Dendrimers in Nanomedicine

Dendrimers in Nanomedicine

Concept, Theory and Regulatory Perspectives

Edited by
Neelesh Kumar Mehra
Keerti Jain

CRC Press
Taylor & Francis Group
Boca Raton London New York

CRC Press is an imprint of the
Taylor & Francis Group, an **informa** business

First edition published 2021
by CRC Press
6000 Broken Sound Parkway NW, Suite 300, Boca Raton, FL 33487-2742

and by CRC Press
2 Park Square, Milton Park, Abingdon, Oxon, OX14 4RN

© 2021 Taylor & Francis Group, LLC

CRC Press is an imprint of Taylor & Francis Group, LLC

Library of Congress Cataloging-in-Publication Data
Names: Mehra, Neelesh K. (Neelesh Kumar), 1949- editor. | Jain, Keerti,
editor.
Title: Dendrimers in nanomedicine : concept, theory and regulatory
perspectives / edited by Neelesh Kumar Mehra and Keerti Jain.
Description: First edition. | Boca Raton : CRC Press, 2021. | Includes
bibliographical references and index.
Identifiers: LCCN 2020041011 | ISBN 9780367466053 (hardback) | ISBN
9781003029915 (ebook)
Subjects: LCSH: Dendrimers. | Nanoscience.
Classification: LCC TP1180.D45 D457 2021 | DDC 615/.6--dc23
LC record available at https://lccn.loc.gov/2020041011

ISBN: 978-0-367-46605-3 (hbk)
ISBN: 978-0-367-68901-8 (pbk)
ISBN: 978-1-003-02991-5 (ebk)

Typeset in Times

Contents

Preface

Dendrimers, hyperbranched macromolecules, emerged just few decades ago but show promising potential as drug delivery nanocarriers, theranostic agents and gene vectors and in the pharmaceutical research and innovation area as well as in other healthcare applications. Although tremendous advancements have been made in dendrimer chemistry and their applications since their emergence, the synthesis, development and design of pure and safe dendrimer-based products have been a major challenge in this area.

This book has exhaustively covered the nanotechnological aspects, concepts, properties, characterisation, application, biofate and regulatory aspects of dendrimers. It includes a contribution of vivid chapters from across the world by renowned formulators, researchers and academicians with their specialisation area of interest in the field of chemistry, biology, pharmacy and nanomedicine.

- It includes dendrimers' advancements and future opportunity in nanomedicine with safety and efficacy in the development of healthcare and biotechnological products.
- It covers the physicochemical aspects, biofate, drug delivery aspects and gene therapy using dendrimers, etc.
- It covers the biomedical application of dendrimers in the field of biological sciences.
- It includes the dendrimer–guest interaction chemistry i.e. how dendrimers can be explored as a template/scaffold to load drug molecules or diagnostic agents or targeting ligands or gene, DNA or siRNA and so on either in their cavities or on their surface.

This book provides the comprehensive overview of the latest research efforts in designing, optimising, development and scale up of dendrimer-mediated delivery systems. It analyses the key challenge, synthesis, design, molecular modelling, fundamental concepts, drug delivery aspects, analytical tools and biological fate and regulatory consideration to the practical use of dendrimer application.

Chapter 1, written by editors, Dr. Keerti Jain and Dr. Neelesh Kumar Mehra and co-authors, comprises the recent nanotechnological advancement including contribution of numerous biomaterials.

Chapters 2 and **14**, written by Dr. Javed Ahmad and co-authors, explains various examples of drug targeting exploiting dendrimer-based formulations to an adequate length through different routes of administration and dendrimeric-based drug delivery potential.

Chapter 3 was written by Dr. Smriti Sharma on the dendrimers' nanomedicine and their history, concept and properties of dendrimers.

Chapter 4, edited by Dr. Kalpana Nagpal, demonstrates the interesting polymeric architecture of dendrimers in terms of their chemistry, classification, types for better understanding and further implementation in different biomedical fields.

Chapter 5 was written by editors on macromolecular architecture and molecular modelling of dendrimers.

Chapter 6 was written by Dr. Abha Sharma on the various synthesis methods of dendrimers.

Chapter 7 was contributed by Dr. Hitesh Kulhari and it is divided into two parts. The first part describes various strategies for functionalisation of different dendrimers. The second part discusses the applications of functionalised dendrimers in diverse areas.

Chapter 8, contributed by Dr. Sougata Jana and co-authors, describes the various analytical characterisation techniques for dendrimers.

Chapter 9 discusses the various interaction mechanisms for dendrimer–guest molecules.

Chapters 10 and 11 were contributed by Dr. Awanish Mishra and describe the dendrimers as a nanovector for gene delivery and diagnostic applications of dendrimers, respectively.

Chapters 12 and 15 were contributed by Dr. Sumit Sharma and describe the various conjugation chemistry approaches with nanomaterials and their possible biomedical applications.

Chapter 13 was contributed by Dr. Vijay Mishra who discusse the possible role and application of dendrimers in cancer therapy.

Chapter 16 was contributed by Miss Anisha D'Souza and Dr. Pratik Kumar Patel on biofate, toxicity and regulatory consideration of dendrimers.

Editors
Neelesh Kumar Mehra, PhD
Keerti Jain, PhD

Editors

Dr. Neelesh Kumar Mehra is working as an Assistant Professor of Pharmaceutics & Biopharmaceutics, at the Department of Pharmaceutics, National Institute of Pharmaceutical Education & Research (NIPER), Hyderabad, India. He earned his PhD with Prof. N.K. Jain from Dr. H. S. Gour University, Sagar and PostDoc from Irma Lerma Rangel College of Pharmacy, Texas A & M Health Science Centre, Kingsville, TX, USA. He served as Manager in Product Development, Sentiss Research Centre, Sentiss Pharma Pvt Ltd. Gurgaon for development, scale-up and technology transfer of complex ophthalmic, inhalation and optic pharmaceutical products. He received the '**TEAM AWARD**' for successful commercialisation of Ophthalmic Suspension product. He has authored more than 60 peer-reviewed publications in high repute international journals and 7 book chapter contributions. He guided PhD and M.S. students for their dissertations/research projects. He has received numerous outstanding awards including Young Scientist and Team Awards for their research output. He is a peer reviewer of various international journals and publishers. Currently, he is editing books on biopharmaceutical and nanotechnology based with CRC Press and Elsevier Pvt Ltd. He has 11 years rich research and teaching experience in the formulation and development of complex, innovative ophthalmic and injectable biopharmaceutical products including micro- and nanotechnologies for regulated market and so on.

Dr. Keerti Jain is working as an Assistant Professor and Scientist in Department of Pharmaceutics, National Institute of Pharmaceutical Education and Research (NIPER), Raebareli, Lucknow, India. Dr. Jain earned her M. Pharm. and Doctorate degree from Dr. H. S. Gour Central University, Sagar. She did her Post Doctorate from M.S. University of Vadodara, India as a SERB-National Post-Doctoral Fellow. She has more than 11 years of research experience on nanomedicine-based drug delivery systems.

Dr. Jain has supervised M. Pharm. and Pharm. D. students for their research works, which have been published in quality journals with a good impact factor. She is the author of more than 60 international manuscripts in peer reviewed-high impact journals, including publications such as *Progress in Polymer Science* having impact factor: 26.932, in biomaterials, Drug Discovery Today, nanomedicine, Current Opinion in pharmacology, antimicrobial agents and chemotherapy and other high impact factor journals. Her review articles entitled 'Dendrimer Toxicity: Let's Meet the Challenge' and 'Dendrimers as Nanocarriers in Drug Delivery' published in International Journal of Pharmaceutics (2010) and Progress in Polymer Science (2014) was listed in the top 25 articles. She has authored of 10 books and book chapters with international publishers. She has been awarded with research fellowships including SERB – Post Doctorate Fellowship, Senior Research Fellowship (CSIR), Junior Research Fellowship (UGC), Fellowship for Training of Young Scientist (MPCST), Travel Grant Award (INSA) and International Travel Fellowship (ICMR and DBT) as well as several other prestigious grants, fellowships and awards. She has been nominated for Ranbaxy Research Scholar Award – 2012 and BioAsia Innovation Award – 2015. She has also been awarded with many best research presentation awards in international conferences like NIPiCON-2014, International Science Congress-2012, APTI – Jaipur – 2018, SPER-Bangkok – 2019 and so on. She has been the recipient of Pharmaceutical Science Alumni Award – 2006 at Dr. H. S. Gour University, Sagar (M.P.). She has been invited to present her research work at several national and seven international conferences including ICYRAM-2012 held at Singapore, Bioencapsulation – 2016 held at Lisbon, Portugal and International Conference – 2019 held at Bangkok. She won the Most Innovative Idea Award in LUFTHANSA impact week, which is a reputed international platform.

Contributors

Ameeduzzafar
Department of Pharmaceutics
College of Pharmacy
Jouf University
Aljouf, Saudi Arabia

Muhammad Afzal
Department of Pharmacology
College of Pharmacy
Jouf University
Aljouf, Saudi Arabia

Javed Ahmad
Department of Pharmaceutics
College of Pharmacy
Najran University
Najran, Saudi Arabia

Khalid Saad Alharbi
Department of Pharmacology
College of Pharmacy
Jouf University
Aljouf, Saudi Arabia

Abuzer Ali
College of Pharmacy
Taif University
Taif-Al-Haweiah, Saudi Arabia

Nabil K. Alruwaili
Department of Pharmaceutics,
College of Pharmacy,
Jouf University, Aljouf, KSA

Saima Amin
Department of Pharmaceutics
School of Pharmaceutical Education and Research
   (SPER)
Hamdard, India

Sonali Batra
Department of Pharmaceutical Sciences
Guru Jambheshwar University of Science and
   Technology
Hisar, India

Dnyaneshwar Baswar
Department of Pharmacology and Toxicology
National Institute of Pharmaceutical Education
   and Research-Raebareli (NIPER-R)
Lucknow, India

Valamla Bhavana
Pharmaceutical Nanotechnology Research
   Laboratory
Department of Pharmaceutics
National Institute of Pharmaceutical Education &
   Research (NIPER)
Hyderabad, India

Thakor Pradip
Pharmaceutical Nanotechnology Research
   Laboratory
Department of Pharmaceutics
National Institute of Pharmaceutical Education &
   Research (NIPER)
Hyderabad, India

Anisha D'Souza
School of Pharmacy
Duquesne University
Pittsburgh, Pennsylvania

Ankita Devi
Department of Pharmacology and Toxicology
National Institute of Pharmaceutical Education
   and Research-Raebareli (NIPER-R)
Lucknow, India

Rahul Gauro
Department of Pharmaceutics
Delhi Pharmaceutical Sciences and Research
   University (DPSRU)
New Delhi, India

Syed Sarim Imam
Department of Pharmaceutics,
College of Pharmacy
King Saud University
Riyadh, Saudi Arabia

Keerti Jain
Department of Pharmaceutics
National Institute of Pharmaceutical Education &
    Research (NIPER)
Rae Bareli, India

Vineet Kumar Jain
Department of Pharmaceutics
Delhi Pharmaceutical Sciences and Research
    University (DPSRU)
New Delhi, India

Sougata Jana
Department of Pharmaceutics
Gupta College of Technological Sciences
Asansol, India
and
Department of Health and Family Welfare
Directorate of Health Services
Kolkata, India

Hitesh Kulhari
School of Nano Sciences
Central University of Gujarat
Gandhinagar, India

Gagandeep Maan
Department of Pharmacology and Toxicology
National Institute of Pharmaceutical Education
    and Research-Raebareli (NIPER-R)
Lucknow, India

Sreejan Manna
Department of Pharmaceutical Technology
Brainware University
Barasat, Kolkata, India

Neelesh Kumar Mehra
Pharmaceutical Nanotechnology Research
    Laboratory
Department of Pharmaceutics
National Institute of Pharmaceutical Education &
    Research (NIPER)
Hyderabad, India

Awanish Mishra
Department of Pharmacology and Toxicology
National Institute of Pharmaceutical Education
    and Research-Raebareli (NIPER-R)
Lucknow, India

Vijay Mishra
School of Pharmaceutical Sciences
Lovely Professional University
Phagwara, India

Abdul Muheem
Department of Pharmaceutics
School of Pharmaceutical Education & Research
Jamia Hamdard, India

Kalpana Nagpal
Amity Institute of Pharmacy
Amity University, Uttar Pradesh
Noida, India

Pallavi Nayak
School of Pharmaceutical Sciences
Lovely Professional University
Phagwara, India

Brijesh Ojha
Department of Pharmaceutics
Delhi Pharmaceutical Sciences and Research
    University (DPSRU)
New Delhi, India

Harvinder Popli
Department of Pharmaceutics
Delhi Pharmaceutical Sciences and Research
    University (DPSRU)
New Delhi, India

Pratikkumar Patel
Department of Chemical Sciences, Synthesis
Bernal Institute
University of Limerick
Limerick, Ireland

Deep Pooja
School of Science
RMIT University
Melbourne, Victoria, Australia

Mohammad Akhlaquer Rahman
Department of Pharmaceutics and Industrial
    Pharmacy
College of Pharmacy
Taif University
Taif-Al-Haweiah, Saudi Arabia

Divya Bharti Rai
School of Nano Sciences
Central University of Gujarat
Gandhinagar, India

Anjali Rajora
Amity Institute of Pharmacy
Amity University
Noida, India

Md. Rizwanullah
Department of Pharmaceutics
School of Pharmaceutical Education and Research
   (SPER)
Jamia Hamdard
New Delhi, India

Abha Sharma
Department of Medicinal Chemistry
National Institute of Pharmaceutical Education
   and Research-Raebareli (NIPER-R)
Lucknow, India

Smriti Sharma
Amity Institute of Pharmacy
Amity University
Noida (UP), India

Sumit Sharma
Chitkara College of Pharmacy
Chitkara University
Punjab, India

Biplab Sikdar
Department of Regulatory Toxicology
National Institute of Pharmaceutical Education
   and Research-Raebareli (NIPER-R)
Lucknow, India

Amit Singh
Doctoral Researcher,
Institute of Molecular Biosciences,
Karl-Franzens-University Graz,
Humboldtstrasse 50/3, A-8010 Graz, Austria

Manvendra Singh
School of Pharmaceutical Sciences
Lovely Professional University
Phagwara, India

Ashima Thakur
Department of Medicinal Chemistry
National Institute of Pharmaceutical Education
   and Research-Raebareli (NIPER-R)
Lucknow, India

Samridhi Thakral
Department of Pharmaceutical Sciences
Guru Jambheshwar University of Science and
   Technology
Hisar, India

Musarrat Husain Warsi
Department of Pharmaceutics and Industrial
   Pharmacy
College of Pharmacy
Taif University
Taif-Al-Haweiah, Saudi Arabia

Mohammad Yusuf
College of Pharmacy
Taif University
Taif-Al-Haweiah, Saudi Arabia

1

Nanotechnology: Introduction and Basic Concepts

Brijesh Ojha, Vineet Kumar Jain, Neelesh Kumar Mehra and Keerti Jain

CONTENTS

1.1 Introduction

Nanotechnology is a fast-growing science of nanosized particle growth and consumption, measuring size in the nanometre range. Additionally, nanotechnology is the practice of systemically characterising, manipulating and arranging matter at the nanometre scale, which has brought about a revolution in science, technology, engineering, drug discovery and therapy. Nanotechnology includes a combination of various sciences, including not only medicine, but also physics, chemistry, biochemistry and molecular biology. It describes and deals with previously unexplored functional characteristics (optical, magnetic, electronic and catalytic) for scientific benefit (El-Sayed, 2001). The nanometre scale is around 1000 times smaller than structures that the human eye might see but still 1000 times bigger than an atom. In the recent era, advancements have been made in methods and techniques that deal and tackle with the nanoscale size range and are classified under nanotechnology. Nanotechnology is defined as the control or reconstruction of matter at atomic and molecular levels of approximately 1–100 nm size range and the science that underlies known as nanoscience (Jain, 2018a, 2018b; Joshi *et al.*, 2019; Bhushan, 2016). Today, there are many operations that take a lot of time and are very costly too. Speedier and much cheaper therapies can be created using nanotechnology in the pharmaceutical industry. Nanotechnology encompassing various nanomaterials and nanocarriers such as nanoemulgel, nanocapsules, carbon nanotubes (CNTs), fullerenes, hyperbranched polymers, dendrimers, etc. is being explored for various applications in medicine and drug delivery to treat severe and fatal diseases such as visceral leishmaniasis, triple negative breast cancer, AIDS, refractory cancers, etc. (Khan *et al.*, 2019; Ahmad *et al.*, 2018; Cuneo *et al.*, 2020; Huang *et al.*, 2020; Fana *et al.*, 2020). Nanotechnology is also being explored in personalised medicine and 3-D printing technologies (Afsana *et al.*, 2019).

The highly contagious infectious respiratory disease COVID-19, caused by a novel coronavirus SARS-CoV-2, attained the global pandemic status in just few months, which started in late 2019. Currently, millions of people have been infected while counting deaths in lakhs worldwide and affecting almost all countries around the world with this viral infection till the writing of this chapter. It is necessary in this global crisis that the scientists and researchers working in the fields of technology, engineering, medicine, healthcare, public health, etc. should join hands to devise a treatment or preventive strategy to stop spread of this disease. Nanotechnology can also be investigated here to develop, formulate and design treatment and vaccine to fight/stop this disease. Nanomaterials and nanotechnology could be explored here on the basis of phenomena that they are being explored for targeting therapeutic agents selectively to cancer cells or infected cells to avoid harm to healthy cells. This phenomenon could be useful in the treatment/management/prevention of this pandemic COVID-19 (Chan, 2020; Sportelli *et al.*, 2020; Jain, 2017; Ahmad *et al.*, 2018).

The nanotechnology is affecting all the fields including healthcare and being explored extensively for various applications. Nanotechnology has led to some important innovations in different fields ranging from aircraft to healthcare and drug delivery to marine science and also has ability to be explored promisingly for fighting with sudden crisis such as COVID-19 pandemic which emerged in late 2019.

1.2 Nanotechnology in Biomedical/Healthcare/Pharmaceutical Science

Drugs normally function through the entire body before entering the site affected by the disease. With these pharmaceutical nanotechnologies, the medication may be delivered at a specific location that will make the therapy even more powerful and reduce the chances of side effects (Misra *et al.*, 2010). Nanoparticles (NPs) have been defined as solid colloidal particles varying in dimensions from 10 to 400 nm (Figure 1.1). These NPs consist of macromolecular substances in which the reactionary agents (bioactives) have been encapsulated or dissolved or entrapped, or to which the active agent is attached or adsorbed (Muthu and Singh, 2009). Pharmaceutical nanotechnology, through early diagnosis, prediction, prevention, personalised treatments and medication, offers a novel strategy and advanced technology for cancer (Kalyankar *et al.*, 2012). Early diagnosis and targeted therapy techniques are the key areas of research where nanotechnology can play an important role (Sutradhar and Amin, 2014). Fiber optic technology is used for

Chem Soc Rev

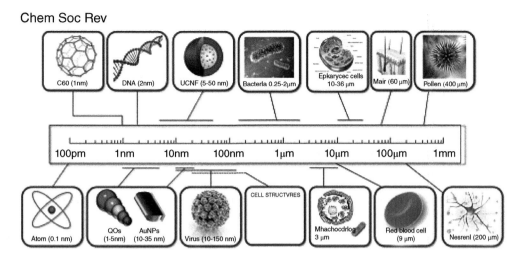

FIGURE 1.1 The size of nanoparticles compared with biologically related molecules and organisms. Reproduced with permission from (Gnach *et al.*, 2014).

tracking diseases. Optical biosensors are used to measure physical parameters such as pH, blood supply volume, blood oxygen concentrations and radiation dosage. Next generation endoscopy can expand its ability from imaging to treatment and therapy using nanofibre technology (Quan and Zhang, 2015).

Nanomaterials and nanodevices are the two basic forms of pharmaceutical nanotechnology and play a key role in pharmaceutical drug delivery and other fields (Mohan and Varshney, 2012). Enormous surface areas and quantum effects are the two important factors that render nanomaterials some unique properties compared to other materials including drug targeting, intracellular drug delivery, theranostic applicability, gene delivery, etc. Nanomaterials are made of biomaterials used in orthopaedic or dental implants or as scaffolds for tissue-based items. Their surface can be updated, or coatings may be developed which enhances biocompatibility with human cells. These are further graded into two forms, namely nanocrystal and nanostructural materials. Nanocrystals are ground in special mills, and the resulting drugs can be applied intravenously or bronchially through an inhaler as nanosuspensions. The small size increases the surface/volume ratio and the bioavailability of insoluble pharmaceutical materials (Clement *et al.*, 2012; Malamatari *et al.*, 2018). Nanosuspensions are colloidal submicron dispersions of condensed nanosized drug particles stabilised by polymers and surfactants. Nanosuspensions are formulated to increase the dissolution rate and mucoadhesion of poorly soluble drugs to increase their bioavailability on administration *via* routes such as oral, pulmonary, ocular, etc. Scientists have been exploring nanosuspensions for the delivery of BCS class II drugs, which are poorly water soluble. Recently, a nanosuspension of fluvoxamine, a selective serotonin reuptake inhibitor, which is commonly used as first line anti-depressant treatment for major depressive disorders, was prepared to improve its bioavailability. A nanosuspension of itraconazole, a broad spectrum anti-fungal drug, has been prepared and converted to inhalable and re-dispersible NPs for deep lung delivery by scientists. A mucoadhesive nanosuspension-based formulation of loratadine, an anti-histaminic anti-allergic drug, was prepared for nasal delivery with improved bioavailability (Tyagi and Madhav, 2020; Wan *et al.*, 2020; Alshweiat *et al.*, 2020; Villanueva-Martínez *et al.*, 2020).

1.3 History of Nanomaterials

The history of nanomaterials began immediately after the big bang, when the early meteorites developed nanostructures. Nature subsequently generated many other nanostructures such as seashells, skeletons, etc. Through the use of fire, early humans developed nanoscale smoked particles. The nanomaterials' science story, however, began much later. One of the first experimental reports is colloidal gold particles synthesised by Michael Faraday as early as 1857. Nanostructured catalysts have also been explored for over 70 years. By the early 1905, Einstein estimated the size of the sugar compound to be 1 nm using the known physical properties of its solution. By the early 1940s, precipitated and fumed silica NPs were produced and marketed in the USA and Germany as substitutes for rubber reinforcements for ultrafine carbon black (Einstein, 1906). In 1974, Norio Taniguchi first coined the term 'nanotechnology'. Scanning tunneling microscopy (STM) was developed by physicist Binnig and Rohrer in 1981 and is a new advance type of microscope (Bayda *et al.*, 2020). For this, they have won the Nobel Prize in Physics in 1986. In the same year, Curl, Kroto and Smalley discovered a spherical stable form of carbon later known as fullerenes or bucky balls (Kroto, 1985). Later in 1991, Sumio Iijima observed hollow tubes using transmission electron microscopy, which were later added as another member of the fullerene family (Iijima, 1991). A few years later in 1990s, Don Eigler of IBM prepared the letter of IBM logo using STM by modifying 35 xenon atoms on a base of nickel (Eigler and Schweizer, 1990). In 2004, Xiaoyou and colleagues accidently discovered a new class of carbon dots using the modification of CNTs (Xu *et al.*, 2004). A number of studies have recently highlighted the enormous potential that nanotechnologies have in biomedicine for the diagnosis and treatment of many human diseases.

1.4 Classification of Nanotechnology

Nanotechnology explored for application in the fields of healthcare, pharmaceutical and biomedical sciences can be broadly classified into two categories, which are nanomedicine and nanodevices, which play a vital role in healthcare, pharmaceutical and biomedical industries. These nanomedicines and nanodevices further can be classified as given below:

1.4.1 Nanomaterials

These consist of biomaterials that are used in orthopaedic or dental implants or tissue engineered products. Their surfaces are coated and modified to enhance the biocompatibility with living cells. They are further classified into nanocrystals and nanostructures.

1.4.1.1 Nanocrystals

The drug to be injected into the cell is generated in nanosize and can act as a carrier of its own. Using polymeric macromolecules and non-ionic surfactants, the drug particle is reduced to the nanosize range and has a stabilised surface (Joshi *et al.*, 2019; Wanigasekara and Witharana, 2016). These materials are also used directly in prosthesis, implants and bone replacement.

1.4.1.2 Nanostructures

These materials are the processed form with special shapes and functions.

   a. **Polymer based**
      i. Liposomes
     ii. Dendrimers
    iii. Micelles
     iv. NPs
      v. Drug conjugates
   b. **Non-polymer based**
      i. CNTs
     ii. Metallic NPs
    iii. Quantum dots (QDs)
     iv. Fullerenes

1.4.2 Nanodevices

These are nanoscale, tiny devices which include nanoelectromechanical systems and/or microelectrome-chanical systems, microfluidics and microassays which include biosensors and detectors used for diagnosis (Pandit *et al.*, 2016).

1.5 Nanomaterials/Nanocarriers for Drug Delivery

Various polymer- and non-polymer-based materials are being explored for drug delivery applications by scientists all over the world. The important features of some extensively explored nanomaterials are discussed below (as shown in Figure 1.2).

1.5.1 Liposomes

Liposomes consist of vesicles made up of bilayers or multilayers containing, or having, phospholipids and cholesterol surrounding an aqueous space. Drug(s) can be trapped within the liposome and released to the surface of the intestinal membrane for absorption from the liposomes. Phospholipids, which are amphiphilic molecules (having a hydrophilic head and hydrophobic tail) consisting of a glycerol molecule attached to a phosphate group (PO_4^{2-}) and two saturated or unsaturated fatty acid chains, form the essential or important component of liposomes (Figure 1.2). The advantages and disadvantages of liposomes in medicine are summarised in Table 1.1 (Anwekar *et al.*, 2011; Pinot *et al.*, 2014). The hydrophilic part is essential, which is mainly phosphoric acid bound to a water-soluble molecule, while the hydrophobic part consists of two fatty acid chains of 10–24 carbon atoms and up to six double bonds in each chain. When these phospholipids are disseminated or scattered in aqueous medium, they form lamellar sheets by arranging them in such a way that the polar head group faces the aqueous zone outwards, while the fatty acid groups face each other and actually form spherical/vesicle structures called liposomes. The polar

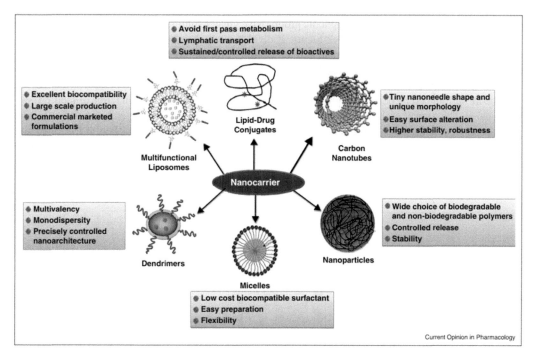

FIGURE 1.2 Nanostructures of different types of nanomaterials with their advantages. Reproduced with permission from (Jain *et al.*, 2014b).

TABLE 1.1

Advantages and disadvantages of liposomes

Advantages	Disadvantages
These are non-toxic, biocompatible, completely biodegradable and non-immunogenic for systemic as well as for non-systemic administration	There may be chances of oxidation and hydrolysis type of reactions of phospholipids
Increase therapeutic index and efficacy of drugs (e.g. actinomycin-D)	Short half life
Increase the stability of potentially instable drugs *via* encapsulation	Solubility issues
Reduction in the toxicity side effects associated with the drug (e.g. amphotericin B, Taxol)	Leakage of the encapsulated drug
Reduce the exposure of sensitive tissues to toxic drugs	Production cost is high

Source: Anwekar et al. (2011).

component remains in contact or touch with the aqueous environment, along with shielding or keeping the non-polar portion free (Kahraman *et al.*, 2017; Gracia *et al.*, 2019).

1.5.2 Dendrimers

Dendrimers are nanosized, radially symmetric molecules with a well-defined, homogeneous and monodisperse design consisting of arms or branches similar to trees. Dendrimers are hyperbranched macromolecules with a precisely designed architecture, and the end-groups (*i.e.* groups approaching the outer periphery) could be functionalised, thereby changing their physicochemical or biological properties (Bosman *et al.*, 1999). Functional end groups present on the surface are responsible for miscibility, high reactivity and high solubility of dendrimers. The existence of surface groupings influences dendrimer's solubility. The solubility of dendrimers in polar solvents is greater because of their hydrophilic groups, while those dendrimers that have hydrophobic groups have greater solubility in non-polar solvents. Based on the growth of branches, dendrimer's structure is divided into generations as shown in Figure 1.3, and the change in molecular weight, radii and surface charge or the number of terminal nitrogen atoms of dendrimers with generation is summarised in Table 1.2.

Dendrimers are structures that are three-dimensional, hyper-branched and monodisperse and have a central core surrounded by peripheral groups. Such characteristics are important in terms of their physicochemical and biological properties. Normally, dendrimers have three distinctive architectural elements, (i) the core, (ii) the branches (an internal layer composed of repeated units attached to the core) and (iii) the terminal groups attached to the branches (Figure 1.2). The advantages of dendrimers as drug

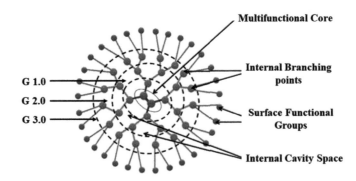

FIGURE 1.3 Schematic presentation of the typical structure of dendrimers comprising three major parts (i) the core, (ii) internal branches and cavity space and (iii) surface functional groups.

TABLE 1.2

Generation of PAMAMs with their molecular weight, radii and surface charge or terminal nitrogen (G-0 to G-10)

Generation	Atoms	Measured Radii (Å)	Molecular Weight	Surface Charge/Terminal Nitrogen Atoms
0	84	30	517	4
1	228	44	1,430	8
2	516	58	3,256	16
3	1,092	72	6,909	32
4	2,244	90	14,215	64
5	4,548	108	28,826	128
6	9,156	134	58,048	256
7	18,372	162	116,493	512
8	36,804	194	233,383	1024
9	73,668	228	467,162	2048
10	147,396	270	934,720	4096

Source: Maiti *et al.* (2004), (www.Dendritech.com, 2017).

TABLE 1.3

Advantages and disadvantages of dendrimers

Advantages	Disadvantages
Enhance the solubility of highly lipophilic drugs	Not appropriate carrier for hydrophilic drugs
Surface functional groups can be tuned finely to tailored chemical and physical properties	Cellular toxicity
Several functional groups for target drug delivery	Metabolism and elimination depend on the generation and material used in dendrimers
Drugs associated covalently	High cost
Also act as a solubility enhancer	High expertise is required to synthesise dendrimers

Source: Kahraman *et al.* (2017).

delivery systems are summarised in Table 1.3. The dendritic polymer arrangement creates internal cavities, in which the drug can be deposited, increasing its solubility and stability. This unique feature of dendrimers makes them ideal macromolecules to be used as pharmaceutical excipients such as drug delivery vehicles, solubilising agents, gene delivery vectors, diagnostic agents, etc. (Jain, 2018b; Kaur *et al.*, 2017; Tiwari *et al.*, 2010). Poly-amidoamines (PAMAMs) were the first synthesised dendrimers also known as starburst dendrimers (Klajnert and Bryszewska, 2001). These nanoscale polymers consist of a centre of ethylenediamine, a regular internal branching structure of amidoamine and a main terminal surface of amines. In an iterative manufacturing process, dendrimers are 'grown' off a central core, with each subsequent stage representing a new generation of dendrimers (Kwun *et al.*, 2018). Growing generations (molecular weight) yield greater molecular radii, twice the number of reactive surface sites and nearly double the previous generation's molecular weight.

1.5.3 Polymeric Micelles (PMs)

PMs are self-assembled core–shell nanostructures formed by amphiphilic block copolymers in an aqueous solution. Micelle formation in aqueous solution occurs when the block copolymer concentration rises above a certain concentration called critical aggregation concentration (CAC) or critical micelle concentration (CMC). Hydrophobic segments of block copolymers begin to interact at the CAC or CMC to reduce the interaction with water molecules, leading to the formation of a micelle in a vesicular or core–shell structure (Xu *et al.,* 2013). Micelle morphology has a major effect on micelles' pharmacokinetic properties. For example, worm-like filomicelles showed ten times longer circulation time compared to the spherical equivalent made of a similar material. Poly(ethylene glycol) (PEG), with a molecular weight of 2–15 kDa, is the most widely used hydrophilic type of micelle for drug delivery. PEG forms a hydrophilic corona on the micelle surface that minimises the non-specific contact with blood components and increases the circulation time (Zhang *et al.*, 2014).

1.5.4 Polymeric NPs

These are colloidal particles composed of synthetic or natural polymers in the size of 10 nm–1 μm. Due to their inherent properties such as biocompatibility, non-immunogenicity, non-toxicity and biodegradability, these NPs provide an alternative to the above listed nanosystems. Polymeric NPs are classified into nanocapsules and nanospheres (Rangasamy, 2011). Nanocapsules are structures in which the drug is constrained to a cavity enclosed by a special polymeric membrane, while in nanospheres, the drug is distributed within the polymer matrix. Natural polymers used for NP preparation include gelatin, albumin and alginate, whereas preformed polymers such as polyesters (polycaprolactone) are used as synthetic polymers for NPs (Deng *et al.*, 2020; Bagre *et al.,* 2013; El-Gogary *et al.*, 2020; Jain *et al.*, 2014a).

1.5.5 Drug Conjugates

The combination of low molecular-weight drugs and polymers induces dramatic changes in drug pharmacokinetic activity in the entire body and at the cellular level. They are intended to increase the overall molecular weight, which promotes their retention in cancer cells by use of an enhanced permeation retention effect using a passive delivery approach.

Polymer–drug conjugates provide many properties advantageous for drug delivery applications, such as (Marasini *et al.*, 2017):

I. Potential to achieve large drug payloads.
II. Enhanced drug solubility.
III. Regulation of drug pharmacokinetics (including extended plasma exposure and enhanced biodistribution behaviour, resulting in increased therapeutic effectiveness).
IV. Reduced local and systemic side effects as a result of highly cytotoxic medicine.
V. Improved *in vivo* drug stability.
VI. Controlled drug release rate and site.

1.5.6 CNTs

CNTs, discovered by the Japanese scientist Sumio Iijima in 1991, are carbon allotropes, made of graphite, and formed in cylindrical tubes with diameter in nanometres and several millimetres in length (Iijima, 1991). Their remarkable structural, mechanical and electronic properties are due to their small size and mass, their immense mechanical ability and their high electrical and thermal conductivity (Jain, 2018; Usui *et al.*, 2012). Having a high surface area, excellent chemical stability and a rich polyaromatic electronic structure, CNTs can adsorb or conjugate a wide range of therapeutic molecules (drugs, proteins, antibodies, DNA, enzymes, etc.) (Sharma *et al.*, 2017; Zhang *et al.*, 2010). Single-walled CNTs are used as a platform to investigate surface–protein and protein–protein binding as well as to build highly sensitive detectors for electronic biomolecules (Jain, 2005).

1.5.7 Metallic NPs

Metal NPs are created specifically through metals. Such NPs exhibit special optoelectrical properties due to the well-known localized surface plasmon resonance characteristics. In the visible zone of the electromagnetic solar spectrum, Cu, Ag and Au have a broad absorption band (Dreadena *et al.*, 2012).

As an alternative to quantum dots, they were used for the active delivery of bioactives, drug discovery, bioassays, identification, imagery and many other applications due to surface functionalisation capability (Borse *et al.*, 2020; Chen *et al.*, 2020; Hu *et al.*, 2020; Rangasamy, 2011).

1.5.8 QDs

QDs are semi-conductive materials consisting of a shell-coated semi-conductor core to enhance optical properties. Their properties derive from their physical size varying in radius from 10 to 100 A^0. QDs are widely used in fluorescence-related biological applications, including DNA array technology, cell biology and immunofluorescence assays, particularly in the immunostaining of proteins, microtubules, actins and nuclear antigens (Michalet *et al.*, 2005; Bajwa *et al.*, 2016a). Cadmium selenide (CdSe), cadmium telluride (CdTe), indium phosphide (InP) and indium arsenide (InAs) are the most widely used QDs. Such particles act as contrast agents in the bioimaging process, offering much greater resolution than current fluorescent colours. These particles can consume white light and re-emit it within nanoseconds, with different energy gaps in the bulk band corresponding to different particle combinations (Bawarski *et al.*, 2008; Bajwa *et al.*, 2016b).

1.5.9 Fullerenes

Carbon is one of the most common elements in nature, and it definitely occupies one of the first positions in the scientific field of nanomaterials. The discovery of buckminsterfullerene (also known as fullerene or buckyball), the third carbon allotrope after graphite and diamond, by Robert F. Curl, Harold W. Kroto and Richard E. Smalle in 1985 prompted the entry into the field of carbon-based nanomaterials (Kroto *et al.*, 1985). Fullerenes are similar to CNTs, as their molecular frame is made entirely of a large π-conjugated carbon skeleton. C60 is a relatively stable compound composed of 60 carbon atoms arranged in a 'soccer cage', 0.72 nm in diameter. The highly delocalised π double bond system contributes to an unusual redox chemical cycle. Thus, C60 has been described as a 'free radical sponge' with several hundred-fold higher antioxidant efficacy than traditional antioxidants, which also play a crucial role in the treatment of disorder related to the cardiovascular system and central nervous system (Wagner *et al.*, 2006; Liu *et al.*, 2014).

1.6 Applications of Nanotechnology and Nanosystems

Using the principle of nanotechnology, the latest approach to pharmaceutical-based therapy is carried out in which drugs were systemically absorbed by the entire body to affect a single specific organ, according to which that organ or diseased portion of it can be treated with molecular precision with little or no side effects. Current pharmaceuticals depend on minor differential selectivity of binding or absorption, and a sufficient dose to be successful against the diseased organ is likely to have substantial deleterious effects on the body as a whole when poor binding and absorption are summed up in the rest of the body. Pharmaceutical nanotechnology offers a wide range of nanosize systems or devices, which offer various advantages. Some of the important applications of these nanosystems are discussed below:

1.6.1 Drug Delivery

Nanotechnology allowed an integrated physical, chemical and biological property drug delivery system that can serve as effective delivery tools for bioactives currently available. Many nanobased carrier systems include polymer NPs, liposomes, dendrimers, PMs, polymer conjugates and antibody–drug conjugates. Drug delivery based on NPs has many advantages, such as improving drug-therapeutic efficacy and pharmacological features. NPs improve the solubility of poorly water-soluble drugs, improve pharmacokinetics, increase the half-life of drugs by reducing immunogenicity, increase specificity towards the target cell or tissue (reducing side effects), improve bioavailability, decrease the metabolism of drugs and allow for more controllable release of therapeutic compounds and simultaneous delivery of two or more drugs (Sanvicens and Marco, 2008).

1.6.2 Gene Therapy

Normal gene is introduced using a carrier molecule instead of an abnormal disease-causing gene. Nanotechnology provided an efficient and promising tool in the treatment of systemic genes. Gene therapy uses chitosan, gelatin, poly-1-lysine and modified silica NPs. These have improved efficiency in transfection and decreased cytotoxicity. Nanotechnology offers perfect gene delivery vectors (Jain, 2005).

1.6.3 Implants

Due to their size and controlled and approximately zero order kinetics, NPs can serve as delivery systems; otherwise, they can cause toxicity compared to I.V. Carriers are liposomes, transferosomes and ethosomes. These help to reduce peak plasma levels and reduce the risk of adverse reactions, allow for a more stable and prolonged course of action, decrease the frequency of re-dose and increase patient acceptance and conformance (Jain and Jain, 2002).

1.6.4 Diagnosis

The combination of NPs and other materials based on nanotechnology has the potential to solve this emerging challenge and provide technologies that allow single cells and single molecules to be diagnosed (Sanvicens and Marco, 2008). Some applications include precise cell and tissue labelling, long-term imaging, multicolour multiplexing, complex visualisation of subcellular structures and fluorescence resonance energy transfer and magnetic resonance imaging (MRI). Nanomaterials such as dendrimers, QDs, CNTs and magnetic NPs replace the MRI agents. Because of their high intensity, photostability, resolution and resistance, they aid in obtaining a very effective, stable, sharp and clearer image.

1.6.5 Tissue Engineering

Nanotechnologies and microtechnologies can be combined with biomaterials to create tissue engineering scaffolds that can sustain and control the actions of cells. The ability to design biomaterials closely emulating the structure of extracellular matrix (ECM) and ECM functionality is crucial to effective tissue regeneration. Recent nanotechnology developments have allowed the design and manufacture of biomimetic microenvironments at a nanoscale, providing an analogue to the native ECM (Shi *et al.*, 2010). The ECM plays a very important role in processing, releasing and activating a wide variety of biological factors, as well as supporting interactions between cell cells and water-soluble factors (Taipale and Keski-Oja, 1997).

1.6.6 Biosensors

Various methods are used to assess the specific pathological proteins and physiological–biochemical markers associated with disease or impaired body metabolism. Biosensors are a measuring device consisting of a probe with a sensitive biological recognition element or bio-receptor, a physiochemical detector part, and a transducer for the amplification and transduction of such signals into measurable shape. Biosensors are used for identification of targets, validation, production of assays, ADME and determination of toxicity (Khopde and Jain, 2001).

1.6.7 Cancer Therapy

Nanotechnology will also be able to increase the number of cancers detected early by improved imaging, and this may lead to better results for cancer patients in combination with more rigorous application of current screening techniques. Methods for the delivery of colloidal drugs such as liposomes and micelles were intensively investigated for their use in cancer therapy.

1.6.8 Drug Discovery

The key advantages of nanotechnology in the pharmaceutical sector are enhanced bioavailability, minimised toxicity, sustained and controlled release, targeted delivery, preventing occluding blood capillaries, being able to easily pass through most physiological barriers, providing successful delivery to the brain and intracellular compartments, protecting fragile medicines or proteins from harsh biological conditions, detecting disease more rapidly, safely and reliably, and feasibility to perform less invasive surgery with more precise, affordable and large-scale manufacturing. Nanotechnology will enhance the drug delivery process through miniaturisation, automation, speed and reliability of assays (Rangasamy, 2011).

1.7 Approved Marketed Products and Products Currently Undergoing Clinical Trials

Over the past two decades, the Food and Drug Administration (FDA) has approved many nanotherapeutics to treat cancer, hepatitis, diabetes, high cholesterol, myeloma and certain infectious diseases (Table 1.4) (Ventola, 2017). In addition, many products based on nanocarriers are currently available in various stages of preclinical and clinical development. As per a research study, *Anselmo and Mitragotri* have uncovered 18 new NPs coming into clinical trials since 2016. Of those 18 NPs, 12 are liposomes and 17 are indicated for cancer (15 are for diagnosis and 2 are for imagery). Only one non-cancer indication is mRNA-1944, two mRNAs that express heavy and light anti-Chikungunya antibody chains formulated in lipid NPs (Anselmo and Mitragotri, 2019) (as listed in Table 1.5).

TABLE 1.4

List of FDA-approved marketed nanoformulations

S. No.	Name	Generic Name	Used in	Advantages of NP
Liposome NPs				
1	Doxil	Doxorubicin HCl liposome	Multiple myeloma	Increases bioavailability and reduced systemic toxicity
2	Abelcet	Liposomal amphotericin B	Fungal infection	Reduces toxicity
3	DepoDur	Lipsomal morphine sulphate	Postoperative analgesia	Extended release
4	Onivyde	Liposomal irinotecam	Pancreatic cancer	Enhances drug delivery and reduces systemic toxicity
Polymeric NPs				
1	Adynovate	PEGylated-Factor VIII (Protein polymer conjugate)	Hemophilia	Enhances the stability of the protein
2	Plegridy	PEGylated-IFN beta-1a (Protein polymer conjugate)	Multiple sclerosis	Enhances the stability of the protein
Micelle NPs				
1	Estrasorb	Micellar estradiol	Vasomotor symptoms in menopause	Controlled delivery
Nanocrystal NPs				
1	Avinza	Morpine sulphate	Psychostimulant	Enhanced drug loading, Extended release
2	Rapamune	Sirolimus	Immunosuppressant	Enhanced bioavailability
Inorganic NPs				
1	Dexferrum	Iron dextran	Iron deficiency	Increased dose
2	Ferrlecit	Sodium ferric gluconate complex in sucrose injection	Iron deficiency	Increased dose
Protein NPs				
1	Abraxane	Albumin bounded paclitaxel NP	Pancreatic cancer	Enhanced solubility and target delivery to tumours
2	Ontak	Denileukin diftitox	T-cell lymphoma	Target T-cell specificity

Source: Ventola (2017).

TABLE 1.5

List of currently undergoing clinical trials as per clinicaltrial.gov

S. No.	Brand Name	Drug Form	Investigated Application	Clinicaltrial.gov Status
Liposomes (cancer)				
1	MM-310-(Merrimack Pharmaceuticals)	Nanoliposomal encapsulated docetaxel and functionalized with antibodies targeted to the EphA2 receptor	Solid tumours	NCT03076372 (Ph I): Recruiting
2	EGFR(V)-EDV-Dox	Bacterially derived minicell encapsulating doxorubicin	Recurrent glioblastoma	NCT02766699 (Ph I): Recruiting
3	Alprostadil liposome-(CSPC ZhongQi Pharmaceutical Technology)	Alprostadil liposome	Safety and tolerability	NCT03669562 (Ph I): Recruiting
4	Liposomal Annamycin-(Moleculin Biotech)	Liposomal Annamycin	Acute myeloid leukemia	NCT03388749 (Ph II): Recruiting NCT03315039 (Ph II): Recruiting
5	FF-10831-(Fujifilm Pharmaceuticals)	Liposomal Gemcitabine	Advanced solid tumours	NCT03440450 (Ph I): Recruiting
6	Anti-EGFR-IL-dox-(Swiss Group for Clinical Cancer Research; University Hospital, Basel, Switzerland)	Doxorubicin-loaded anti-EGFR immunoliposomes	Advanced triple negative EGFR positive breast cancer	NCT02833766 (Ph II): Recruiting NCT03603379 (Ph I): Recruiting
7	TLD-1/Talidox-(InnoMedica)	A new formulation of liposomal doxorubicin	Advanced solid tumours	NCT03387917 (Ph I): Recruiting
8	NC-6300-(NanoCarrier)	Micelle encapsulated epirubicin	Advanced solid tumours or soft tissue sarcoma	NCT03168061 (Ph II): Recruiting
Liposomes (gene therapy: Cancer)				
1	MRT5201-(Translate Bio)	mRNA encapsulated in PEGylated liposomes	Ornithine transcarbamylase deficiency	NCT03767270 (Ph I): Not yet recruiting
2	Lipo-MERIT-(Biontech RNA Pharmaceuticals)	Four naked ribonucleic acid (RNA)-drug products formulated with liposomes	Cancer vaccine for advanced melanoma	NCT02410733 (Ph I): Recruiting

(Continued)

Table 1.5 *(Continued)*
List of currently undergoing clinical trials as per clinicaltrial.gov

S. No.	Brand Name	Drug Form	Investigated Application	Clinicaltrial.gov Status
Liposomes (immunotherapy: Cancer)				
1	IVAC_W_bre1_uID	Patient-specific liposomes (specificity for antigen-expression on a patient's tumour) complexed RNA	Triple negative breast cancer	NCT02316457 (Ph 1): Recruiting
Liposomes (gene therapy: Vaccine)				
1	mRNA-1944-(Moderna)	Two mRNAs that encode heavy and light chains of anti-Chikungunya antibody formulated in Moderna's proprietary lipid nanoparticle technology	Safety, tolerability, pharmacokinetics and pharmacodynamics towards the prevention of Chikungunya virus infection	NCT03829384 (Ph 1): Recruiting
Micelles (cancer)				
1	MTL-CEBPA-(Mina alpha)	Double stranded RNA formulated into SMARTICLES amphoteric liposomes	Advanced liver cancer	NCT02716012 (Ph 1): Recruiting
2	Imx-110-(Immix Biopharma Australia)	Micelle encapsulating a Stat3/NF-kB/ poly-tyrosine kinase inhibitor and low-dose doxorubicin	Advanced solid tumours	NCT03382340 (Ph 1): Recruiting
3	IT-141-(Intezyne Technologies)	Micelle formulation of SN-38	Advanced cancer	NCT03096340 (Ph 1): Recruiting
Inorganic NPs (cancer)				
1	NU-0129-(Northwestern)	Spherical nucleic acid platform consisting of nucleic acids arranged on the two surfaces of a spherical gold NP	Glioblastoma	NCT03020017 (Ph 1): Active, not recruiting
NPs for imaging applications				
1	AguIX-(National Cancer Institute, France)	Polysiloxane Gd-chelate-based NPs	Advanced cervical cancer	NCT03308604 (Ph 1): Recruiting
2	ONM-100-(OncoNano Medicine)	Micelle covalently conjugated to indocyanine green	Intraoperative detection of cancer	NCT03735680 (Ph II): Not yet recruiting

Source: Anselmo and Mitragotri (2019).

1.8 Conclusions

Nanotechnology has been the cornerstone of impressive manufacturing technologies and rapid development over a period of century. For example, nanotechnology has a significant effect in the pharmaceutical areas of practice in drug delivery (liposomes, dendrimers, carbon nanotubes, NPs, fullerenes, etc.) and in medical devices such as diagnostic biosensors, drug delivery systems and imaging samples. It provides new technologies, possibilities and scope, which through its nanoengineered technologies are expected to have a great effect on many areas of illness, diagnostics, prognosis and illness care. Pharmaceutical nanotechnology offers opportunities for improving materials and medical devices and helping to develop new technologies where existing and more traditional technologies can meet their limits.

Abbreviations

ADME	Absorption, distribution, metabolism and excretion
BCS	Biopharmaceutical classification system
CAC	Critical aggregation concentration
CMC	Critical micelle concentration
CNT	Carbon nanotubes
COVID	Corona virus disease
ECM	Extra cellular matrix
EPR	Enhanced permeation retention
FDA	Food and Drug Administration
FRET	Fluorescence resonance energy transfer
LSPR	Localized surface plasmon resonance
MEMS	Micro electromechanical systems
NEMS	Nano electromechanical systems
NP	Nano particles
PAMAM	Poly- amidoamines
PEG	Poly ethylene glycol
PM	Polymeric micelle
QD	Quantum dots
SARS	Severe acute respiratory syndrome
SSRI	Selective serotonin reuptake inhibitor
STM	Scanning tunneling microscopy

ACKNOWLEDGEMENT

Dr. Keerti Jain is thankful to National Institute of Pharmaceutical Education and Research (NIPER), Raebareli for extending the facilities to write this book chapter.

The NIPER, Raebareli communication number for this manuscript is NIPER-R/Communication/139.

DISCLOSURE

The authors report no conflict of interest related to this article.

REFERENCES

Afsana, J. V., Haider, N., & Jain, K. 2019. 3D printing in personalized drug delivery. *Current Pharmaceutical Design*, *24*(42), 5062–5071.

Ahmad, J., Gautam, A., Komath, S., Bano, M., Garg, A., & Jain, K. 2018. Topical nano-emulgel for skin disorders: Formulation approach and characterization. *Recent Patents on Anti-Infective Drug Discovery*, *14*(1), 36–48.

Alshweiat, A., Csóka, I. I., Tömösi, F., et al. 2020. Nasal delivery of nanosuspension-based mucoadhesive formulation with improved bioavailability of loratadine: Preparation, characterization, and in vivo evaluation. *International Journal of Pharmaceutics*, *579*, IJP-119166.

Anselmo, A. C., & Mitragotri, S. 2019. Nanoparticles in the clinic: An update. *Bioengineering & Translational Medicine*, *4*(3), 1–16.

Anwekar, H., Patel, S., & Singhai, A. K. 2011. Liposome-as drug carriers. *International Journal of Pharmacy & Life Science*, *2*(7), 1–7.

Bagre, A. P., Jain, K., & Jain, N. K. 2013. Alginate coated chitosan core shell nanoparticles for oral delivery of enoxaparin: In vitro and in vivo assessment. *International Journal of Pharmaceutics*, *456*(1), 31–40.

Bajwa, N., Kumar Mehra, N., Jain, K., & Kumar Jain, N. 2016a. Targeted anticancer drug delivery through anthracycline antibiotic bearing functionalized quantum dots. *Artificial Cells, Nanomedicine and Biotechnology*, *44*(7), 1774–1782.

Bajwa, N., Mehra, N. K., Jain, K., & Jain, N. K. 2016b. Pharmaceutical and biomedical applications of quantum dots. *Artificial Cells, Nanomedicine and Biotechnology*, *44*(3), 758–768.

Bawarski, W. E., Chidlowsky, E., Bharali, D. J., & Mousa, S. A. 2008. Emerging nanopharmaceuticals. *Nanomedicine: Nanotechnology, Biology, and Medicine*, *4*, 273–282.

Bayda, S., Adeel, M., Tuccinardi, T., & Cordani, M. 2020. The history of nanoscience and nanotechnology: From chemical – physical applications. *Molecules*, *25*, 1–15.

Bhushan, B. 2016. *Introduction to nanotechnology: History, status, and importance of nanoscience and nanotechnology education*. In *Global Perspectives of Nanoscience and Engineering Education, Springer International Publishing Switzerland 2016*, pp. 1–31.

Borse, V. B., Konwar, A. N., Jayant, R. D., & Patil, P. O. 2020. Perspectives of characterization and bioconjugation of gold nanoparticles and their application in lateral flow immunosensing. *Drug Delivery and Translational Research*, *10*, 878–902.

Bosman, A. W., Janssen, H. M., & Meijer, E. W. 1999. About dendrimers: Structure, physical properties, and applications. *Chemical Reviews*, *99*(7), 1665–1688.

Chan, W. C. W. (2020). Nano Research for COVID-19. *American Chemical Society Nano*, 14, 3719–3720.

Chen, H., Luo, C., Yang, M., Li, J., Ma, P., & Zhang, X. 2020. Intracellular uptake of and sensing with SERS-active hybrid exosomes: Insight into a role of metal nanoparticles. *Nanomedicine*, *15*(9), 913–926.

Cuneo, T., Wang, X., Shi, Y., & Gao, H. 2020. Synthesis of hyperbranched polymers via metal-free ATRP in solution and microemulsion. *Macromolecular Chemistry and Physics*, *221*(6), 1–8.

Deng, S., Gigliobianco, M. R., Censi, R., & Di Martino, P. 2020. Polymeric nanocapsules as nanotechnological alternative for drug delivery system: Current status, challenges and opportunities. *Nanomaterials*, *10*(5), 1–39.

Dreadena, E. C., Alkilanyb, A. M., Huangc, X., Murphy, C. J., & Mostafa, A. E.-S. 2012. The Golden Age: Gold nanoparticles for biomedicine. *Department of Health and Human Sciences*, *4*(7), 2740–2779.

Eigler, D. M., & Schweizer, E. K. 1990. Positioning single atoms with a scanning tunnelling microscope. *Nature*, *344*, 524–526.

El-Gogary, R. I., Khattab, M. A., & Abd-Allah, H. 2020. Intra-articular Multifunctional Celecoxib Loaded Hyaluronan Nanocapsules For The Suppression of Inflammation In An Osteoarthritic Rat Model. *International Journal of Pharmaceutics*, IJ119378.

El-Sayed, M. A. (2001). Some interesting properties of metals confined in time and nanometer space of different shapes. *Accounts of Chemical Research*, *34*(4), 257–264.

Fana, M., Gallien, J., Srinageshwar, B., Dunbar, G. L., & Rossignol, J. 2020. PAMAM dendrimer nanomolecules utilized as drug delivery systems for potential treatment of glioblastoma: A systematic review. *International Journal of Nanomedicine*, *15*, 2789–2808.

Gnach, A., Lipinski, T., Bednarkiewicz, A., & Capobianco, J. A. 2014. Upconverting nanoparticles: Assessing the toxicity. *Chemical Society Reviews*, *44*, 1561–1584.

Gracia, B., Nano, C., Gracia, E. B., et al. 2019. Nanomedicine review: Clinical developments in liposomal applications. *Cancer Nanotechnology*.

Hu, X., Zhu, Z., & Dong, H., 2020. Inorganic and metal-organic nanocomposites for cascade-responsive imaging and photochemical synergistic effects. *Inorganic Chemistry*, *59*(7), 4617–4625.

Huang, H. J., Kraevaya, O. A., Voronov, I. I., Troshin, P. A., & Hsu, S. H. 2020. Fullerene derivatives as lung cancer cell inhibitors: Investigation of potential descriptors using qsar approaches. *International Journal of Nanomedicine*, *15*, 2485–2499.

Iijima, S. 1991. Helical microtubules of graphitic carbon. *Nature*, *354*, 56–58.

Jain, K. 2017. Dendrimers: Smart nanoengineered polymers for bioinspired applications in drug delivery. In *Biopolymer-Based Composites: Drug Delivery and Biomedical Applications*, Jana, S., Maiti, S., & Jana, S. (Eds.), London: Elsevier Ltd.

Jain, K. K. 2005. The role of nanobiotechnology in drug discovery. *Drug Discovery Today*, *10*.

Jain, S., & Jain, N. K. 2002. Liposomes as drug carrier. In *Controlled and novel drug delivery*, Jain, S. & Jain, N. K. (Eds.), 2nd edition, CBS Publisher (pp. 304–352).

Jain, K., Mehra, N. K., & Jain, N. K. 2014a. Potentials and emerging trends in nanopharmacology. *Current Opinion in Pharmacology*, *15*, 97–106.

Jain, V., Gupta, A., Pawar, V. K., et al. 2014b. Chitosan-assisted immunotherapy for intervention of experimental leishmaniasis via amphotericin B-loaded solid lipid nanoparticles. *Applied Biochemistry and Biotechnology*, *174*(4), 1309–1330.

Jain, K. 2018a. Dendrimers: Emerging anti-infective nanomedicines. In *Sequel volume of Multivolume International Book "NanoBioMedicine"*, Singh, B., Ho, R. J. Y., & Kanwar, J. R. (Eds.), CRC Press (pp. 121–138). eBook ISBN 9781351138666.

Jain, K. 2018b. Nanohybrids of dendrimers and carbon nanotubes: A benefaction or forfeit in drug delivery? *Nanoscience & Nanotechnology-Asia*, *9*, 21–29.

Joshi, K., Chandra, A., Jain, K., & Talegaonkar, S. 2019. Nanocrystalization: An emerging technology to enhance the bioavailability of poorly soluble drugs. *Pharmaceutical Nanotechnology*, *7*(4), 259–278.

Kahraman, E., Güngör, S., & Özsoy, Y. 2017. Potential enhancement and targeting strategies of polymeric and lipid-based nanocarriers in dermal drug delivery. *Therapeutic Delivery*, *8*(11), 967–985.

Kalyankar, T. M., Butle, S. R., & Chamwad, G. N. 2012. Application of nanotechnology in cancer treatment. *Research Journal of Pharmacy and Technology*, *5*(9), 1161–1167.

Kaur, A., Jain, K., Mehra, N. K., & Jain, N. K. 2017. Development and characterization of surface engineered PPI dendrimers for targeted drug delivery. *Artificial Cells, Nanomedicine and Biotechnology*, *45*(3), 414–425.

Khan, M. A., Jain, V. K., Rizwanullah, M., Ahmad, J., & Jain, K. 2019. PI3K/AKT/mTOR pathway inhibitors in triple-negative breast cancer: A review on drug discovery and future challenges. *Drug Discovery Today*, *24*(11), 2181–2191.

Khopde, A. J., & Jain, N. K. 2001. Dendrimer as potential delivery system for bioactive. In *Advances in controlled and novel drug delivery*, Jain, N. K. (Ed.), CBS Publisher, New Delhi (pp. 361–380).

Klajnert, B., & Bryszewska, M. 2001. Dendrimers: Properties and applications. *Acta Biochimica Polonica*, *48*(1), 199–208.

Kroto, H. W. (1985). Buckminsterfullerene. *This Week's Citation Classic, Nature*, *318*, 162–163.

Kroto, H.W., Health, J. R., O'Brien, S. C., Curl, R. F., & Smalley, R. E. 1985. Buckminsterfullerene. *Nature Publishing Group*, *318*, 162–163.

Kwun, Y. C., Farooq, A., Nazeer, W., Noreen, S., Kang, S. M., & Zahid, Z. 2018. Computations of the M-polynomials and degree-based topological indices for dendrimers and polyomino chains. *International Journal of Analytical Chemistry*.

Maiti, P. K., Tahir, C., Wang, G., & Goddard, W. A. 2004. Structure of PAMAM dendrimers: Generations 1 through 11. *Macromolecules*, (*37*), 6236–6254.

Malamatari, M., Taylor, K. M. G., Malamataris, S., Douroumis, D., & Kachrimanis, K. 2018. Pharmaceutical nanocrystals: production by wet milling and applications. *Drug Discovery Today*, *23*(3), 534–547.

Marasini, N., Haque, S., & Kaminskas, L. M. 2017. Polymer-drug conjugates as inhalable drug delivery systems: A review. *Current Opinion in Colloid and Interface Science*, *31*, 18–29.

Michalet, X., Pinaud, F. F., Bentolila, L. A., et al. 2005. Quantum dots for live cells, in vivo imaging, and diagnostics. *National Institute of Health*, *307*(5709), 538–544.

Misra, R., Acharya, S., & Sahoo, S. K. 2010. Cancer nanotechnology: Application of nanotechnology in cancer therapy. *Drug Discovery Today*, *15*(19–20), 842–850.

Mohan, S., & Varshney, H. M. 2012. "Nanotechnology" current status in pharmaceutical science: A review. *International Journal of Therapeutic Applications*, *6*(April), 14–24.

Muthu, M. S., & Singh, S. 2009. Targeted nanomedicines: Effective treatment modalities for cancer, AIDS and brain disorders. *Nanomedicine*, *4*(1), 105–118.

Pandit, S., Dasgupta, D., Dewan, N., & Ahmed, P. 2016. Nanotechnology based biosensors and its application. *The Pharma Innovation Journal*, *5*(6), 18–25.

Pinot, M., Vanni, S., Pagnotta, S., et al. 2014. Polyunsaturated phospholipids facilitate membrane deformation and fission by endocytic proteins. *Lipid Cell Biology*, *345*(6197), 693–697.

Quan, Q., & Zhang, Y. 2015. Lab-on-a-Tip (LOT): Where nanotechnology can revolutionize fibre optics. *Nanobiomedicine*, *2*, 1–5.

Rangasamy, M. 2011. Nano technology: A review. *Journal of Applied Pharmaceutical Science*, *1*(2), 8–16.

Sanvicens, N., & Marco, M. P. 2008. Multifunctional nanoparticles – Properties and prospects for their use in human medicine. *Trends in Biotechnology*, *26*(8), 425–433.

Sharma, P., Jain, K., Jain, N. K., et al. 2017. Ex vivo and in vivo performance of anti-cancer drug loaded carbon nanotubes. *Journal of Drug Delivery Science and Technology*, *41*, 134–143.

Shi, J., Votruba, A. R., Farokhzad, O. C., & Langer, R. 2010. Nanotechnology in drug delivery and tissue engineering: From discovery to applications. *NIH Public Nano Letters*, *10*(9), 3223–3230.

Sportelli, M. C., Izzi, M., Kukushkina, E. A., et al. 2020. Can nanotechnology and materials science help the fight against SARS-CoV-2? *Nanomaterials*, *10*(4), 802.

Sutradhar, K. B., & Amin, M. L. 2014. Nanotechnology in cancer drug delivery and selective targeting. *ISRN Nanotechnology*, *2014*, 1–12.

Taipale, J., & Keski-Oja, J. 1997. Growth factors in the extracellular matrix. *The FASEB Journal*, *11*(1), 51–59.

Tiwari, A. K., Gajbhiye, V., Sharma, R., & Jain, N. K. 2010. Carrier mediated protein and peptide stabilization. *Drug Delivery*, *17*(8), 605–616.

Tyagi, Y., & Madhav, N. S. 2020. Inbuilt novel bioretardant feature of biopolymer isolated from cucumis sativa for designing drug loaded bionanosuspension. *Journal of Drug Assessment*, *9*(1), 72–81.

Usui, Y., Haniu, H., Tsuruoka, S., & Saito, N. 2012. Medicinal chemistry carbon nanotubes innovate on medical technology. *Medicinal Chemistry*, *2*, 1–6.

Ventola, C. L. 2017. Progress in nanomedicine: Approved and investigational nanodrugs. *Pharmacy and Therapeutics*, *42*(12), 742–755.

Villanueva-Martínez, A., Hernández-Rizo, L., & Ganem-Rondero, A. 2020. Evaluating two nanocarrier systems for the transdermal delivery of sodium alendronate. *International Journal of Pharmaceutics*, *582*, IPJ-119312.

Wagner, V., Husing, B., Gaisser, S., & Bock, A.-K. 2006. Nanomedicine: Drivers for development and possible impacts. *European Commission's Joint Research Centre Report No. 46744*, 1–116.

Wan, K. Y., Weng, J., Wong, S. N., Kwok, P. C. L., Chow, S. F., & Chow, A. H. L. 2020. Converting nanosuspension into inhalable and redispersible nanoparticles by combined in-situ thermal gelation and spray drying. *European Journal of Pharmaceutics and Biopharmaceutics*, *149*(January), 238–247.

Wanigasekara, J., & Witharana, C. 2016. Applications of nanotechnology in drug delivery and design – An insight. *Current Trends in Biotechnology and Pharmacy*, *10*(1), 78–91.

www.dendritech.com. 2017. *PAMAM Dendrimers*. Available from http://www.dendritech.com/pamam.html

Xu, W., Ling, P., & Zhang, T. 2013. Polymeric Micelles, a promising drug delivery system to enhance bioavailability of poorly water-soluble drugs. *Journal of Drug Delivery*, *2013*(1), 1–15.

Xu, X., Ray, R., Gu, Y., et al. 2004. Electrophoretic analysis and purification of fluorescent single-walled carbon nanotube fragments. *Journal of American Chemical Society*, *(126)*, 12736–12737.

Zhang, Y., Bai, Y., & Yan, B. 2010. Functionalized carbon nanotubes for potential medicinal applications. *Drug Discovery Today*, *15*(11–12), 428–435.

Zhang, Y., Huang, Y., & Li, S. 2014. Polymeric micelles: Nanocarriers for cancer-targeted drug delivery. *AAPS PharmSciTech*, *15*(4), 862–871.

2

Fundamental Aspects of Targeted Drug Delivery

Ameeduzzafar, Nabil K. Alruwaili, Md. Rizwanullah, Syed Sarim Imam, Muhammad Afzal, Khalid Saad Alharbi and Javed Ahmad

CONTENTS

2.1 Introduction

For many years, a disease (acute and chronic) has been treated by many formulations such as tablets, capsules, cream, paste, ointment, pills, syrup, aerosol, etc. These conventional dosage forms are primarily prescribed pharmaceutical products. Treatment of any disease by the conventional systemic delivery system is a significant challenge. Conventional systemic drug delivery systems distribute the drug to all parts of the body and are deficient in drug-specific affinity towards a treatment site (pathological organs, tissues or cells). They require more doses to achieve the desired concentration for therapeutic responses. The drug is deposited into any healthy organ or tissue, which is not involved in the treatment process. In addition, these systems require a large amount of drug for therapeutic activity and cause undesirable side or toxic effects (Haider et al. 2020).

Drug targeting provides new and novel approaches for solving all problems related to conventional systemic drug therapy. Drug targeting means the delivery of drugs only to the particular organ selectively and quantitatively. In drug targeting, most of the drug accumulated only in effective organs and tissue and an insignificant concentration level in other organs and tissue, which prevents the unnecessary side or toxic effect (Rizwanullah et al. 2018; Rehman et al. 2019). Drug targeting can be done by considering the points such as (a) specific properties of target cells, (b) nature of markers or transport carriers or vehicles, which transport the drug to a specific target site and (c) ligands and physically modulated components. For these purposes, so many strategies, including individualizing drug therapy, nanoparticle-based drug delivery systems (polymeric nanoparticles and lipid nanoparticles), drug conjugates, therapeutic drug monitoring and stimuli-sensitive targeted therapy, are used intensely (Akhter et al. 2018).

Many of the conventional and gene-based drugs can be transported selectively to a particular target site of the body, such as organs, tissues and cells. Liposomal formulation Doxil® (Doxorubicin) (Wang et al. 2013) and Abraxane® nanoparticles (Paclitaxel) have been approved by the US FDA as new target drug delivery systems. Tumour-targeting was also performed with specific monoclonal antibodies such as Erbitux® (Cetuximab), Panitumumab (Vectibix®), Trastuzumab (Herceptin®), Imatinib (Gleevec®), Erlotinib (Tarveca®), Sorafenib (Nexavar®) and Sunitinib (Sutent®). These drugs are clinically FDA approved and used in clinics (Emery et al. 2009; Zhou et al. 2009; Tsai et al. 2017).

The objectives of drug targeting are the selective convey, absorption and dispersion of the therapeutic agent to the target site, minimizing the dose of the drug, minimizing the undesirable side effects and achieving the optimal pharmacological response (Rizwanullah et al. 2020)

2.2 Advantages of Drug Targeting

The potential advantages of the drug targeting approach are as follows:

- Reducing the frequent dosing.
- Enhanced bioavailability.
- Improving the fluctuations in plasma levels.
- Lowering the risk of unwanted side effects.
- Showing better targeting.
- Reducing the unwanted drug deposition to other organs.
- Constant rate delivery of drugs and giving a required therapeutic response.
- Minimizing the exposure of the drug to the biological environment.

- Enhancing the patient compliance.
- Improving the physical, chemical and biological stability of therapeutics.
- Enhanced permeability and retention (EPR) effect.

2.3 Pharmacokinetic Consideration in Drug Targeting

The pharmacokinetics can be explained as the study of drug absorption, distribution, metabolism and excretion in the body with respect to time. It has largely been used to describe how a drug is handled by the body. The pharmacokinetic data are used to correlate the therapeutic/toxicological effects of the drug, i.e., its pharmacodynamics (Ahmad et al. 2015b).

The development of controlled release formulation is not only for control or delayed release of the drug, but also to formulate to maintain the desired concentration level in a particular target site/blood. Pharmacokinetics is important for developing new carrier systems suitable for targeted drug delivery. The physiologically based pharmacokinetic modelling was first proposed by Bischoff and Dedrick (Bischoff and Dedrick, 1968).

In early years, researchers have focussed on the compartmental analysis to study the pharmacokinetics. Compartmental analysis has been assessed by considering the human body as a series of black boxes or compartments. The ideal compartmental model for the drug acting at the target site, manifesting its toxicity at the toxic site and drug distribution through the remainder of the body, is shown in Figure 2.1A. The drug distributes between the two macroscopic compartments other than the response site and toxic site: (i) a series of accessible tissues that are in rapid equilibrium with the plasma or central compartment and (ii) a series of less accessible tissues that are not in ready equilibrium with the plasma or tissue compartment. The ideal drug is one whose greatest affinity is for its response on the target site and which has minimal contact with the non-toxic site.

For better understanding of pharmacokinetics of the drug, targeting requires considering the points such as clearance analysis for tissue distribution of drug carriers, prolongation of retention time in blood, reduction of glomerular filtration, reduction of reticuloendothelial uptake and EPR effect for tumours (Maeda et al. 2000; Maeda, 2001). Moreover, the drug is released in the target site and then eliminated and not distributed to other organs or tissues (Lukyanov et al. 2004). When the dosage form of the drug (nanoparticles, liposomes, dendrimers, drug conjugates, etc.) reaches the target region, many additional processes take place such drug release, drug elimination, drug–receptor binding on the target cell surface and receptor binding, lysosome uptake, nuclear targeting, drug elimination, drug binding, etc., in the intracellular target region (Petrak, 2005). The conventional dosage form is rapidly absorbed in the body. The frequency of the dose regimen is determined through data of the biological half-life and therapeutic index.

$$\text{Therapeutic index} = \frac{\text{Maximum tolerated toxic dose}\left(\text{MTTD}\right)}{\text{Minimum effective concentration}\left(\text{MEC}\right)}$$

The main drawback of conventional drug delivery systems is their inabilities to exhibit the therapeutic efficacy of those drugs having low therapeutic index and short elimination half-life. This drawback can be short by targeted drug delivery approach.

The following pharmacokinetic behaviour of target delivery was identified.

- Maintenance of therapeutic concentration with minimal fluctuation over a prolonged period throughout a dosing interval.
- Keeping the drug concentration within the therapeutic range at a steady state.
- Maintaining the release of the drug in such a way that it gives at least 80% bioavailability as compared to conventional therapy.
- Increasing the elimination half-life by the design of a suitable dosage regimen.
- Better patient compliance and lower incidence of toxicity.

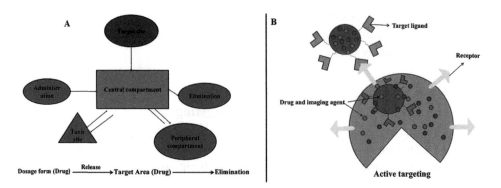

FIGURE 2.1 Illustration showing the concept of drug delivery and targeting. (A) Compartmental distribution of the drug after administration. (B) Transport of the therapeutic agent through active targeting.

The pharmacokinetics of targeted (controlled release) delivery can be divided into two release patterns, zero order and first order. The following assumption considered the release pattern of the drug from the targeted delivery.

- Drug absorption, metabolism and excretion are first-order rate processes.
- Drug absorption and elimination are irreversible.
- The drug released after oral administration is completely absorbed.
- The rate-limiting step is the absorption process.

For the development of target drug delivery, the following points should also be taken into consideration.

2.3.1 Rate of Elimination

There must be controlled removal of the drug carrier conjugate from the systemic circulation. The drug–carrier conjugate and the systemic compartment environment need to be eliminated during the design and development of targeted delivery systems. There must be non-specific interaction between the conjugate and compartment. The carrier should have the ability to restrict all unwanted interactions between the drug and the physiological environment (Morgan et al. 1990; Opanasopit et al. 2002).

2.3.2 Release Rate to the Non-Target Site

The drug release at the non-target site could lead to obtaining benefits, and release of the drug at the target site might help to obtain maximum therapeutic activity. The drug release at non-target sites causes toxicity due to the presence of higher drug concentration.

2.3.3 Delivery Rate to the Target Site

The desired therapeutic effect at the site of action can never be achieved if the drug conjugates reach the target site very slowly. The free drug concentration must reach the therapeutic level (i.e., the area under the curve in a drug concentration versus time plot for the target site) to obtain the desired pharmacological action. The delivery to the target organ by the drug carrier conjugate does not guarantee the availability of the free drug at the actual target site.

2.3.4 Rate of Removal of the Free Drug from the Target Site

Targeted drug delivery systems are those that deliver the drug at the desired target site in maximum concentration. The drug should be encapsulated for selective targeted delivery, and the delivery system should not have performed poorly. DNA in gene therapy needs to be delivered into the targeted area cytoplasm, so it will be preferred to release the drug within the cells and enhance retention of the drug to the proximity to its target area. The high elimination rate of the free drug from the central compartment tends to increase the required rate of input (of the drug carrier) to maintain a therapeutic effect (Boddy et al. 1989).

2.3.5 Elimination Rate of the Drug–Carrier Conjugate and Free Drug from the Body

The elimination of a complete drug–carrier system should be minimum for the maximum drug targeting. The elimination of these delivery systems from kidneys is difficult due to their large size (Petrak and Goddard, 1989). The removal of drug conjugates from the circulation is mainly done by the liver. The elimination rate of the free drug from systemic circulation should be rapid and fast. It is related to the free drug transfer from the target site to the central compartment of the body. In this way, the drug-delivery system will achieve a decrease in drug-associated toxicity.

2.4 Rationale for Drug Targeting

The basic rationale of drug targeting is the manipulation of the pharmacodynamics of the therapeutic agent by enhancing the pharmacokinetics. Targeted drug delivery systems can diminish the unfavourable effects by changes in the unfortunate disposition of the therapeutic agent and decreasing its occurrence in the non-targeted region. It also enhances the therapeutic efficacy of the API by preventing its deactivation by chemical or enzymatic action into the body as well as no degradation before accomplishment of the site of a target with a reduction of the dose (Rizwanullah et al. 2016; Barkat et al. 2020). This rationale can be achieved by the development of novel drug delivery such as nanoparticulate systems, liposomes, solid lipid nanoparticles, nanostructured lipid carriers, dendrimers, etc. in the form of active targeting or passive or self-programmed approaches.

The most crucial advantage of the nanoparticle-based system is the ability to encapsulate more than one drug with different physicochemical properties. Two drugs with different physicochemical properties can be co-delivered for a combined effect and to avoid multi-drug resistance (Ahmad et al. 2016). Therapeutic proteins, peptides and nucleic acid can also be co-delivered with nanoparticles. Nanoparticles exhibit significantly higher drug loading ability, higher storage stability and controlled release of the drug (Ahmad et al. 2017). The controlled drug release profile of nanoparticles is because of the slow degradation of polymers/lipids present in the core. In addition, targeting ligands can be conjugated to the surface of nanoparticles. By surface decoration with different targeting ligands, nanoparticles can be actively targeted to the site of action. Furthermore, the PEGylation of nanoparticles enhances their hydrophilicity and increases their biological half-life by decreasing clearance by the reticuloendothelial system (Shen et al. 2016). The diagrammatic representation of the surface-decorated nanoparticulate system is illustrated in Figure 2.2.

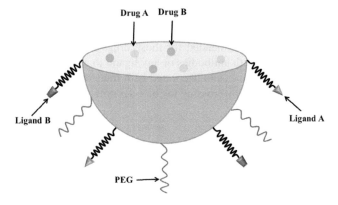

FIGURE 2.2 Illustration showing the surface decorated nanoparticulate system.

2.5 Factors Influencing Drug Targeting

Different factors influencing the drug targeting are discussed below.

2.5.1 Low Absorption of Therapeutic Agents

Many therapeutic agents such as alendronate, gentamycin, vancomycin, paclitaxel, docetaxel, topotecan and peptides and proteins specifically suffer from low absorption due to the enzymatic barriers and poor permeation across the biological membrane (Meerum et al. 1999; Reif et al. 2002; Chu et al. 2008; Park et al. 2015). Peptides that have a high molecular weight (MW) (>6000 Da) and have low permeability across the biological membrane, and hence are not suitable for oral administration and require more frequent doses on parenteral administration. Targeted delivery achieved the desired concentration in the target region and enhanced the absorption of such types of therapeutic agents (Renukuntla et al., 2013; Choonara al., 2014).

2.5.2 Short Half-Life

The half-life of the drug is also an essential parameter to be considered in targeted drug delivery. The half-life of the therapeutic agents or gene, which is low or up to 4 h, is an excellent candidate for targeted delivery because it reduces the dose, frequency and fluctuation of the drug in the response surface graph. The mRNA genes which have short half-life can be regulated by targeting polymerase to RNA binding proteins (Pérez-Herrero and Fernández-Medarde, 2015).

2.5.3 Volume of Distribution

The volume of distribution indicated the drug distributed in the body, which means the volume of body fluid in which the therapeutic agent dissolved. The drug has a large volume of distribution which indicated the therapeutic agent present in body tissue at desired concentration action did not reach the target site, whereas less volume of distribution indicated that the drug is present in the systemic circulation. The drug (acidic drug) has a plasma protein-bound smaller volume of distribution, whereas basic drugs have a large volume of distribution because they are bound to the extravascular region (De Buck et al. 2007; Wooten, 2012; Holt et al. 2019). Moreover, the volume of distribution is influenced by the state of disease of the patient. Fewer serum albumins are present for binding with the drug in case of liver failure, allowing a higher volume of distribution. The drug-like propranolol and digoxin have a large volume of distribution, necessary to make a target delivery for delivering the therapeutic drug to the target site for therapeutic action and prevent undesirable side effects.

2.5.4 Specificity of Drugs towards the Target Site

Lack of target specificity and selectivity of the drug towards the target site than the normal organ/ cells/tissues makes inadequate deposition, leading to little or zero therapeutic activity. Delivery of drugs with receptor targeting is recommended for those drugs that have low specificity. It can be achieved by conjugating the drug with antibodies or any ligand to augment target specificity. Protein or peptide loaded PEGylated liposomes and decorated with transferrin showed a significant concentration of drug that reached the brain by crossing the blood–brain barrier (BBB) (Kamaly et al. 2012; Guan et al., 2019).

2.5.5 Low Therapeutic Index

Drugs with a low therapeutic index require a more amount of dose or increase the frequency for desired pharmacological conditions but increase the side effect or toxic effects. Drug targeting increases the therapeutic index by transporting all the drug to the target site or receptor site without accumulation of the drug in the normal cell and hence reduces the dose frequency as well as toxicity. Most of the anti-cancer drugs such as methotrexate, 5-fluorouracil and carboplatin show higher toxicity due to their low therapeutic index (Muller and Milton, 2012; Paci et al. 2014).

2.5.6 Concentration of Drugs

Drugs at required concentration at the receptor site produced the pharmacological action; it means the drug concentration must be retained within the therapeutic window. If administered a single dose in a large quantity, it may be beyond the therapeutic window (maximum safe concentration) and lead to an undesirable side effect; in addition, it abruptly declines below the minimum effective concentration of the blood plasma profile. However, in multiple dosing, the variation in drug levels can be reduced, but it reduces patient compliance. Hence, the targeted delivery transports the desired therapeutic concentration to the target site/organ/cells/tissue without affecting the normal cells (Tekade et al. 2015). Stock and his associates studied the concentration-dependent binding in the membrane proteins with the ligands. They showed a significant concentration of ligand-bound protein between therapeutic windows, which showed significant activity in cancer and neurological diseases (Stock et al. 2017).

2.5.7 Particulate Location and Distribution

Any therapeutic agent administered into the body moved towards the target site (protein/macromolecule) and bound with the active site. After that, pharmacological action takes place due to the change in the function of the binding molecule. In this case, the transporter or carriers play an important role. The transporter is a membrane-bound protein; it conveys the therapeutic agent across membranes or inside the target region. The carrier is a secreted protein that carries the bound drug to the cell transporters and increases the concentration of therapeutic molecules to the target site (Chomoucka et al. 2010). Target delivery decreased the clearance rate because most of the drug reaches the target site, enhances the half-life and prevents the toxic effect on other organs. Receptor sink is the term used for such target-dependent clearance, which plays an important role in the internalization of the therapeutics/biological through cell membrane receptors. Selection of novel techniques is also important for targeting to avoid the toxic effect. Delivery of the drug to the body for the management of any diseases is not easy because it was affected by various parameters such as serum protein interaction, types of nanocarriers, clearance and deposition in the non-targeted organ. This can be prevented by manipulation in the formulation like a modification on the surface. Liposomes are well-known formulations and exhibit a site-avoidance mechanism which means not to come in contact with organ/tissue/cells such as kidneys, brain or heart. Nanoparticles loaded with docetaxel showed significant anti-tumour activity in breast cancer in vivo in a targeted form (Tao et al. 2013).

2.5.8 Molecular Weight (MW) of the Drug

MW of the drug is an essential parameter for drug targeting. Low MW (<500 Da) means easy penetration into the cell membrane through a diffusion mechanism like a chemotherapeutic drug. The MW of the drug (>1000 kDa) decreases the penetration across the cellular membrane along the gastrointestinal tract (Mohammed et al. 2017).

Lysozymes are low MW proteins that are suitable carriers to target drugs to kidneys. They are quickly reabsorbed and catabolized by the proximal tubular cells of the kidneys and release the drug into the kidneys and give a proper pharmacological action. Low MW chitosan is a valuable carrier for renal targeting of prednisolone in the proximal tubules by receptor-mediated endocytosis. Heparin is also an example of low MW protein for targeting of lymph node metastasis (Ye et al. 2015). A 5-fluorouracil loaded Au-NPs/chitosan nanoparticle delivery (25 kD) showed significant accumulation in the cancer cell and increased the half-life of the drug without affecting normal cells (Salem et al. 2018).

2.5.9 Physiochemical Properties

Solubility, ionization and lipophilicity are the key factors for drug targeting. Lipophilicity is the main factor for crossing of a drug to the BBB (diacetylmorphine/heroin) (Veiseh et al. 2010). Ionization means the development of charge on the molecules due to hydronation or dehydronation (a functional group at pH 1.5–7.8). The binding of a drug to the active site also depends upon ionization. The unionized state is involved in the hydrogen bond, whereas the ionized state provides the strength of salt bridges or H-bonds. The acidity or basicity of the therapeutic molecule is responsible for controlling the absorption or distribution (Balamuralidhara et al. 2011; Rizwanullah et al. 2017).

Ionization of drug molecules also participate as an approach for the targeting. Most of the cell membranes of the body have a negative charge (BBB, cornea). The molecule has a negative charge due to which it easily crossed the membrane. The molecules have a negative charge, or neutral charge first make it as positively charge by using any carriers such polymer or antibody and then helpful in efficient transportation. Macromolecules such as protein and peptide cationization by conjugation of positively charged albumin or antibodies are administered as targeted delivery like in the BBB due to the electrostatic process. The solubility of a drug increases the dissolution profile, hence giving it an improved possibility to attain a higher volume of distribution and to reach more drugs to target organs/tissue/cells. The injectable route varies solubility, i.e., intravenous, subcutaneous and intravenous routes. The soluble drug was mostly administered through the intravenous route. However, other parenteral routes have limitations such as the volume of injection (SC-2.5 mL. IM-5 mL).

Protection of drug degradation in the body environment is the effect on targeting because the maximum amount of drugs reached the target area. There are many techniques that prevent the degradation of the drug in which the PEGylation techniques can protect against proteolytic degradation as well as increase the solubility of therapeutic agents (Remaut et al. 2007; Veronese and Pasut, 2008; Mishra et al. 2016). Most of the anti-cancer therapies were proved by many researchers by PEGylation (Tekade and Sun, 2017).

2.5.10 pH of the Environment

Targeting of the drug is influenced by the pH of the environment. The targeting of the drug to the colon depends on pH. Treatment of tumours by passive targeting is performed at low pH because internal stimulation takes place in the pH gradient (Duan et al. 2013; Soni et al. 2018; Harshita et al. 2019). Cellular internalization, like endocytosis, is influenced by pH, i.e., early endosomes at acidic pH, while late endosomes are mildly acidic pH (5.5), which results in liposome formation. The acidic pH system will permit entry into the endosomes and help to release the drug and diffuse into the cytoplasm (Liu et al. 2014a).

2.6 Levels and Approaches of Drug Targeting

The therapeutic efficacy of the targeted drug delivery system would not only enhance the therapeutic efficacy of drugs but also reduce the toxicity associated with the drug to allow the use of lower doses of the drug in therapeutic treatment. There are two approaches used for drug targeting such as active targeting and passive targeting.

2.6.1 Passive Targeting

Passive targeting includes the deposition of the drug or drug–carrier system at a specific site due to physicochemical or pharmacological factors, without involving any targeting ligand which enhanced permeability and retention due to direct entry into leaky vasculature. It facilitates the advancement of a targeted nanocarrier structure loaded with therapeutic agents for an improved effective drug profile with insignificant toxic effects (Ahmad et al. 2015a; Yu et al. 2015). The drug or drug carrier nanosystems can be passively targeted, making use of the pathophysiological and anatomical opportunities. Examples include the targeting of anti-malarial drugs for the treatment of leishmaniasis, brucellosis, candidiasis and cancer.

2.6.2 Active Targeting

Active targeting employs specific modification or surface engineering of drug/drug carrier systems with active agents having a selective affinity for recognizing and interacting with a specific cell, tissue or organ

(Figure 2.1B) in the body after circulation and extravasations (Choi and Kim, 2007; Jhaveri and Torchilin, 2016; Anarjan, 2019). The direct pairing of drugs to targeting ligands restricts the pairing capacity to a few drug molecules. In contrast, the union of drug carrier nanosystems and ligands allows the introduction of thousands of drug molecules by means of one receptor-targeted ligand (Danhier et al. 2010; Ahmad et al. 2020). Xi and his associates developed curcumin-loaded self-assembled micelles using alendronate-hyaluronic acid-octadecanoic acid (ALN-HA-C18) for osteosarcoma therapy. Curcumin-loaded ALN-HA-C18 micelles showed a high affinity to the bone. Curcumin-loaded ALN-HA-C18 micelles showed much higher cytotoxic activity against MG-63 cells compared to free curcumin (Xi et al. 2019). A significant amount of doxorubicin was accumulated in the bone metastatic cancer cell after administration of intravenous administration of doxorubicin-loaded DOX-hyd-PEG-ALN micelle-targeted delivery (Ye et al. 2015). Significant cytotoxicity of vancomycin was observed on the administration of vancomycin in the form of alendronate-decorated biodegradable polymeric micelles in bone-targeting with significant anti-bacterial activity (Cong et al. 2015). The active targeting delivery approach can be further categorized into three different levels of targeting.

2.6.2.1 First-order targeting

It is also known as organ level targeting. It is defined as to the limited distribution of the drug–carrier systems to the capillary bed of a programmed target area like organs or tissue. An example includes compartmental targeting in the peritoneal cavity, joints, pleural cavity, lymphatics, cerebral ventricles, eyes, etc. Carboplatin-loaded polycaprolactone nanoparticles showed significant deposition into the brain via nose delivery by crossing the BBB as well as showed significant improvement in cytotoxicity in LN229 cells (Alex et al. 2016).

2.6.2.2 Second-order Targeting

Second-order targeting refers to site-specific delivery of therapeutics to definite cell types such as tumour cells and not deliver to the normal cells. Targeted delivery of drugs in the kupffer cells of liver is an example of second-order targeting. Targeted delivery of doxorubicin with the galactosamine ligand showed significantly more effectiveness in killing HepG2 cells than doxorubicin (Shen et al. 2011). Primaquine (PQ) emulsion showed more accumulation in the hepatocyte cell of the liver for the treatment of vivax malaria (Dierling and Cui, 2005). PLGA nanoparticles of paclitaxel conjugated with galactose ligands showed significant accumulation into the liver cell and significant anti-cancer efficacy with insignificant toxicity to other cells or tissues or organs (Sakhrani and Padh, 2013). Melgert et al. (2001) investigated that when a significant amount of dexamethasone accumulated in Kupffer cell of liver and observed on administration, it conjugated with mannosylated albumin as a ligand.

2.6.2.3 Third-order Targeting

In this, the targeting of the therapeutic agent takes place in intracellular targeted site/cells by the endocytosis mechanism or by using non-toxic ligands. Biswas and his associates developed surface engineering liposomes of paclitaxel for mitochondrial targeting. The formulation showed significant deposition of drugs in mitochondria of cancer cells as well as significant anti-cancer activity into cell culture as compared to unmodified liposomes (Biswas et al. 2012). Yang and his associates developed exosomes of linezolid for targeting of intracellular MRSA infection and showed more effectiveness against intracellular MRSA infections in vitro and in vivo than the free linezolid (Yang et al. 2018).

2.6.2.4 Ligand-Mediated Targeting

Targeting of drugs with the ligand is the novel tool for conveying a therapeutic agent to the target organ without any effect on other organs. This targeting is also called active targeting. Ligands are molecules that are attached to the surface of a therapeutic agent or any nanocarrier system such as liposomes and recognized the specific receptor site and transport them (Bouillon et al. 2010) because they had appropriate functional groups. Peptides, proteins, antibodies, oligosaccharides, folate and transferrin are the example of bioactive ligands (Disney and Angelbello, 2016; Warner et al. 2018). Mostly, such coordination is at the surface of the molecule, so surface modification leads to better recognition and interactions (Liu et al. 2012; Tekade and Sun, 2017). Some examples of ligand-mediated drug targeting are shown in Table 2.1.

2.6.3 Physical Targeting

2.6.3.1 pH-Sensitive Systems

Release of the drug from the carriers or delivery systems depends upon the pH (Figure 2.3A). These are cross-linked nanoparticles constituting acidic or basic moieties that participate in swelling of polymers based on pH in the body. Appropriate selection of polymer depends upon physiological conditions of the target region (Yuan et al. 2010). Change in the pH enhances the repulsion within the network of the polymer and releases the drug into the target region. Most of the colonic delivery was done by the pH-sensitive polymer. The polymers which are used as a pH-sensitive polymer are polyacrylic acids, methacrylic acid (MAA), polyethyleneimine (PEI) and cellulose acetate phthalate (Liu et al. 2014a; Mukhopadhyay et al. 2018). Woraphatphadung and his associates developed pH-sensitive polymeric micelles with curcumin colon-targeting. They showed a different release profile, i.e., 20% release in stomach pH and 50–55% in intestinal fluid and 60–70% in colon fluid (Woraphatphadung et al. 2018).

Fan and his associates developed pH-sensitive liposomes using oleic acid (OA), cholesteryl hemisuccinate (CHEMS) and linoleic acid (LA) and compared them with cholesterol and phosphatidyl ethanolamine (PE). They showed enhanced responsiveness at the tumour microenvironmental level as well as exhibited tumour-targeting with controlled drug release (Fan et al. 2017). The pH-sensitive dendrimers are used in biomedical applications. The pH-sensitive dendrimers have many advantages, such as increased efficacy, high payload of a drug, reduced toxicity, controlled release and high drug targeting. Kono and his associates developed dendrimers using pH-sensitive polymer poly(L-glutamic acid) and coated them with fully PEGylated doxorubicin. The dendrimers exhibited fast release in a weak acid environment with higher cytotoxicity than free drug (Kono et al. 2008).

Qi and his associates developed pH-sensitive carboxymethyl chitosan-modified polyamidoamine dendrimers and showed significant drug release in the tumour at a specific pH (Qi et al. 2016). Nguyen and

TABLE 2.1

Ligand-mediated drug targeting

Ligand	Carrier	Therapeutic Agent	Receptor/ Target Site	Significance	Ref.
Folate (Folic acid)	Liposomes	Mitomycin C and doxorubicin	Folate	Overexpressed on prostate cancer cells significantly enhanced cytotoxicity when compared to non-folate liposome	Patil et al. (2018)
Transferrin	Nanoparticle	Paclitaxel	Transferrin	Overexpressing in Transferrin receptor. Significantly promote drug delivery to cancer cells. higher intracellular uptake especially in nuclei	Nag et al. (2016)
New p32/gC1qR peptide	Nanoparticle	LyP-1	P-32 peptide	p32-expressing breast tumours in mice	Paasonen et al. (2016)
Galactosamine	Nanoparticle	Docetaxel	Galactosamine	Significantly reduced tumour size most significantly on hepatoma-bearing nude mice	Zhu et al. (2016)

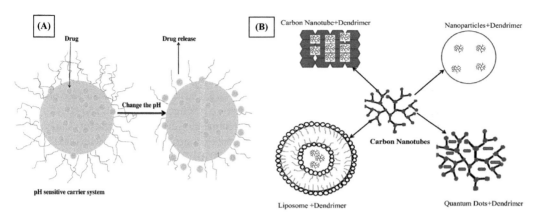

FIGURE 2.3 Illustration showing physical targeting and a hybrid system for drug targeting. (A) pH-Sensitive system for drug targeting. (B) Hybrid-dendrimer system for drug targeting.

his associates developed a redox and pH dual-responsive drug delivery system heparin (Hep)-conjugated poly (amidoamine) dendrimer (P-SS-Hep) via a redox-sensitive disulphide bond. P-SS-Hep has been shown to be an effective drug-loading carrier along with the ability to deliver the loaded drug to the target site (Nguyen et al. 2017).

Inapagolla and his associates developed methylprednisolone conjugated poly(amidoamine) (PAMAM) dendrimers and exhibited reducing allergen-induced airway inflammatory response (87% and 67% reduction) (Inapagolla et al. 2010).

2.6.3.2 Temperature-sensitive Systems

These open-loop delivery systems are commonly referred to as 'pulsatile' or 'externally regulated' systems, and they are frequently explored for targeted delivery as well as a controlled release system. They also increase circulation time and passive targets. They have a property to improve cell response to alteration of temperature in the external environment. Mostly, temperature-sensitive polymers were employed for all drug delivery systems and worked in the physiological range as well as the tumour-targeted range involving hyperthermia (42°C). They have a reversible volume phase transition and sol–gel phase transition in response to temperature at the target site. The temperature-responsive dendrimer system can be thermo-sensitive dendrimers which can counter on alteration in temperature in the extracellular environment, resulting in a significant change in its characteristics such as conformation, hydrophilic or hydrophobic balance and solubility. They also exhibit the lower critical solution temperature (LCST) phenomenon, above which the dendrimer solution is separated in phases due to the aggregation and collapse of the dendrimer chains and expunged water (Needham and Dewhirst, 2001). Zhang and his associates developed temperature-responsive dendrimers using polyamidoaminepoly (N-isopropylacrylamide). They showed a rapid shrinking rate at the temperature above the LCST and exhibited a higher equilibrium swelling ratio at room temperature (Zhang et al. 2004).

2.7 Double/Dual Targeting

Complex diseases such as cancer or other inflammatory disorders are generally multifactorial in character, involving a severance of disease-mediating ligands and receptors. Some proinflammatory cytokines such as TNF, IL-1 and IL-6 have been recognized as key players in inflammatory diseases. In cancer, tumour cells often regulated diverse growth-promoting receptors that can act either independently or crosstalk intracellularly through signalling networks. Therefore, treatment with mAbs that target only a singular antigen has limitations and does not act on others (Dong et al. 2011). The barrier of multiple targets on one target should result in enhanced therapeutic efficacy. This can be achieved by combination therapy with mAbs but also other therapeutic compounds.

Dual targeting with bispecific antibodies is a novel approach for combination therapy. The idea of dual targeting with bispecific antibodies is based on the targeting of multiple disease-modifying molecules with one drug. It is divided into two categories (i) directly act on target structures, e.g., cell surface receptors and (ii) delivery (retargeting) of a therapeutically active moiety, e.g., effector molecules and effector cells. The double targeting strategies are mainly applied for the treatment of cancer, inflammatory and various infectious diseases. The bispecific antibodies targeting different epitopes on disease may increase binding as well as enhanced neutralization (Muller and Kontermann, 2010). They increase the anti-proliferative effect and help to prevent the development of resistance. Various bispecific molecules such as EGFR and IGF-1R are developed for the targeting (Emanuel et al. 2011).

Ni and co-workers developed curcumin and 5-fluorouracil nanoparticles for dual targeting for the treatment of hepatocarcinoma. They exhibited high cellular uptake and cytotoxicity to tumour cells (Ni et al. 2019). Feng and his associates developed liposomes of redox sensitivity for dual targeting for bone and CD44 for the treatment of orthotopic osteosarcoma. Alendronate and doxorubicin are drugs used for dual targeting. They exhibited significantly high cytotoxicity of the human cell line as well as rapid cellular uptake against other reference liposomes. They also exhibited a significant reduction of OS nude mouse model tumours as well as endurance time (Feng et al. 2019). In another study, Huang and his associates developed mesoporous silica/polyglutamic acid peptide dendrimers for dual targeting in thrombus disease using anti-thrombus drug nattokinase (Huang et al. 2019). They exhibited strong thrombolysis activity confirmed by the in vitro and in vivo study. Dong and his associates developed an amphiphilic dendrimer for dual targeting gene (siRNA) for cancer therapy. The amphiphilic dendrimer prevents the degradation of siRNA and maintains the openness site for cancer cells. It exhibited higher anti-cancer activity and enhanced delivery of siRNA without any acute toxicity than other conventional dendrimers (Dong et al. 2018). Furthermore, Liu and his associates delivered the siRNA by using dual targeting peptide decorated dendrimers (Liu et al. 2014b). The siRNA peptide dendrimer complex showed enhanced cell targeting by binding with integrins and cell penetration by binding with neutrophilin-1 receptor and thereby improved gene silencing and anti-tumour activity.

2.8 Carriers for Drug Targeting

Carriers are substances which are or any other systems accountable for the successful shipping of a therapeutic agent to the active/receptor site (organ/tissues/cell). Carriers are transporters and specifically designed in a different form for arresting of therapeutic agents inside by encapsulation or adsorbed on the surface of formulation (Kumari et al. 2010). Carriers improve the physical stability, safety and effectiveness as well as therapeutic efficacy of the therapeutic agent.

2.8.1 Pharmaceutical Carrier

Pharmaceutical carriers include delivery of therapeutic agents with polymeric and lipid carriers which include nanoparticles, microparticles, nanocapsules, solid lipid nanoparticles, nanostructured lipid carriers, liposomes, niosomes, ethosomes and invasomes. These carriers allow high litheness in both structural and physicochemical properties.

2.8.2 Endogenous Carrier

It is a carrier that transports the lipid and cholesterol into the bloodstream. These carriers are not recognized by the reticuloendothelial system and reach the target site. Serum albumin, resealed erythrocytes and lipoproteins (chylomicrons, very-low-density lipoprotein, low-density lipoprotein and high-density lipoprotein) are well-known examples of endogenous carriers (Joshi et al. 2012).

2.8.3 Cellular Carrier

These carriers are present in the human body and have a property to transport the drug to the target site. These cellular carriers include erythrocytes, serum albumin, antibodies, platelets and leukocytes which are some of the cellular carriers (Ghaffarian and Muro, 2013).

2.8.4 Macromolecular Carriers

They are large biological molecules and used in drug targeting. They are decorated or surface engineered on formulation and release the formulation of therapeutic agents into the target site. These examples include antibodies, polysaccharides, plasmid, DNA and oligonucleotides for the gene delivery (Raucher et al. 2018).

2.9 Dendrimer-based Targeted Drug Delivery

Dendrimers are the nanocarrier system having 3-D structure synthetic macromolecules. They are a monodispersed system having the branches of layer extended from the centre of core molecules. Dendrimers have been widely investigated predominantly in therapeutics and diagnosis, particularly in cancer because of definite nanosize, branched 3-D structure and surface modification. Dendrimers are a more responsive delivery system and can efficiently identify circulating tumour cells when conjugated with precise antibodies. They improve the bioavailability as well as therapeutic efficacy (Gilani et al. 2018). Surface-modified dendrimers such as Lewis X antibody (aSlex)-conjugated PAMAM dendrimers were effectively approached for targeting of therapeutic and transported maximum DOX to the target region. The formulation was tested on murine C-26 colon carcinoma cells. Dendrimers showed sustained release of DOX at the target site and a remarkable effect compared to normal formulation. The summary of dendrimer-based drug delivery and targeting is shown in Table 2.2.

Dendrimer-based formulations produced the toxicity due to their surface cationic charge. However, many researchers reduced this toxicity by developing new biodegradable polymeric dendrimers such as melamine dendrimers (Abd-El-Aziz et al. 2016) and poly-L-lysine dendrimers (Rahimi et al. 2016). The drug toxicity was also reduced by the surface modification of the dendrimers such as carbohydrate coating (Xia et al. 2019), PEGylation (Somani et al. 2018), amino acid and peptide conjugation (Heitz et al. 2017) and acetylation (Yan et al. 2017).

Currently novel approaches are investigated to reduce the toxicity of dendrimers as well as drug loading by developing the hybrid dendrimers (Figure 2.3B). In these cases, the dendrimers were conjugated with other nanocarrier systems such as microspheres, liposomes, quantum dots (QDs) and carbon nanotubes (CNTs) (Kesharwani et al. 2018). In hybrid dendrimer liposomes, the entrapment efficiency of the drug increases upon the encapsulation of dendrimers into a liposome. The dendrimer encapsulated with opposite charge liposomes. This enhanced the drug loading of methotrexate several times in an acidic environment and because of charge interaction change the pH, which leads to solubility and concentration gradient influx and prolonged-release (Khopade et al. 2002). CNTs are carbon-based nanomaterials and used in biomedical applications. However, they have many disadvantages, such as toxicity, which have limited application nowadays. When hybridized with dendrimers, the toxicity is reduced, and they are used prominently (Kesharwani et al. 2012). Dendrimers exhibit more hydrophilicity to CNTs. The hybrid dendrimers containing silver nanoparticles with multi-walled carbon nanotubes enhanced the anti-microbial activity against Gram-positive and negative bacteria (Murugan and Vimala, 2011).

QDs are also used as targeting agents for imaging of theranostic agents, but they have the drawbacks such as toxicity and hydrophobicity. Dendrimer–QD hybrids have established a system to reduce the toxicity of QDs as well as increase aqueous solubility and quantum yield (Larson et al. 2003). PAMAM dendrimer hybrids with QDs improved the cellular uptake, cytosolic distribution in primary cultured mesenchymal stem cells and prolonged intracellular fluorescence intensity (Higuchi et al. 2011).

TABLE 2.2

Dendrimer-based drug delivery and targeting

Formulation	Therapeutics	Outcomes	Ref.
Poly(propylene) imine (5.0G PPI) dendrimer	Docetaxel	Significantly reduced haemolytic toxicity	Mody et al. (2014)
Dendrimer stabilized smart nanoparticles	Paclitaxel	Stable at physiological pH and showed rapid release in the tumour environment (pH5.5)	Tekade et al. (2015)
Poly(L-glutamic acid)	Doxorubicin	Significantly different release of doxorubicin at pH-5 compared to pH-7 due to cleavage of hydrazine bonds	Yuan et al. (2010)
Lipid dendrimer hybrid nanoparticles	Vancomycin	It showed significantly faster release and anti-bacterial activity (8-fold) at pH6 than pH-7.4	Maji et al. (2019)
PEG-PAMAM	Doxorubicin	It significantly enhanced the anti-tumour toxicity and reduced the systemic toxicity at pH-5	Hu et al. (2016)
PEG-chitosan-folic acid PAMAM dendrimers	pDNA	The formulation was stable in the acidic environment compared to physiological pH.	Wang et al. (2018)
Folate-anchored dendrimer	Retinoic acid, Methotrexate	Significant cargo on folate receptor of cancer cells in high concentration without affecting other normal cells	Tekade et al. (2008)
PAMAM dendrimer	16 Biotin	enhanced internalization into two cancer cell lines via receptor-mediated endocytosis	Wang et al. (2018)
PAMAM with d-glucoheptono-1,4-lactone	Pyridoxal and biotin	lower cytotoxicity and 3–4 times higher cellular uptake in cancer glioblastoma (U-118 MG) and squamous carcinoma (SCC-15)	Uram et al. (2017)
Polyamidoamine dendrimer	Pyridoxal and biotin	Significantly reached higher concentration to the target site confirmed by cytotoxicity assays (MTT, neutral red and crystal violet) and estimation of apoptosis by confocal microscopy detection	Uram et al. (2013)
Ligands (sialic acid, glucosamine)-anchored dendrimers	Paclitaxel	biodistribution studies in rats showed a significantly higher accumulation of PTX in the brain as compared to free PTX	Patel et al. (2016)
Polysorbate 80 (P80) anchored PPI dendrimer	Docetaxel	Gamma scintigraphy and biodistribution studies further confirmed the targeting efficiency and higher biodistribution of ligand-conjugated dendrimers into the brain	Gajbhiye and Jain (2011)
PAMAM dendrimer complex attached to liposomes	Doxorubicin	Significant release of doxorubicin	Papagiannaros et al. (2005)
PAMAM dendrimers	siRNA delivery to target Hsp27	Silencing of the hsp27 gene led to induction of caspase-3/7-dependent apoptosis and inhibition of PC-3 cell growth in vitro	Liu et al. (2009)
Hybrid dendrimers (mPEG-b-PCL) and G1-PEA	Vancomycin	It showed sustained release (68% in 48 h), with 16 fold minimum inhibitory concentration against MRSA as compared to free drug	Omolo et al. (2018)
PPI dendrimers	Amphotericin B	High intracellular uptake as wells as a significant reduction in toxicity as compared to pure drug	Jain et al. (2015)
Folate-PEG-Appended Dendrimer	Doxorubicin	Enhanced cellular accumulation and cytotoxic activity of DOX in folate receptor-α (FR-α)-overexpressing KB cells	Mohammed et al. (2019)
Folate-PEG-appended dendrimer	siRNA	High accumulation in tumour tissues after intravenous injection into the mice without any toxicity into other cells	Arima et al. (2012)

(Continued)

TABLE 2.2 (*Continued*)
Dendrimer-based drug delivery and targeting

Formulation	Therapeutics	Outcomes	Ref.
Single-walled carbon nanotubes- (PAMAM dendrimers		The significant cytotoxic effect in C2C12 cells	Cancino et al. (2015)
Dendrimer-magnetic NPs	Iron	capture 86% ± 5% of tumour cells and isolate a rare number of tumour cells from spiked whole blood samples in only 15 min without the damage of cell viability	Zhang et al. (2016)

2.10 Conclusions and Future Directions

There is a model move from conventional to targeted drug delivery, where the selected drug reached the desired target site, and its pharmacological effect is shown with minimized side effects to other regions. At a higher dose of the drug, it gives a adverse effect, and targeted delivery may reduce the dose as well as minimizing the effect over the non-targeted region. Using the level of targeted strategy, the organelles can be avoided. The targeted delivery system gives a comprehensive release of drug in a sustained and controlled pattern for enhancing the therapeutic response and increases patient compliance. Molecular pharmacology of the therapeutic agent and carriers can be mutual with the disease state, and the microenvironment in the target region (disease) to give information about receptors may enhance the concept of targeted drug delivery. Dendrimers are promising nanoformulations for delivery of a therapeutic agent to the target region with precise size and molecular weight, and they are a 3-D macromolecular branched architect form in which the therapeutic moiety encapsulated in guest form and interacts with the cell receptor. Furthermore, the targeting enhanced by surface engineering manipulation of formulation enhanced the interaction with the receptor and enhanced the pharmacological response. The dendrimer has many advantages but has toxicity due to its cationic charge on the surface; for this, it is limited in clinical application and has a high cost. Therefore, hybrid dendrimers are a novel tool to minimize the drawback of dendrimers and enhanced clinical application of the therapeutic agent. In this chapter, the concept of drug targeting and various hybrid dendrimers have been explored for a wide range of biomedical applications.

Abbreviations

API	Active Pharmaceutical Agents
BBB	Blood-brain barrier
DOX	Doxorubicin
EPR	Enhanced permeability and retention
HDL	High-density lipoprotein
LCST	Lower critical solution temperature
LDL	Low-density lipoprotein
MDR	Multi-drug resistance
MMA	Methacrylic acid
MW	Molecular weight
PCL	Polycaprolactone
PEI	Polyethyleneimine
PLGA	Poly(lactic-co-glycolic acid)
PQ	Primaquine
PTX	Paclitaxel
QDs	Quantum dots
USFDA	US Food and Drug Administration
VLDL	Very-low-density lipoprotein

REFERENCES

Abd-El-Aziz, A.S., Abdelghani, A.A., El-Sadany, S.K., Overy D.P., Kerr, R.G. 2016. Antimicrobial and anti-cancer activities of organoiron melamine dendrimers capped with piperazine moieties. *European Polymer Journal 82*: 307–323.

Ahmad, J., Amin, S., Rahman, M. et al. 2015b. Solid matrix based lipidic nanoparticles in oral cancer chemotherapy: Applications and pharmacokinetics. *Current Drug Metabolism 16*(8): 633–644.

Ahmad, J., Akhter, S., Rizwanullah, M. et al. 2015a. Nanotechnology-based inhalation treatments for lung cancer: State of the art. *Nanotechnology, Science and Applications 8*: 55–66.

Ahmad, J., Akhter, S., Greig, N.H. et al. 2016. Engineered nanoparticles against MDR in cancer: The state of the art and its prospective. *Current Pharmaceutical Design 22*(28): 4360–4373.

Ahmad, J., Singhal, M., Amin, S. et al. 2017. Bile salt stabilized vesicles (Bilosomes): A novel nano-pharmaceutical design for oral delivery of proteins and peptides. *Current Pharmaceutical Design 23*(11): 1575–1588.

Ahmad, J., Ameeduzzafar, Ahmad, M.Z., Akhter, H. 2020. Surface-engineered cancer nanomedicine: Rational design and recent progress. *Current Pharmaceutical Design 26*(11): 1181–1190.

Akhter, M.H., Rizwanullah, M., Ahmad, J. et al. 2018. Nanocarriers in advanced drug targeting: Setting novel paradigm in cancer therapeutics. *Artificial Cells, Nanomedicine, and Biotechnology 46*(5): 873–884.

Alex, A.T., Joseph, A., Shavi, G., Rao, J.V., Udupa, N. 2016. Development and evaluation of carboplatin-loaded PCL nanoparticles for intranasal delivery. *Drug Delivery 23*(7): 2144–2153.

Anarjan, F.S. 2019. Active targeting drug delivery nanocarriers: Ligands. *Nano-Structures & Nano-Objects 19*: 100370.

Arima, H., Yoshimatsu, A., Ikeda, H. et al. 2012. Folate-PEG-appended dendrimer conjugate with α-cyclodextrin as a novel cancer cell-selective siRNA delivery carrier. *Molecular Pharmaceutics 9*(9): 2591–2604.

Balamuralidhara, V., Pramodkumar, T.M., Srujana, N. et al. 2011. pH Sensitive Drug Delivery Systems: A Review. *American Journal of Drug Discovery and Development 1*(1) 24–48.

Barkat, M.A., Harshita, Rizwanullah, M. et al. 2020. Therapeutic nanoemulsion: Concept to delivery. *Current Pharmaceutical Design 26*(11): 1145–1166.

Biswas, S., Dodwadkar, N.S., Deshpande, P. P., Torchilin, V. P. 2012. Liposomes loaded with paclitaxel and modified with novel triphenylphosphonium-PEG-PE conjugate possess low toxicity, target mitochondria and demonstrate enhanced antitumor effects in vitro and in vivo. *Journal of Controlled Release 159*(3): 393–402.

Boddy. A., Aarons, L., Petrak, K. 1989. Efficiency of drug targeting: Steady-state considerations using a three-compartment model. *Pharmaceutical Research 6*(5): 367–372.

Bouillon, C., Tintaru A., Monnier, V. 2010. Synthesis of poly(amino) ester dendrimers via active cyanomethyl ester intermediates. *Journal of Organic Chemistry 75*: 8685–8688.

Cancino, J., Paino, I.M., Micocci, K.C., Selistre-de-Araujo, H.S., Zucolotto, V. 2015. In vitro nanotoxicity of single-walled carbon nanotube-dendrimer nanocomplexes against murine myoblast cells. *Toxicology Letters 219*(1): 18–25.

Choi, S.W., Kim, J.H. 2007. Design of surface-modified poly (D,L-lactide-co-glycolide) nanoparticles for targeted drug delivery to bone. *Journal of Controlled Release 122*(1): 24–30.

Chomoucka, J., Drbohlavova, J., Huska, D., et al. 2010. Magnetic nanoparticles and targeted drug delivering. *Pharmacological Research 62*(2) 144–149.

Choonara, B.F., Choonara, Y.E., Kumar, P. et al. 2014. A review of advanced oral drug delivery technologies facilitating the protection and absorption of protein and peptide molecules. *Biotechnology Advances 32*(7): 1269–1282.

Chu, Z., Chen, J.S., Liau C.T. et al. 2008. Oral bioavailability of a novel paclitaxel formulation (Genetaxyl) administered with cyclosporine A in cancer patients. *Anticancer Drugs 19*(3): 275–281.

Cong, Y., Quan, C., Liu, M. et al. 2015. Alendronate-decorated biodegradable polymeric micelles for potential bone-targeted delivery of vancomycin. *Journal of Biomaterials Science, Polymer Edition 26*(11): 629–643.

Danhier, F., Feron, O., Préat, V. 2010. To exploit the tumor microenvironment: Passive and active tumor targeting of nanocarriers for anti-cancer drug delivery. *Journal of Controlled Release 148*(2):135–146.

De Buck, S.S., Sinha, V.K., Fenu, L.A. 2007. The prediction of drug metabolism, tissue distribution, and bioavailability of 50 structurally diverse compounds in rat using mechanism-based absorption, distribution, and metabolism prediction tools. *Drug Metabolism and Disposition 35*(4):649–659.

Disney, M.D., Angelbello, A.J. 2016. Rational design of small molecules targeting oncogenic noncoding RNAs from sequence. *Accounts of Chemical Research 49*(12): 2698–2704.

Dong, J., Sereno, A., Aivazian, D. et al. 2011. A stable IgG-like bispecific antibody targeting the epidermal growth factor receptor and the type I insulin-like growth factor receptor demonstrates superior anti-tumor activity. *MAbs* 3(3): 273–288.

Dong, Y., Yu, T., Ding, L. et al. 2018. A dual targeting dendrimer-mediated siRNA delivery system for effective gene silencing in cancer therapy. *Journal of the American Chemical Society* 140(47): 16264–16274.

Duan, Q., Cao, Y., Li, Y. et al. 2013. pH-responsive supramolecular vesicles based on water-soluble pillar [6] arene and ferrocene derivative for drug delivery. *Journal of the American Chemical Society* 135(28): 10542–10549.

Emanuel, S.L., Engle, L.J., Chao, G. et al. 2011. A fibronectin scaffold approach to bispecific inhibitors of epidermal growth factor receptor and insulin-like growth factor-I receptor. *MAbs* 3(1): 38–48.

Emery, I.F., Battelli, C., Auclair, P.L., Carrier, K., Hayes, D.M. 2009. Response to gefitinib and erlotinib in Non-small cell lung cancer: A restrospective study. *BMC Cancer* 9: 333.

Fan, Y., Chen, C., Huang, Y., Zhang, F., Lin, G. 2017. Study of the pH-sensitive mechanism of tumor-targeting liposomes. *Colloids and Surfaces B: Biointerfaces* 151: 19–25.

Feng, S., Wu, Z.X., Zhao, Z. et al. 2019. Engineering of bone and CD44-dual-targeting redox-sensitive liposomes for the treatment of orthotopic osteosarcoma. *ACS Applied Materials & Interfaces* 11(7): 7357–7368.

Gajbhiye, V., Jain, N. K. 2011. The treatment of Glioblastoma Xenografts by surfactant conjugated dendritic nanoconjugates. *Biomaterials* 32(26): 6213–6225.

Ghaffarian, R., Muro, S. 2013. Models and methods to evaluate transport of drug delivery systems across cellular barriers. *Journal of Visualized Experiments* 80: e50638.

Gilani, S.J., Jahangir, M.A., Chandrakala, et al. 2018. Nano-based therapy for treatment of skin cancer. *Recent Patents on Anti-Infective Drug Discovery* 13(2): 151–163.

Guan. J., Jiang, Z., Wang, M. 2019. Short peptide-mediated brain-targeted drug delivery with enhanced immunocompatibility. *Molecular Pharmaceutics* 16(2): 907–913.

Haider, N., Fatima, S., Taha, M. et al. 2020. Nanomedicines in diagnosis and treatment of cancer: An update. *Current Pharmaceutical Design* 26(11): 1216–1231.

Harshita, Barkat, M.A., Rizwanullah, M. et al. 2019. Paclitaxel-loaded nanolipidic carriers with improved oral bioavailability and anticancer activity against human liver carcinoma. *AAPS PharmSciTech* 20(2): 87. doi: https://doi.org/10.1208/s12249-019-1304-4

Heitz, M., Kwok, A., Eggimann, G. A. 2017. Peptide dendrimer-lipid conjugates as DNA and siRNA transfection reagents: Role of charge distribution across generations. *Chimia (Aarau).* 71(4): 220–225.

Higuchi, Y., Wu, C., Chang, K.L. et al. 2011. Polyamidoaminedendrimer-conjugated quantum dots for efficient labelling of primary cultured mesenchymal stem cells. *Biomaterials* 32, 6676–6682.

Holt, K., Ye, M., Nagar, S., Korzekwa, K. 2019. Prediction of tissue-plasma partition coefficients using microsomal partitioning: Incorporation into physiologically based pharmacokinetic models and steady-state volume of distribution predictions. *Drug Metabolism and Disposition* 47(10): 1050–1060.

Hu, W., Qiu, L., Cheng, L. et al. 2016. Redox and pH dual responsive poly(amidoamine) dendrimer-poly(ethylene glycol) conjugates for intracellular delivery of doxorubicin. *Acta Biomaterialia* 36: 241–253.

Huang, M., Zhang, S.F., Lu, S. et al. 2019. Synthesis of mesoporous silica/polyglutamic acid peptide dendrimer with dual targeting and its application in dissolving thrombus. *Journal of Biomedical Materials Research Part A* 107(8): 1824–1831.

Inapagolla, R., Guru, B.R., Kurtoglu, Y. et al. 2010. In vivo efficacy of dendrimeremethylprednisolone conjugate formulation for the treatment of lung inflammation. *International Journal of Pharmaceutics* 399(1–2): 140–147.

Jain, K., Verma, A.K., Mishra, P.R., Jain, N.K. 2015. Surface-engineered dendrimeric nanoconjugates for macrophage-targeted delivery of amphotericin B: Formulation development and in vitro and in vivo evaluation. *Antimicrobial Agents and Chemotherapy* 59(5): 2479–2487.

Jhaveri, A., Torchilin, V. 2016. Intracellular delivery of nanocarriers and targeting to subcellular organelles. *Expert Opinion on Drug Delivery* 13(1): 49–70.

Joshi, M.D., Unger, W.J., Storm, G., van Kooyk, Y., Mastrobattista, E., 2012. Targeting tumor antigens to dendritic cells using particulate carriers. *Journal of Controlled Release* 161(1): 25–37.

Kamaly, N., Xiao, Z., Valencia, P.M., Radovic-Moreno, A.F., Farokhzad, O.C., 2012. Targeted polymeric therapeutic nanoparticles: Design, development and clinical translation. *Chemical Society Reviews* 41(7): 2971–3010.

Kesharwani, P., Ghanghoria, R., Jain, N.K. 2012 Carbon nanotube exploration in cancer cell lines. *Drug Discovery Today* 17: 1023–1030.

Kesharwani, P., Gothwal, A., Iyer, A.K. et al. 2018. Dendrimer nanohybrid carrier systems: An expanding horizon for targeted drug and gene delivery. *Drug Discovery Today 23*(2): 300–314.

Khopade, A.J., Caruso, F., Tripathi, P., Nagaich, S., Jain, N.K. 2002. Effect of dendrimer on entrapment and release of bioactive from liposomes. *International Journal of Pharmaceutics 232*(1–2): 157–162.

Kono, K., Kojima, C., Hayashi, N. et al. 2008. Preparation and cytotoxic activity of poly (ethylene glycol)-modified poly (amidoamine) dendrimers bearing adriamycin. *Biomaterials 29*(11): 1664–1675.

Kumari, A., Yadav, S.K., Yadav, S.C. 2010. Biodegradable polymeric nanoparticles based drug delivery systems. *Colloids and Surfaces B: Biointerfaces 75*(1): 1–18.

Larson, D.R., Zipfel, W.R., Williams, R.M. et al. 2003. Water-soluble quantum dots for multiphoton fluorescence imaging in vivo. *Science 300*(5624): 1434–1436.

Liu, M., Li, Z.H., Xu, F.J. et al., 2012. An oligopeptide ligand-mediated therapeutic gene nanocomplex for liver cancer-targeted therapy. *Biomaterials 33*(7): 2240–2250.

Liu, X.X., Rocchi, P., Qu, F.Q. et al. 2009. PAMAM dendrimers mediate siRNA delivery to target Hsp27 and produce potent antiproliferative effects on prostate cancer cells. *ChemMedChem 4*(8): 1302–1310.

Liu, J., Huang, Y., Kumar, A. et al., 2014a. pH-sensitive nano-systems for drug delivery in cancer therapy. *Biotechnology Advances 32*(4): 693–710.

Liu, X., Liu, C., Chen, C. et al. 2014b. Targeted delivery of Dicer-substrate siRNAs using a dual targeting peptide decorated dendrimer delivery system. *Nanomedicine 10*(8): 1627–1636.

Lukyanov, A.N., Elbayoumi, T.A., Chakilam, A.R., Torchilin, V.P. 2004. Tumor-targeted liposomes: Doxorubicin-loaded long-circulating liposomes modified with anti-cancer antibody. *Journal of Controlled Release 100*(1):135–144.

Maeda, H. 2001. The enhanced permeability and retention (EPR) effect in tumor vasculature: The key role of tumor-selective macromolecular drug targeting. *Advances in Enzyme Regulation 41*: 189–207.

Maeda, H., Wu, J., Sawa, T., Matsumura, Y., Hori, K. 2000. Tumor vascular permeability and the EPR effect in macromolecular therapeutics: A review. *Journal of Controlled Release 65*(1–2): 271–284.

Maji, R., Omolo, C.A., Agrawal, N. et al. 2019. pH-Responsive lipid-dendrimer hybrid nanoparticles: An approach to target and eliminate intracellular pathogens. *Molecular Pharmaceutics 16*(11): 4594–4609.

Meerum, T.J.M., Malingre, M.M., Beijnen, J.H. et al. 1999. Coadministration of oral cyclosporin A enables oral therapy with paclitaxel. *Clinical Cancer Research 5*(11): 3379–3384.

Melgert, B.N., Olinga, P., Van Der Laan, J.M.S. et al. 2001. Targeting dexamethasone to Kupffer cells: Effects on liver inflammation and fibrosis in rats. *Hepatology 34*(4): 719–728.

Mishra, P., Nayak, B., Dey, R.K. 2016. PEGylation in anti-cancer therapy: An overview. *Asian Journal of Pharmaceutical Sciences 11*(3) 337–348.

Mody, N., Tekade, R.K., Mehra, N.K., Chopdey, P., Jain, N.K. 2014. Dendrimer, Liposomes, Carbon Nanotubes and PLGA Nanoparticles: One Platform Assessment of Drug Delivery Potential. *AAPS PharmSciTech 15*(2): 388–399.

Mohammed, M.A., Syeda, J.T.M., Wasan, K.M., Wasan, E.K. 2017. An Overview of Chitosan Nanoparticles and Its Application in Non-Parenteral Drug Delivery. *Pharmaceutics 9*(4): 53. doi: https://doi.org/10.3390/pharmaceutics9040053

Mohammed, A.F.A., Higashi, T., Motoyama, K. et al. 2019. In vitro and in vivo co-delivery of siRNA and doxorubicin by folate-PEG-appended dendrimer/glucuronylglucosyl-β-cyclodextrin conjugate. *AAPS Journal 21*(4): 54.

Morgan. P.J., Harding, S.E., Petrak, K. 1990. Interactions of a model block copolymer drug delivery system with two serum proteins and myoglobin. *Biochemical Society Transactions 18*(5): 1021–1022.

Mukhopadhyay, P., Maity, S., Mandal, S. et al. 2018. Preparation, characterization and in vivo evaluation of pH sensitive, safe quercetin-succinylated chitosan-alginate core-shell-corona nanoparticle for diabetes treatment. *Carbohydrate Polymer 182*: 42–51.

Muller, D., Kontermann, R.E. 2010. Bispecific antibodies for cancer immunotherapy: Current perspectives. *Bio Drugs 24*(2): 89–98.

Muller, P.Y., Milton, M.N. 2012. The determination and interpretation of the therapeutic index in drug development. *Nature Reviews Drug Discovery 11*(10): 751–761.

Murugan, E., Vimala, G. 2011. Effective functionalization of multiwalled carbon nanotube with amphiphilicpoly(propyleneimine) dendrimer carrying silver nanoparticles for better dispersability and antimicrobial activity. *Journal of Colloid and Interface Science 357*(2): 354–365.

Nag, M., Gajbhiye, V., Kesharwani, P., Jain, N.K. 2016. Transferrin functionalized chitosan-PEG nanoparticles for targeted delivery of paclitaxel to cancer cells. *Colloids and Surfaces B: Biointerfaces 148*: 363–370.

Needham, D., Dewhirst, M.W. 2001. The development and testing of a new temperature-sensitive drug delivery system for the treatment of solid tumors. *Advanced Drug Delivery Reviews 53*(3): 285–305.

Nguyen, T.L., Nguyen, T.H., Nguyen, C.K., Nguyen, D.H. 2017. Redox and pH responsive poly (amidoamine) dendrimer-heparin conjugates via disulfide linkages for letrozole delivery. *Biomed Research International 2017*: e8589212.

Ni, W., Li, Z., Liu, Z. et al. 2019. Dual-targeting nanoparticles: Codelivery of curcumin and 5-fluorouracil for synergistic treatment of hepatocarcinoma. *Journal of Pharmaceutical Sciences 108*(3): 1284–1295.

Omolo, C.A., Kalhapure, R.S., Agrawal, N. et al. 2018. A hybrid of mPEG-b-PCL and G1-PEA dendrimer for enhancing delivery of antibiotics. *Journal of Controlled Release 290*: 112–128.

Opanasopit, P., Nishikawa, M., Hashida, M. 2002. Factors affecting drug and gene delivery: Effects of interaction with blood components. *Critical Reviews in Therapeutic Drug Carrier Systems 19*(3):191–233.

Paasonen, L., Sharma, S., Braun, G.B. et al. 2016. New p32/gC1qR ligands for targeted tumor drug delivery. *Chembiochem 17*(7): 570–575.

Paci, A., Veal, G., Bardin, C. et al. 2014. Review of therapeutic drug monitoring of anticancer drugs part 1 – Cytotoxics. *European Journal of Cancer 50*(12): 2010–2019.

Papagiannaros, A., Dimas, K., Papaioannou, G. T., Demetzos, C. 2005. Doxorubicin-PAMAM dendrimer complex attached to liposomes: Cytotoxic studies against human cancer cell lines. *International Journal of Pharmaceutics 302*(1–2): 29–38.

Park, J.W., Kim, S.J., Kwag, D.S. et al. 2015. Multifunctional delivery systems for advanced oral uptake of peptide/protein drugs. *Current Pharmaceutical Design 21*(22): 3097–3110.

Patel, H.K., Gajbhiye, V., Kesharwani, P., Jain, N.K. 2016. Ligand anchored poly(propyleneimine) dendrimers for brain targeting: Comparative in vitro and in vivo assessment. *Journal of Colloid and Interface Science 482*: 142–150.

Patil, Y., Shmeeda, H., Amitay, Y. et al. 2018. Targeting of folate-conjugated liposomes with co-entrapped drugs to prostate cancer cells via prostate- specific membrane antigen (PSMA). *Nanomedicine 14*(4):1407–1416.

Pérez-Herrero, E., Fernández-Medarde, A. 2015. Advanced targeted therapies in cancer: Drug nanocarriers, the future of chemotherapy. *European Journal of Pharmaceutics and Biopharmaceutics 93*: 52–79.

Petrak, K. 2005. Essential properties of drug-targeting delivery systems. *Drug Discovery Today 10*(23–24): 1667–1673.

Petrak, K., Goddard, P. 1989. Transport of macromolecules across the capillary walls. *Advanced Drug Delivery Reviews 3*(2): 191–214.

Qi, X., Qin, J., Fan, Y. et al. 2016. Carboxymethyl chitosan-modified polyamidoamine dendrimer enables progressive drug targeting of tumors via pH-sensitive charge inversion. *Journal of Biomedical Nanotechnology 12*(4): 667–678.

Rahimi, A., Amjad-Iranagh, S., Modarress, H. 2016. Molecular dynamics simulation of coarse-grained poly(L-lysine) dendrimers. *Journal of Molecular Modeling 22*(3): e59.

Raucher, D., Dragojevic, S., Ryu. J. 2018. Macromolecular drug carriers for targeted glioblastoma therapy: Preclinical studies, challenges, and future perspectives. *Frontiers in Oncology 8*: e624.

Rehman, S., Nabi, B., Zafar, A., Baboota, S., Ali, J. 2019. Intranasal delivery of mucoadhesive nanocarriers: A viable option for Parkinson's disease treatment? *Expert Opinion on Drug Delivery 16*(12): 1355–1366.

Reif, S., Nicolson, M.C., Bisset, D. et al. 2002. Effect of grapefruit juice intake on etoposide bioavailability. *European Journal of Clinical Pharmacology 58*(7):491–494.

Remaut, K., Lucas, B., Braeckmans, K. et al. 2007. PEGylation of liposomes favours the endosomal degradation of the delivered phosphodiester oligonucleotides. *Journal of Controlled Release 117*: 256–266.

Renukuntla, J., Vadlapudi, A.D., Patel, A., Boddu, S.H., Mitra, A.K. 2013. Approaches for enhancing oral bioavailability of peptides and proteins. *International Journal of Pharmaceutics 447*(1–2): 75–93.

Rizwanullah, M., Ahmad, J., Amin, S. 2016. Nanostructured lipid carriers: A novel platform for chemotherapeutics. *Current Drug Delivery 13*(1): 4–26.

Rizwanullah, M., Amin, S., Ahmad, J. 2017. Improved pharmacokinetics and antihyperlipidemic efficacy of rosuvastatin-loaded nanostructured lipid carriers. *Journal Drug Targeting 25*(1): 58–74.

Rizwanullah, M., Amin, S., Mir, S.R., Fakhri, K.U., Rizvi, M.M.A. 2018. Phytochemical based nanomedicines against cancer: Current status and future prospects. *Journal of Drug Targeting 26*(9): 731–752.

Rizwanullah, M., Harshita, A.M. et al. 2020. Polymer-lipid hybrid nanoparticles: A next-generation nanocarrier for targeted treatment of solid tumors. *Current Pharmaceutical Design 26*(11): 1206–1215.

Sakhrani, N.M., Padh, H. 2013. Organelle targeting: Third level of drug targeting. *Drug Design, Development and Therapy 7*: 585–599.

Salem, D.S., Sliem, M.A., El-Sesy, M., Shouman, S.A., Badr, Y. 2018. Improved chemo-photothermal therapy of hepatocellular carcinoma using chitosan-coated gold nanoparticles. *Journal of Photochemistry and Photobiology B: Biology 182*: 92–99.

Shen, Z., Wei, W., Tanaka, H. et al. 2011. A galactosamine-mediated drug delivery carrier for targeted liver cancer therapy. *Pharmacological Research 64*(4): 410–419.

Shen, Z., Nieh, M.P., Li, Y. 2016. Decorating nanoparticle surface for targeted drug delivery: Opportunities and challenges. *Polymers (Basel) 8*(3). pii: E83. doi: https://doi.org/10.3390/polym8030083

Somani, S., Laskar, P., Altwaijry, N. et al. 2018. PEGylation of polypropylenimine dendrimers: Effects on cytotoxicity, DNA condensation, gene delivery and expression in cancer cells. *Scientific Reports 8*(1): e9410.

Soni, K., Rizwanullah, M., Kohli, K. 2018. Development and optimization of sulforaphane-loaded nanostructured lipid carriers by the Box-Behnken design for improved oral efficacy against cancer: In vitro, ex vivo and in vivo assessments. *Artificial Cells, Nanomedicine, and Biotechnology 46*(sup1): 15–31.

Stock, L., Hosoume, J., Treptow, W. 2017. Concentration-Dependent Binding of Small Ligands to Multiple Saturable Sites in Membrane Proteins. *Scientific Reports 7*(1): e5734.

Tao, W., Zeng, X., Liu, T. et al. 2013. Docetaxel-loaded nanoparticles based on star-shaped mannitol-core PLGA-TPGS diblock copolymer for breast cancer therapy. *Acta Biomaterialia 9*(11): 8910–8920.

Tekade, R.K., Dutta, T., Tyagi, A. et al. 2008. Surface-engineered dendrimers for dual drug delivery: A receptor up-regulation and enhanced cancer targeting strategy. *Journal of Drug Targeting 16*(10): 758–772.

Tekade, R.K., Tekade, M., Kumar, M., Chauhan, A.S. 2015. Dendrimer-stabilized smart-nanoparticle (DSSN) platform for targeted delivery of hydrophobic antitumor therapeutics. *Pharmaceutical Research 32*(3):910–928.

Tekade, R.K., Sun, X. 2017. The Warburg effect and glucose-derived cancer theranostics. *Drug Discovery Today 22*: 1637–1653.

Tsai, M.H., Pan, C.H., Peng, C.L., Shieh, M.J. 2017. Panitumumab-conjugated Pt-drug nanomedicine for enhanced efficacy of combination targeted chemotherapy against colorectal cancer. *Advanced Healthcare Materials 6*(13): 1700111. doi: https://doi.org/10.1002/adhm.201700111

Uram, Ł., Szuster, M., Gargasz, K. et al. 2013. In vitro cytotoxicity of the ternary PAMAM G3-pyridoxal-biotin bioconjugate. *International Journal of Nanomedicine 8*: 4707–4720.

Uram, T., Szuster, M., Filipowicz, A. et al. 2017. Cellular uptake of glucoheptoamidatedpoly(amidoamine) PAMAM G3 dendrimer with amide-conjugated biotin, a potential carrier of anticancer drugs. *Bioorganic & Medicinal Chemistry 25*(2):706–713.

Veiseh, O., Gunn, J.W., Zhang, M. 2010. Design and fabrication of magnetic nanoparticles for targeted drug delivery and imaging. *Advanced Drug Delivery Reviews 62*(3): 284–304.

Veronese F.M., Pasut G. 2008. PEGylation: Posttranslational bioengineering of protein biotherapeutics. *Drug Discovery Today: Technologies 5*(2–3): 57–64.

Wang, R., Billone, P.S., Mullett, W.M. 2013. Nanomedicine in Action: An Overview of Cancer Nanomedicine on the Market and in Clinical Trials. *Journal of Nanomaterials 2013*. doi: https://doi.org/10.1155/2013/629681

Wang, T., Zhang, Y., Wei, L. et al. 2018. Design, synthesis, and biological evaluations of asymmetric bow-tie PAMAM dendrimer-based conjugates for tumor-targeted drug delivery. *ACS Omega 3*(4): 3717–3736.

Warner, K.D., Hajdin, C.E., Weeks, K.M. 2018. Principles for targeting RNA with drug-like small molecules. *Nature Review Drug Discovery 17*(8): 547–558.

Wooten, J.M. 2012. Pharmacotherapy considerations in elderly adults. *Southern Medical Journal 105*(8): 437–445.

Woraphatphadung, T., Sajomsang, W., Rojanarata, T. et al. 2018. Development of chitosan-based pH-sensitive polymeric micelles containing curcumin for colon-targeted drug delivery. *AAPS PharmSciTech 19*(3): 991–1000.

Xi, Y., Jiang, T., Yu, Y. et al. 2019. Dual targeting curcumin loaded alendronate-hyaluronan- octadecanoic acid micelles for improving osteosarcoma therapy. *International Journal of Nanomedicine 14*: 6425–6437.

Xia, C., Yin, S., Xu, S. et al. 2019. Low molecular weight heparin-coated and dendrimer-based core-shell nanoplatform with enhanced immune activation and multiple anti-metastatic effects for melanoma treatment. *Theranostics 9*(2): 337–354.

Yan, C., Gu, J., Lv, Y., Shi, W., Jing, H. 2017. Improved intestinal absorption of water-soluble drugs by acetylation of G2 PAMAM dendrimer nanocomplexes in rat. *Drug Delivery and Translational Research 7*(3): 408–415.

Yang, X., Shi, G., Guo, J., Wang, C., He, Y.2018. Exosome-encapsulated antibiotic against intracellular infections of methicillin-resistant *Staphylococcus aureus*. *International Journal of Nanomedicine 13*: 8095–8104.

Ye, W.L., Zhao, Y.P., Li, H.Q. et al. 2015. Doxorubicin-poly (ethylene glycol)-alendronate self-assembled micelles for targeted therapy of bone metastatic cancer. *Scientific Reports 5*: e14614.

Yu, H., Tang, Z., Zhang, D. et al. 2015. Pharmacokinetics, biodistribution and in vivo efficacy of cisplatin loaded poly(L-glutamic acid)-g-methoxy poly(ethylene glycol) complex nanoparticles for tumor therapy. *Journal of Controlled Release 205*:89–97.

Yuan, H., Luo, K., Lai, Y. et al. 2010. A novel poly(l-glutamic acid) dendrimer based drug delivery system with both pH-sensitive and targeting functions. *Molecular Pharmaceutics 7*(4): 953–962.

Zhang, J.T., Huang, S.W., Zhuo, R.X. 2004. Temperature-sensitive polyamidoaminedendrimer/poly(N-isopropylacrylamide) hydrogels with improved responsive properties. *Macromolecular Bioscience 4*(6): 575–578.

Zhang, P., Zhang, Y., Gao, M., Zhang, X. 2016. Dendrimer-assisted hydrophilic magnetic nanoparticles as sensitive substrates for rapid recognition and enhanced isolation of target tumor cells. *Talanta 161*: 925–931.

Zhou, S.F., Zhou, Z.W., Yang, L.P., Cai, J.P. 2009. Substrates, inducers, inhibitors and structure-activity relationships of human Cytochrome P450 2C9 and implications in drug development. *Current Medicinal Chemistry 16*(27): 3480–3675.

Zhu, D., Tao, W., Zhang, H. et al. 2016. Docetaxel (DTX)-loaded polydopamine-modified TPGS-PLA nanoparticles as a targeted drug delivery system for the treatment of liver cancer. *Acta Biomaterialia 30*:144–154.

3

Dendrimers in Nanomedicine: History, Concept and Properties of Dendrimers

Smriti Sharma

CONTENTS

3.1 Introduction

Since the last two decades, science and technology has been at the hub of human effort, for creating new tools and products. It is natural to imagine that technology can take us further from the present technologies such as cell phones, laptop, computers, regenerative medicine, targeted drugs, environmentally friendly equipment, etc. Nanotechnology is one of the emerging concepts for the current and future possibilities. It has applications not only in electronics and communications but also in food, agricultural systems, energy, environment, nanomedicines, etc. (Thiruvengadam et al. 2018; Wong et al. 2019; Davidovits 2019).

Dendrimers are distinct nanostructures/nanoparticles/nanomotifs with 'onion skin-like' branched layers (Kannan et al. 2014). They are frequently utilised for enhancing and exposing the particular function. One of the nature's examples is tree, which utilised the dendritic pattern of leaves for the sunlight. The roots of the tree provide a dendritic pattern and help in collecting water from the soil (Figure 3.1). One such example is found in the human body also, during breathing air into the lungs, the air passes through a dendritic network of bronchioles and alveoli for providing a maximum surface for the transfer of oxygen into the blood stream (Mehta et al. 2019). Likewise, the arterial network possesses a dendritic pattern for the transportation of oxidised blood to the different organs. The brain and central nervous system has a large amount of cells, in the dendritic pattern for the exchange of information with the surrounding tissues (Figure 3.1) (Kweon et al. 2017).

Nanostructure growth occurs in concentric layers which are known as generations. The outer generation of each dendrimer presents a specific number of functional groups, which may act as a monodispersed

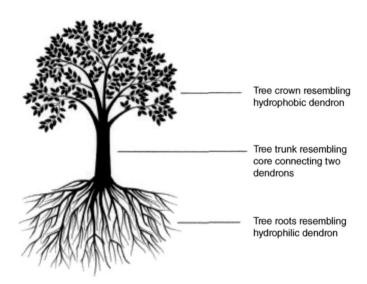

Tree crown resembling
hydrophobic dendron

Tree trunk resembling
core connecting two
dendrons

Tree roots resembling
hydrophilic dendron

FIGURE 3.1   Dendritic network of roots and leaves (Reproduced with copyright permission from Sikwal et al., 2017).

platform for the suitable interaction between nanoparticle–drug and nanoparticle–tissue interactions (Mehta et al. 2019). These features have gained attention in healthcare as nanostructures are utilised for conventional small drugs, proteins and DNA/RNA and in few examples as active nanoscale drugs. Dendrimer-based drugs as diagnostic agents are upcoming promising candidates for many nanomedicine applications (Baig et al. 2015; Wu et al. 2015). This chapter provides a brief discussion on the historical aspects of dendrimers and an overview of the architectural concept as well as their different physicochemical properties.

3.2 Historical Perspective

So far, one of the most persistent technologies observed on our Earth is considerably a dendritic architecture. A number of examples of these prototypes may be found in both the abiotic system and biotic systems. Various light prototypes and crystals of snow are examples of abiotic systems, whereas tree branches, roots, animal and plant vascular systems and neurons are examples of biotic systems. The dendritic patterns/prototypes are found in different dimensional lengths; for example, in trees the dendritic size is found in metres, in fungi, the size varies from millimetres (mm) to centimetres (cm), whereas in neurons, the size is in microns (μ). Still the reason for such a wide imitation of these dendritic technologies is not clear. However, it is speculated that these evolutionary architectures provide maximum interfaces for optimum energy.

For the first time, the concept of dendritic complex nature was suggested by Flory in the 1970s (Tomalia 2016). He explained the synthesis of abiotic macromolecules (dendrimers) without using any of the biological systems. The technique utilised in the synthesis is slightly different from the traditional polymerisation technique. Then in 1978, Buhleier et al. reported the concept of repetitive growth for the synthesis of low molecular weight amines (Buhleier et al. 1978; Karthikeyan et al. 2016). After few years, the Tomalia group coined the term dendrimers and described in detail the synthesis of polyamidoamine (PAMAM) derivatives (Figure 3.2) (Chauhan 2018; Tomalia et al. 1985). Two methodologies came into limelight for the synthesis of dendrimers. One is divergent methodology and the other is convergent methodology (Walter and Malkoch 2012). In divergent methodology, the core molecule expands outward in a differing fashion with an increasing number of different coupling steps. In convergent methodology, the periphery of the molecule proceeds inward to afford building blocks (dendrons). Then they coupled into a branching monomer through the reaction of a single reactive group located at its focal point (Tang 2017).

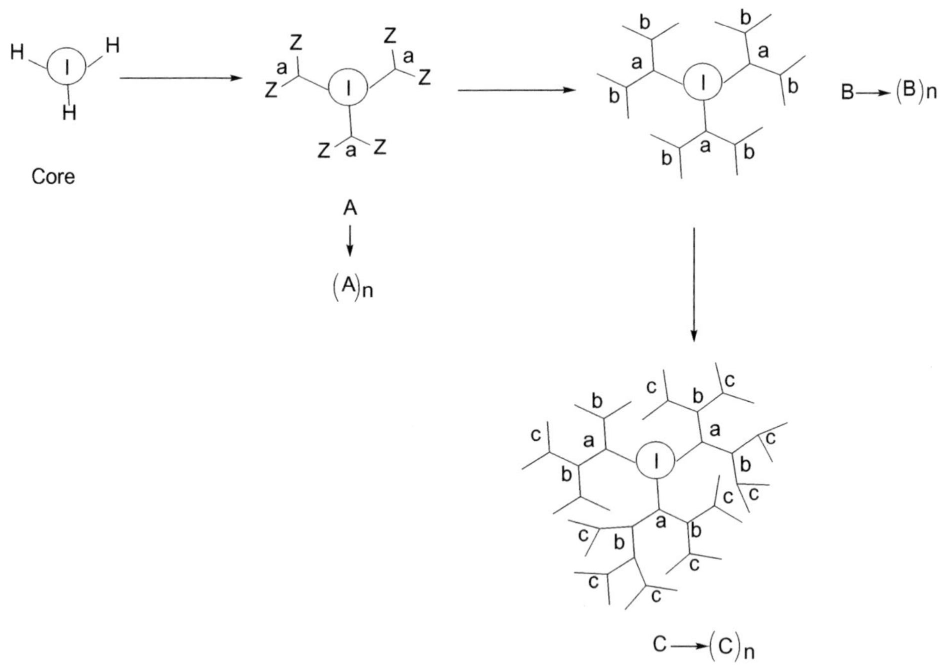

FIGURE 3.2 Flow chart for the formation of PAMAM derivatives from the (initiator core) (I) to branches {generations $(A)_n$, $(B)_n$ and (C)n}.

Divergent methodology based on acrylate monomers was discovered in 1979. It was further developed in Dow's laboratories with purity and good yield, while divergent iterative methodology utilised acrylonitrile which suffered from low yield and difficult product isolation and cannot be used to produce large molecules (Buhleier et al. 1978). After almost one and a half decade, two research groups (Worner and Mulhaupt and de Brabander-van den Berg & Meijer (DSM)) were able to enhance the Voegtle approach for the preparation of poly(propyleneimine) (PPI) dendrimers (Worner and Muhlhaupt 1993; Wallace et al. 1995; Newkome et al. 1994; Frechet et al 2000). In the year 1988, Fréchet et al. notably expanded the area of poly(ether) dendrimers with the discovery of convergent synthesis, which leads to globular macromolecules. Some other researchers also utilised a convergent approach for the preparation of dendrimers with different functional groups. The dendritic macromolecules and hybrid macromolecules were prepared by using one or two dendrimers (Hawker and Frechet 1990; Wooley et al. 1991). Further double stage convergent synthesis was also developed.

The convergent synthesis approach leads to the first solid phase synthesis of a dendritic molecule as well as a hyperbranched polyester. The structure elucidation was done through mass spectrometry techniques. This technique was utilised for the exact determination of protein molecular weight (up to 1 M Dalton) through different techniques such as electrospray ionisation, matrix-assisted laser desorption/ionisation, time-of-flight techniques, etc. (Kallos et al. 1991; Esfand and Tomalia 2001; Tomalia and Durst 1993). The dendritic growth and amplification have been carried out by utilising different techniques such as size exclusion chromatography, light scattering/viscosity measurement, photophysical property measurement, electron microscopy, gel/capillary electrophoresis, atomic force microscopy, etc. (Fréchet and Tomalia 2002; Glicksman 2012).

Dendritic polymers are now known as the fourth major class of polymeric architecture, which consist of different subsets: (1) random hyperbranched polymers, (2) dendrigraft polymers, (3) dendrons and (4) dendrimers. During the twentieth century, polymers were discovered and led to significant impact on the industries and economy (Wallace et al. 1995). (Hermes 1996).The macromolecular architectures have been divided into four major classes: class first, which has linear macromolecules, and class two which has cross-linked architectures. Both the classes explain the origination of traditional polymers as well as

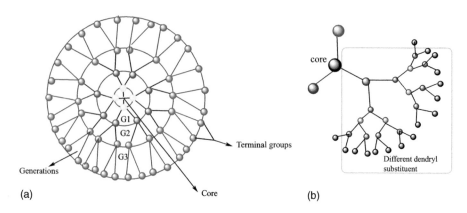

FIGURE 3.3   (a) Basic Structure of dendrimers and their elaborate form (b).

their properties. The third class includes branched molecules and is considered to be the focus point for the growth of the new polyenes. The class fourth includes dendritic structure-controlled macro-molecules, for example, dendrimers, dendrigrafts, etc. that are useful for the biological polymers and assemblies (Kallos et al. 1991).

3.3 Architecture of Dendrimers

Dendrimer is a Greek term and made of two words 'dendron' which means tree and 'meros' meaning part (Klajnert and Bryszewska 2001). These molecules are globular in shape and nanometre in size and possess discrete molecular architecture. Dendrimers are divided into three parts: core, branches and terminal groups. The central part of the dendrimers known as core comprises a single atom or a group of atoms, which provides linkage to the branches (Figure 3.3) (Klajnert and Bryszewska 2001).

The branches emerging from the core comprise repeating units in radially concentric layers, known as generations (G), and the terminal functional groups form the outer surface of the macromolecules, which are utilised for the drug encapsulation and nucleic acid complexation (Tomalia et al. 1985). In addition, their number of groups can be calculated with the following Equation 3.1.

$$n_G = F_K \left(F_V - 1 \right) G \tag{3.1}$$

n_G: number of terminal groups in the G-th generation; F_k: number of functional groups in the core; F_v: number of functional groups emerging from the branches; G: number of generations.

This equation explains the directly proportional relation to the number of terminal groups and the functional groups of the core. The number of branches increases exponentially with the number of generations.

The conformational nature of the branching units and the terminal groups provides mechanical stability to the dendrimers. The density of the dendrimers can be changed by the modification of the branching units. Accordingly, dendrimers can be more rigid and more flexible, for example, polyphenylene and PAMAM (Klajnert and Bryszewska 2001).

3.4 Types of Dendrimers

On the basis of the number of branches, dendritic molecules can be divided into three classes of molecules: (i) cascadanes, (ii) dendrimers and (iii) hyperbranched molecules. Among all the dendrimers, cascadanes are the most perfect ones, as they are of the same types and same molecular weight with regular branches. As compared to cascadanes, dendrimers are less perfect molecules, whereas hyperbranched molecules are the total group of imperfect molecules (Sowinska and Lipkowska 2014; Fischer and Vogtle 1999).

3.5 Nomenclature of Dendrimers

The IUPAC-nomenclature facilitates chemists to name molecules in a systematic way. However for complicated structures, nomenclature becomes difficult, and different rules were formed to enhance the traditional IUPAC-nomenclature. Initially, nodal nomenclature was used for naming the dendritic structures. The limitations of the nodal nomenclature lead to the development of new rules Newkome nomenclature and more refined Cascadane nomenclature.

3.5.1 Newkome Nomenclature

For the first time in 1993, Newkome suggested a nomenclature for the dendrimers. According to Newkome's theory, the priority is given to the peripheral terminal groups of the dendritic molecules and their fragments like dendrons and dendryl/cascadyl substituents. Then, the class was titled with the term 'cascade', and the individual branches are specified, which started from the core to the individual generations that are separated by the colons. Later, the terminal groups are than characterised (Newkome et al. 1993, 1994). Finally, the nomenclature is done according to the given equation:

$$Z \text{ cascade} = \text{core building block} \left[N_{core} \right] : \left(\text{branching unit} \right)^{G} : \text{end groups} \quad (3.2)$$

Z: number of terminal groups
N_{core}: core multiplicity
G: number of generations with branching building blocks

3.5.2 Cascadane Nomenclature

A more defined and detailed nomenclature assigned to complex dendrimers is known as cascadane nomenclature (Friedhofen and Vögtle 2006). The generations correlate with the branches, and their number is mentioned in the superscript and subscript form. Furthermore, the terminal groups are assigned in subscript. Then, the term cascadane is used in the subscript. The location points of the two branches are given in brackets. For example in the poly(propyleneimine) [PPI or POPAM] dendrimer, the generations G1.0 and G2.0 with the subsequent number of branches 4 and 8 are specifically indicated as superscripts and subscripts. The number of terminal groups (16) is shown in subscript form, and the class designation 'cascadane' reveals at the end of the nomenclature of compound. The locus 4,4 of the two branch atoms is given in brackets (Friedhofen and Vögtle 2006). It is shown in the following example:

1, 4-Diaminobutane[N,N,N′,N′] : {4-azabutyl(4,4)}$^{G1.0, G2.0}$ 4n, 8n : 3-aminopropyl$_{16}$-cascadane

Stepwise nomenclature of the dendritic molecules:

Step 1: All the dendritic structures have self-similarity.
Step 2: Broadly dendritic structure is divided into two parts: core and a dendritic part (Figure 3.3).
Step 3: Collection of different dendritic parts is known as dendrons.
Step 4: The suffix '-cascadane' is used for the molecules having seven to nine core units with two minimum similar dendryl substituents, whereas for different dendryl units, the suffix used is '-cascadyl' (Figure 3.4).

The compound which has three dendryl substituents and the dendron D1 region emerges out twice and describes the compound as a cascadane and hence it is a dendrimer. Furthermore, the D1 region is connected with a six-membered chain at 1 and 6 positions. The dendryl D2 substituent is considered as a cascadyl substituent. So the name of the core unit is: 5-(A-methyl)-3-B-1-(C-methyl)-2-(D2-cascadyl)-hexane(1,6) (Figure 3.5a).

Step 5: The particular part of the dendryl substituent from which branches arises is known as generations.

Step 6: The listing of different dendrons begins with the longest chain in the first generation. If the chain lengths are the same, then the priority will be given to the next generation.

Step 7: The core unit of the dendrimers is named according to the shortest chain, which joins different dendrons of the same types. After that, the position at which dendrons connected to the core unit is represented in brackets as shown in Figure 3.5b.

Step 8: Next numbering of the atoms in branches begins from the core to the terminal groups.

Step 9: The end of a generation is represented by the junction, and the terminal groups are not involved in the formation of a generation.

Step 10: First the naming is done for the scaffold part then to the core part. These names are represented in curly brackets. After that, the generation numbers are represented in the superscript form in and then they are kept in the curly brackets. The subscript stands for the total number of the units in the relevant generation. The particular generations are split from one another and also from the terminal groups by colons. The terminal group is indicated by the subscript. It is explained by the following example and is shown in Figure 3.5.

$$\text{Core}(1,1):\{\text{branch A } (A,A)\}^{G1.0}{}_{2n} : \{\text{branch B } (B,B)\}^{G2.0}{}_{4n}:\text{end}_8\text{-cascadane}$$

Step 11: Sometimes scaffold units are repeated in different generations; they are placed in the superscript form after the curly bracket and then differentiated by commas.

Step 12: If recurrence of units occurs in generations, which do not follow each other according to the step 11, and then the term registered for the generation will precede the terminal groups.

Step 13: Instead of curly brackets, angle brackets are utilised for naming the scaffold units, which have symmetrical branches in a particular generation. After the angle brackets, a subscript represents the total number of scaffold units in the generation.

$$\text{Core}(1,1):\{\text{branch A}(1,1)\}G1_2,G3_8. :\{\text{branch B}(1,1)\}G2_4:\{<\text{branchB}(1,1)>_{14}. <\text{branch C}(1,1)>2^*\}G4_{16}: \text{end}_{32}\text{.cascadane}.$$

If the scaffold has unsymmetrical branches, it is divided into smaller fragments, so that the cascadyl substituent becomes symmetrical again. It is explained in the following example:

$$(A(1,1):\{\text{branch A}(1,1)\}^{G2.0}{}_4:\{\text{branch B}(1,1)\}^{G1.0}{}_2,{}^{G3.0}{}_8: \text{end}_{16}\text{-cascadyl})-(A(1,1):\{\text{branch B}(1,1)\}^{G1.0}{}_2: \{\text{branch A}(1,1)\}^{G2.0}{}_4:\{<\text{branchB}(1,1)>_6. <\text{branch C}(1,1)>_2.\}^{G3.0}{}_8: \text{end}_{16}\text{-cascadyl})\text{-core}.$$

Step14: According to the nodal nomenclature, aromatic hydrocarbon parts are considered as a chain number, and then ring atoms are numbered accordingly.

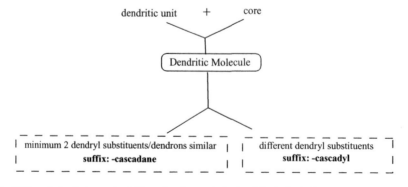

FIGURE 3.4 Differentiation between dendritic molecules on the basis of dendryl substituents.

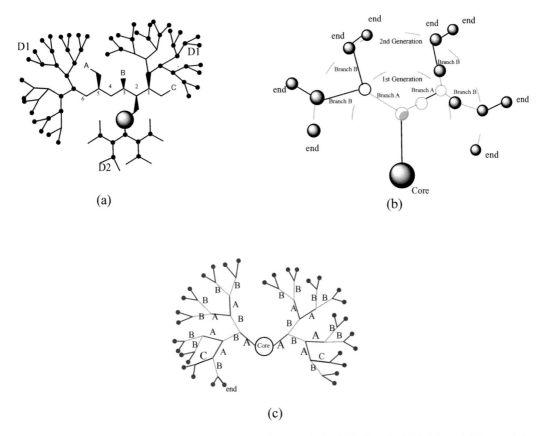

FIGURE 3.5 (a) Diagrammatic representation of compound 5-(A-methyl)-3-B-1-(C-methyl)-2-(D2-cascadyl)-hexane(1,6), (b) classification of a dendritic unit and (c) dendritic structure with different branches in one generation.

3.6 Properties of Dendrimers

Dendrimers are branched and nanosize macromolecules because of their molecular architecture. Their properties are outlined below:

1. **Monodisperity:** The dendritic polymers are monodisperse in nature because of the well-defined molecular structure and it is achieved by controlled synthesis (Dvornic and Tomalia 1996).
2. **Molecular size and mass:** The size and molecular mass of the dendrimers are specific. This leads to refined molecular architecture with better physical and chemical properties and also it is equivalent to medium sized proteins. With the increase in the molecular size, the intrinsic viscosity reaches the maximum level at the fourth generation and then it declines. Such behaviour is absent in linear polymers. (Table 3.1) (Bhattacharya et al. 2013)
3. **Rheological properties:** Dendrimers in comparison to linear polymers in solutions form tightly packed balls and impacted rheological properties.
4. **Solubility:** Properties such as high solubility, miscibility and high reactivity are found in polymers having high chain ends. Dendrimers having lipophobic groups are soluble in polar solvents, and at the same time, hydrophobic end groups are soluble in non-polar solvents. A major difference is also seen in the case of chemical reactivity. Upon catalytic hydrogenolysis, dendritic polymers can be debenzylated, whereas linear polyester remains unreactive (Fréchet 1994).
5. **Pharmacokinetic properties:** Dendrimers have some exclusive properties due to the presence of globular shape and internal cavities. They can encapsulate target molecules inside the cavity. Few

TABLE 3.1

Relationship between dendrimers and different biological entities

Different Types of Molecule	Generation	Molecular Mass	Radius	Variety of Surface Functional Group
PAMAM	G3	2411	1.1 nm	12 primary amines
PAMAM	G6	28.788	3.2 nm	128 primary amines
PAMAM-OH	G6	28.913	—	128 hydroxyls
Ovalbumin (Medium sized protein)	—	43.000	2.5 nm	20 primary amines, 10 phenol groups, four thiols and seven imidazoles
Hemocyanin (Large protein)	—	~5.000.000	—	$\approx$ 2000 primary amines, 700 thiols and 1900 phenols

scientists like Meijer and co-workers experimented by trapping small molecules such as rose bengal or p-nitrobenzoic acid inside the 'dendritic box' of PPI dendrimers, with 64 branches onto the periphery (Jansen et al. 1994, 1995). The outer part of the box can be hydrolysed and can liberate target molecules. The shape of the target and the architecture of the box and its cavities are utilised for determining the number of molecules that can be entrapped. The other group developed one of the methods in which boxes could be opened by the use of UV light (Araújo et al. 2018). For example, a fourth generation PPI dendrimer was utilised for terminating the 32 end groups in azobenzene groups. The E isomer can be turned into the Z form by using a wavelength of 313 nm and again can be converted back into the E form by using a wavelength of 254 nm light (Araújo et al. 2018). Many of the dendrimers are used for biomedical applications, for example, cationic dendrimers used for haemolysis and cytotoxicity purposes, whereas anionic dendrimers with carboxylate are not cytotoxic (Araújo et al. 2018).

6. **Multivalency:** Due to the hyperbranched arrangement in the dendrimers, the multivalent surfaces are created, which reveals a large number of functional groups on the surface and they provide exclusive adsorption or aggregation behaviours. The functional groups of proteins having the same molecular size are less than the dendrimers (Table 3.1).

7. **Electrostatic nature:** Dendrimers interact with oppositely charged end groups of the membrane and may increase the cell permeability for the delivery of the drug.

8. **Cytotoxicity:** The number of generation, surface groups and the nature of terminal moieties such as anionic, neutral, cationic, etc. affect the cytotoxicity. The cytotoxicity increases with the higher generation of dendrimers and also with the positive charge on the surface.

9. **Luminescence:** The dendrimers also possess luminescence properties and among them, PAMAM dendrimers produce strong blue luminescence. The reason for luminescence is the presence of tertiary amine and oxygen atoms. Due to the electrochemical activation, the lone pair electron excited from the ground state to higher energy state and absorbs visible light and then the phenomenon fluorescence-relaxation leads to blue colour.

3.6.1 Features Influencing the Dendrimers' Properties

1. **pH:** The pH value affects the conformation and shape of the dendrimers. At low pH (pH < 4), the inner surface of the dendrimers becomes hollow with the increasing generation number, whereas at neutral pH, back-folding occurs. At the higher pH (pH > 10), the dendrimer contracted, due to the neutral charge and acquired a globular-like structure.

2. **Solvent:** The solvent plays a very important role in estimating the conformational state of a dendrimers. The decreased solvation power leads to the back-folding in all generations of dendrimers.

3. **Salt:** High salt concentration favours the contracted conformation of dendrimers and especially to PPI dendrimers, whereas in a low salt environment the conformation further extended to minimise the repulsive force between the structures.

3.7 Conclusions and Future Perspective

Dendrimers create exhilarating opportunities for scientists to generate macromolecule structures with distinct functions. The hyperbranched macromolecules of the dendrimers present an inimitable architecture and have a spherical shape, multiple functional group opportunities and uniform size. Initially, focus was on the exploration of different methods of synthesis and investigating the properties of the dendrimers, but now it shifted towards their applications.

Abbreviations

AFM	Atomic force microscopy
cm	Centimetres
DNA	Deoxyribonucleic acid
EM	Electron microscopy
ESI	Electrospray ionization
MALDI	Matrix assisted laser desorption/ionization
mm	Millimetres
MS	Mass spectrometric
PAMAM	Poly (amido) amine
PPE	Polyphenylene
PPI	Poly-propylene-imine
RNA	Ribonucleic acid
SEC	Size exclusion chromatography
TOF	Time-of-flight techniques
UV	Ultraviolet

ACKNOWLEDGEMENT

The author would like to thank the Amity University, Noida, during writing this chapter.

REFERENCES

Araújo, R.V., Santos, S.S., Ferreira, E.I., Giarolla, J. 2018. New advances in general biomedical applications of PAMAM Dendrimers. *Molecules 23*(11): 2849.

Baig, T., Nayak, J., Dwivedi, V., Singh, A., Srivastava, A. Tripathi, P.K. 2015. A review about Dendrimers: synthesis, types, characterization and applications. *International Journal of Advances in Pharmacy, Biology and Chemistry 4*(1): 44–59.

Bhattacharya, P., Geitner, N.K., Sarupria, S., Ke, P.C. 2013. Exploiting the physicochemical properties of dendritic polymers for environmental and biological applications. *Physics Chemistry Chemical Physics 15*(13): 4477–4490.

Buhleier, E., Wehner, W., Vogtle, F. 1978. Cascade and nonskid-chain-like syntheses of molecular cavity topologies. *Synthesis, 1978*(2): 155–158.

Chauhan, A.S. 2018. Dendrimers for drug delivery. *Molecules 23*(4): 938.

Davidovits, P. 2019. Nanotechnology in biology and medicine. In *Physics in Biology and Medicine, 5th Edition*, 293–305. Elsevier.

Dvornic, P.R., Tomalia, D.A. 1996. Recent advances in dendritic polymers. *Current Opinion in Colloid & Interface Science 1*: 221–235.

Esfand, R., Tomalia, D.A. 2001. Poly(amidoamine) (PAMAM) Dendrimers: from biomimicry to drug delivery and biomedical applications. *Drug Discovery Today 6*(8): 427–436.

Fischer, M., Vogtle, F. 1999. Dendrimers: from design to application-A progress report. *Angewandte Chemie International Edition 38* (7): 884–905.

Fréchet, J. M. 1994. Functional polymers and Dendrimers: reactivity, molecular architecture, and interfacial energy. *Science 263*(5154): 1710–1715.

Fréchet, J.M.J., Tomalia, D.A. 2002. Introduction to the dendritic state: Dendrimers and other dendritic polymers. *In Dendrimers and other dendritic polymers. John Wiley & Sons Ltd*, 1–44.

Friedhofen, J.H., Vögtle, F. 2006. Detailed nomenclature for dendritic molecules. *New Journal of Chemistry 30*(1): 32–43.

Glicksman, M.E. 2012. Mechanism of dendritic branching. *Metallurgical and Materials Transactions 43A*: 391–404.

Grayson, S.M., Frechet, J.M.J. 2001. Convergent dendrons and Dendrimers: from synthesis to applications. *Chemical Review, 101*(12): 3819–3868.

Hawker, C.J., Frechet, J.M.J. 1990. Control of surface functionality in the synthesis of dendritic macromolecules using the convergent-growth approach. *Macromolecules 23*(21): 4726–4729.

Hermes, M.E. 1996. *Enough for One Lifetime*. American Chemical Society, Washington, DC. XVII, 345.

Jansen, J.F., de Brabander-van den Berg, E. M. M., Meijer, E.W. 1994. Encapsulation of guest molecules into a dendritic box. *Science 266*(5188): 1226–1229.

Jansen, J.F., de Brabander-van den Berg, E. M. M., Meijer, E.W. 1995, The dendritic box: shape-selective liberation of encapsulated guests. *Journal of American Chemical Society 117*(15): 4417–4418.

Kallos, G.J., Tomalia, D.A., Hedstrand, D.M., Lewis, S., Zhou, J. 1991. Molecular weight determination of a polyamidoamine Starburst polymer by electrospray ionization mass spectrometry. *Rapid Communications in Mass Spectrometry 5*(19): 383–386.

Kannan, R.M., Nance, E., Kannan, S., Tomalia, D.A. 2014. Emerging concepts in dendrimer-based nanomedicine: from design principles to clinical applications. *Journal of Internal Medicine 276*(6): 579–617.

Karthikeyan, R., Kumar, P.V., Koushik, O.S. 2016. Dendrimeric biocides - A tool for effective antimicrobial therapy. *Journal of Nanomedicine & Nanotechnology 7*(2): 1000359.

Klajnert, B., and Bryszewska, M. 2001. Dendrimers: properties and applications. *Acta Biochimica Polonica 48*(1): 199–208.

Kweon, J.H., Kim, S., Lee. S.B. 2017. The cellular basis of dendrite pathology in neurodegenerative diseases. *BMB Report 50*(1): 5–11.

Mehta, P., Kadam, S., Pawar, A., Bothiraja, C. 2019. Dendrimers for pulmonary delivery: current perspectives and future challenges. *New Journal of Chemistry 43*(22): 8396—8409.

Newkome, G.R., Baker, G.R., Young, J.K., Traynham, J.G. 1993. A systematic nomenclature for cascade polymers. *Journal of Polymer Science Part A, 31*(3): 641–651.

Newkome, G.R., Morrefield, C.N., Keith, J.M., Baker, G.R. and Escamilla, G.H. 1994. Chemistry within a Unimolecular Micelle Precursor: Boron Superclusters by Site- and Depth-Specific Transformations of Dendrimers. *Angew Chem Int Ed Engl, 33*(6): 663–665.

Sikwal, D.R., Kalhapure, R.S., Govender, T. 2017. An emerging class of amphiphilic Dendrimers for pharmaceutical and biomedical applications: Janus Amphiphilic Dendrimers. *European Journal of Pharmaceutical Sciences, 97*: 113–134.

Sowinska, M., Lipkowska, Z.U. 2014. Advances in the chemistry of Dendrimers. *New Journal of Chemistry 38*: 2168–2203.

Tang, Z. 2017. Research progress on synthesis and characteristic about Dendrimers. *IOP Conference Series: Earth and Environmental Sciences 100*(1): 2024.

Thiruvengadam, M., Rajakumar, G., Chung, lll-M. 2018. Nanotechnology: current uses and future applications in the food industry. *3Biotech 8*(1): 1–13.

Tomalia, D.A. 2016. Functional Dendrimers. *Molecules 21*(8): 1035.

Tomalia, D.A., Baker, H., Dewald, J., Hall, M., Kallos, G. 1985. A new class of polymers: starburstdendritic macromolecules. *Polymer Journal 17*: 117–132.

Tomalia, D.A., Durst, H.D. 1993. Supramolecular chemistry I—Directed synthesis and molecular recognition. In *Topics in Current Chemistry 165*: 193–313.

Walter, M.V. Malkoch, M. 2012. Simplifying the synthesis of Dendrimers: accelerated approaches. *Chemical Society Reviews 41*(13): 4593–4609.

Wallace, W.E., van Zanten, J.H., Wu, W.L. 1995. Influence of an impenetrable interface on a polymer glass-transition temperature. *Physical Review E, 52*(4) R3329–R3332.

Wong, J.K.H., Tan, H.K., Laua, S.Y., Yapa, P.S., Danquahb, M.K. 2019. Potential and challenges of enzyme incorporated nanotechnology in dye wastewater treatment: a review. *Journal of Environmental Chemical Engineering 7*(4): 103261.

Wooley, K.L., Hawker, C.J., Fréchet, J.M.J. 1991. Polymers with controlled molecular architecture: control of surface functionality in the synthesis of dendritic hyperbranched macromolecules using the convergent approach. *Journal of the Chemical Society, Perkin Transactions 1* (5): 1059–1076.

Worner, C, Muhlhaupt, R. 1993. Poly nitrile and polyamine functional poly(trimethylene imine) Dendrimers. *Angewandte Chemie International Edition 32*(9): 1306–1308.

Wu, L., Ficker, M., Christensen, J.B., Trohopoulos, P.N., Moghimi, S.M. 2015. Dendrimers in medicine: therapeutic concepts and pharmaceutical challenges. *Bioconjugate Chemistry 26*(7): 1198–1211.

4

Classification and Types of Dendrimers: Introduction and Chemistry

Anjali Rajora and Kalpana Nagpal

CONTENTS

4.1 Introduction

The term dendrimeris derived from 'dendron' (Greek word), which means 'tree' and from the suffix 'mer' which means 'segment' (Tomalia 2009). It refers to a synthetic three-dimensional structure, which has branching parts (Singh et al. 2008, Akbarzadeh et al. 2018). Dendrimers contain certain macromolecule groups consisting of a highly ordered and predictable structure and are synthesised from the core with repetitive occurrence of chemical reactions (Dias et al. 2020). Donald Tomalia was a polymer chemist who coined the term 'dendrimers' at Dow Chemical in Midland, Michigan. Fritz Vogtle's group synthesised dendrimers in 1978 at the University of Bonn for the very first time when they came across certain multifunctional branches (Nagpal et al. 2018). Their development as highly branched multifunctional polymeric systems resulted from different innovations and advances of polymer science and architectural chemistry. These are defined structures whose properties resemble those of biomolecules, which offer high functionality and versatility in drug delivery (Singh et al. 2008). These are well organised, branched three-dimensional nanoscopic macromolecules (i.e., molecular weight of 5000–500,000 g/mol) that have a low polydispersity index and play an essential role in the field of nanomedicines (Nimesh 2013). The comparison of the size of megamers, dendrimers, dendrons, monomers and atoms is shown in Figure 4.1a. They have the capability to entrap and conjugate with hydrophobic as well as hydrophilic molecules with high molecular weight either by host–guest interactions or covalent bonding (Madaan et al. 2014; Tomalia 2005). The nanosize is very important for their intracellular drug delivery as well as for evading them from the reticulo endothelial system (RES) of the human body (Dutta et al. 2008). This is the major class after linear, branched and cross-linked architectures (Thatikonda et al. 2013). 'Dentromers' have evolved recently as a new class of dendrimers, which were synthesised by triple branching (Astruc et al. 2018).

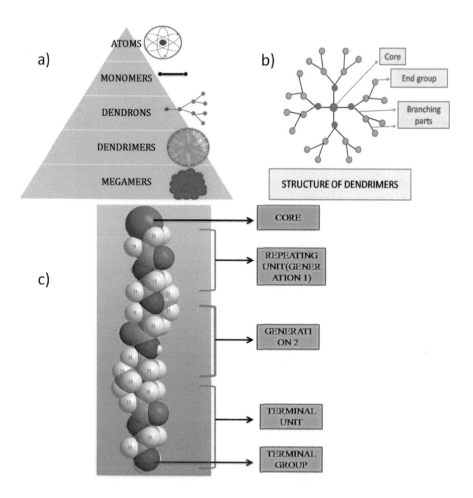

FIGURE 4.1   (a) Comparison of megamers, dendrimers, dendrons, monomers and atoms, (b) a typical structural unit of dendrimers and (c) chemical perspective of dendrimer's structure defining the overall sequence of the main core; the repeating units form different generations and the terminal unit.

These are 'The Polymers of 21st Century', and their structure influences their applications (Sohail et al. 2020). First, Vogtle synthesised dendrimers, and then, a cascade synthesis was performed with Michael type addition of acrylonitrile to amine groups. Furthermore, reduction of nitrile groups to primary amines was followed and repeated in an infinite order to produce highly branched macromolecular ligands theoretically. These highly branched polymers with a very low polydispersity index were made by stepwise polymerisation by the research groups of Denkewalter and Tomalia in the 1980s. Newkome's research group also reported similar macromolecule synthesis at the same time and referred to them as 'arborols' from the latin word, i.e., 'arbor', which stands for 'tree' (Abbasi et al. 2014). Because of their irregular arrangement and easy controllable properties, they are thought to be as the crossroad between associating polymers, traditional colloids and biosystems (Lombardo 2014). Thus, dendrimers give an excellent platform to attach genes, drugs and targeting moieties as they have multi-valent periphery with a well-defined structure, but at the same time toxicity caused by them is also an important issue. These molecules have found great applications in medicine, the biomedical field and technology. They have several advantages in comparison to other nanosized polymers and have also led to different studies for drugs to increase biomedical applications such as molecular labels, probe moieties and gene transfection (Tsai and Imae 2011). Advanced mass spectrometry studies have been performed to determine the purity of dendrimers. Thus, dendrimers have been proved to be flexible monodisperse molecules (Bosman et al. 1999).

4.2 Structure of Dendrimers

Dendrimers are nanoscale molecules often called 'cascade molecules', which are spheroid or globular in shape, highly branched and versatile and provide a high degree of surface functionality in the size range of 1–10 nm as shown in Figure 4.1b (Abbasi et al. 2014; Luna et al. 2014). Figure 4.1c describes the chemical perspective of dendrimer's structure defining the overall sequence of the main core; the repeating units form different generations and the terminal unit. These structures are designed precisely such that they encapsulate molecules in their interior voids or attach them to their surface to carry therapeutically active molecules (Thatikonda et al. 2013). Dendrimers have a wide range of applications in different fields of supramolecular chemistry especially in the self-assembly process and host–guest reactions. These monodispersed artificial molecules are very much defined and have a large number of functional groups, which form a highly compact structure with many applications (Abbasi et al. 2014; Santos et al. 2019). Dendrimers have certain distinct features due to their hyperbranched design and well-ordered synthesis. The main parts of dendrimers are:

(a) a central core (having a single atom or a group of atoms),
(b) generations: the building blocks having many layers of repeating units also known as branches and
(c) numerous functional groups, i.e., terminal groups on the surface which play an important role in deciding their properties (Sherje et al. 2018).

4.3 Generation of Dendrimers

Dendritic polymers are open and covalent assembly of branched cells (Tomalia 2005). They are synthesised by certain repeated reactions, and each reaction leads to the formation of a generation, which leads to doubling of the surface end groups and increases molecular weight. The size of these dendrimers is actually dependent upon their generations, which corresponds to the branching layers (Tsai and Imae 2011). Lower generation dendrimers (G0.0, G1.0 and G2.0) are more open and asymmetric as compared to those of higher generation (above G2.0). When the chains originating from the core become longer and further branched (in G4.0 and higher generations), they become globular in shape. As dendrimers extend out to peripheral groups, they form a closed densely packed membrane like structures. Once a critical branched state is reached, dendrimers are unable to grow further due to lacking space; this is called the 'starbust effect'. The size of dendrimers is dependent upon the generation, which corresponds to the branching layers (Tsai and Imae 2011). They have unique characteristics, which increases their applicability in technology, treatment and research in various areas such as industrial and biomedical fields (Klajnert and Bryszewska 2001). Toxicity of dendrimers is mainly dependent upon the terminal groups' nature. The anionic or neutral macromolecules have a lesser cytotoxic and haemolytic effect than cationic ones (Szulc et al. 2016).

4.4 Properties of Dendrimers

On comparison of dendrimers with other nanosized structures (e.g., carbon nanotubes, bucky balls or traditional polymers), they are either extremely non-defined and have incomplete structural diversity (Abbasi et al. 2014). Because of the presence of terminal groups on them in large quantity, they interact with solvents, surfaces or other molecules. The binding properties, solubility and reactivity of dendrimersare dependent upon highly interactive functionalities present on them (Nimesh 2013). They have adequate biocompatibility so that they can cross biological membranes or barriers easily when required. Their functional groups bear positive, negative or neutral charge, which is important for dendrimers to act as drug delivery vehicles. Due to their nanosize, they act with different cellular components to increase toxicity, which is dependent upon the number of functional groups and surface charge present on them (Kesharwani et al. 2014).

4.5 Chemistry of Different Dendrimers

Certain new methods have been developed for the synthesis of dendrimers such as Lego chemistry and click chemistry using reactions such as the Diels–Alder (DA) reaction, copper-assisted azidealkyne cycloaddition (CuAAC), thiolene and thiolyneclick reaction, etc.

4.5.1 Lego Chemistry

This is a new method of dendrimer production which uses branched monomers, i.e., DB_2 and CA_2. Each one generation is produced in a single quantitative step, where only nitrogen and water are formed as by-products, and thus, generation 3.0 is produced in three steps. The end groups, on the other hand, are hydrazines and phosphines, and their reactivity was explained (Maraval et al. 2003). Similarly, the reaction of generation 4.0 and CD_5 monomer leads to additional branching and ending group number in a single pace from 48 to 250. The characteristic of 'Lego chemistry' is that it requires minimum solvent and is single for each generation and effortless purification.

4.5.2 Click Chemistry

This concept was initially introduced in 2001 by Barry Sharpless and co-workers (Kolb et al. 2001). It is a group of chemical reactions, which are versatile in joining diverse structures without requiring the protection steps: majorly cycloadditions, nucleophilic ring opening, non-aldol carbonyl chemistry and carbon multiple bond addition reactions. These different reactions from stable products with no or few by-products. Reactions belonging to this family of click reactions must be wide in scope and modular, give high yields and be tolerant to many functional groups, leaving harmless byproducts, which results in products that can be purified by non-chromatographic techniques. The reaction needs to be stereospecific and is carried out under mild reaction conditions. Kolb et al. (2001) added that an internal driving force is present in chemical reactions, which pushes these reactions to full conversion, both on large and small scales. These reactions are carried out by high thermodynamic forces usually 20 kcal/mol and are called 'spring loaded'. The characteristics which define click chemistry are its high chemoselectivity, high reaction enthalpy, environment safety and atom economy, high chemoselectivity and reaction enthalpy, which offer ability to produce new structures which would not have been possible otherwise (Kolb et al. 2001).

4.5.2.1 The DA Reaction

This reaction has several advantages given by click reactions including high conversion, atom economy and good chemical and region selectivity. Another characteristic of this reaction is the option to decide whether we wish a reversible or irreversible reaction by cautiously selecting precursors. The utilisation of cyclopentadienone helps to build carbon-rich dendrimers as the diene was made popular by Morgenroth and Müllen (1997). In this reaction, retro-Diels–Alder is prevented, and carbon monoxide molecules are lost. However, conditions for carrying out this reaction have not been evolved clearly except applying microwave irradiation to attain 300 °C. DA was much better as compared to organo-metallic coupling. The materials rich in carbon commonly attain melting points of 350 °C. Thus, the elevated temperature does not become an issue (Arseneault et al. 2015).

4.5.2.2 CuAAC

This reaction is a modification of cycloaddition among azide and alkyne catalysed with the help of copper (Cu) under mild conditions. It contains both hydrophobic and hydrophilic substrates, which is a valuable feature of amphiphilic molecules. It can work in a wide range of pH, i.e., 5–12 and can continue at room temperature. The catalyst is relatively gentle and inexpensive when compared to a variety of organo-metallic compounds. Because of all such advantages, these conditions are widely used for the formation of dendrimers (Arseneault et al. 2015).

4.5.2.3 The Thiolene and Thiolyneclick Reaction (TEC)

The TEC origin is even more dispersed than CuAAC according to some research studies, but now it has developed into a definite set of conditions. The TEC includes the reaction among thiol and terminal alkenes which can proceed either through a nucleophilic one or free-radical mechanism. The latter is more popular because it can be photocatalysed. Classic conditions are orthogonal and gentle to a large scale of reactions, and even CuAAC and Michael addition (Arseneault et al. 2015).

4.6 Types of Dendrimers

Depending on the chemistry of different dendrimers, they are of many types as mentioned below:

4.6.1 Polyamidoamine (PAMAM) Dendrimers

These are also known as PAMAM dendrimers and were introduced in 1985. Donald A Tomalia named them 'starburst polymers', which were considered as a novel class of polymers (Chauhan 2018). Their core part is made up of a linear sequence of molecules having primary amines. The compounds used commonly as the core component are cystamine (having four core multiplicity, i.e., upto four branches), ammonia (three core multiplicity) and ethylenediamine(four core multiplicity) (Araújo et al. 2018). The schematic diagram of PAMAM dendrimers is shown in Figure 4.2a.

The divergent method is used for their synthesis, which starts from ethylenediamine and ammonia as core reagents. The products upto the 10th generation are obtained and have molecular weight upto 9,30,000 gm/mol and apolydispersity index of around 1.01. The polydispersity index of PAMAM dendrimers is low, which means that particle size distribution is homogeneous for each generation (Kesharwani et al. 2014). Another method used for the synthesis of EDA core–PAMAM dendrimers involves Michael addition of EDA to methacrylate and then, amidation of formed multiester with EDA as shown in Figure 4.2b. The G 0.5 dendrimer formed leads to an increase in its generation on addition with EDA (Peterson et al. 2001). Methyl acrylate and EDA are added according to the number of generations desired G0.0, G1.0, G2.0, G3.0, G4.0, etc. It is also possible to have generations like G0.5 by addition of methyl acrylate and then terminating the reaction sequence leading to terminal carboxylate groups (Abedi-Gaballu et al. 2018; Das et al. 2018). Drugs are encapsulated within PAMAM's internal cavity, throughout the structure or on the surface of dendrimers to prevent their degradation. Cationic PAMAM dendrimers are gaining more attention as they provide better drug delivery and nanoparticle formation (Smith et al. 2012).

They are appropriate for passive targeting and accumulation of drugs within the tumor site by the enhanced permeation and retention effect, which reduces the side effects of loaded drugs. However, toxicity associated with charge limits the use of full generation amine-terminated PAMAM dendrimers (Vu et al. 2019). They find applications such as drug delivery to molecule encapsulation and form building blocks of nanostructures to micelles, which acts as a contamination removing agent (Maiti et al. 2004). The star-like pattern is seen while observing the structure of high-generation dendrimers of these categories in two dimensions and thus named 'Starbust' (Tomalia et al. 1985). They are used in computer toners, materials science, etc. (Shahi et al. 2015). G5.0 PAMAM dendrimers have a size of 5 nm approximately due to which they are capable of moving through biological tissues and may increase the blood circulation time of conjugated or entrapped drug (Van Dongen et al. 2013).

4.6.2 Hyperbranced Polyethylene Glycol (PEG) Dendrimers

They are inexpensive as they are prepared easily but are not symmetrical in shape and are polydispersed. Both these classes forma nanocavity, in the core of which different molecules such as bioactive compounds have been encapsulated. Primarily on their surface, bioactive compounds are also added at the dendritic scaffold. This set of polymers bear several functional groups on their outer surface. Several functional groups on the surface like PEGylation have often been applied broadly to other nanoparticles and liposomal carriers. Also, alteration of interior groups of dendrimers can affect their solubility behavior, which makes encapsulation of different drugs possible. Dendritic polymernanocavities or addition of drugs into their peripheral region has been performed to elicit drug release by changing the bio-environment at the site suited for action. The risk of triggering premature release of drug in the extracellular fluid should be considered while scheming a targeted PEGylated drug delivery system. Thus, it is advised to study drug release behaviour from PEGylated dendrimers with the help of virtual experiments prior to proceeding to any *in vitro* and *in vivo* tests. The introduction of the PEG chains increases the solubilisation efficiency of dendrimers (Paleos et al. 2010). The chemical structure of hyperbranched PEGpseudo dendrimers with hydroxyl (–OH) end groups is shown in Figure 4.3 (Morshed et al. 2019).

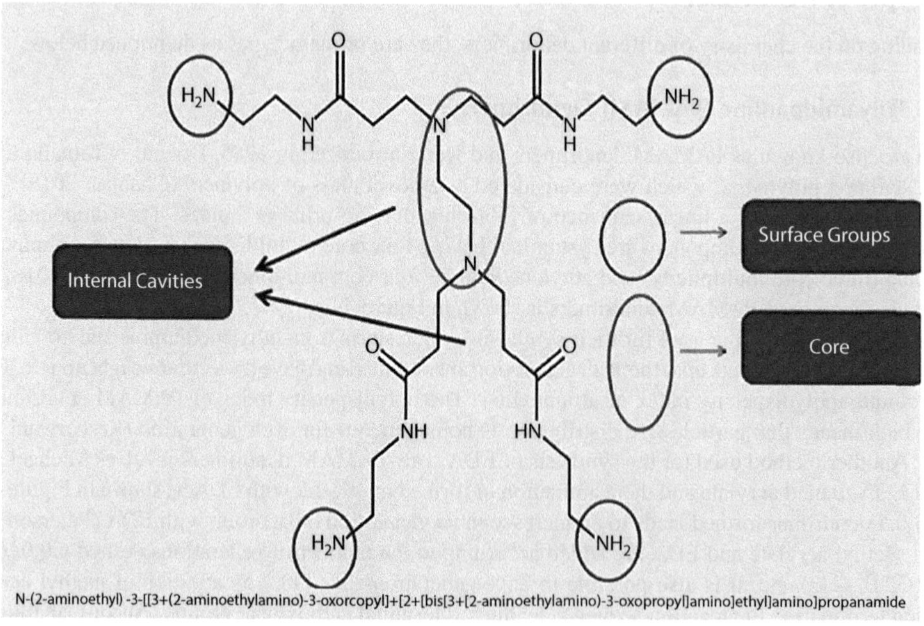

H_2N
NH_2
O
O
N H
N
N H
N
Surface Groups
Internal Cavities
Core
O
O
NH
HN
H_2N
NH_2
N-(2-aminoethyl) -3-[[3+(2-aminoethylamino)-3-oxorcopyl]+[2+[bis[3+[2-aminoethylamino)-3-oxopropyl]amino]ethyl]amino]propanamide

(a)

FIGURE 4.2 (a) Schematic representation of PAMAM dendrimers. *(Continued)*

(b)

FIGURE 4.2 (Continued) (b) Diagram of the synthesis of ethylenediamine (EDA)core–PAMAM dendrimers. Reprinted from "Synthesis and CZE analysis of PAMAM dendrimers with an ethylenediamine core," *Proc. Estonian Acad. Sci. Chem.*, 2001, 50(3), 156–166. Copyright [2001] with permission from The Estonian Academy Publishers.

4.6.3 Polypropylene Imines (PPIs)/Diamino Butane(DAB)/Polypropylene Amine

Theyare the oldest dendrimers which were developed originally by Vögtle; their structure contains primary amine as the end group and certain tertiary trispropylene amine as the interior part. They are generally available up to G5.0 and often called DAB dendrimers where DAB is the core structure,. The method

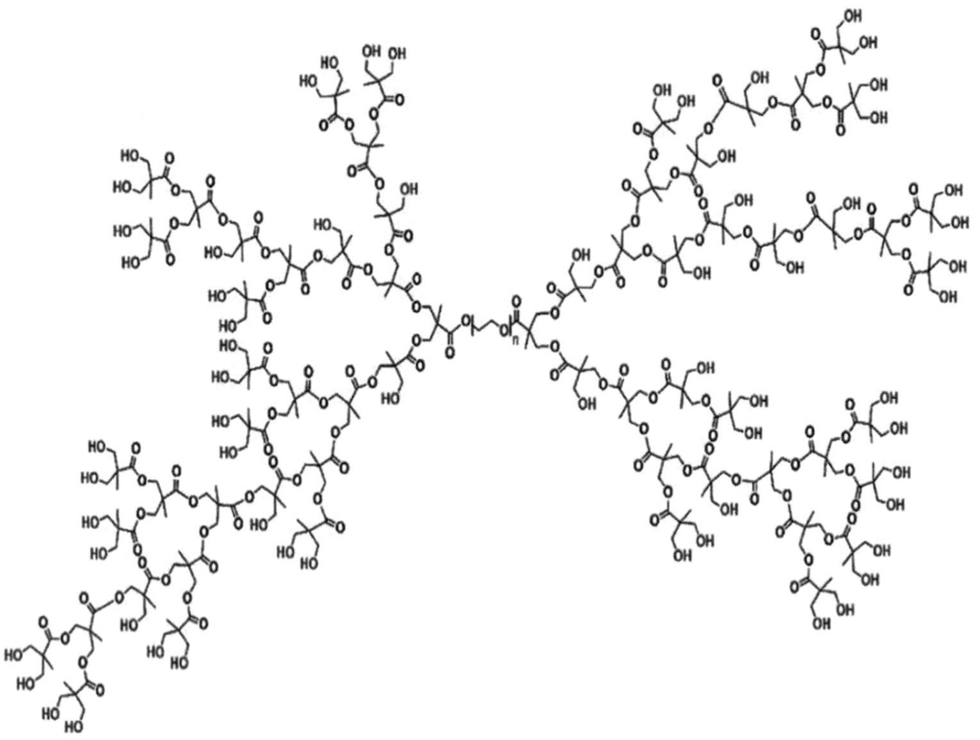

FIGURE 4.3   Chemical structure of hyperbranched PEGpseudo dendrimers with–OH end groups. Reproduced Mohammad Neaz Morshed et al. (2019), "Surface modification of polyester fabric using plasma-dendrimer for robust immobilisation of glucose oxidase enzyme." Copyright [2019], Springer Nature.

commonly used for the synthesis is divergent, and these dendrimers are widely used in materials science and biology. The example is Asramol preparation using Diagnostic and Statistical Manual of Mental Disorders (Shahi et al. 2015; Dutta et al. 2008). It is synthesised from 1,4 diamino butane and then developed by a repetitive chain having (a) double Michael addition of acrylonitrile to primary amino groups then and (b) hydrogenation by keeping Raney cobalt under pressure (Chavda 2007). The PPI dendrimers are shown in Figure 4.4 (Wu et al. 2013).

4.6.4 Carbohydrate-Containing Dendrimers

These types of dendrimers are also called 'Glyco dendrimers', and their synthesis is shown in Figure 4.5a; these dendrimers are covered with carbohydrates on their external face. Both convergent and divergent methods are used for their synthesis. Certain grouped glycosides in the form of well-branched oligosaccharides can also act as dendritic wedges in the expansion of carbohydrate dendrimers performed subsequently. It is estimated that these novel saccharide-containing polymers, which are soluble in water and immensely branched, will also discover applications in the biological area as well and in the framework of new materials (Jayaraman et al. 1997).

(a) Carbohydrate-Coated Dendrimers: The creation of these types of glycoden drimers is carried out either by the convergent approach or by the divergent approach by doing end group alterations in a pre-existing non-carbohydrate dendritic core. As the chemical treatment with free saccharides is a delicate task, the employment of O-protected saccharides is beneficial here provided that the dendritic core is secure under the conditions necessary for the protection (Jayaraman et al. 1997).

(b) Carbohydrate-Coated Dendritic Wedges and Cluster Glycosides: With the development in the production of carbohydrate-containing dendrimers, attention is paid to the expansion of methods to

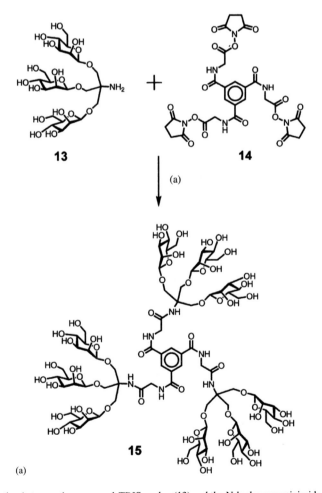

FIGURE 4.4 PPI G3 dendrimer. Reproduced Wu et al. (2013), "Dendrimers as carriers for siRNA delivery and gene silencing: a review."

FIGURE 4.5 (a) Reaction between the mannosyl-TRIS wedge (13) and the N-hydroxysuccinimide-activated core (14) can be conducted in Ž. The absence of protecting groups on the saccharide residues to give the dendrimer (15). Reagents: (a) C_5H_5N/DMF/60°C. Reprinted from "Design and synthesis of glyco dendrimers," by W. Bruce Turnbull and J. Fraser Stoddart, 2002, *Rev. Molecular Biotechnol.*, 90, 231–255. Copyright [2002], with permission from Elsevier." *(Continued)*

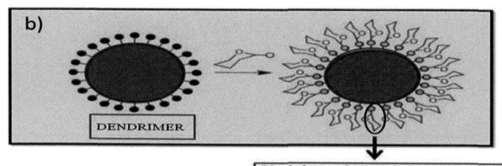

Black dots and open circles represent reactive functionalities located at the dendrimer outer-surface and at the saccharide aglycone. These functionalities allow the saccharides to get attached to the dendrimer.

FIGURE 4.5 (Continued) (b) Representation of carbohydrate-coated dendrimers by modification of non-carbohydrate dendritic matrices.

multivalent cluster glycosides, based on dendritic wedge-like carriers. The main motivating force to do research was the development of binding efficiency of some non-synthetic carbohydrates towards carbohydrate-binding proteins like lectins, which result from the grouping of carbohydrate units. To achieve good bonding between protein receptors and saccharide ligands leading to irreparable blocking of these types of receptors and to inhibit their interaction with natural ligands, lots of small carbohydrate groups are produced along with the growth of neoglyco conjugates, produced by the bonding of carbohydrates to proteins or synthetic polymers like carriers (Jayaraman et al. 1997).

(c) Full Carbohydrate Dendrimers: Synthetic full carbohydrate dendrimers are analogues of polysaccharides, which can imitate a wide range of polysaccharide structures and their properties. The three-dimensional strongly packed design of extremely branched dendritic oligosaccharides is responsible for inclusion properties linked with the interior parts of molecules. The full carbohydrate dendrimers which are water soluble have internal voids like cyclodextrins which have the capability to encapsulate and solubilise lipophilic organic molecules (Jayaraman et al. 1997).

4.6.5 PEG Core Dendrimers

These dendrimers along with biocompatible polymers like the PEG chain as the core show various advantages such as less cytotoxicity of dendrimers and better biodistribution credited to this PEG core and provide extra probability for surface modification of dendrimers because of their less spatial hindrance among neighbouring surface functionalities (Guillaudeu et al. 2008; Kim et al. 2004) effectively made such dendrimers with three-arm PEG as the central core and 2,2-bis-(hydroxymethyl) propanoic acid as repeating units as shown in Figure 4.6 (Cheng et al. 2011). In this type, AB_2/CD_2 monomers are used for the synthesis of four-arm-based hybrid copolymers containing the PEG core using orthogonal amine/epoxy and thiol-yne chemistry. These four-arm hybrid dendrimers are a good platform for gene transfection as they induce endosomal escape and also show better DNA binding than two-arm hybrid dendrimers. The ability of PEG dendrimers for encapsulating drugs increases with increased dendrimer's generation and chain size of PEG grafts (Kojima et al. 2000). Biological activity of four-arm-based and two-arm-based dendrimers reveals a greater dendritic effect and low cytotoxicity with the increase in internalisation of cells related to the increase in amine end groups. The production of this type of dendrimer is also time-consuming and requires different processes for purification. Therefore, PEG is used as a mono or di-functional core for divergent synthesis. The strategy to perform purification of PEG dendrimer hybrids simply by precipitation dialysis relies upon the solubility of the PEG core. However, studies revealed that PAMAM-PEG-PAMAM copolymers have reduced toxicity and increased

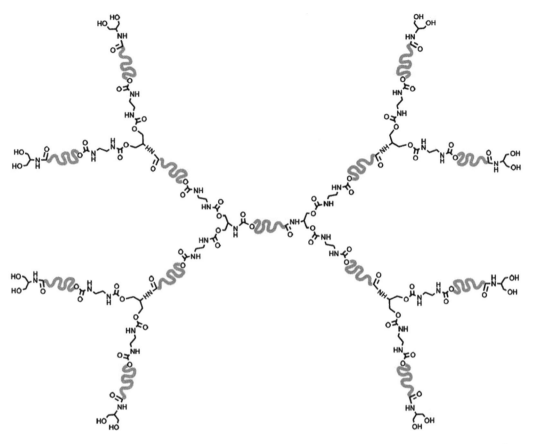

FIGURE 4.6 Biocompatible dendrimer consisted of the PEG core and repeated units. The green line represents the PEG chain. Reprinted from "Design of biocompatible dendrimers for cancer diagnosis and therapy: current status and future perspectives," by Yiyun Cheng, Libo Zhao, Yiwen Li and Tongwen Xu, 2011, *Chem. Soc. Rev.*, 40, 2673–2703. Copyright [2011] by the Royal Society of Chemistry. Reprinted with permission.

colloidal stability in comparison to PAMAM dendrimers. On the basis of the bi-functional PEG core, fourth generation dendrimers having 32 orthogonal surface groups and 20 inner functionalities were produced in just four steps. The victory of this approach provoked the discovery of different hybrid dendritic architectures in order to know their structure–property relationship concerning biological activity (Albertazzi et al. 2012).

4.6.6 Amine-Terminated PEG Core Dendrimers

Amine-terminated PAMAM dendrimers are considered as a perfect class of building blocks for producing multifunctional vectors. This is because they have well-defined extremely branched structures and various surface groups accessible for gathering of many different types of functional entities and intrinsic properties advantageous for gene delivery (Morshed et al. 2019; Yuan et al. 2010). The production of these dendrimers is carried out in four steps based on a four-arm PEG star core using dialysis and precipitation as purification methods. Comparatively simple purification, easy production and compatibility of alkyne groups of third generation with thiolyne and copper-catalysed azide alkyne click chemistry make amine dendrimers very smart scaffolds for a variety of biomedical applications. The role of several end groups and structural design in shaping the relations of these four arm- and two arm-based hybrid cationic dendrimers containing living cells was determined with respect to the internalisation pathway, membrane affinity and toxicity (Albertazzi et al. 2012). The mechanical and physicochemical property layers of these dendrimers depend upon the fabrication method (Martin and Vermette 2006). Amine-terminated

PEG core dendrimers for full generations EDA were used to produce amine-terminated dendrimers while for half generations like 0.5–4.5 methyl acrylate was chosen as adendron (Gürbüz et al. 2016).

4.6.7 Phosporous Dendrimers

This is a novel class of cascade molecules in which both the core and its subsequent branches are quaternary phosphonium ion sites (Majoraland Caminade 1999). These dendrimers are synthesised on reaction of hydroxyl benzaldehyde under alkaline conditions with a core having P-Cl function, followed by reduction of the aldehyde groups with a phosphorhydrazide. These two steps were continued until highest generation (generation no. 12) is obtained. The sequence of these reactions is started from the core having aldehyde or P-Cl functions. The existence of PCl_2 or aldehyde end-groups in each step allows building a wide variety of reactions. Other than this method described above, a new method was developed recently using AB_5 and CD_5 monomers for multiplying functional groups. This method produced layered dendrimers containing two types of recurring units. The phosphorus-containing dendrimer with phosphonate surface groups is shown in Figure 4.7 (Cheng et al. 2011). Their generations do not affect thermal stability

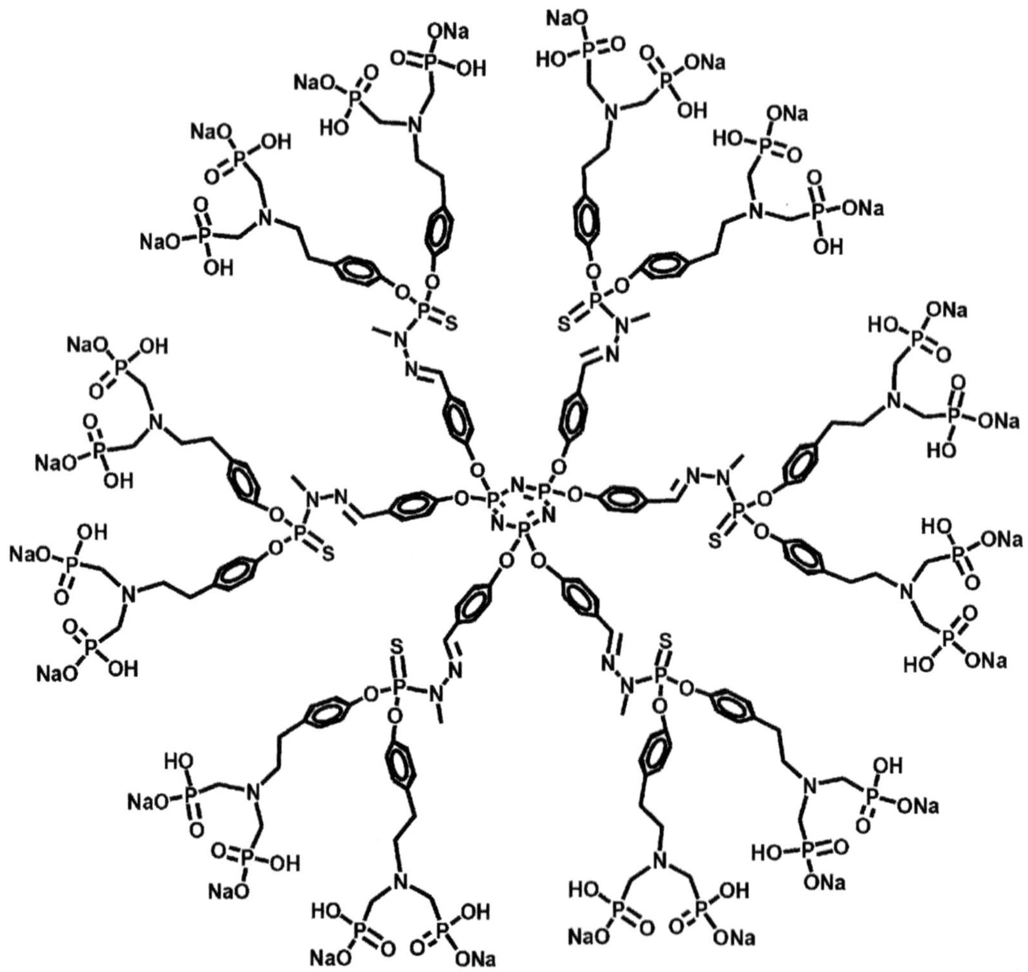

FIGURE 4.7   Phosphorus-containing dendrimers with phosphonate surface groups. Reprinted from "Design of biocompatible dendrimers for cancer diagnosis and therapy: current status and future perspectives," by Yiyun Cheng, Libo Zhao, Yiwen Li and Tongwen Xu, 2011, *Chem. Soc. Rev.*, 40, 2673–2703. Copyright [2011] by the Royal Society of Chemistry. Reprinted with permission.

while types of end groups do. Their very first application was in catalysis where they allow reusage of dendrimer catalysts; they are also used in nanoelectronics, immobilisation of protein, electrochemical sensors, materials science and biology (Caminade et al. 2003).

4.6.8 Polyphenylene Dendrimers

Polyphenylene dendrimers are considered incredibly promising templates because of their perfect structure and specific spatial definition of functional units in centre in the dendritic scaffold or at periphery (Alikhani et al. 2014). Furthermore, in the living organism environment they are chemically inert, which is another important requirement for biologically oriented applications (Mihov et al. 2005). They must be stiff as their rotation is possible only around those C–C bonds, which are in the ring contributing high thermal and chemical stability. On comparing with linear polyparaphenylenes, branched polyphenylene and corresponding dendrimers must have similarity in thermal stability but should have more solubility due to branching of polymers; the packing phenomenon seen in linear poly paraphenylenes is prevented. The DA reaction among tetraphenylcyclopentadienone derivatives and ethynyl-substituted phenylene allows the easy and quick production of monodispersed polyphenyleneden drimers until fourth generation with molecular masses more than 20,000 g/mol (Nguyen et al. 2013; Wiesler et al. 2001).

4.6.9 Dendrimers Containing Heteroatoms

The other heteroatoms such as Si, B, Ge or Bi are also incorporated into dendrimers. *Silicon-containing dendrimers* were the initial heteroatom-bearing dendrimers produced. Three kinds of linkage have been used at branching points, i.e., siloxane (Si-O), carbosilane (Si-C) or silane (Si-Si) groups. Siloxane-containing dendrimers were produced by the recurrence of two steps, which starts from a trifunctional core like methyltrichlorosilane: the first step includes a nucleophilic substitution of chlorosilyl groups by diethoxyhydroxymethylsilane sodium salt while the next step involves the reaction of ethoxysilane end groups with $SOCl_2$ to yield Si-Cl end groups. Carbosilane (CBS) dendrimers were produced in extremely good quantity by using alternate alkenylation with Grignard reagents and hydrosilylation (Arif et al. 2013). The CBS dendrimers are shown in Figure 4.8 (Wu et al. 2013). These two steps are repeated to obtain fourth generation (G1.0–G4.0), which contains 48 ethoxy end groups. Side reactions such as dehydrogenation R-addition of hydrosilylation are mostly prevented by carefully controlling temperature and choosing the right solvent during the hydrosilylation. *Boron-containing dendrimers* include the inclusion of boron on dendrimers and take benefit of the tendency of alkyne moieties to react with decaborane, i.e., $B_{10}H_{14}$ to afford 1,2-dicarba-closo-dodecaboranes. In *Germanium-containing dendrimers*, the initial dendrimers containing germanium at branching points have been explained by Huc et al. (1996) where they used both divergent and convergent approaches for their synthesis. In *Bismuth-containing dendrimers*, the first exceptional dendritic bismuthane 183-G2 was produced up to the second generation by Suzuki and his colleagues (Majoraland Caminade 1999).

4.6.10 Amino Acid Dendrimers (poly-L-lysine and Polyarginine Dendrimers)

Amino acids are protein components and their naturally occurring metabolites. However, arginine and lysine which are 'amino acid dendrimers' are low biocompatible than expected. Surface charges greatly influence their biodistribution behaviour. Anionic and neutral dendrimers have perfect blood distribution time and are not much recognised by the RES. Different types of amino acids like cationic, neutral, anionic or hydrophilic/hydrophobic were placed on the PPI dendrimers and PAMAM surface. To increase gene transfection and reduce toxicity G4.0, the PAMAM dendrimer surface was functionlised with arginine. They transfected 40% of cells effectively with target plasmid and showed improved gene delivery and low toxicity compared to PAMAM dendrimers grafted with lysine. The gene delivery and cytotoxicity of amino acid-coated PAMAM dendrimers depend on the type of amino acid used. Neutral amino acid (leucine and phenylalanine) and anionic amino acid (glutamate)-coated PAMAM dendrimers are found

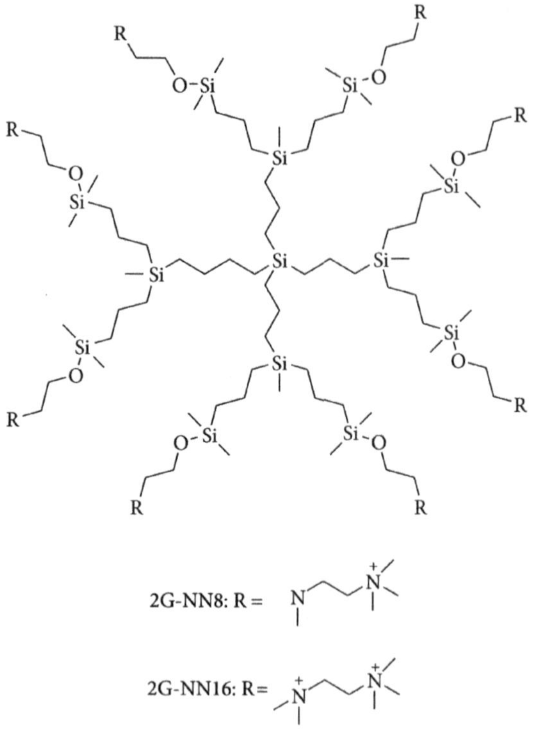

FIGURE 4.8 CBS dendrimers. Reproduced from "Dendrimers as carriers for siRNA delivery and gene silencing: a review," by Jiangyu Wu et al. (2013).

to be less harmful than cationic amino acids (arginine or lysine). Hydrophobhic and neutral amino acid dendrimers due to their high permeation efficiency particles through the cell membrane have high gene delivery capacity (Cheng et al. 2011).

(a) **Poly(L-lysine) dendrimers:** Their construction units are L-lysines, which occur naturally, and their main advantage is their biodegradable and biocompatible behaviour in the presence of enzymes and acids (Fox et al. 2009). The structure of poly(L-lysine) dendrimers of generation 3.0 is shown in Figure 4.9 (Liu et al. 2012). When combined with a Endo-Porter (weak-alkaline peptide, i.e., ampiphilic), these dendrimers show improved transfection because of ampiphilic and weak alkalinity of Endo-Porter. This in turn increases the gene silencing efficiency and delivery by promoting siRNA release in the intracellular area (Liu et al. 2012). The half inhibitory action of these dendrimers in comparison to PAMAM dendrimers of similar generation is found to be only two-fold lower. Poly(L-ornithine) and poly(L-lysine) dendrimers clear from blood circulation quickly and get accumulated in kidneys and liver. But if anionic amino acids are used such as glutamate, biocompatility of these dendrimers can be improved.

(b) **Poly(arginine) dendrimers:** The polyhedral poly(arginine) dendrimer (G2.0 and G3.0, *in vitro* model) and poly(lysine) dendrimer (G6.0, *in vivo* model) with a biocompatible poly(glutamic acid) dendrimer were found to have antiangiogenic activity systemically and often suggest the use of amino acid dendrimers in cancer (Cheng et al. 2011).

4.6.11 Surface-Engineered Dendrimers

In this, a dendrimer is considered as the core and tailored with phosphate, fluorine, folate, lipids and specific antibodies. These engineered dendrimers provide improved siRNA loading protecting siRNA from degradation by different enzymes like RNasefor better transfection and permeation, endosomal escape and good

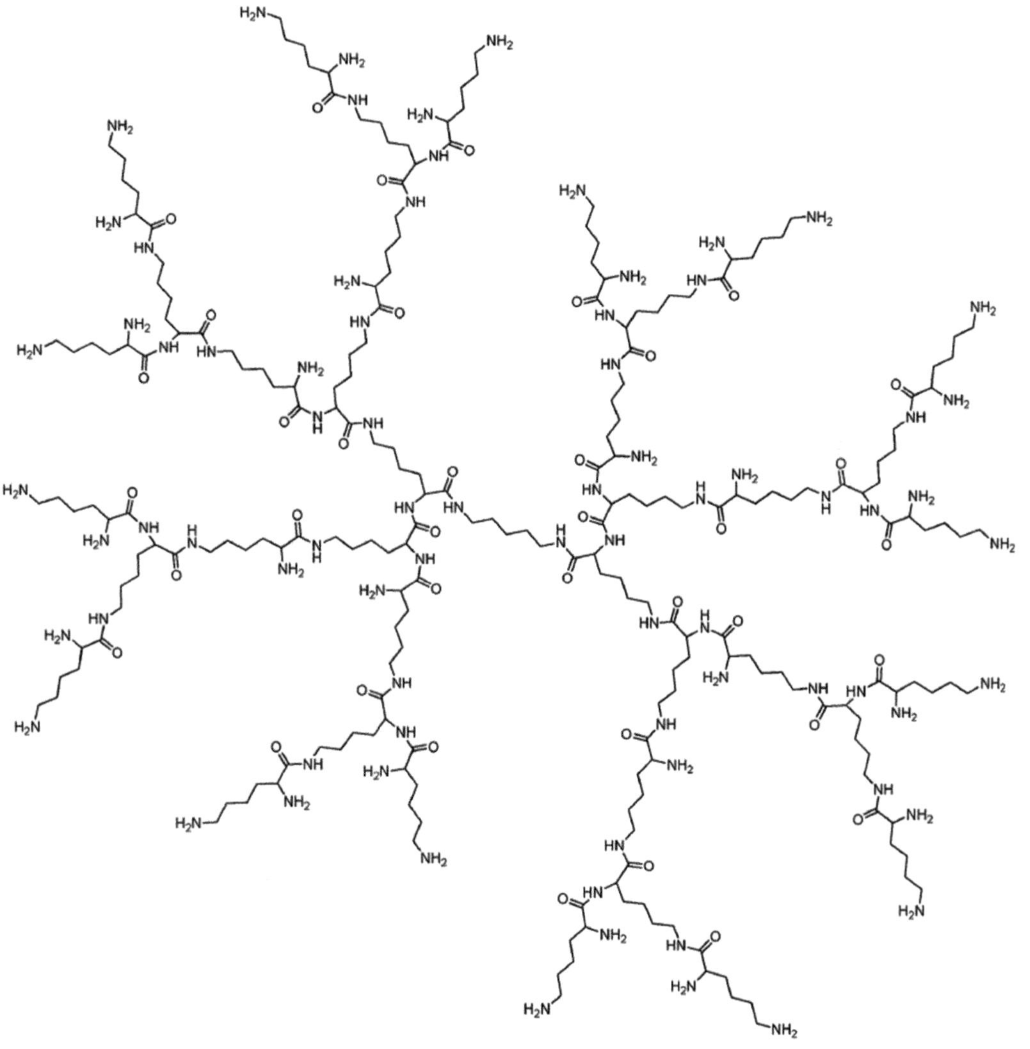

FIGURE 4.9 Dendritic poly(L-lysine) dendrimer of generation 3. Reprinted from "Dendrimers as non-viral vectors for siRNA delivery," by Xiaoxuan Liu, Palma Rocchi and Ling Peng, 2012, *New J. Chem.*, 36, 256–263. Copyright [2012] by the Royal Society of Chemistry. Reprinted with permission.

targeting. These modifications are done to increase the targeting potential of dendrimers, lower toxicity and improve siRNA loading. Different surface-engineered dendrimers have been available like lipid-modified dendrimers, fluorinated dendrimers, amino acid-modified dendrimers, saccharide-modified dendrimers, cationic-moiety-modified dendrimers, other ligand-modified dendrimers, protein- and peptide-modified dendrimers, polymer-modified dendrimers and nanoparticle-modified dendrimers (Figure 4.10; Tambea et al. 2017; Yang et al. 2015; Wan and Alewood 2016).

4.6.12 Miscellaneous Dendrimers

(a) **Organosilicon dendrimers:** These are unimolecular inverted micelles, which consist of hydrophobic organosilicon externally and nucleophilic hydrophilic PAMAM internally and are referred to as 'Organosilicon dendrimers'. They provide exclusive potential in chemical catalysis,

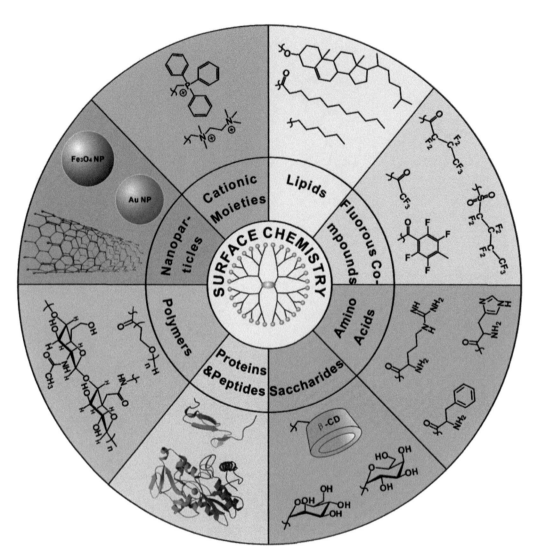

FIGURE 4.10 Surface-engineered dendrimers in gene delivery. Reprinted (adapted) with permission from Yang J, Zhang WQ, Chang H, Cheng Y, "Surface-engineered dendrimers in gene delivery," *Chem. Rev.*, 2015, 115(11), 5274–5300. Copyright [2015] American Chemical Society.

electronics, photonics and nanolithography because of their properties such as structure constancy, ability to encapsulate various guest species with nanoscopic topological accuracy and capability to make complexes (Kesharwani et al. 2014). These are extremely valuable precursors for preparing a honeycomb-like network with nanoscopic PAMAM and the OS domain (Chavda 2007; Dwivedi and Singh 2014). The introduction of organosilicon branch-cell layers such as (3-acryloxypropyl) trimethoxy-silane or (3-acryloxypropyl) methyldimethoxysilane around the PAMAM interiors is often achieved by the Michael addition reaction (Dvornic et al. 2005).

(b) **Core–shell (tecto) dendrimers:** They have extremely ordered polymeric structural design which is obtained because of their dendrimers having controlled covalent bonding. These dendrimers contain a core often having therapeutic agents further enclosed by dendrimers (Figure 4.11a). Folate is used as a targeting moiety, and fluorescein is used as the core reagent for detection in the synthesis of these dendrimers. (Kesharwani et al. 2014).

(c) **Glycerol dendrimers:** They have 100% DB (degree of branching) and consist of two types of structural units, i.e., terminal and dendritic, while supplementary linear groups like primary or secondary hydroxy groups are present in hyperbranched polyglycerol, which leads to the decrease of their DB, i.e., about 60% (Haag et al. 2000). The large oxygen content present in them suggests them to be used as synthetic ionophores and hyperbranched polymers, and cross-linked dendrimers possibly are more ionophoric than their precursors. Cross-linked as well as non-cross-linked glycerol dendrimers extract ammonium, europiumand guanidiumpicrates in chloroform effectively with less sensitivity (Zimmerman et al. 2007). The release rate of the micelles and drug loading capacity increases with the increase in the generation of glycerol dendrimers. Figure 4.11b shows the divergent synthesis of poly(glycerol-succinic acid) dendrimers [Gn]-PGLSA where hydrogenolysis in THF/methanol yielded the G3 dendrimer [G3]-PGLSA-OH (11) (95% yield). [G4]-PGLSA-bzld, i.e., benzylideneacetal (12) and [G4]-PGLSA-OH (13). (Carnahan and Grinstaff 2001; Li et al. 2014).

(d) **Triazine dendrimers:** They are made up of 1,3,5-triazine rings as branching units with spacer having amine groups at both ends. They were found to have capacity of delivering siRNA when different types of generations, cores and surface functionalities were evaluated. Both the end group and core structure affect delivery capacity. Dendrimers having an inflexible structure along with lipophilic terminals and arginine-like terminals show improved siRNA gene silencing property on Hela cells in the luciferase model (Liu et al. 2012).

(e) **Chiral dendrimers:** They are made up of branches which are chemically alike to the chiral core but constitutionally dissimilar. They are a unique class having certain applications in chiral molecular identification and in asymmetrical catalysis as they are non-racemic dendrimers, which are chiral and have non well-defined stereochemistry (Chavda 2007).

FIGURE 4.11 (a) PAMAM core–shell tecto-dendrimer. Reprinted (adapted) with permission from "A review of Dendrimer-based approach to novel drug delivery systems," by Navneet Kumar Verma, Gulzar Alam and J. N. Mishra, 2015, *Int. J. Pharm. Sci. Nanotech.*, 8(3), 2906–2918. *(Continued)*

11; R = H; [G3]-PGLSA-OH

a

12; R = bzld; [G4]-PGLSA-bzld

b

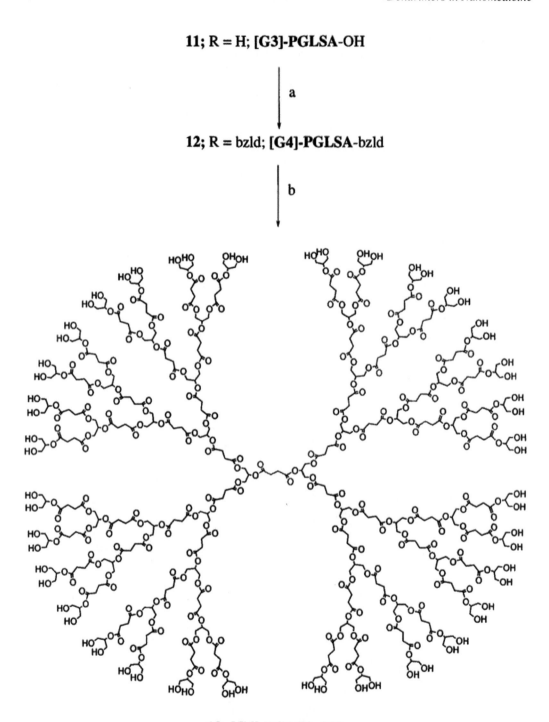

13; [G4]-PGLSA-OH

FIGURE 4.11 (Continued) (b) Divergent synthesis of G4 dendrimers composed of glycerol and succinic acid. Reagents and conditions: (a) 4, DPTS, DCC, THF:DMF (10:1), 25 °C, 14 h, 73% yield; (b) 50 psi of H2, Pd/C, THF:MeOH (9:1), 25 °C, 10 h, 93% yield. Here DPTS = 4-(dimethylamino) pyridinium 4-toluenesulfate, DCC = dicyclohexylcarbodiimid, THF = tetrahydrofuran, DMF = N,N-dimethylformamide, Pd/C = 10% palladium on activated carbon. Reprinted with permission from "Synthesis and characterization of poly(glycerol–succinic acid) Dendrimers," by Michael A. Carnahan and Mark W. Grinstaff, *Macromolecules*, 34(22), 7648–7655. Copyright [2001] American Chemical Society.

4.7 Conclusions

Dendrimers, highly branched systematic structures of nanosize range, are mainly the result of polymerisation reactions. The concept of generation of dendrimers and the effect of changes in the structure of dendrimers with increasing generations is also described. Their characteristic properties such as narrow size distribution and a low polydispersity index offer easy passage through the biological membrane; the dendrimers find applicability for delivery of drugs, genes and diagnostics. This chapter summarises the chemistry of different types of dendrimers along with their methods of synthesis (Table 4.1). The recent advancements in dendrimers such as the inclusion of heteroatoms, 'click chemistry' and 'Lego chemistry' have been discussed. The narrow polydispersity and size allow easy passage of dendrimers through biological membranes. However, due to certain limitations like tedious, cost and time-consuming synthesis methods, their difficult translation to clinical applications in medicine is limited. So, comparatively easier preparation synthesis methods may be researched for their better application in the service of mankind.

TABLE 4.1

Summary of specific application for different types of dendrimers

S.No.	Types	Application Area	Reference
1	PAMAM dendrimers	Gene delivery, gene transfection tumour targeting and delivery of diagnostic agents	Santos et al. (2019)
2	Hyperbranced PEG dendrimers	Polymer science and technology, sustained and controlled drug delivery, oral delivery and better transfection efficiency	Jain et al. (2010)
3	PPI dendrimers	Immunobio sensors, chemical biosensors and enzymatic biosensors	Olayiwola et al. (2020)
4	Carbohydrate-containing dendrimers	Coating agent, molecular recognition and biomedical materials	Jayaram et al. (1997)
5	PEG core dendrimers	Advanced drug delivery systems or as targeting vessels and in gene delivery	Cheng et al. (2011) and Santos et al. (2019
6	Amine-terminated PEG core	Clinical gene therapy for the treatment of mucosal diseases and targeted gene delivery	Yang et al. (2015)
7	Phosporous dendrimers	*In vitro* drug delivery and for elaboration of highly sensitive bio-sensors (transfection agents)	Cheng et al. (2011)
8	Polyphenylene dendrimers	Nanocarriers for biological drug delivery, as hosts for guest molecules due to large stable cavities	Alikhani et al. (2014)
9	Dendrimers Containing heteroatoms	Targeted and controlled release drug delivery	Majoraland Caminade (1999)
10	Amino acid dendrimers	*Protein* mimetics and biomedical diagnostic reagents	Cheng et al. (2011)
11	Surface-engineered dendrimers	Localized and systemic delivery to diseased tissues	Yang et al. (2010)
12	Organosilicon dendrimers	Nanolithography, electronics, photonics and chemical catalysis	Verma et al. (2015)
13	Core shell (tecto) dendrimers	Diseased cell recognition and diagnosis of disease state drug delivery	Verma et al. (2015)
14	Glycerol dendrimers	Carriers for drug delivery and coatings for tissue repair	Haag et al. (2000)
15	Triazine dendrimers	Delivery of antitumor agents	Yang et al. (2015)

Abbreviations

CO	Carbon monoxide
Cu	Copper
CuAAC	Copper assisted azide alkyne cycloaddition
DAB	Diamino butane
DB	Degree of branching
DMOMS	Methyldimethoxysilane
DSM	Diagnostic and statistical manual of mental disorders
EDA	Ethylenediamine
EPR	Enhanced permeation and retention
OS	Organosilicon
PAMAM	Poly (amido) amine
PEG	Polyethylene glycol
POPAM	Polypropylene amine
PPI	Poly-propylene imines
rDA	Retro diels-alder
RES	Reticuloendothelial system
Si-C	Carbosilane
Si-O	Siloxane
Si-Si	Silane
TEC	Thiolene click reaction
TMOS	Trimethoxy-silane

REFERENCES

Abbasi E, Aval SF, Akbarzadeh A, et al. 2014. Dendrimers: synthesis, applications, and properties. *Nanoscale Research Letters 9*(1): 247.

Abedi-Gaballu F, Dehghan G, Ghaffari M, et al. 2018. PAMAM dendrimers as efficient drug and gene delivery nanosystems for cancer therapy. *Applied Materials Today 12*: 177–190.

Akbarzadeh A, Khalilov R, Mostafavi E, et al. 2018. Role of dendrimers in advanced drug delivery and biomedical applications: a review. *Experimental Oncology 40*(3): 178–183.

Albertazzi L, Mickler FM, Pavan GM, et al. 2012. Enhanced bioactivity of internally functionalized cationic dendrimers with PEG cores. *Biomacromolecules 13*(12): 4089–4097.

Alikhani S, Hasni R, Arif NE. 2014. On the atom-bond connectivity index of some families of dendrimers. *Journal of Computational and Theoretical Nanoscience 11*: 1–4.

Araújo RV, Santos SDS, Igne Ferreira E, Giarolla J. 2018. New advances in general biomedical applications of PAMAM dendrimers. *Molecules 23*(11): 2849.

Arif NE, Hasni R, Khalaf AM. 2013. Chromatic polynomials of POPAM and siloxane dendrimers. *Journal of Computational and Theoretical Nanoscience 10*(2): 285–287.

Arseneault M, Wafer C, Morin JF. 2015. Recent advances in click chemistry applied to dendrimer synthesis. *Molecules 20*(5): 9263–9294.

Astruc D, Deraedt C, Djeda R, et al. 2018. *Dentromers*, a family of super dendrimers with specific properties and applications. *Molecules 23*(4): 966.

Bosman AW, Janssen HM, Meijer EW. 1999. About dendrimers: structure, physical properties, and applications. *Chemical Reviews 99*(7): 1665–1688.

Caminade A-M, Maraval V, Laurent R, Turrin C-O, Sutra P, Leclaire J, … Majoral J-P. 2003. Phosphorus dendrimers: from synthesis to applications. *Comptes Rendus Chimie 6*(8–10): 791–801.

Carnahan MA, Grinstaff MW. 2001. Synthesis and characterization of poly(glycerol–succinic acid) dendrimers. *Macromolecules 34*(22): 7648–7655.

Chauhan AS. 2018. Dendrimers for drug delivery. *Molecules 23*(4): 938.

Chavda HV. 2007. Dendrimers: polymers of 21st century. *Pharmaceutical Reviews 5*(1).

Cheng Y, Zhao L, Li Y, Xu T. 2011. Design of biocompatible dendrimers for cancer diagnosis and therapy: current status and future perspectives. *Chemical Society Reviews 40*(5): 2673–2703.

Das I, Borah JH, Sarma D, Hazarika S. 2018. Synthesis of PAMAM dendrime and its derivative PAMOL: determination of thermophysical properties by DFT. *Journal of Macromolecular Science 7*: 544–551.

Dias AP, da Silva Santos S, da Silva JV, et al. 2020. Dendrimers in the context of nanomedicine. *International Journal of Pharmaceutics 573*: 118814.

Dutta T, Garg M, Jain NK. 2008. Poly(propyleneimine) dendrimer and dendrosome mediated genetic immunization against hepatitis B. *Vaccine 26*(27–28): 3389–3394.

Dvornic PR, Bubeck RA, Reeves SD, Li J, Hoffman LW. 2005. Nano-scale templating using honeycomblike poly(amidoamine-organosilicon) (PAMAMOS) dendrimer networks. *Silicon Chemistry 2*(5–6): 207–216.

Dwivedi D, Singh A. 2014. Dendrimers: a novel carrier system for drug delivery. *Journal of Drug Delivery & Therapeutics 4*(5): 1–6.

Fox ME, Guillaudeu S, Frechet JMJ, Jerger K, Macaraeg N, Szoka FC. 2009. Synthesis and in vivo antitumor efficacy of PEGylated Poly(l-lysine) Dendrimer–Camptothecin conjugates. *Molecular Pharmaceutics 6*: 1562–1572.

Guillaudeu SJ, Fox ME, Haidar YM, Dy EE, Szoka FC, Fréchet JMJ. 2008. PEGylated Dendrimers with core functionality for biological applications. *Bioconjugate Chemistry 2008*(19): 461–469.

Gürbüz MU, Öztürk K, Ertürk AS, Yoyen-Ermis D, Esendagli G, Çalış S, Tülü M. 2016. Cytotoxicity and biodistribution studies on PEGylated EDA and PEG cored PAMAM dendrimers. *Journal of Biomaterials Science 27*(16): 1645–1658.

Haag R, Sunder A, Stumbé J-F. 2000. An approach to glycerol dendrimers and pseudo-dendritic polyglycerols. *Journal of the American Chemical Society 122*(12): 2954–2955.

Huc V, Boussaguet, P, Mazerolles PJ. 1996. *Journal of Organometallic Chemistry 512*: 253.

Jain K, Kesharwani P, Gupta U, Jain NK. 2010. Dendrimer toxicity: let's meet the challenge. *International Journal of Pharmaceutics 394*(1–2): 122–142.

Jayaraman N, Nepogodiev SA, Stoddart JF. 1997. Synthetic carbohydrate-containing dendrimers. *Chemistry – A European Journal 3*(8): 1193–1199.

Kesharwani P, Jain K, Jain NK. 2014. Dendrimer as nanocarrier for drug delivery. *Progress in Polymer Science 39*(2): 268–307.

Kim TI, Seo HJ, Choi JS, et al. 2004. PAMAM-PEG-PAMAM: novel triblock copolymer as a biocompatible and efficient gene delivery carrier. *Biomacromolecules 5*(6): 2487–2492.

Klajnert B, Bryszewska M. 2001. Dendrimers: properties and applications. *Acta Biochimica Polonica 48*: 199–208.

Kojima C, Kono K, Maruyama K, Takagishi T. 2000. Synthesis of polyamidoamine dendrimers having poly(ethylene glycol) grafts and their ability to encapsulate anticancer drugs. *Bioconjugate Chemistry 11*(6): 910–917.

Kolb HC, Finn MG, Sharpless KB. 2001. Click chemistry: diverse chemical function from a few good reactions. *Angewandte Chemie International Edition 40*(11): 2004–2021.

Li Y, Su T, Li S, Lai Y, He B, Gu Z. 2014. Polymeric micelles with π–π conjugated moiety on glycerol dendrimer as lipophilic segments for anticancer drug delivery. *Biomaterials Science 2*(5): 775–783.

Liu X, Rocchi P, Peng L. 2012. Dendrimers as non-viral vectors for siRNA delivery. *New Journal of Chemistry 36*(2): 256–263.

Lombardo D. 2014. Modeling dendrimers charge interaction in solution: relevance in biosystems. *Biochemistry Research International 2014*: 837651.

Luna N, Godínez LA, Rodríguez FJ, Rodríguez A, Larrea GZLD, Ferreyra CFS, Curiel RFM, Manríquez J, Bustos E. 2014. Applications of dendrimers in drug delivery agents, diagnosis, therapy, and detection. *Jounal of Nanomaterials 2014*: 1–19.

Madaan K, Kumar S, Poonia N, Lather V, Pandita D. 2014. Dendrimers in drug delivery and targeting: drug-dendrimer interactions and toxicity issues. *Journal of Pharmacy and Bioallied Sciences 3*: 139–150.

Maiti PK, Cagin T, Wang G, Goddard WA. 2004. Structure of PAMAM Dendrimers: generations 1 through 11. *Macromolecules 37*: 6236–6254.

Majoral JP, Caminade AM. 1999. Dendrimers containing heteroatoms (Si, P, B, Ge, or Bi). *Chemical Reviews 99*(3): 845–880.

Maraval V, Pyzowski J, Caminade AM, Majoral JP. 2003. "Lego" chemistry for the straightforward synthesis of dendrimers. *The Journal of Organic Chemistry 68*(15): 6043–6046.

Martin Y, Vermette P. 2006. Low-fouling amine-terminated poly(ethylene glycol) thin layers and effect of immobilization conditions on their mechanical and physicochemical properties. *Macromolecules 39*(23): 8083–8091.

Mihov G, Grebel-Koehler D, Lübbert A, et al. 2005. Polyphenylene dendrimers as scaffolds for shape-persistent multiple peptide conjugates. *Bioconjugate Chemistry 16*(2): 283–293.

Morgenroth F, Müllen K. 1997. Dendritic and hyperbranched polyphenylenes via a simple Diels-Alder route. *Tetrahedron 53*(45): 15349–15366.

Morshed MN, Behary N, Bouazizi N, Guan J, Chen G, Nierstrasz V. 2019. Surface modification of polyester fabric using plasma-dendrimer for robust immobilization of glucose oxidase enzyme. *Scientific Reports 2019*(9): 15730.

Nagpal K, Mohan A, Thakur S, Kumar P. 2018. Dendritic platforms for biomimicry and biotechnological applications. *Artificial Cells, Nanomedicine, and Biotechnology 46*(Suppl.): 861–875.

Nguyen T-T-T, Baumgarten M, Rouhanipour A, Räder HJ, Lieberwirth I, Müllen K. 2013. Extending the limits of precision polymer synthesis: giant polyphenylene dendrimers in the megadalton mass range approaching structural perfection. *Journal of the American Chemical Society 135*: 4183–4186.

Nimesh S. 2013. Dendrimers. In *Gene Therapy Potential Applications of Biotechnology*. ed. S. Nimesh. 259–285, *Woodhead Publishing Series in Biomedicine*. doi: 10.1533/9781908818645.259

Olayiwola IA, Mamba B, Feleni U. 2020. Poly (propylene imine) Dendrimer: a potential nanomaterial for electrochemical application. *Materials Chemistry and Physics* 122641.

Paleos CM, Tsiourvas D, Sideratou Z, Tziveleka LA. 2010. Drug delivery using multifunctional dendrimers and hyperbranched polymers. *Expert Opinion on Drug Delivery 7*(12): 1387–1398.

Peterson J, Ebber A, Allikmaa V, Lopp M. 2001. Synthesis and CZE analysis of PAMAM dendrimers with an ethylenediamine core. *Proceedings-Estonian Academy of Sciences Chemistry 50*(3): 156–166.

Santos A, Veiga F, Figueiras A. 2019. Dendrimers as pharmaceutical excipients: synthesis, properties, toxicity and biomedical applications. *Materials 13*(1): 65.

Shahi S, Kulkarni MS, Karva GS, Giram P, Gugulkar RR. 2015. Review article: dendrimers. *International Journal of Pharmaceutical Sciences and Research 33*(1): 187–198.

Sherje AP, Jadhav M, Dravyakar BR, Kadam D. 2018. Dendrimers: a versatile nanocarrier for drug delivery and targeting. *International Journal of Pharmaceutics 548*(1): 707–720.

Singh I, Rehni AK, Kalra R, Joshi G, Kumar M. 2008. Dendrimers and their pharmaceutical applications--a review. *Pharmazie 63*(7): 491–496.

Smith SA, Choi SH, Collins JNR, Travers RJ, Cooley BC, Morrissey JH. 2012. Inhibition of polyphosphate as a novel strategy for preventing thrombosis and inflammation. *Blood 120*(26): 5103–5110.

Sohail I, Bhatti IA, Ashar A, Sarimc FM, Mohsina M, Naveeda R, Yasir M, Iqbal M, Nazir A. 2020. Polyamidoamine (PAMAM) dendrimers synthesis, characterization and adsorptive removal of nickel ions from aqueous solution. *Journal of Materials Research and Technology 9*(1): 498–506.

Szulc A, Pulaski L, Appelhans D, Voit B, Klajnert-Maculewicz B. 2016. Sugar-modified poly(propylene imine) dendrimers as drug delivery agents for cytarabine to overcome drug resistance. *International Journal of Pharmaceutics 513*(1–2): 572–583.

Tambea V, Thakkara S, Ravala N, Sharmab D, Kaliac K, Tekade RK. 2017. Surface engineered dendrimers in siRNA delivery and gene silencing, *Current Pharmaceutical Design 2017*(23): 1–24.

Thatikonda S, Yellanki SK, Charan DS, Arjun D, Balaji A. 2013. Dendrimers – a new class of Polymers. *International Journal of Pharmaceutical Sciences Review and Research 6*: 2174–2183.

Tomalia D, Baker H, Dewald J. et al. 1985. A new class of polymers: starburst-dendritic macromolecules. *Polymer – Journal 17*: 117–132.

Tomalia DA. 2005. Birth of a new macromolecular architecture: dendrimers as quantized building blocks for nanoscale synthetic polymer chemistry. *Progress in Polymer Science 30*(3–4): 294–324.

Tomalia DA. 2009. In quest of a systematic framework for unifying and defining nanoscience. *Journal of Nanoparticle Research 11*(6): 1251–1310.

Tsai HC, Imae T. 2011. Fabrication of dendrimers toward biological application. *Progress in Molecular Biology and Translational Science 104*: 101–140.

Van Dongen MA, Vaidyanathan S, Banaszak Holl MM. 2013. PAMAM dendrimers as quantized building blocks for novel nanostructures. *Soft Matter 9*(47): 11188.

Verma NK, Alam G, Mishra JN. 2015. A review of dendrimer-based approach to novel drug delivery systems. *International Journal of Pharmaceutical Sciences and Nanotechnology 8*(3): 2906–2918

Vu MT, Bach LG, Nguyen DC, et al. 2019. Modified carboxyl-terminated PAMAM dendrimers as great cytocompatiblenano-based drug delivery system. *International Journal of Molecular Sciences 20*(8): 2016.

Wan J, Alewood PF. 2016. Peptide-decorated dendrimers and their bioapplications. *Angewandte Chemie International Edition 55*(17): 5124–5134.

Wiesler U-M, Berresheim AJ, Morgenroth F, Lieser G, Müllen K. 2001. Divergent synthesis of polyphenylene dendrimers: the role of core and branching reagents upon size and shape. *Macromolecules 34*(2): 187–199.

Wu J, Huang W, He Z. 2013. Dendrimers as carriers for siRNA delivery and gene silencing: a review. *The Scientific World Journal 2013*: 1–16.

Yang, J, Zhang, Q, Chang, H, Cheng, Y. 2015. Surface-engineered dendrimers in gene delivery. *Chemical Reviews 115*(11): 5274–5300.

Yuan Q, Yeudall WA, Yang H. 2010. PEGylated polyamidoamine dendrimers with bis-aryl hydrazone linkages for enhanced gene delivery. *Biomacromolecules 11*(8): 1940–1947.

Zimmerman SC, Quinn JR, Burakowska E, Haag R. 2007. Cross-linked glycerol dendrimers and hyperbranched polymers as ionophoric, organic nanoparticles soluble in water and organic solvents. *Angewandte Chemie International Edition 46*(43): 8164–8167.

5

Macromolecular Architecture and Molecular Modelling of Dendrimers

Rahul Gauro, Keerti Jain, Vineet Kumar Jain, Neelesh Kumar Mehra and Harvinder Popli

CONTENTS

5.1 Introduction

Drug design is an iterative process that starts when a chemist discovers a compound that shows an interesting biological profile and completes when the new chemical entity activity profile and chemical synthesis are optimised. Traditional drug development methods rely on a step-by-step synthesis and screening system for a large number of compounds to optimise the profiles of operation. Scientists have used computer models of new chemical entities over the last ten to twenty years to help identify activity profiles, geometries and reactivities (Sharma and Gupta 2009).

Molecular modelling is one of the fastest growing sectors in sciences. It can differ from the design and visualisation of molecules. It encounters all types of molecules in molecular modelling such as structural (graphic, ball and string and wire), phenomenological (homology and secondary structure prediction) and statistical (computer simulations). The emergence of computerised molecular modelling in the late 1960s had made traditional models less desirable. Not only are computers capable of drawing and manipulating three-dimensional molecules, they are also powerful tools for predicting the molecular spatial structure through energy minimisation (EM) calculations based on quantum mechanics (Sharma and Gupta 2009).

5.1.1 Molecular Modelling of Dendrimers

The architecture of dendrimers in aqueous media is influenced by several critical process parameters. These include the generation number (size), monomer form (e.g. spacer length, density groups, void space and branching units) and surface terminal groups (charge and hydrophobicity). The presence of other molecules and valence of counter-ions in the solution, the concentration of ion salts or pH also play a role. Many experimental methods, including nuclear magnetic resonance (NMR), infrared spectroscopy, mass spectroscopy, fluorescence and small angle neutron scattering, are used to test the dendrimer structure (Caminade, Laurent and Majoral 2005). Although these methods provide valuable information on the size and constituents of molecules, they have certain limitations. For example, it can be challenging to determine the spatial configuration and geometric characterisation (Lee and Larson 2009). Because dendrimers have been synthesised using repetitive series of monomers, sampling configurations are limited due to low signal differentiation. In addition, versatile dendrimers may have a range of acceptable solution configurations with rapid exchange between them (Javor and Reymond 2009).

Sometimes, specific groups of chemicals can be used to test their local environment if they are different. For example, amide protons in poly(L-glutamic acid) dendrimers had separate chemical NMR shifts, and these were used to check their solvent exposure in two different generations by adjusting the temperature (Ranganathan and Kurur 1997), as well as obtaining information on flexibility and the interaction of lipidic peptide G3.0 dendrons (Zloh et al. 2005).

Computational techniques can provide valuable insight into studying and exploring the properties of complex systems like dendrimers. Molecular modelling may be used to calculate conformational analysis, molecular interaction between drug and biological, and for experimental data validation. In the end, molecular modelling approaches have the ability to provide valuable data that can help to reduce laboratory experiments that are often laborious, time consuming and costly. We believe that if driven by sound, quality by design and design of experimentation can be made much more effective easier in scale-up (Martinho et al. 2014).

5.2 Applications of Dendrimers

Dendritic materials were discovered and produced in a variety of applications due to their interesting structural characteristics such as globular morphology, nanosize, interior void spaces, multiple surface flexibility and specific properties such as excellent solubility, mono-dispersity, biocompatibility, minimal toxicities and so on.

The dynamic design of dendritic architecture results in novel dendritic structures with different applications. Now there are over fifty dendrimer families with each having its own structure and property. It makes dendrimers a suitable candidate for specific superior applications. Because of their internal void space and peripheral functional groups, they behave similar to proteins (Mariyam et al. 2018; Liu et al. 2010). So they are called artificial proteins.

A large number of peripheral groups may be modified in order to reduce the toxicity of dendrimers and to allow for the fixation of targeting moieties. Modified dendrimers with lower toxicity may act as nanodrugs against viruses, bacteria, tumours and infectious diseases. The encapsulation efficiency of dendrimers makes them perfect nanovectors for active biomaterials in therapeutic applications (Mariyam et al. 2018). Dendrimer applications in the biomedical field are summarised in Figure 5.1.

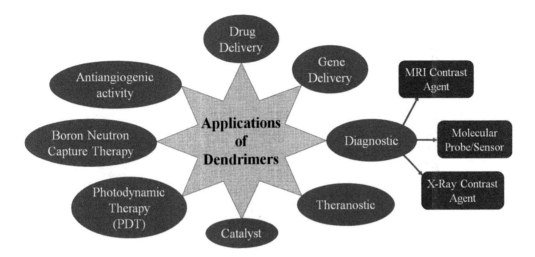

FIGURE 5.1 Biomedical applications of dendrimers.

5.3 Core–shell Patterns That Influence the Modular Reactivity of Dendrimers

Dendritic species, having an unsaturated outer monomer shell consisting of amine and ester domains, exhibited self-reactive behaviour. They were often encountered if either amine or ester groups were not fully saturated. These species, which include missing branch structures, result in the formation of mono-dendrimers containing both macro-cyclic terminal groups and moderate amounts of megamers (i.e. dimeric, trimeric, etc.) (Kallos et al. 1991).

Ideal dendrimer structures (i.e. saturated-outer-monomer shell products) could, however, be separated from such side products by thin-layer chromatography isolation and silica gel column chromatography techniques (Hummelen, Van Dongen and Meijer 1997). Ideal dendrimer structures with mathematically predictable masses, as well as unsaturated monomer shell products with mass defects, were easily characterised by matrix-assisted laser desorption/ionisation-time of flight mass spectrometry and electrospray ionisation (Huang et al. 2019).

Tomalia and co-workers reported that the unfilled-outer-monomer shell species are self-reactive intermediate species that do indeed lead to megamer formation. In general, poly(amidoamine) (PAMAM) dendrimers (i.e. all-ester-group or all-amine-saturated surfaces) are very robust species (i.e. are analogous to inert gas configuration observed at the atomic level). In this regard, authors suggest that the dendrimers do not have self-reactive characteristics (Figure 5.2). Such samples may be stored for months or years without modifications. On the other hand, PAMAM dendrimer samples containing unfilled monomer shells (i.e. ester and amine group domains on the dendrimer surface) are notorious by exhibiting self-reactive properties leading to terminal looping (i.e. macrocycle formation) and megamer formation (Tomalia 2005).

Remarkably, these self-reactivity patterns are also observed for dimensionally larger core–shell dendrimer architectures. For example, shell-saturated, core–shell tecto (dendrimers) architectures do not exhibit self-reactivity, whereas shell-saturated, core–shell tecto (dendrimer) architectures exhibit profound self-reactivity, unless pacified by reagents with orthogonally reactive functionality. This behaviour is comparable to the behaviour of atoms and basic dendrimers (Tomalia 2005).

	Atoms	Dendrimers	Core–Shell Tecto(dendrimers)
Dimensions	0.05–0.6 nm	1–15 nm	5.0 = 100 nm
Valency (Reactivity)	Unfilled Outer Electron Shell	Unfilled Outer Branch Cell Shell	Unfilled Outside Dendrimer Shell
(Core–Shell) Architecture Induced Reactivity (Unfilled Shells)	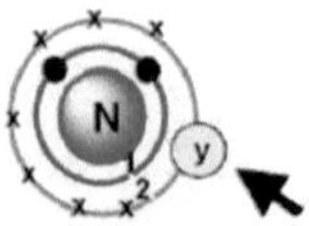 (e.g., fluorine) Unfilled Shell (x)	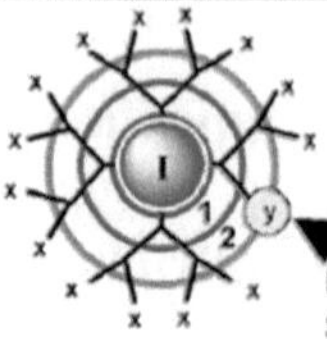 Unfilled Shell (x)	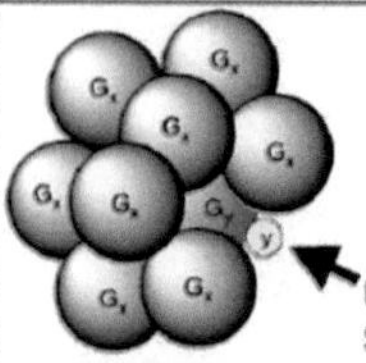 Unfilled Shell (x)
Functional Components Directing Valency	Missing One Electron (y) in Outer Shell (x) Penultimate to Saturated Noble Gas Configuration	Missing One Terminal Branch Cell in Outer Shell (x) Exposing Functionality (y)	Missing One Dendrimer Shell Reagent Exposing Functionality (y)

FIGURE 5.2 Quantified module (building block) reactivity patterns at subnanoscale (atoms), lower nanoscale (dendrimers) and higher nanoscale [core–shell tecto(dendrimers)] rates involving unsaturated electron, monomer or dendrimer shells (Tomalia 2005).

5.4 New Nanosynthetic Methods for Organic Chemists

Description of the hierarchical complexity of precise, regulated nanostructures clearly shows the value of quantified building blocks for viable bottom-up synthetic strategies. The value of atom modules (0.1–0.6 nm) for the age of small molecules (traditional chemistry) and monomers (0.5–1.0 nm) for the age of macromolecules (polymers) clearly indicates the significant role of dendrimers (1.0–20 nm) that is expected to play as sufficiently large, quantified building blocks for the synthesis of well established, more complex nanostructures. Experimental research has already demonstrated the ability to regulate the shape, size and chemical functionality of a wide range of dendrimer structures. The first step has been taken to demonstrate the use of such engineered dendrimeric structures as basic building blocks for the synthesis of well-defined nanostructures beyond dendrimers (i.e. megamers), specifically the recent new core–shell tecto (dendrimers) class (Tomalia 2010; Uppuluri et al. 2000).

5.5 Classical to Dendrimer-Based Chemistry

Historically, it is well known that Dalton proposed the use of atom modules for the synthesis of higher chemical complexity in his New System of Chemical Philosophy (1808), and Staudinger's Catenation of Monomers for the development of macromolecules (Mülhaupt 2004) provided crucial enabling building blocks and thus synthetic frameworks, for the very important fields they pioneered.

Such historical events met with resistance at their inception and, in some cases, they encountered strong peer scrutiny (Mülhaupt 2004). In view of recent design presentations, Tomalia made a bold proposal that dendrimers are the emerging science of synthetic nanochemistry (Figure 5.3) (Tomalia 1994). The potential use of dendrimers as reactive building blocks is expected to provide the enabling framework in the routine synthesis of well-defined organic, inorganic and hybridised biomolecular nanostructures (Figure 5.4) (Tomalia 2005). The significant role that synthetic organic and polymer chemists are presently playing in the development of this new field is readily apparent from a recent issue of Tetrahedron Symposia-in-Print (Smith 2003).

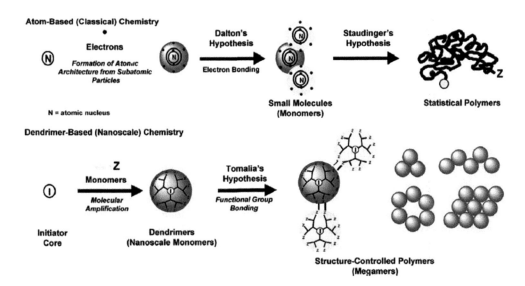

FIGURE 5.3 A comparison of atom-based classical chemistry (Dalton and Staudinger hypotheses) with dendrimer-based nanoscale chemistry (Tomalia hypothesis) for the synthesis of higher-complexity structures (Tomalia 2005).

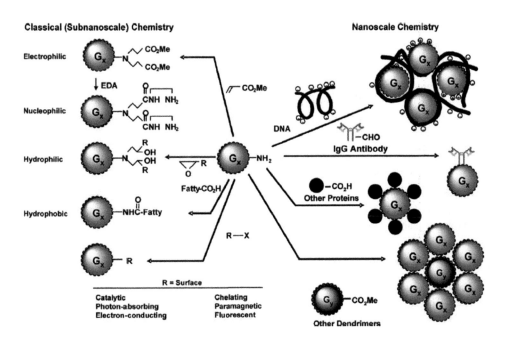

FIGURE 5.4 Options to adjust amine-terminated dendrimers using classical subnanoscale and nanoscale reagents (Tomalia 2005).

5.6 Computational Molecular Models

Computational molecular models are the products of mathematical equations that predict the positions and reactions of electrons and nuclei. Mathematical models are divided into traditional mechanical and quantum dynamics approaches. Traditional mechanics considers molecules to be a set of atoms and bonds that are viewed as balls and springs. Data including atomic radii and spring stiffness are used to find the 'right' position of the atoms. It is a simple and accurate method for locating the stable geometry of a molecule. Quantum mechanical methods solve the Schrodinger equation in two different ways: *ab initio* (from the beginning) and semi-empirically. Semi-empirical approaches use experimental data to simplify the solution of the Schrodinger equation and resolve more easily. Several methods have been developed and available for this simplification including Huckel, Extended Huckel, Indian National Database of Seafarers and Modified Neglect of Diatomic Overlap. Each of them has a parameter set based on experimental measurements for multiple compounds. On the other hand, *ab initio* methods use only standard mathematical approximation. Although these methods are technically 'pure,' they are computationally intensive and rely on massive, ultra-high-speed computers (Sharma and Gupta 2009).

Molecular modelling is the analysis of molecules and molecular structures using computer models. The most popular model for a molecule is a group of atoms, considered to be a point mass with a position in space and probably other properties. All the atoms in the molecular system interact with each other; these interactions are represented by the energy interaction that is the property of the atoms and function of the positions. Interaction is the central quantity of molecular modelling, and the creation of an effective mechanism is one of the major problems. Recipes are referred to as force fields (Sharma and Gupta 2009).

Construction and manipulation of these molecules use biology (for functions, genes, genetics and modifications), chemistry (for the structure of the components and the knowledge of what binds them together), physics (for principles of heat, forces, statistical mechanics and thermodynamics) and computer science (for the representation, simulation, visualisation and organisation of models) (Sharma and Gupta 2009). In order to use molecular modelling in a comprehensive way, it is important to understand the relationship between the 3D structure of the system and its energy. A number of mathematical models link the structure to energy.

5.7 Important Aspects of Molecular Modelling

Molecular dynamics (MD) simulations can be used to understand the structure and dynamics of any drug carrier or polymer. This information regarding structure and dynamics may be used to study interactions of carriers with various drug molecules, proteins, peptides, nucleic acids, etc. In the following section of review, we discussed about some different aspects of molecular modelling.

5.7.1 Quantum Mechanics

Molecular orbital theory/linear arrangement of atomic orbits and orbitals (electrons) specifically treated in quantum mechanics. Quantum mechanics is speculated to be more precise results in fundamental constants and related quantities determination (Ramachandran et al. 2008).

5.7.2 Molecular Mechanics (MM)

MM is a computational method used to model the conformation behaviour and the energetic properties of molecules. The molecule is being treated at the atomic level, that is, electrons are not being treated directly. MM uses an energy function described in such a way that, given a particular conformation (i.e. given a set of spatial coordinates for all atoms), the energy of the molecule can be measured. The energy function is empirical; that is, it is not completely dependent on systematic theories. Typically, a mixture of quantum mechanical equations and experimental data is used to create the energy function (Kitchen et al. 2004).

The energy feature distinguishes between 'bonded' and 'non-bonded' interactions. Bonded interactions occur between atoms which are bound by no more than three bonds. Non-bonded interactions occur between atoms, which are either not bound at all (e.g., atoms on two residues which are apart, far from each other, in a polypeptide chain) or are only linked by more than two bonds (e.g. the first and last n-pentane carbon atoms) (Sharma and Gupta 2009).

5.7.3 Bonding Interactions

Bonding is the interaction among two atoms directly bound to each other and is thought to be harmonic (imagine a spring connecting the two atoms – preventing atoms from becoming too close and too far away). The bond angle is the contact between the three related atoms, which is often believed to be harmonic (imagine a spring that holds the angle between the two bonds at some point, e.g. 109°) for an angle of H-C-H in methane. Dihedral (torsion) angles are observed in the case of four atoms bound by three bonds, and we look straight down to the second bond, the dihedral angle is the angle between the first and the third bond. The energies associated with dihedral angles are classified using a cosine series: this allows the angle to have a variety of desired values. For example, in ethane, the angle of dihedral H-C-C-H prefers to be 60°, 180° or 240° (i.e. −60°) (Sharma and Gupta 2009).

5.7.4 Non-Bonding Interactions

Non-bonded interactions act between atoms in the same molecule and those in other molecules. Force fields usually divide non-bonded interactions into two: electrostatic interactions and van der Waals interactions. They attract long ranges (due to the London dispersion forces) but are highly repulsive in a short range (because atoms cannot overlap). Coulombic interactions operate for all the atoms that bear a charge. Clearly Na^+ has a charge of $+1e$. Methanol (CH_3OH) has a minimum charge of $0e$ (zero electrons). However, oxygen is a much more electronegative element than either carbon or hydrogen, so that it is in fact slightly negatively charged, and therefore, a partial charge of say −0.5e would be given, with + 0.4e being assigned to the hydroxyl hydrogen and + 0.1e being assigned to the carbon (Sharma and Gupta 2009).

5.7.5 Energy Function

This method, referred to as a potential method, measures the molecular potential energy as a sum of energy terms representing the variance of bond lengths, bond angles and torsional angles from the equilibrium values, plus the non-bonded pairs of atoms representing van der Waals and electrostatic interactions.

$$E = E_{bonds} + E_{angle} + E_{dihedral} + E_{non\text{-}bonded}$$

The set of parameters consisting of bond angles, partial charge values, equilibrium bond lengths, force constants and van der Waals parameters are collectively referred as the force field. Different MM implementations use slightly different mathematical formulas and thus different constants for the potential function. The different strength fields widely used in biomolecular simulations are assisted model building and energy refinement), GROningen molecular simulation and Chemistry at Harvard Macromolecular Mechanics specifically developed to represent the most important biological units, such as amino acids and nucleic acids (Kamp et al. 2008; Loschwitz et al. 2020).

5.7.6 EM

EM techniques are used to iterate the conformation of a molecule in order to reduce its energy. EM is especially useful for the removal of bad clashes between atoms that may have formed while constructing or modelling a molecule (e.g., a protein-modelled homology structure). There are several approaches that can be used; all have their own strengths and weaknesses. They are steepest descent, second order approaches and conjugate gradient. Both EM strategies concentrate on down-hill searching – so they aim to locate the nearest local minimum on the energy sheet. When a much deeper (*i.e.* better) energy minimum is close, however, a high-energy barrier separates it from the starting point, it will not be located. EM is therefore not capable of finding a global energy minimum; the absolute best molecular conformation can only be found if we start very close to it (Chatzieleftheriou, Adendor and Lagaros 2016).

5.8 MD

MD incorporates energy estimates from the strength field approach with the laws of Newtonian (as opposed to quantum) mechanics. The simulation of MD models is used to model complex behaviour of molecular systems. Starting with an acceptable initial conformation of our system, we used Newton's motion equations over small time steps (usually 10–15 sec) to determine how the system evolves over time. The simulation is initialised by specifying the position and assigning a force vector to each atom in the molecule. The acceleration of each atom is then determined from the equation a = F/m, where F is the negative gradient of the potential energy function and m is the mass of the atom. For example, a long MD simulation of the protein provides insight into its conformational flexibility. Simulation will tell you what conformational states are accessible to proteins and the timing of these conformational fluctuations. The Verlet algorithm is used to measure the velocity of the atoms from the position of the forces and the atoms. When the velocity is determined, new atom positions and the assembly temperature can be measured. Such values are then used to measure the paths or time-dependent positions for each atom. The trajectory developed by the MD simulation can also be used to measure thermodynamic properties (e.g. heat capacity) (Adcock and McCammon 2006).

The main benefit of MC and MD computer simulation methods over different experimental techniques lies in their ability to explain the atomistic detail and the behaviour of individual molecules in exquisite details. These methods are also capable of tracking the time evolution of the structure and interactions of molecules from the femto to the microsecond scale. This level of sophistication allows the study of almost any imaginable property of single molecules or bulk materials (Hollingsworth and Dror 2018).

5.9 The Molecular Model Design

The molecular model is constructed by fitting a known sequence into an estimated electron density, derived from some phase model. This method is subject to mistakes that are corrected as the processing proceeds. Usually, the model is designed with strong local geometry, but global geometry issues are cleared by refinement. The following errors can occur during the molecular model design (Maximova et al., 2016; McGreevy et al., 2016).

a. The right atom in the wrong density. Density can be misinterpreted in such a way that a wrong piece of structure is built into it. It can be especially popular with regard to the positioning of solvent and surface side chains. Atoms with no supporting density at all. This will happen for a number of causes, the most common of which is disorder.

b. Atoms appear where no atoms are supposed to occur. Noise exists in three dimensions, and noise peaks can look like atoms.

c. Errors in chain registration. The electron density does not hold any marks, so it is simple to extract a few residues while following the chain. This is especially probable after tracing through a surface loop or another region of low density.

d. Misleading the main chain. Frequently, the chain is traced in a few bits, which must then be linked. It is possible to misconnect the bits. A length of chain may also be traced backwards (Sharma and Gupta 2009).

5.10 Molecular Modelling of Dendrimers for Nanoscale Applications

Dendrimers are well described, highly branched macromolecules that radiate from the central core and are synthesised by a stepwise, repetitive sequence of reactions that guarantees complete shells for each generation, resulting in monodisperse polymers. Synthetic procedures developed for dendrimer preparation require almost complete control of essential molecular design parameters, such as shape, size, surface/interior chemistry, flexibility and topology.

Recent findings indicate that dendritic polymers will provide the key to the production of a reliable and economical manufacturing and manufacturing method for usable nanoscale materials with specific properties (electronic, magnetic, optical, optoelectronic, chemical or biological). In addition, they may be used in the creation of modern nanoscale devices (Cagin, Wang, Martin, Breen and Goddard 2000). For example, PAMAM dendrimers have been used to attract copper (II) ions within the macromolecules, where they are subsequently reacted by solubilising H_2S to form metal sulphides (Naylor and Goddard Hi 1989).

These organic/inorganic, dendrimer-based hybrid organisms have been referred to as 'nanocomposites' and exhibit unusual properties. For example, the solubility of nanocomposites is determined by the characteristics of host dendrimer molecules. This helps the inorganic guest compounds to be solubilised in conditions where they are naturally insoluble. As it has been identified that there is no covalent bond between the host and the guest, these findings suggest that the inorganics are physically and spatially constrained by the dendrimer shell. This structure is not checked yet. Here, we use 3D structure building methods based on the CCBB MC approach and MD techniques to explore the structural features of the PAMAM dendrimers (Tomalia, 1991).

5.10.1 Docking Based on Pharmacophore

In the field of molecular modelling, docking is a technique that predicts the preferred orientation of one molecule to another when bound together to form a stable complex. Knowledge of the desired orientation, in effect, can be used to predict the strength of the interaction or binding affinity between the two molecules by using, for example, the role of score. Associations between biologically related molecules, such as proteins, carbohydrates, nucleic acids and lipids, play a central role in the transmission of signals. In addition, the relative orientation of the two interactive partners can influence the type of signal produced (e.g. agonism vs. antagonism). Docking is therefore useful for predicting both the intensity and the form of signal generated. Docking is often used to predict the binding orientation of small molecule drug candidates to their protein targets in order to predict the affinity and behaviour of small molecules. Docking therefore plays a significant role in the ethical production of drugs. Considering the biological and pharmaceutical importance of molecular docking, significant efforts have been made to develop the methods used to predict docking (Joseph-mccarthy et al. 2003).

5.10.2 Computer-aided Production of Retro-Metabolic Products: Soft Drugs

Soft drug design approaches aim to develop new therapeutic agents that undergo simple, ideally hydrolytic metabolism to produce inactive metabolites. This approach is general and can be used in virtually any therapeutic region, particularly where the desired activity is localised, relatively short-lived or subject to easy titration. In most cases, the purpose of this approach is to develop near steric and electronic analogues for an existing drug that serves as the lead in the development of the drug. Computer programs that can create virtual libraries of potential analogues and can provide quantitative tools to rank them on the basis of the similarity of their properties to the original lead are therefore of special interest (Buchwald 2007).

5.10.3 Determination of Drug Excipient Activity

The molecular modelling technique has become common in the study of drug–excipient interaction, which helps to simulate the form and location of interaction on a computer monitor. It was stated in a study that seven glucose units were combined to provide well-formed energy to reduce conformation. The depth of the cavity and the diameter of the wider and narrower rim were measured and compared to the values of the literature using the DTMM kit. Likewise, ciprofloxacin, norfloxacin and other structures have been built to reduce energy conformation. The dimensions of these molecules have been measured and compared to the values in the literature.

The drug molecules were allowed to enter the cavity, and the likelihood of penetration was observed. Progress in the development of a beta-cyclodextrin inclusion complex with ciprofloxacin, norfloxacin, tinidazole and methotrexate was finally registered (Nadendla 2004).

5.10.4 Quantitative Structure Activity Relationship (QSAR) Analysis

QSAR is a method used to measure the interaction between structural and biological properties. The most general form of QSAR can be expressed by the following equation: biological activity = f (physicochemical and/or structural parameters). Physicochemical descriptors include parameters that account for topology, electronic properties, hydrophobicity and steric effects, which are empirically calculated using analytical methods. The procedure used in QSAR includes chemical analyses and biological experiments. Researchers have been trying to create drugs based on QSAR for several years. The example of QSAR in modelling is the 1-(X-phenyl)-3,3-dialkyl triazene sequence. These compounds are of concern for their anti-tumour activity, but they are also mutagenic. QSAR is used to explain how the structure can be adjusted to minimise mutagenicity without drastically reducing the anti-tumour function. In a quantitative activity relationship analysis, the anti-leishmanial activity of the substituted pyrimidine and pyrazolo pyrimidine analogues was calculated using physicochemical and steric descriptions (hydrophobicity, molar refractivity, Supton resonance, Verloop steric parameters and van der Waals volumes of the substituent groups) of the different substituents. The study of pyrimidine analogues suggested the need for unsubstituted pyrimidine for anti-leishmanial action. Linear multiple regression analysis using the least squares method was used to establish a correlation (Ghasemi et al. 2018).

5.10.5 Lead Generation

Lead is any chemical compound that displays biological activity. It is not the same as a drug molecule, but its production is a significant step in the process of drug discovery. It is a method to classify potential drug compounds or leads that interact with a target with appropriate potency and selectivity. Lead generation is a complex process involving two basic steps; lead discovery is a step towards finding a chemical compound with the desired biological activity; lead optimisation involves developing around the basic lead structure all the desirable properties, such as protection, solubility and so on, with the aid of molecular modelling (Carlson 2002).

5.11 Conclusions

Dendrimers are a new class of 3D, man-made molecules formed by an uncommon synthetic route that combines repetitive branching sequences to create a special novel architecture. The advantages of dendrimers include regulated size, shape and differentiated functionality, their ability to produce both isotropic and anisotropic assemblies, their compatibility with many other nanoscale building blocks such as metal nanocrystals, DNA and nanotubes and their capacity for organised self-assembly. Their ability to combine both organic and inorganic components and their capacity to encapsulate or grow into single-molecular functional devices make dendrimers exceptionally versatile among existing nanoscale building blocks and materials. Dendritic polymers, in particular dendrones and dendrimers, are expected to play an important role as fundamental modules for nanoscale synthesis. Through this viewpoint, it is reasonable to be positive about the future of this new big polymer class, the dendritic state.

ACKNOWLEDGEMENT

The authors would like to thank National Institute of Pharmaceutical Education and Research (NIPER), Raebareli; Delhi Pharmaceutical Sciences and Research University (DPSRU), New Delhi; and National Institute of Pharmaceutical Education and Research (NIPER), Hyderabad for extending the facilities to write this chapter.

The NIPER, Raebareli communication number for this manuscript is NIPER-R/Communication/140.

CONFLICT OF INTEREST

The authors reported no conflict of interest related to this chapter.

REFERENCES

Adcock, Stewart A, and J Andrew McCammon. 2006. "Molecular Dynamics: Survey of Methods for Simulating the Activity of Proteins." *Chem. Rev.*, *106*(5) (May): 1589–1615.

Buchwald, Peter. 2007. "Computer-aided Retrometabolic Drug Design: Soft Drugs." *Mol. Cell. Pharmacol.*, 2, 923–933.

Cagin, Tcagin, Guofeng Wang, Ryan Martin, Nicholas Breen, and William A Goddard. 2000. "Molecular Modelling of Dendrimers for Nanoscale Applications." *Nanotechnology*, *11*: 77–84.

Caminade, Anne Marie, Régis Laurent, and Jean Pierre Majoral. 2005. "Characterization of Dendrimers." *Adv. Drug Deliv. Rev.* 57(15): 2130–2146.

Carlson, Heather A. 2002. "Protein Flexibility and Drug Design: How to Hit a Moving Target." *Curr. Opin. Chem. Biol.*, *6*(4), 447–452.

Chatzieleftheriou, Stavros, Matthew R Adendor, and Nikos D Lagaros. 2016. "Generalized Potential Energy Finite Elements for Modelling Molecular Nanostructures." *J. Chem. Inf. Model.* 56(10).

Ghasemi, Fahimeh, Alireza Mehridehnavi, Alfonso Pérez-Garrido, and Horacio Pérez-Sánchez. 2018. "PT SC." *Drug Discov. Today* doi:10.1016/j.drudis.2018.06.016.

Hollingsworth, Scott A, and Ron O Dror. 2018. "Review Molecular Dynamics Simulation for All." *Neuron* 99(6): 1129–1143. doi:10.1016/j.neuron.2018.08.011.

Huang, Zhihao et al. 2019. "Binary Tree-Inspired Digital Dendrimer." *Nat. Commun.* 10(1): 1–7. doi:10.1038/s41467-019-09957-6.

Hummelen, Jan C, Joost LJ Van Dongen, and EW Meijer. 1997. "Electrospray Mass Spectrometry of Poly(Propylene Imine) Dendrimers-The Issue of Dendritic Purity or Polydispersity." *Chem. Eur. J.* 3(9): 1489–1493.

Javor, Sacha, and Jean Louis Reymond. 2009. "Molecular Dynamics and Docking Studies of Single Site Esterase Peptide Dendrimers." *J. Org. Chem.* 74(10): 3665–3674.

Joseph-McCarthy, Diane et al. 2003. "Pharmacophore-Based Molecular Docking to Account for Ligand Flexibility." *188*(2001): 172–188.

Kallos, GJ et al. 1991. "Molecular Weight Determination of a Polyamidoamine Starburst Polymer by Electrospray Ionization Mass Spectrometry." *Rapid Commun. Mass Spectrom.* 5(9): 383–386.

Kitchen, Douglas B, Hélène Decornez, John R Furr, and Jürgen Bajorath. 2004. "Docking and Scoring in Virtual Screening for Drug Discovery: Methods and Applications." *3*(November).

Lee, Hwankyu, and Ronald G Larson. 2009. "Multiscale Modelling of Dendrimers and Their Interactions with Bilayers and Polyelectrolytes." *Molecules* 14(1): 423–438.

Liu, Jinyao et al. 2010. "The In Vitro Biocompatibility of Self-Assembled Hyperbranched Copolyphosphate Nanocarriers." *Biomaterials* 31(21): 5643–5651. doi:10.1016/j.biomaterials.2010.03.068.

Loschwitz, Jennifer, Olujide O Olubiyi, Jochen S Hub, Birgit Strodel, and Chetan S Poojarid. 2020. Computer Simulations of Protein–Membrane Systems. *Prog. Mol. Biol. Transl. Sci. 170*: 273–403.

Mariyam, Merina et al. 2018. Dendrimers: General Aspects, Applications and Structural Exploitations as Prodrug/Drug-delivery Vehicles in Current Medicine. *Mini-Rev. Med. Chem.. 18*(5), 439–457.

Martinho, Nuno et al. 2014. "Molecular Modelling to Study Dendrimers for Biomedical Applications." *Molecules*, *19*(12): 20424–20467.

Maximova, T, R Moffatt, B Ma, R Nussinov, and A Shehu. 2016. "Principles and Overview of Sampling Methods for Modeling Macromolecular Structure and Dynamics." *PLoS Comput. Biol. 12*(4): e1004619.

McGreevy, Ryan, Ivan Teo, Abhishek Singharoy, and Klaus Schulten. 2016. "Advances in the Molecular Dynamics Flexible Fitting Method for Cryo-EM Modelling." *Methods 100*: 50–60.

Mülhaupt, Rolf. 2004. "Hermann Staudinger and the Origin of Macromolecular Chemistry." *Angew. Chem. Int. Ed., 43*: 1054–1063.

Nadendla, Rama Rao. 2004. "Molecular Modelling : A Powerful Tool for Drug Design and Molecular Docking." *Resonance 9* (May): 51–60.

Naylor, Adel M, William A Goddard Hi, GE Kiefer, and DA Tomalia. 1989. Starburst dendrimers. 5. Molecular shape control. *J. Am. Chem. Soc. 111*(6): 2339–2341.

Ramachandran, KI, G Deepa, and PK Krishnan Namboori. 2008. *Computational Chemistry and Molecular Modeling: Principles and Applications*. Germany: Springer-Verlag GmbH.

Ranganathan, Darshan, and Sunita Kurur. 1997. "Synthesis of Totally Chiral, Multiple Armed, Poly Glu and Poly Asp Scaffoldings on Bifunctional Adamantane Core." *Tetrahedron Lett. 38*(7): 1265–1268.

Sharma, Aarti, and Himanshu Gupta. 2009. "Molecular Modelling." *J. Pharm. Bioallied Sci., 1*(1), 23–29.

Smith, DK. 2003: Tetrahedron Symposia-in-Print. In *Recent Developments in Dendrimer Chemistry*, ed. DK Smith, Vol. *59*. Amsterdam: Elsevier Science, pp. 3787–4024.

Tomalia, Donald A. 1994. "Starburst/Cascade Dendrimers: Fundamental Building Blocks for a New Nanoscopic Chemistry Set." *Adv. Mater., 6*(7–8): 529–539.

Tomalia, Donald A. 2005. "Birth of a New Macromolecular Architecture: Dendrimers as Quantized Building Blocks for Nanoscale Synthetic Polymer Chemistry." *Prog. Polym. Sci., 30*(3–4): 294–324.

Tomalia, Donald A. 2010. "Dendrons/Dendrimers: Quantized, Nano-Element like Building Blocks for Soft-Soft and Soft-Hard Nano-Compound Synthesis." *Soft Matter 6*(3): 456–474.

Tomalia, Donald A, David M Hedstrand, and Michael S Ferritto. 1991. "Comb-Burst Dendrimer Topology: New Macromolecular Architecture Derived from Dendritic Grafting." *Macromolecules 24*(6): 1435–1438.

Uppuluri, Srinivas et al. 2000. "Core-Shell Tecto (Dendrimers): I. Synthesis and Characterization of Saturated Shell Models." *Adv. Mater. 12*(11): 796–800.

van der Kamp, MW, Katherine E Shaw, Christopher J Woods, and Adrian J Mulholland. 2008. "Biomolecular Simulation and Modelling: Status, Progress and Prospects." *J. R. Soc. Interface 5* (Suppl 3): S173–S190.

Zloh, Mire et al. 2005. "Investigation of the Association and Flexibility of Cationic Lipidic Peptide Dendrons by NMR Spectroscopy." *Magn. Reson. Chem. 43*(1): 47–52.

6

Synthesis of Dendrimers

Ashima Thakur and Abha Sharma

CONTENTS

6.1 Introduction

Dendrimers (arborols or cascade molecules) are considered as polymers of the twenty-first century. Dendrimers are nanosized, radially highly branched, homogeneous and monodisperse macromolecules. After three known classes of traditional polymers, i.e., linear, cross-linked and branched, dendrimers are the fourth most important class of macromolecules (Abasi, Fekri Aval and Akbarzadeh et al. 2014). The word "dendrimer" was derived from the Latin term "dendron" and "meros", which signifies tree and parts or units, respectively, and appears as a tree with branching units (Augustus, Allen, Nimibofa et al. 2017). The first example of dendrimers was given by Fritz Vogtle et al. (1978), who synthesised dendrimers via the divergent approach in a cascade manner that have able to bind with the guest or molecules in a host-guest interaction (Newkome, Yao, Baker et al. 1985). In 1981, the synthesis of poly-lysine dendrimers was patented by Denkewalter et al., and other scientists who synthesised dendrimers were Donald Tomalia at Dow Chemicals in 1983 (Tomalia, Christensen and Boas 2012), George Newkome and his group in 1985 (Newkome, Yao, Baker et al. 1985). Furthermore, Jean Frechet and Craig Hawker in 1990 developed a new approach named the convergent synthetic approach for the synthesis of dendrimers

TABLE 6.1

Gradual evolution of dendrimers with the advent of several years

S. No.	Types of Dendrimers	Discoverer	Year of the Synthesis
1.	Poly(propylene imine)	F. Vogtle et al.	1978
2.	PAMAM	D. Tomalia et al.	1983,1985
3.	Arbosols	G. Newkome et al.	1985
5.	Poly(arylether)	J. Frechet and C. Hawker	1990
6.	Polylysine	R.G. Denkewalter et al.	1981
7.	Polyether	J. Frenchet and S. Grayson	2001

(Table 6.1). Dendrimers have a unique branched structure with multivalent properties, which make them a functional material to be used for a variety of applications, for instance, in medicine, chemistry, physics and materials science (Astruc, Boisselier and Ornelas 2010; Sherje, Jadhav and Dravyakar et al. 2018). The general structure of a dendrimer comprises organised AB_n monomers where A and B are two diverse functionalities and n is the number equal to 2 or higher. They are synthesised throughout a linear polymer core or a small molecule by the sequence of repetitive steps with more activated functional groups on its surface, which can be further functionalised (Tang 2017; Abd-El-Aziz and Agatemor 2017). This means that dendrimers have a multifunctional core moiety, usually di-, tri- or tetra-functional, containing many layers of branched monomers which are designated as generation (G) (Walter and Malkoch 2012).

6.2 Structure of Dendrimers

Dendrimers are generally globular branched macromolecules with a void and various types of functional groups on their surface, which are responsible for carrying active molecules (Astruc, Boisselier and Ornelas 2010). Every element of dendrimers plays a particular role and describes some different properties as it develops from generation to generation (Arseneault, Levesque and Morin 2012). Its construction generally contains three main elements as given in Figure 6.1.

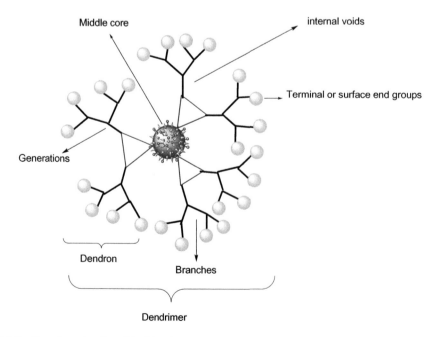

FIGURE 6.1 General construction of dendrimers.

1. A middle core having an atom or an atomic group responsible for chemical functions (Tiwari, Gajbhiye and Sharma et al. 2010). A lot of molecular information about shape, multiplicity and size is expressed via a direct connection to the functional groups present at the surface (Kono 2012).
2. Second is the branches that generally originate from a middle core, with repetitive units arranged in an organised manner. This amplification of repetitive units is called generations. They explained about the nature and capacity of the void regions in the inner shell (Samad, Alam and Saxena 2009).
3. After the generations come the surface functional groups which are generally present at the periphery of the dendrimers (Santos, Veiga, and Figueiras 2019). These groups are reactive or passive terminal groups, which can play different functions responsible for different properties (Sherje, Jadhav, Dravyakar et al. 2018).

6.3 Synthetic Approaches

Dendrimers are synthesised under well-controlled conditions to achieve monodisperse and tree-like branched structures. Alterations of the functionality in every element of the dendrimers (focal point, branches and functional groups) result in a change in their properties, such as thermal stability and solubility achieved by adding active molecules for specific roles (Mignani, Bryszewska and Klajnert-Maculewicz et al. 2015, 14). There is an outward extension from a core by a series of Michael addition reactions for the synthesis of dendrimers. Every single step of the reaction has to be done in this way to prevent side reactions (trailing generations) [9]. These side reactions are responsible for the change in their functionality, and apart from this, the purification of dendrimers is very difficult because the difference between the desired product and side product is very insignificant. A large number of strategies are available in the literature for the synthesis of dendrimers [15]. The selection of method for the dendrimer synthesis mainly depends upon the end goal function. It is mainly divided into three main approaches:

1. Divergent approach
2. Convergent approach
3. Combination of divergent and convergent approaches

6.3.1 Divergent Approach

The divergent approach is one of the traditional and popular ways to synthesise dendrimers, which was used in the early days. Tomalia and Newkome initiated this approach in the 1980s. In this approach, dendrimer synthesis takes place in a step by step fashion starting from the central core to periphery via two fundamental actions. These are

a) Monomer coupling
b) Modification on terminal functional groups to generate deprotected groups, i.e., free functional groups that can react or coupled with new monomers.

The aforesaid two steps were replicated several times until the desired generation of the dendrimer is synthesised. The functionality present in the core reacts with the one reactive group of the monomer and two dormant groups on each site and results in first-generation dendrimers (G1.0) as shown in Figure 6.2. To avoid uncontrolled hyperbranched polymerisation, each terminal functionality on the monomer was planned to be inert to focal monomer functionality. Once the synthesis of first-generation dendrimers (G1.0) is completed, these two dormant functionalities are further activated to get coupled with new peripheral groups present on the monomer (Grayson and Frechet 2001). This activation step of the dormant functionalities present on the monomer includes its deprotection, modification of more reactive moieties and coupling with a new monomer molecule. A huge amount of reagents are mandatory because the steps

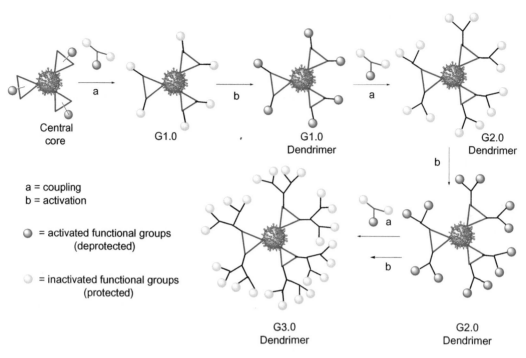

FIGURE 6.2   Divergent route for dendrimer synthesis.

required for completing the reaction increase at the surface. Ultrafiltration, precipitation or simple distillation processes are generally used to purify the macromolecule from side products due to differences in molecular weight. Large scale/various generations of dendrimer synthesis can be done by using this approach if right coupling reagents and proper reaction conditions were maintained. As the generation of dendrimers increases with an increase in coupling and activation steps, it also resulted in the generation of side products through side reactions, for instance, missing repeating units, intramolecular/intermolecular cyclisation or retro-Michael addition and unfinished functionalisation. Due to these reasons, high generation dendrimers contain a large number of structural errors. An example of this is the divergent route for the synthesis of dendrimers using a multifunctional core, i.e., ethylenediamine (EDA). The Michael addition reaction is carried out on two nitrogen atoms of EDA, which resulted in four arms as shown in Figure 6.3. Again EDA is added and further reacted to these four arms to give an amide product. These two processes are repeated all over again to form different generations of dendrimers until the desired generation is synthesised. To prevent the side reactions or errors at the higher generations, a large amount of EDA was used in dendrimer synthesis (Gupta and Nayak 2015).

6.3.1.1 Major Errors Occurring in Divergent Route Synthesis Due To the Following Reasons

a) **Missing Repeat Units**

For the synthesis of dendrimers, the Michael donor (EDA) reacts with Michael acceptor methyl acrylate via the Michael addition reaction in which both the hydrogen atoms of amine react to give a coupled product and are further activated until the desired generation is achieved (Figure 6.4). However, in another case, if one of the hydrogen atoms in amine fails to react, then this missing unit produces defected dendrimers (Figure 6.4).

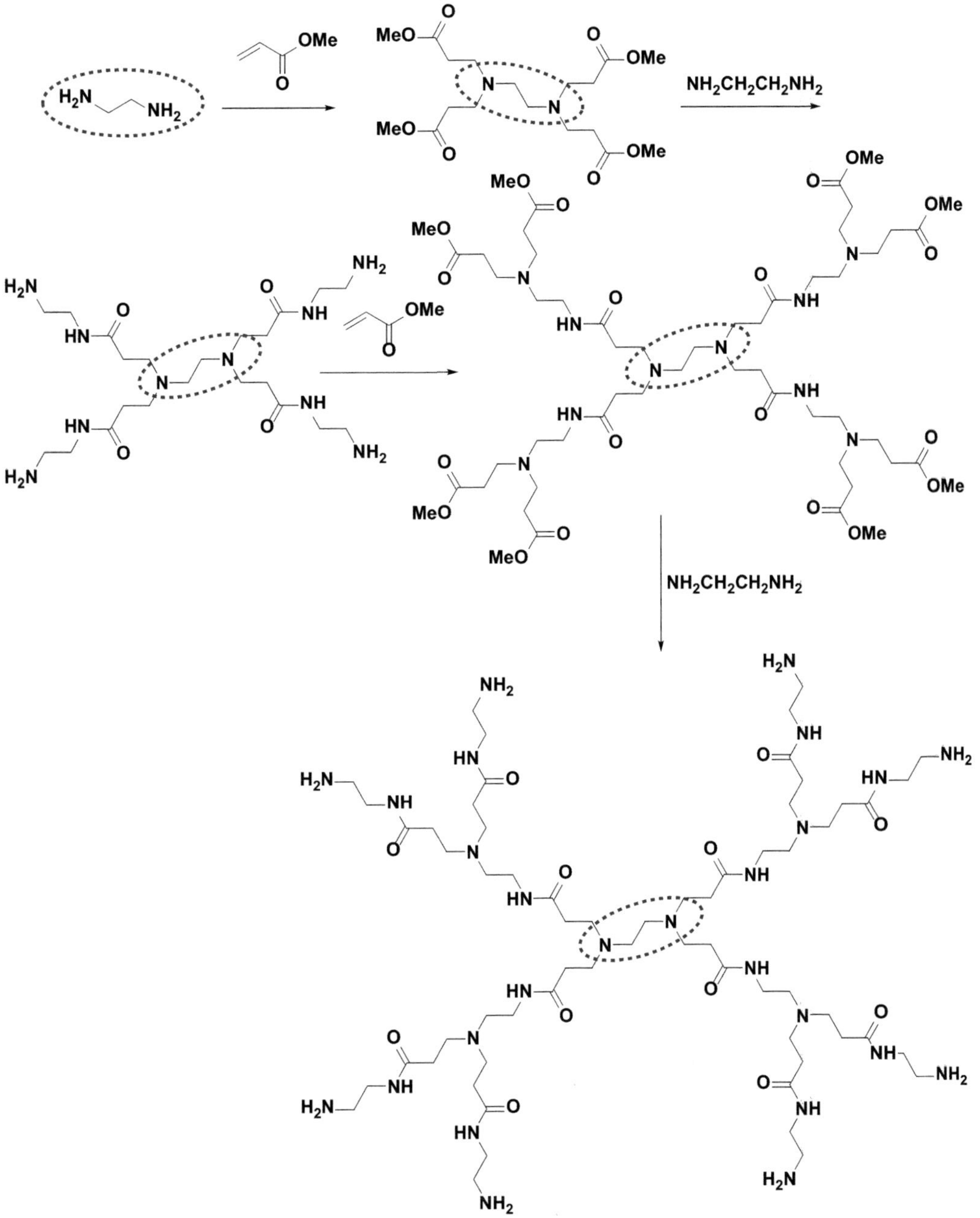

FIGURE 6.3 Synthesis of dendrimers containing the EDA core via the divergent route.

b) **Intramolecular and Intermolecular Cyclisation**

Intramolecular cyclisation may occur due to amidation steps that occur happens in only one arm of the same molecule. Therefore, further growth of the dendrimer is stopped and ends with defected dendrimers. Likewise, a single molecule of EDA reacts with different dendrimers resulting in intermolecular cyclisation which ends with dimer products (Figure 6.5) and therefore stops the synthesis of a further generation of dendrimers.

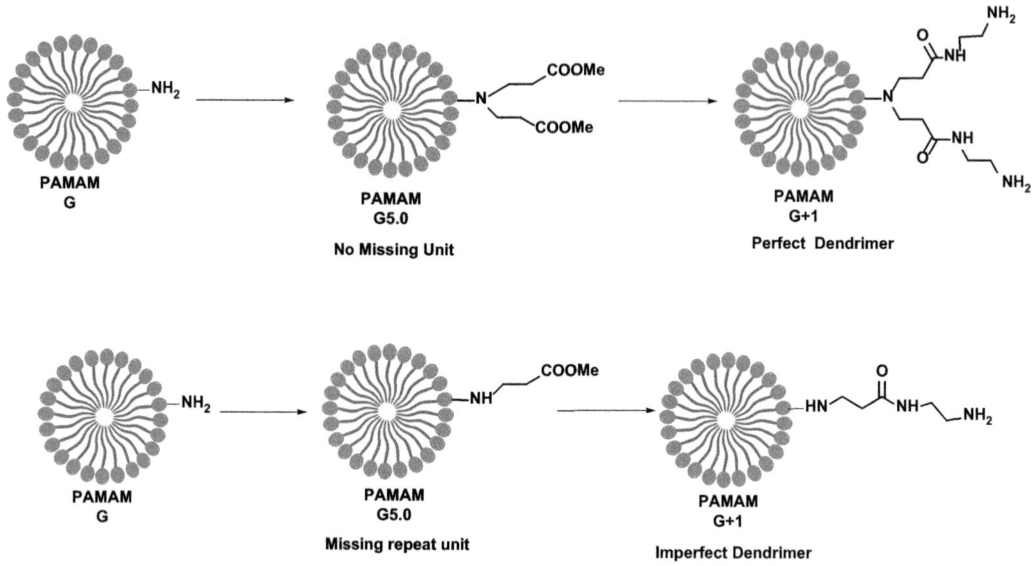

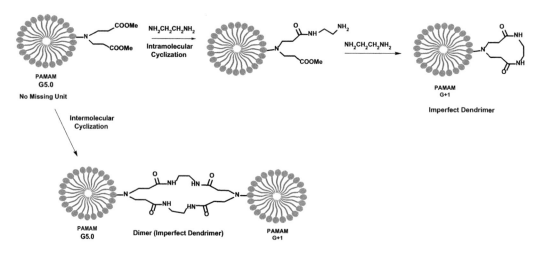

FIGURE 6.4   Missing units during the dendrimer synthesis.

FIGURE 6.5   Intramolecular and intermolecular cyclisation occurs during the synthesis of dendrimers.

c) **Ester Hydrolysis**

Methyl acrylate is an ester, i.e., used as a Michael acceptor in dendrimer synthesis. Ester is hydrolysed very easily throughout dendrimer synthesis and generates acid groups. This acid group creates a problem and reacts with an amine to form amide in the reaction mixture. So, the hydrolysis of methylacrylate in reaction medium resulted in defected dendrimers as given in Figure 6.6.

d) **Retro-Michael reaction**

Additionally, the retro-Michael reaction may occur in the reaction medium, which is the reverse of Michael addition. In this case, degradation of synthesised arms occurs due to which free NH may react to create impure dendrimers (Figure 6.7). Every step of dendrimer synthesis through this divergent route requires purification. Purification is generally based on the molecular weight, size and chemical behaviour. As the generation of dendrimers increases, purification becomes tough, resulting in higher generation dendrimers with lesser purity.

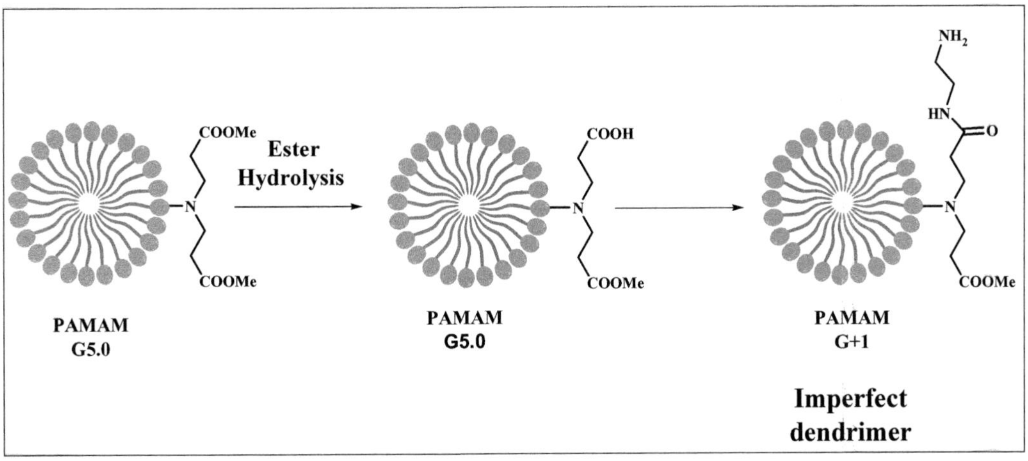

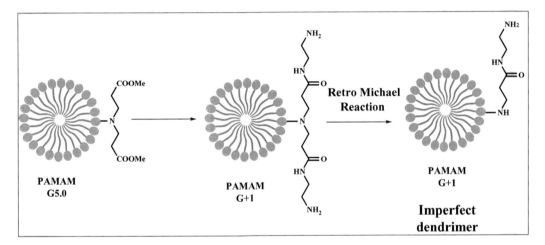

FIGURE 6.6 Ester hydrolysis in divergent route synthesis.

FIGURE 6.7 Defected dendrimers via the retro-Michael reaction.

6.3.2 Click Chemistry in Divergent Route Synthesis of Dendrimers

The term "Click chemistry" was given by Sharpless et al. in the year 2001, which produces the desired final product in excellent yields. Some of the usual reactions are S_N1 and S_N2, i.e., nucleophilic substitution reactions generally for ring-opening and addition reactions which also include 1,3 cycloaddition reactions (Dhaval, Rinkesh and Pravinkumur 2014). Synthesis of various dendrimers (high generation) is achieved by using a divergent approach but this old approach suffers from various limitations like tedious, time-consuming, unfinished reactions (the impaired/defected dendrimer formation) and purification problems. So, click chemistry overcomes the drawbacks of the divergent approach by using a readily available substrate, reagents, shorter reaction time and well-organised and generally by-product free reactions. Gonzaga et al. developed a metal-free click method for the generation of triazole-based dendrimers with high purity and less or no by-products formation. Starting with the synthesis of monomer (bis(bromopropyl) but-2-yne dioate) from a readily available starting material (Figure 6.8), but-2-yne dioic acid with 2-bromopropanol gave (1) with a 96 % yield. The Huisgen 1,3 dipolar cycloaddition reaction (metal-free) of the monomer with 1,4-bisazidomethylbenzene at room temperature afforded (2a) first-generation bromo-terminated dendrimers (G1.0). Then, the nucleophilic substitution of the bromo group by N_3 in DMSO as a solvent resulted in the synthesis of tetra azido dendrimers (G1.0)2b (95 %).

Furthermore, repeating these steps leads to the synthesis of high generation (G2.0 (3a), G3.0 (4a) and G4.0(5a)) dendrimers (Gonzaga, Sadowski and Rambarran et al. 2013).

Another author Arseneault has reported the synthesis of triazole containing third (G3.0) and fourth-generation dendrimers (G4.0) via the Huisgen 1,3-dipolar cycloaddition reaction (Figure 6.9). The metal-free click reaction between the terminal azide and alkynes (disubstituted) resulted in high yield pure dendrimers (Arseneault, Levesque and Morin 2012).

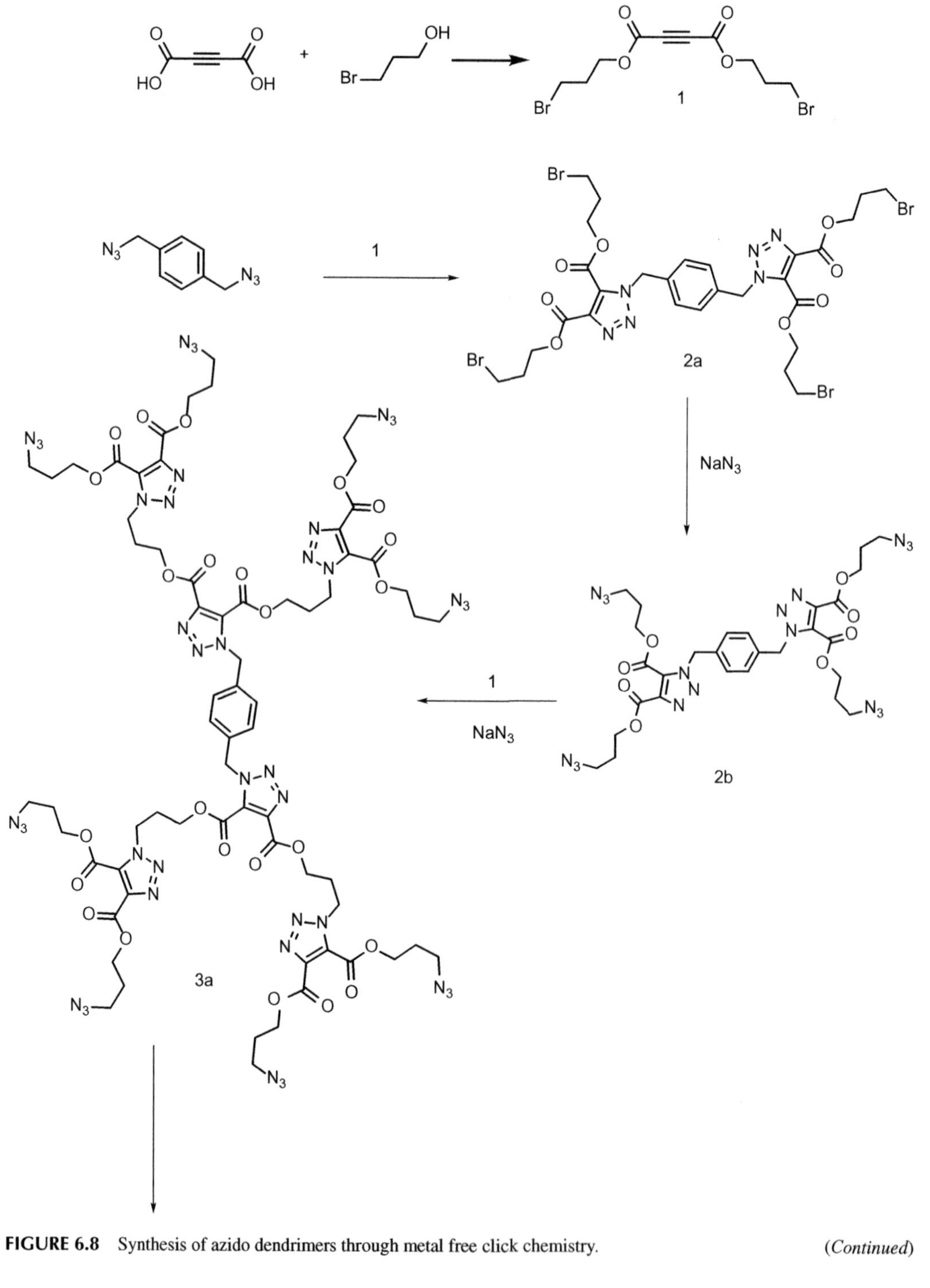

FIGURE 6.8 Synthesis of azido dendrimers through metal free click chemistry.   *(Continued)*

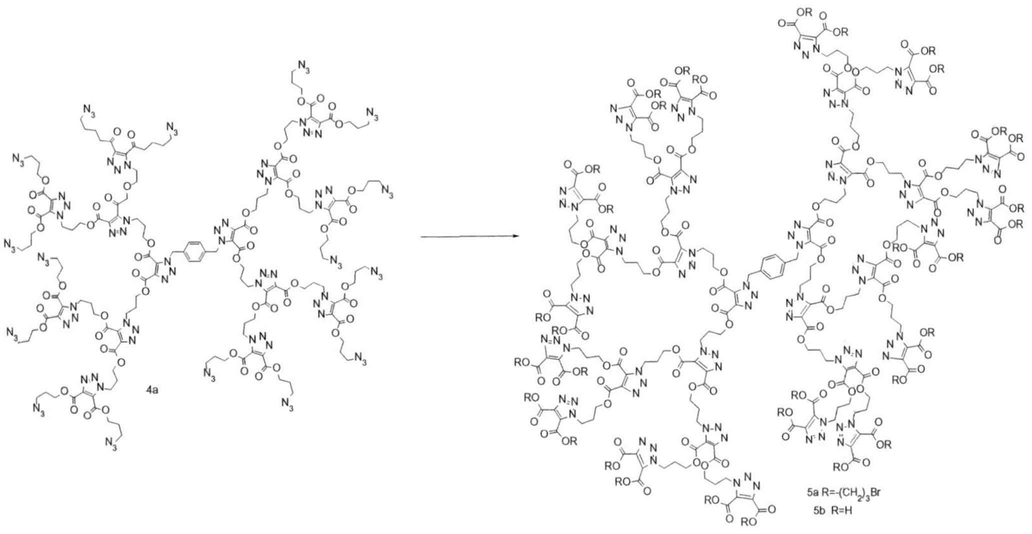

FIGURE 6.8 (Continued) Synthesis of azido dendrimers through metal free click chemistry.

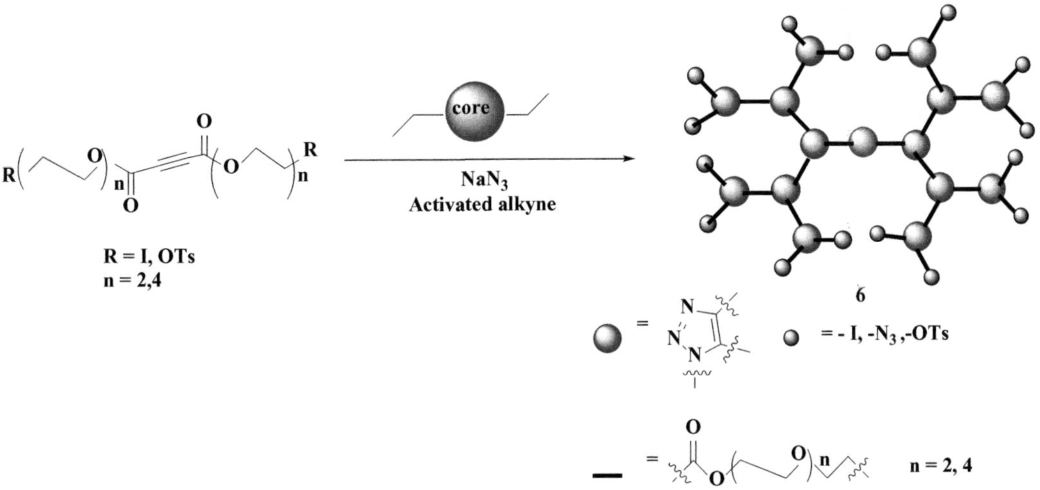

FIGURE 6.9 Synthesis of dendrimers containing ethylene oxide through metal free click chemistry.

6.4 Convergent Approach

A new approach was established to avoid the side reactions and other impurities that arise during the synthesis of dendrimers via the divergent approach. C. Hawker and J. Frechet introduced a convergent method for the synthesis of dendrimers with more purity. This outer part is first synthesised and then linked to the inner core, i.e., the just reverse of the divergent approach. The outer functional groups coupled with each segment of a monomer called dendrons. These functional groups are activated at the focal tip of the dendron and further coupled with more segments of monomer resulting in a bigger dendron, i.e., having more molecular weight than the previous one (Tomalia, Baker and Dewald et al. 1985). These two steps, i.e., coupling and activation are repeated until desired dendrons are synthesised and the dendrons are further coupled to the multifunctional central core resulting in dendrimer synthesis (Figure 6.10). The reaction rate is faster in the case of a convergent approach when compared to a divergent approach

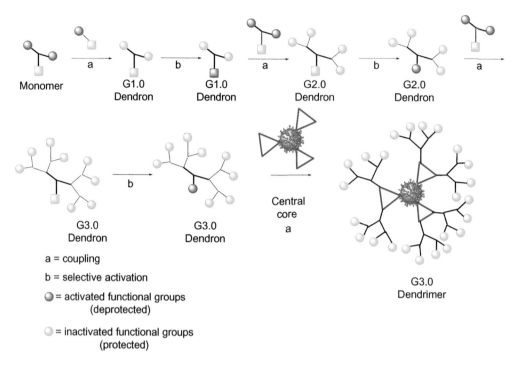

FIGURE 6.10   Convergent route for dendrimer synthesis.

because it involves fewer reactions. A variety of monomers are attached to dendrimers that possess different structural qualities (Gupta and Nayak 2015).

After each coupling step, the molecular mass of the dendron increases with the increase in the number of generations. Uniform and more symmetric dendrimers can be synthesised employing this approach. The overall yield of the dendrimers is reduced as the generation of dendrimers increases (above sixth) due to steric hindrance. This approach is generally used for the lab-scale synthesis of dendrimers and not for industrial scale-up. Purification in this approach is easy as compared to the divergent approach. This approach has an advantage for the synthesis of dendrimers containing diverse functional cores (Augustus, Allen, Nimibofa et al. 2017). This approach mainly highlights two features over the divergent approach, i.e., structural purity and synthetic versatility.

In the previous years, various dendrimers with different cores and dendrons were synthesised via convergent routes (Figure 6.11). Some of the convergent synthesis approaches were efficient and versatile while the remaining ones are imaginary (Gilat, Adronov and Fréchet 1999). Convergent synthesis is generally used for poly-aryl ether (Frechet et al.) and poly-aryl alkyne (Moore et al.) dendrimers. Convergent synthesis can be further classified as follows:

6.4.1 Single-stage Convergent Synthesis

A monomer is essential for the effective synthesis of dendrimers that go through various coupling and activation steps resulting in a higher generation of dendrimers. These coupling steps can complete the synthesis even if bulky dendrons are present.

a) **Poly-aryl ether dendrimers**
   Here, 3,5-dihydroxybenzyl alcohol was used as a monomer for the synthesis of poly-benzyl ether dendrimers. This synthesis was introduced by C. Hawker and J. Frechet in 1990s (Hawker and Fréchet 1990). Second-generation benzyl alcohol (G2.0, 8) was synthesised from the coupling reaction of two phenolic groups of the monomer with 3,5-o-benzyl bromide (7) by using potassium

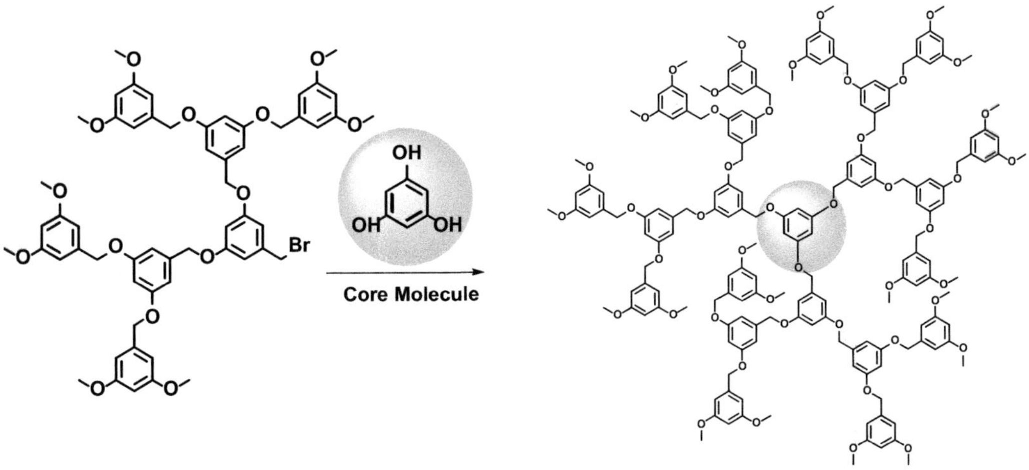

FIGURE 6.11 Dendrimer synthesis containing benzene 1, 3, 5 triol as core.

carbonate and 18-crown-6 with 91 % yield. After this, activation of end groups of benzylic alcohol to afford the next coupling reaction with carbon tetrabromide and triphenyl phosphine generated brominated dendron (9). Then the third generation (G3.0) benzylic alcohol (10) was synthesised by repeating the coupling reaction with 1 equivalent of monomer and 2 equivalents of activated dendron (9). Desired dendrimers with sixth generation (G6.0) were obtained via continuously repeating Williamson coupling and activation steps. This route of synthesis was planned to obtain more yields after each coupling and activation step. For the synthesis of higher generation dendrimers, one of the limitations in this approach could be the steric hindrance of bulky dendrons during the coupling step. (Grayson and Frechet 2001) As in the synthesis of poly-benzyl ether dendrimers (Figure 6.12), 90 % yield was obtained up to the fourth generation (G4.0) dendrimers. However, the yield would fall down to 85 % in the fifth generation (G5.0) and 78 % in the sixth generation (G6.0). Tridendron dendrimers (19–25) are also synthesised by coupling the generation one (G1.0) to sixth (G6.0) with a triphenolic core (22) and have fewer yields due to the steric factor (Figure 6.13). These dendrons were also known as Frechet type dendrons and it is one type of convergent synthesis that can generate dendrimers up to sixth-generation (G6.0) (Sowinska and Urbanczyk-Lipkowska 2014).

b) **Poly-aryl alkyne dendrimers**

Poly-aryl alkyne dendrimers were prepared by Moore and his co-workers using dibromophenylacetylene. Their synthesis includes a palladium-catalysed reaction followed by deprotection of trimethylsilyl (TMS) groups. They modify the surface end groups by converting for enhancing the solubility of phenylacetylene monomers. The modification of surface end groups by using 4-*tert*-butylphenyl peripheral units overcomes their solubility issues only up to the third-generation (G3.0). Furthermore, their modification with 3,5-di-*tert*-butylbenzene groups provided the desired solubility, which resulted in fourth-generation (G4.0) dendrons. In addition to that, they overcome the problem of steric hindrance in the synthesis of a large generation of dendrimers using extended monomer units (Moore and Xu 1991). A new method was developed by Moore and his co-workers to avoid the use of these elongated monomers. The synthesis of a large generation of dendrimers was achieved by inverting the functionalities present on monomers by using dialkyl triazene moieties for focal iodo functionality (Bharathi, Patel, Kawaguchi et al. 1995). Diethynyl monomer (26) with triazene as protecting groups was used for the synthesis of dendrimers (Figure 6.14). First-generation (G1.0) aryl halide dendrons were coupled with end alkyne groups at the monomer surface via two Pd-catalysed reactions yielding a second-generation dendron (G2.0) (30). Furthermore, activation was performed by halogenation of the triazene group to give aryl halide (31) which was coupled with more monomer units (Gorman and Smith 2000). This method was generally able to synthesise dendrons up to the fifth-generation (G5.0) with relatively higher

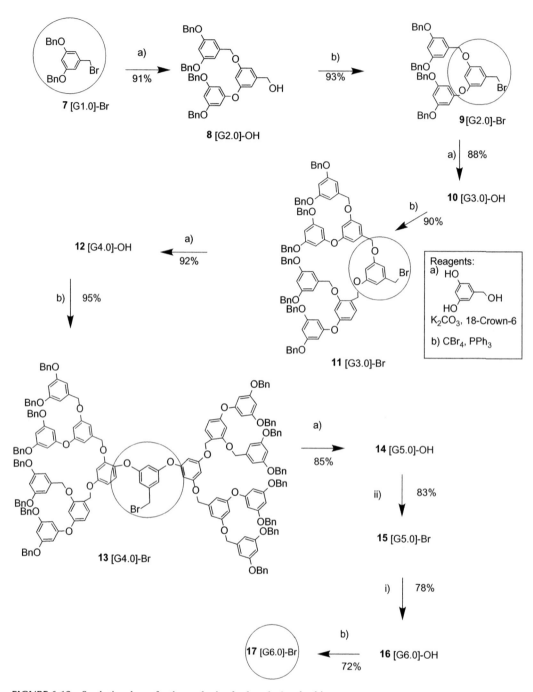

FIGURE 6.12   Synthetic scheme for the synthesis of polyaryl ether dendrimers.

yields using solid-phase synthesis. Phenylacetylene macromolecules are the rigid units, and these dendrimers include segments with different morphologies and properties (Devadoss, Bharathi and Moore 1996).

c) **Polyphenylene dendrimers**

Polyphenylene dendrimers are synthesised either by the divergent or convergent route. Miller and Neenan described the synthesis of poly 1,3,5-phenylene dendrimers by the convergent route in which monomer (36) 3,5-dibromo-1-(trimethylsilyl) benzene is coupled with arylboronic acids

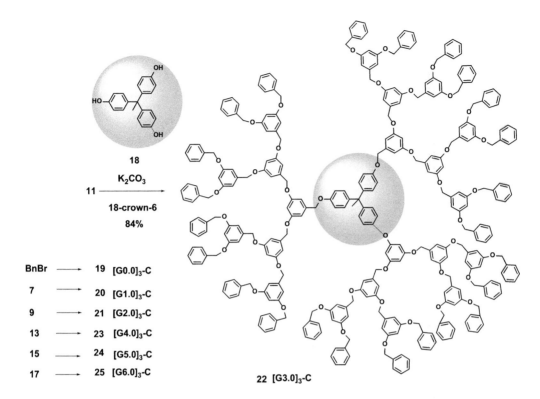

FIGURE 6.13 Synthetic scheme for the synthesis of tridendron dendrimers using the triphenolic core.

34 or 35 via the Suzuki coupling reaction. Deprotection and modification of the TMS protecting group of 37 and 38 give rise to boronic acid moieties 39 and 40 which are further able to couple with the monomer to form third-generation dendrimers (G3.0) 45–48 (Figure 6.15). Synthesis of polyphenylene dendrimers (Figure 6.16) via the convergent route was also reported by Mullen and his co-workers. They have used the pericyclic Diels–Alder reaction between diene and dienophile, i.e., tetrasubstituted cyclopentadienone (49) and dialkynyl monomer (50), respectively, to give dione (51). Furthermore, cyclopentadienone (53) was synthesised via Knoevenagel condensation of (51) dione with 1,3 diphenylacetone (52). Dendrimers synthesised using the convergent route are up to second-generation (G2.0) due to steric complexity present at the wedges of bulky groups (54); further generation of dendrimers was not synthesised. They also have synthesised poly-phenylene pyridyl dendrimers with different surface groups using the Diels–Alder reaction up to the fourth generation (G4.0) via the divergent route (Shifrina, Rajadurai and Firsova et al. 2005).

d) **Poly(alkyl ester) dendrimers**
Hult and his team synthesised poly(alkyl ester) dendrimers by the convergent route using 2,2-bis(hydroxymethyl) propanoic acid as the core moiety (Ihre, Hult and Söderlind 1996). Coupling of the monomer (55) containing 2,2-bis (hydroxyl methyl) propanoate with an acid chloride-containing functional group (56) is performed. Furthermore, there is a deprotection in which benzyl ester is removed by hydrogenolysis. After this, the acid group is converted by oxalyl chloride to acid chloride (58) (Figure 6.17). Fourth-generation poly(alkyl ester) dendrimers (G4.0) were synthesised by continuously repeating these coupling and deprotection steps. Under acidic conditions, these types of dendrimers are more stable because of the protection of ester from any type of nucleophilic attack. Pahovnik et al., also reported the synthesis of fourth-generation (G4.0) polyester dendrimers using 2,2, bis(hydroxymethyl) propanoic acid and glycine as monomer units and further coupled via coupling methods either N, N'-dicyclohexylcarbodiimide (DCC)/4-(dimethylamino) pyridinium 4-toulene sulphonate (DPTS) or 1-ethyl-3-(3-dimethylaminopropyl) carbodiimide (EDC)/4-dimethylaminopyridine (DMAP) (Pahovnik, Čusak, Reven et al. 2014).

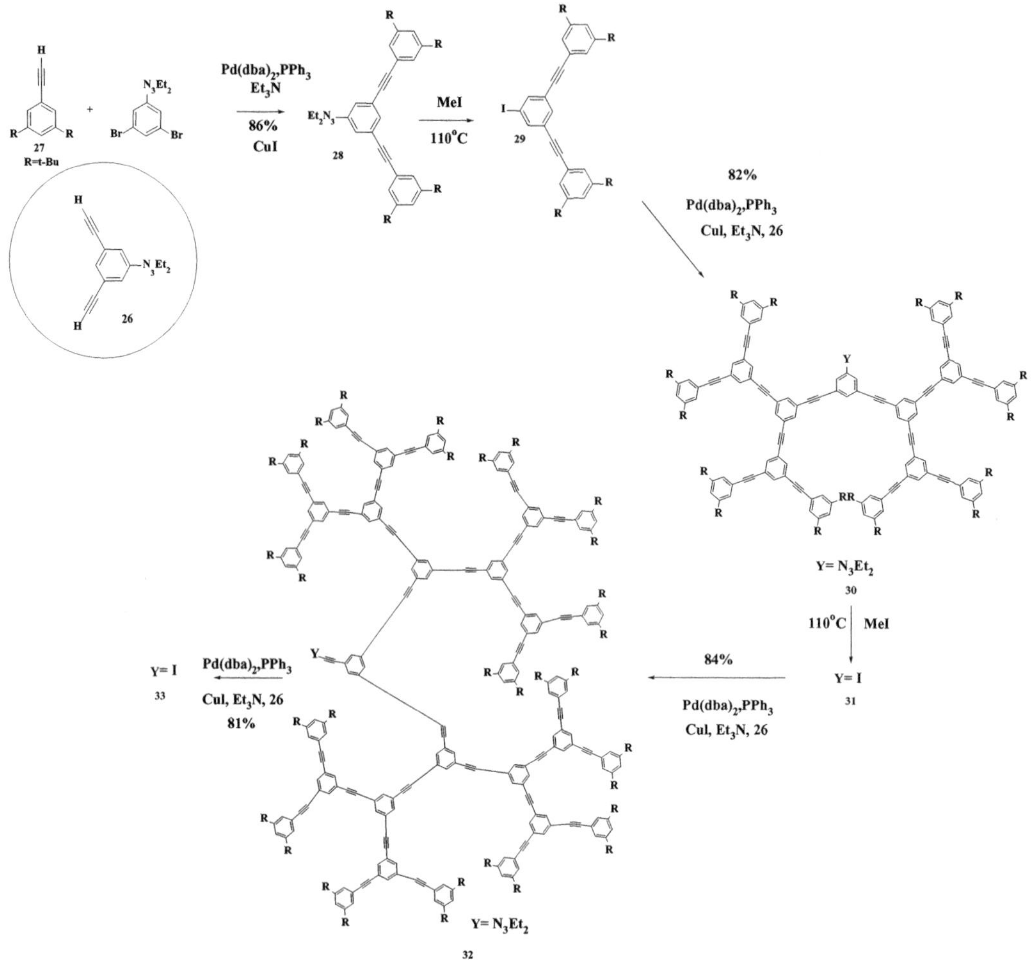

FIGURE 6.14   Synthesis of polyaryl alkene dendrimers.

e) Poly(aryl alkene) dendrimers

These types of dendrimers were reported by Meier, Burn and their co-worker, who described their synthesis by the convergent route. Meier and his team used Horner–Wadsworth–Emmons coupling of aldehydes 62 with bis(phosphite) as monomer units 63 (Meier and Lehmann 1998). Removal of dimethoxy acetal end groups of 64 resulted in activated aldehydes 65 with low yields and longer reaction time for the synthesis of final dendrimers up to the fourth generation (G4.0) 68 (Figure 6.18) (Lehmann, Schartel and Hennecke et al. 1999). In addition to this, Burn and his group utilised two named reactions Heck coupling and Wittig reaction for the synthesis of third-generation dendrimers (G3.0) 74. They have first reacted styrene (69) with 3,5-dibromobenzaldehyde as monomer (70) units followed by the Wittig reaction with methyl triphenylphosphonium iodide (Figure 6.19) (Pillow, Halim and Lupton et al. 1999).

f) Poly(alkyl ether) dendrimers

Frechet et. al., reported the synthesis of poly(alkyl ether) dendrimers which are an aliphatic modification of poly(aryl ether) dendrimers (Figure 6.20). Synthesis of dendrimers starting with monomer 75(3-chloro-2-chloromethyl-propene) is coupled to the hydroxyl group via the Williamson reaction to form ether linkage. In this coupling, bisprotected triol 76 or 77 is coupled with the allylic chloride moiety of the monomer. There are three main roles of this double bond; the first one is the removal of various side products via a reduction in side-chain reactions. The second

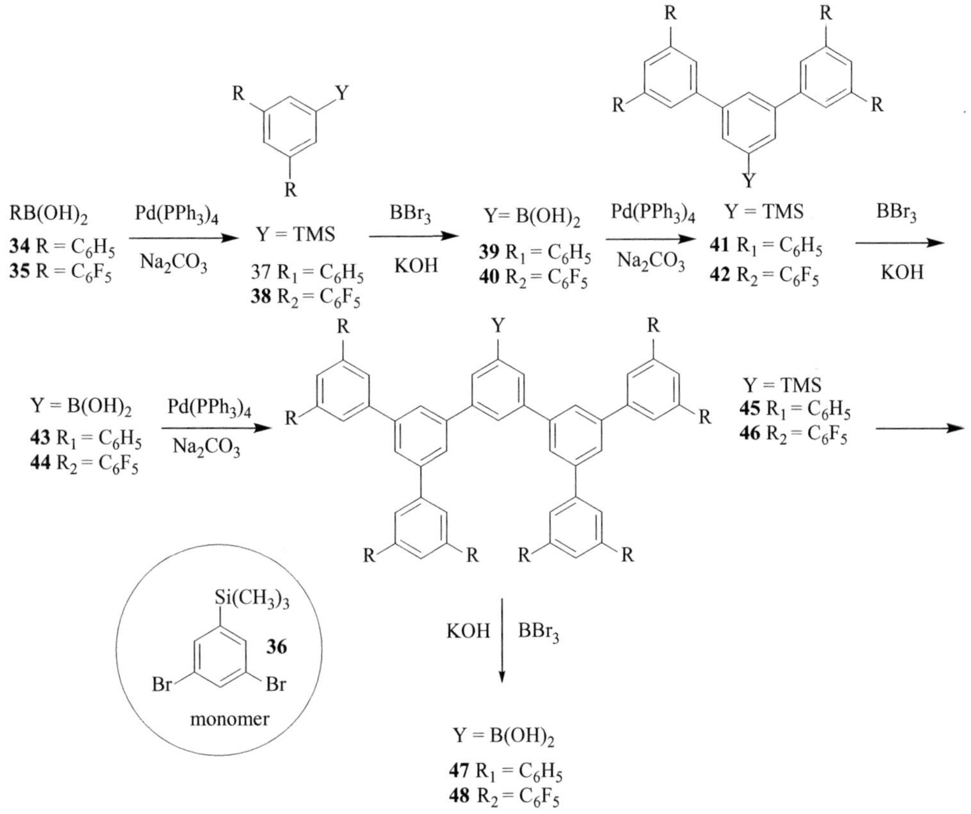

FIGURE 6.15 Polyphenylene dendrimer synthesis using 3,5-dibromo-1-(trimethylsilyl) benzene as the monomer unit.

role is the stimulation of the allylic halide group for the coupling reaction and the third one is the source of hydroxyl groups which can be used later for the development of next generation of dendrimers. During the synthesis of dendrimers, hydroboration and then oxidation reactions were used to convert the double bond of compound 78 and 79 to alcohol 80 and 81. The resulting molecules are ready for more coupling reactions until the desired dendrimers are not synthesised. Therefore, these steps can be repeated to obtain dendrimers (fifth-generation (G5.0)) in high yields.

6.4.2 Other Synthetic Approaches

a) **Lego chemistry**

Lego chemistry is employed for the synthesis of phosphorous dendrimers using branched monomers and highly well-designed functionalised cores. Dendrimers are synthesised using approaches like double exponential growth, hyper monomers and other processes that have several limitations, i.e., more numbers of steps are required for synthesis and requirement of activating agents. Dendrimer synthesis via Lego chemistry overcomes the aforesaid limitations and uses a green protocol like producing environmentally friendly by-products such as N_2 and H_2O, easy purification and fewer amounts of solvents are required for the synthesis. Maraval and his group utilised this Lego chemistry for the synthesis of the fourth- generation (G4.0) of phosphorous dendrimers (Maraval, Pyzowski, Caminade et al. 2003) through a one pot method using CD_2 and AB_2 monomers for the preparation of fourth-generation dendrimers (G4.0). Furthermore, they described another direct synthesis of phosphorous dendrimers using CA_2 and DB_2 monomers in which the functional groups are reversed. Synthesis includes three components, i.e., CA_2 and DB_2 as monomers and B_3 as a core. Initially, three equivalents of CA_2 were reacted with one equivalent of B_3

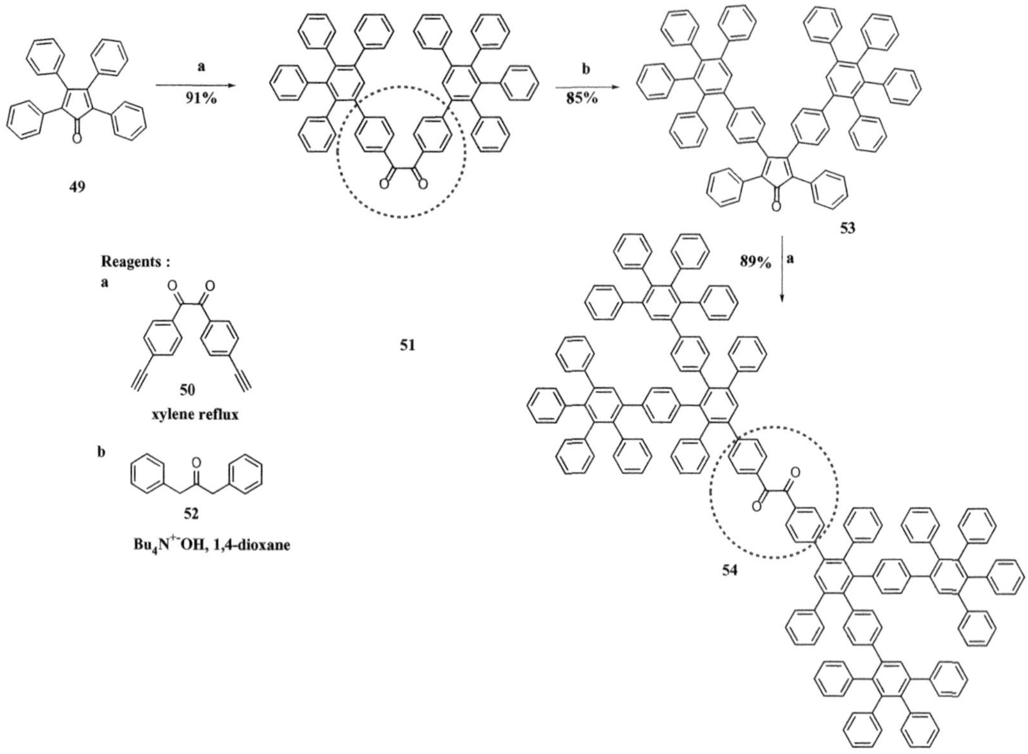

FIGURE 6.16   Synthesis of poly(phenylene- pyridyl) dendrimers.

as the core to synthesise the first generation dendrimers (G1.0) (Figure 6.21). After this, 6 equivalents of DB$_2$ were added into the above mixture to produce second-generation dendrimers (G2.0).
Furthermore, 12 equivalents of CA$_2$ were added to give third-generation dendrimers (G3.0). To generate fourth-generation dendrimers (G4.0), 24 equivalents of DB$_2$ were added to terminal groups
containing phosphines or hydrazines (Dhaval, Rinkesh and Pravinkumur 2014). In the Lego chemistry strategy, highly functionalised cores and branched monomers are applied to prepare phosphorus dendrimers. The end groups are generally phosphines and hydrazines. Fourth-generation
dendrimers (G4.0) are synthesised in only four steps and the number of surface groups increases
from 48 to 250. The synthesis required a minimum volume of solvent, allowed facile purification
and produced environmentally benign by-products such as water and nitrogen (Villalonga-Barber,
Micha-Screttas, Steele et al. 2008).

b) **Synthesis of convergent dendrimers using click chemistry**
   Dendrimers are synthesised by continuously replicating, coupling and activation reactions either
   via divergent or convergent synthesis. Generally, polyether, i.e., Frechet type and poly(amidoamine) (PAMAM), i.e., Tomalia type dendrimers are most widely prepared. The divergent route is
   mainly used for the synthesis of PAMAM type dendrimers. However, this approach has some
   drawbacks, particularly, to obtain pure dendrimers. Therefore, to overcome divergent route drawbacks, Frechet and his group synthesised dendrimers via the convergent route. In this, coupling and
   activation of monomers were performed to obtain different dendron generation, and finally the dendron was coupled with the core but the synthetic procedure was very complex. A new methodology
   was used to simplify the procedure, i.e., incorporating "click chemistry". Click chemistry was
   introduced by Sharpless in 2001 for the synthesis of dendrimers, and it offers various advantages
   over the existing synthetic routes. Lee and his co-workers synthesised PAMAM dendrimers having
   a carbazole core via incorporating click chemistry in the convergent route (Lee, Han, Yun et al.

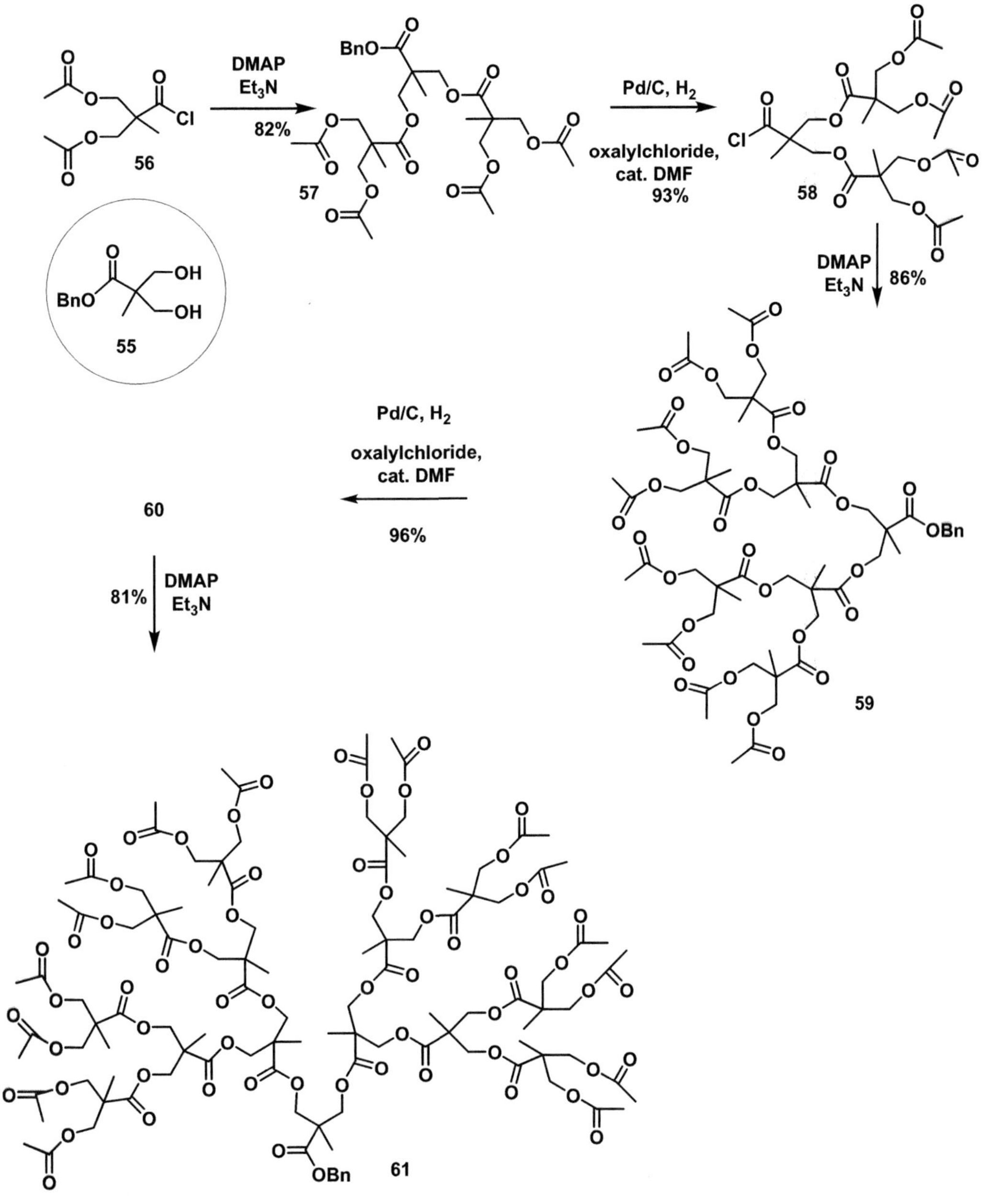

FIGURE 6.17 Synthesis of poly(alkyl ester) dendrimers.

2013). N-octyl-3,6-dibromocarbazole was treated with NaN_3 in the presence of N,N-dimethyl ethylene diamine, Cu (II) and sodium ascorbate for the synthesis of core N-octyl-3,6-diazidocarbazole (94). Furthermore, the core was reacted with PAMAM dendrons using click chemistry to obtain third-generation (G3.0) PAMAM dendrimers (Figure 6.22). Han et al., described the protocol for the synthesis of PAMAM dendrimers including tetra(ethylene oxide) as the core via the convergent approach using click chemistry (Han, Kim, and Lee 2012). They reported the first-time convergent synthesis of triblock (ABA) PAMAM dendrimers in which the 'A' block is various generations

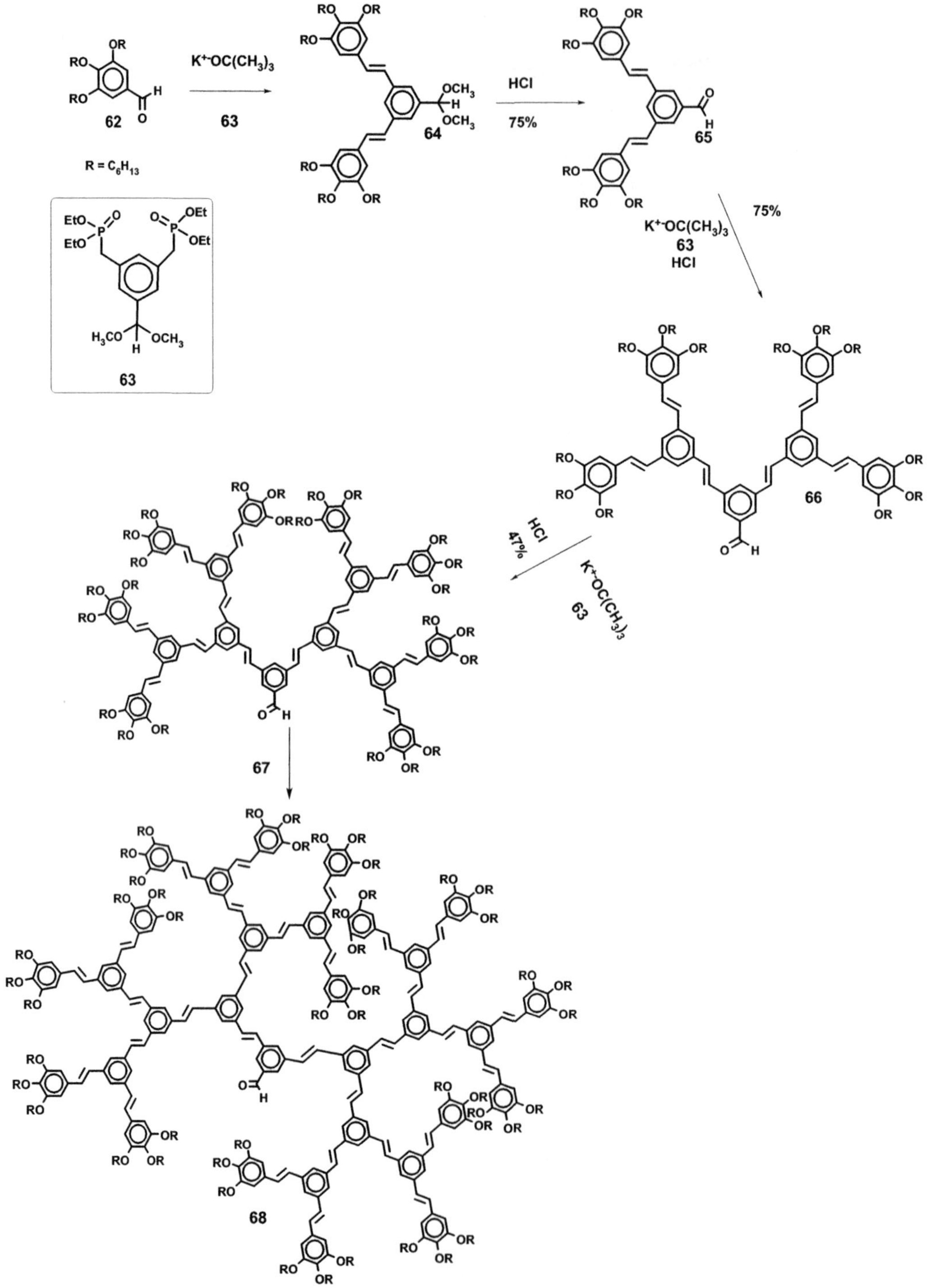

FIGURE 6.18 Synthesis of poly(aryl alkene) dendrimers using bis(phosphite) as the monomer.

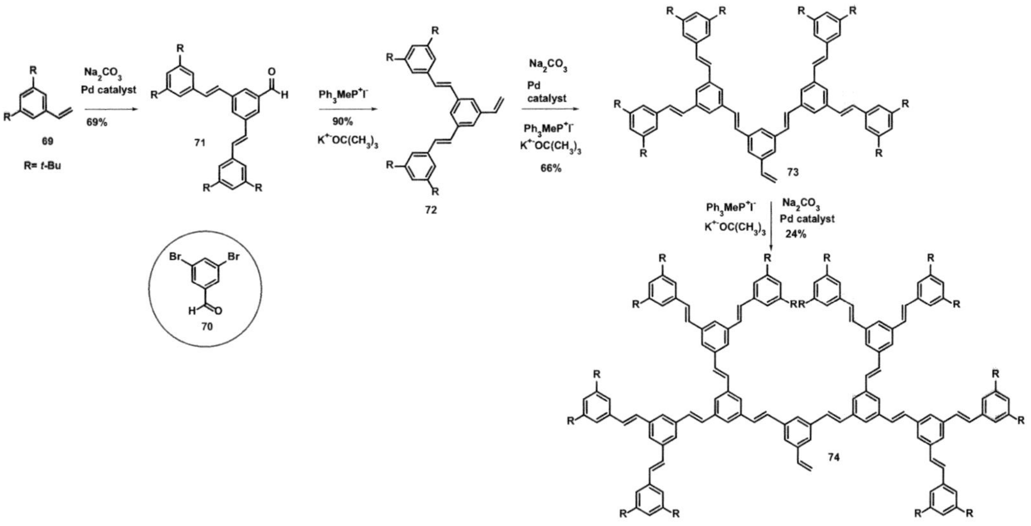

FIGURE 6.19 Synthesis of poly(aryl alkene) dendrimers using 3,5-di bromobenzaldehyde as monomer units.

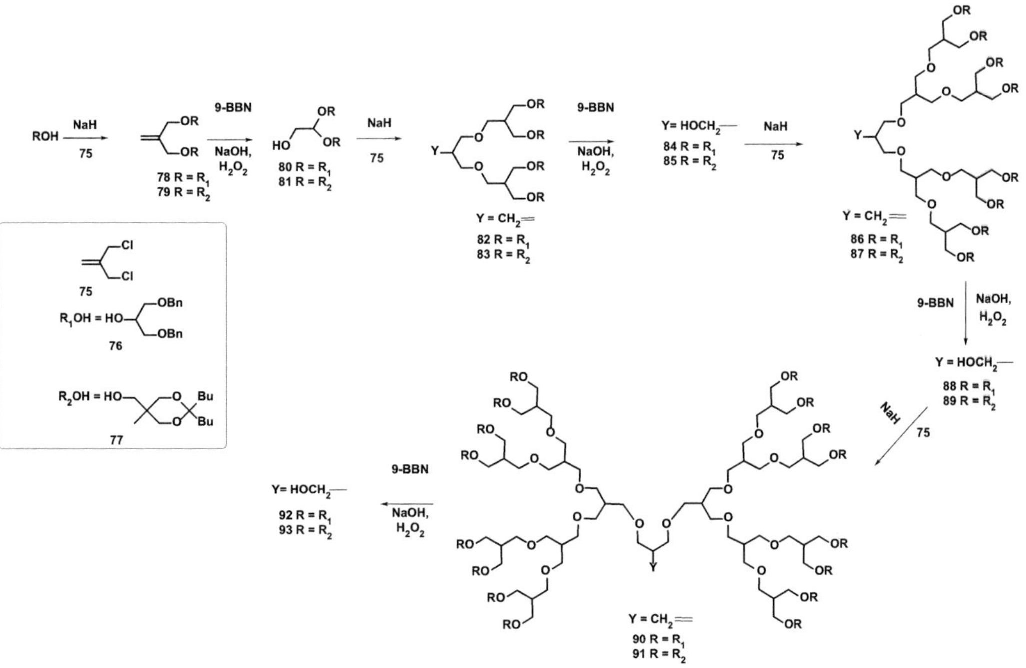

FIGURE 6.20 Synthesis of poly(alkyl ether) dendrimers.

of PAMAM and 'B' is tetra(ethylene oxide). Starting with the synthesis of compound 95, i.e., tetramethylene glycol diazide was obtained by the mesylation reaction of tetraethyleneglycol with methanesulfonyl chloride in the presence of NaN_3 and triethylamine. After this, the reaction of tetraethylene glycol diazide and PAMAM dendrons resulted in triblock type PAMAM dendrimers containing a tetraethylene oxide core in high yield as described in Figure 6.23.

Different Branched Monomers

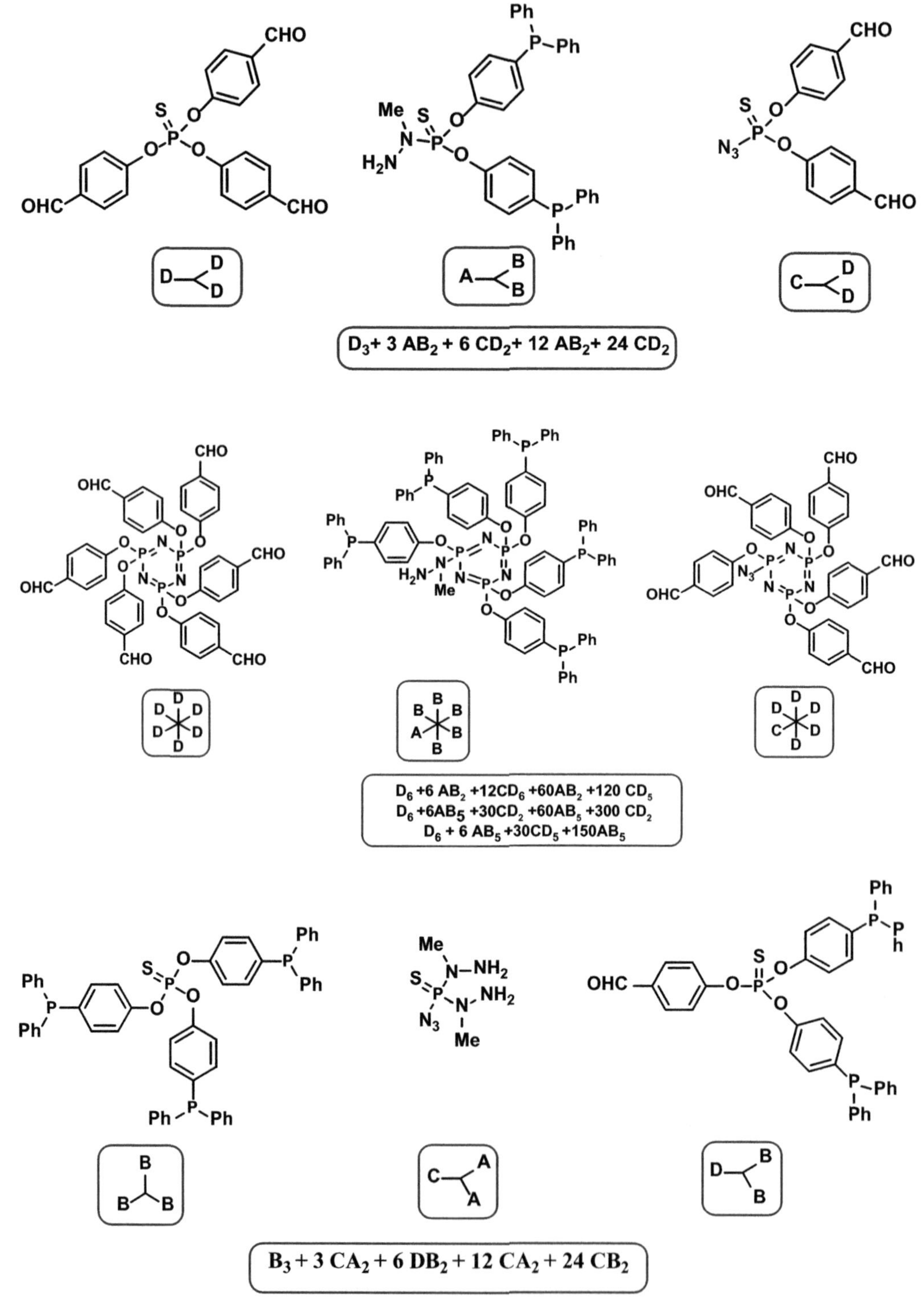

$$D_3 + 3\ AB_2 + 6\ CD_2 + 12\ AB_2 + 24\ CD_2$$

$$D_6 + 6\ AB_2 + 12CD_6 + 60AB_2 + 120\ CD_5$$
$$D_6 + 6AB_5 + 30CD_2 + 60AB_5 + 300\ CD_2$$
$$D_6 + 6\ AB_5 + 30CD_5 + 150AB_5$$

$$B_3 + 3\ CA_2 + 6\ DB_2 + 12\ CA_2 + 24\ CB_2$$

FIGURE 6.21 Synthesis of dendrimers using Lego Chemistry.   *(Continued)*

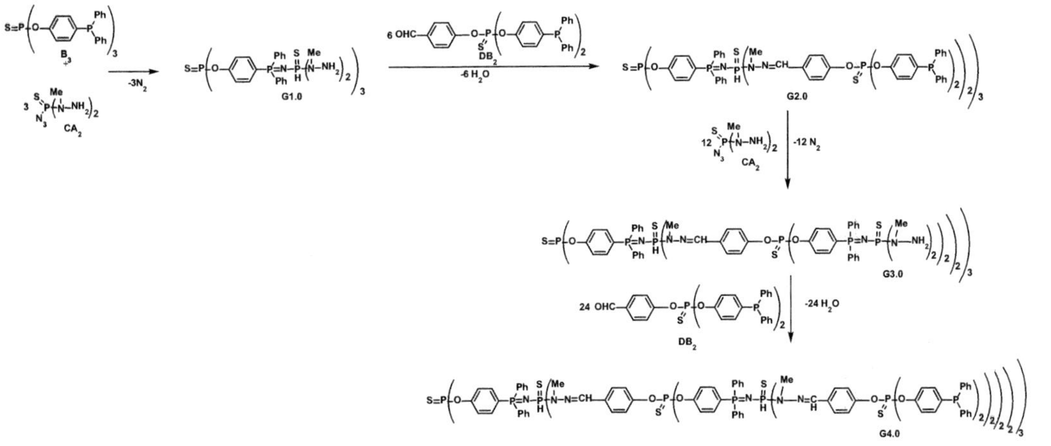

FIGURE 6.21 (Continued) Synthesis of dendrimers using Lego Chemistry.

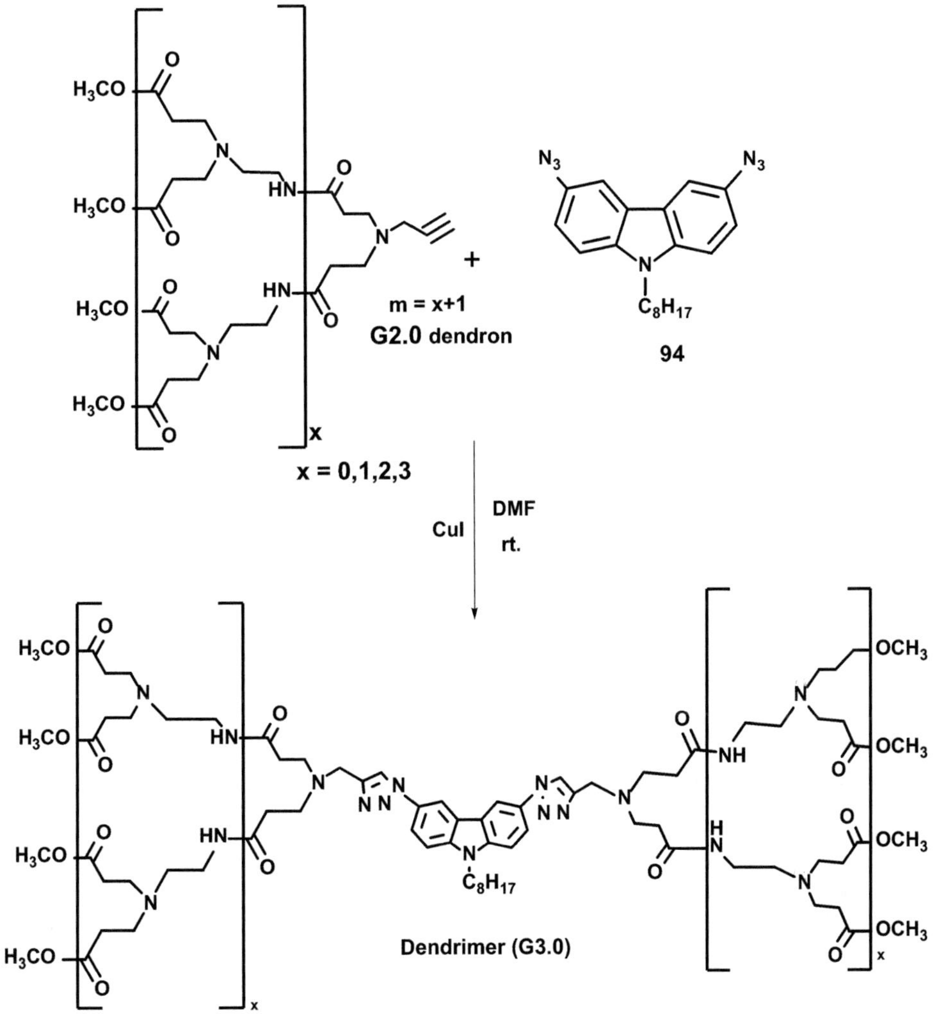

FIGURE 6.22 Synthesis of PAMAM dendrimers containing carbazole as the core.

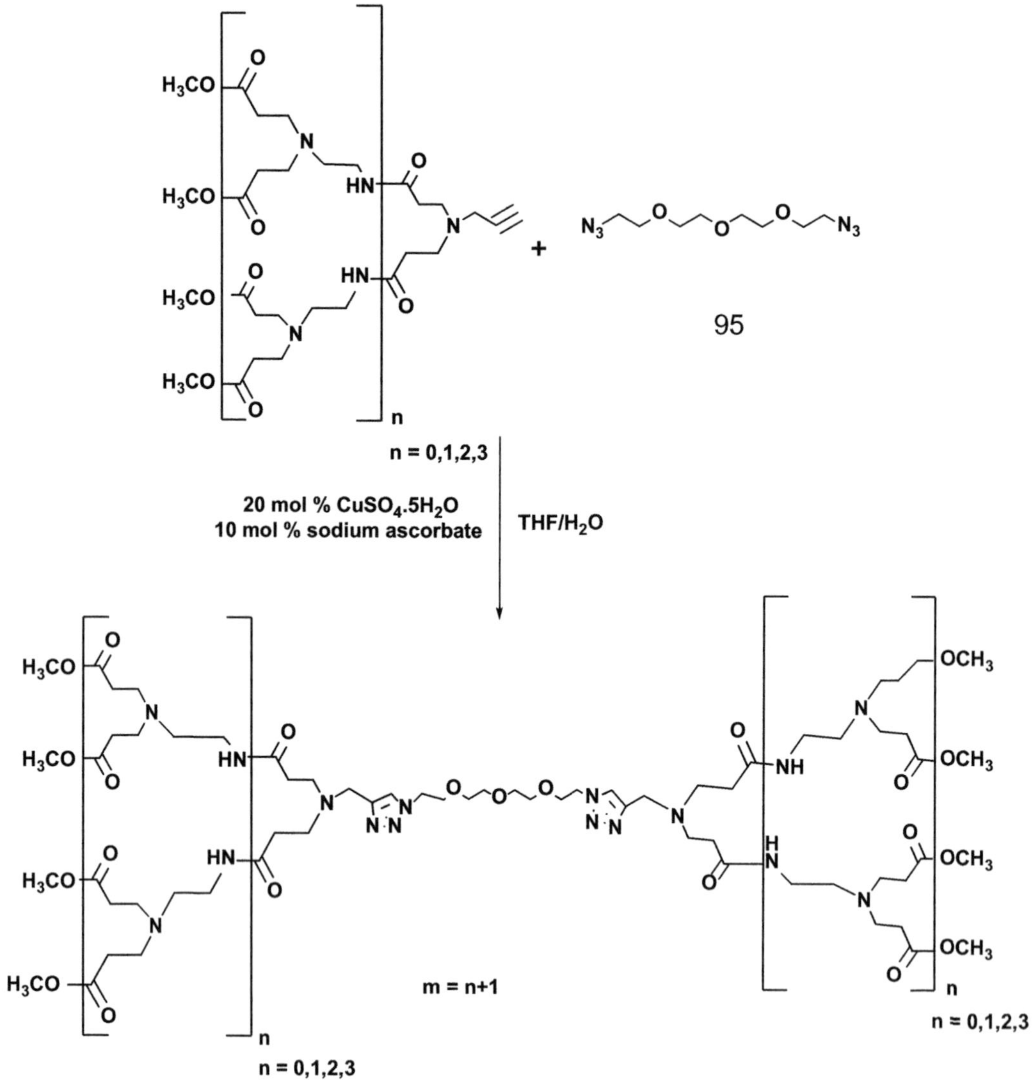

FIGURE 6.23 Synthesis of PAMAM dendrimers containing tetraethylene oxide as the core.

6.5 Combination of Divergent and Convergent Approaches (Accelerated Approaches)

Several approaches have been developed in recent years for the synthesis of dendrimers. However, their limitations such as the formation of side products, long reaction time, multi-step process, etc. bring the concept of accelerated synthesis of dendrimers. Accelerating the synthesis of dendrimers is defined as obtaining a higher branching and/or several functional groups by the use of lesser reaction steps. The major advantage over the existing approaches is that accelerated approaches required a few steps and comparatively lesser time to synthesise desired dendrimers (Montanez, Campos, Antoni et al. 2010). From a scientific perspective, accelerated approaches will make dendrimer synthesis less challenging due to a reduction in the number of steps. Acceleration approaches decrease the production cost of the dendrimers, and eco-friendly synthesis makes its use worldwide. At present, numerous parameters have been documented in the literature that should be considered to accelerate the synthesis of dendrimers.

- Selection of building-blocks
- Numbers of reactions
- One-pot synthesis

In the past decades, researchers much talked about the accelerated approaches for the synthesis of dendrimers. The concept of accelerated synthesis is a combination of convergent and divergent strategies. In the divergent approach, the synthesis of the dendrimer is begun with a core molecule that grows outwards, whereas, in the case of a convergent approach, dendrons are synthesised first and then they are connected to the core molecule inward. Both approaches generally preserve the versatility and product monodispersity of the dendrimer, which is provided by the conventional convergent method; however, it decreases the number of linear synthetic steps that require achieving a larger dendritic material. The procedures include multigenerational coupling that can be classified as follows: (Sowinska and Urbanczyk-Lipkowska 2014).

6.5.1 Hypercore or Double-Stage convergent method

In this approach, hypercores (low generation dendrimer) and low generation dendrons are synthesised in parallel and in the last step, the dendron is coupled with the hypercore which gives rise to desired higher generation dendrimers as shown in Figure 6.24. In comparison to the traditional convergent synthesis, the hypercore method has noteworthy advantages. It depends upon the type of hypercore, which confines the crisis of steric hindrance and thus facilitates access to higher generation monodisperse dendrimers. The synthesis of the hypercore and dendrons is a comparatively more time-consuming process, and this becomes a major demerit of the double-stage convergent method.

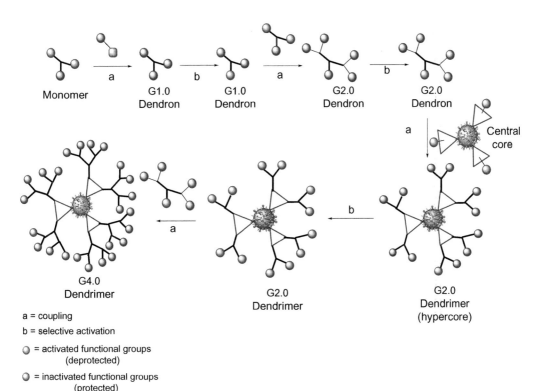

a = coupling
b = selective activation
◐ = activated functional groups
         (deprotected)
◐ = inactivated functional groups
         (protected)

FIGURE 6.24 Synthesis of dendrimers using the hypercore approach.

The introduction of this approach by Frechet and his group is mainly to modify the last step yields of dendrimer synthesis that are less in existing synthetic methodology of dendrimers due to sterically hindered coupling reactions. The main aim of this approach is to join the dendron with functional end groups of the core. 4,4-bis (4'-hydroxyphenyl) pentanol was taken as a monomer for the synthesis of dendrimers up to third-generation (G3.0) via the convergent route. Deprotection of benzyl ether present on the surface results in the activation of surface groups. Furthermore, the hypercore (polyols) when coupled with benzylic bromides results in third-generation (G3.0) hypercore synthesis. In addition to the above, third-generation hypercore 97 (G3.0) coupling with fourth-generation dendron (G4.0) affords seventh-generation dendrimers (98) (G7.0) in 61 % yield (Figure 6.25).

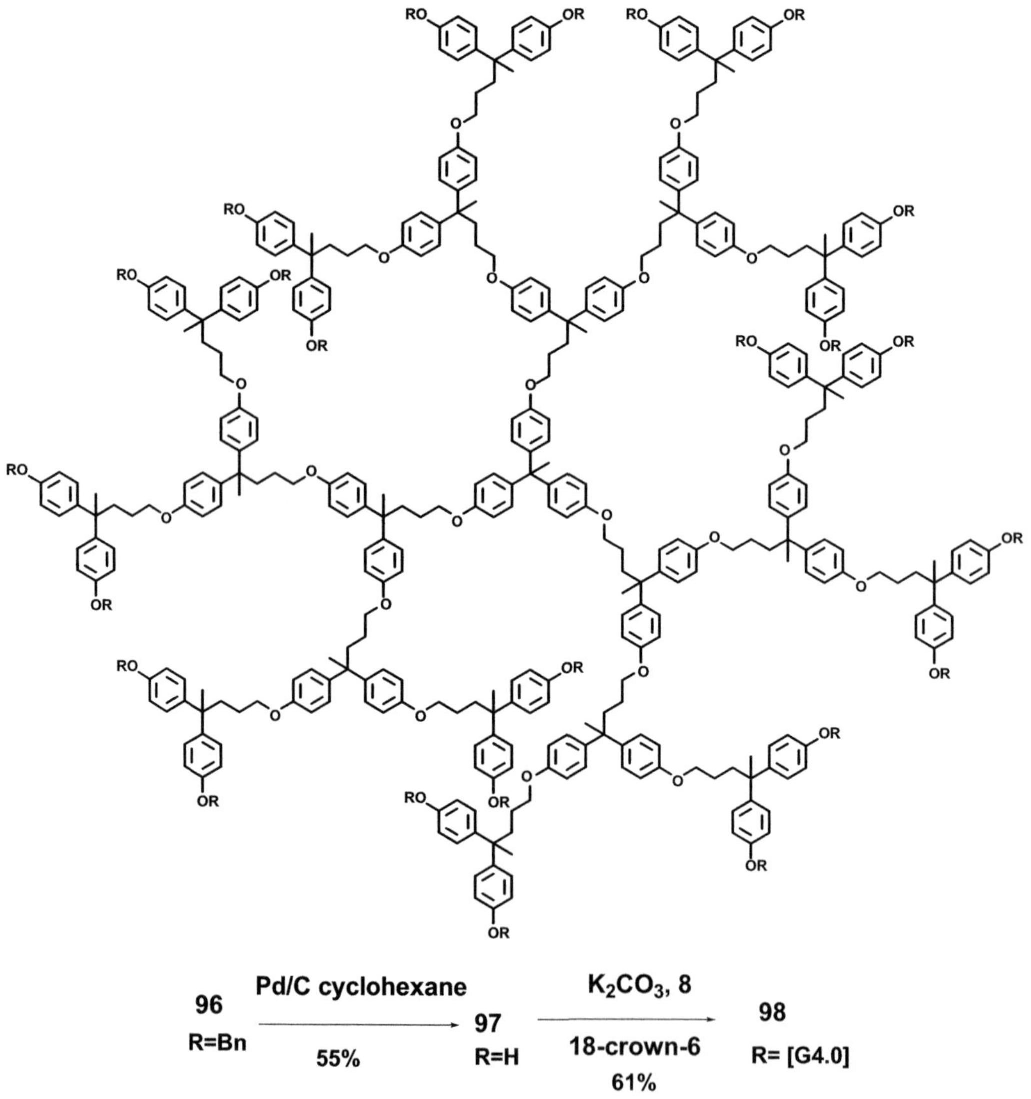

FIGURE 6.25 Synthesis of fourth generation (G4.0) dendrimers via the hypercore approach.

6.5.2 Hypermonomer Method or Branched Monomer Approach

The hypermonomer method is also recognised by the branched monomer approach. In the hypermonomer strategy, generally, monomers containing higher numbers of functional groups are being used compared to conventional monomers. Conventional monomers take more time as well as multiple steps to make desired higher generation dendrimers. Dendrimers with a similar number or a higher number of functional groups can be prepared in the comparatively few steps with the help of hypermonomers (typically of type AB_4 and AB_8) described in Figure 6.26 (Sowinska and Urbanczyk-Lipkowska 2014).

Wooley et al. synthesised fourth-generation dendron 102 (G4.0) via hypermonomer 99 containing a carboxylic acid group. Also, Labbe and co-workers described the synthesis of poly(benzylether) dendrons by using tert-butyldiphenyl silyl protected second (G2.0)- and third-generation (G3.0) hypermonomer 103 and 104 (Figure 6.27). Further activation and then coupling with benzyl bromides resulted in fifth-generation dendrons (G5.0) (80 %yields). Wooley et al. were first to use this approach for the synthesis of fifth-generation dendrimers (G5.0) via coupling of the AB_8 hypermonomer with AB_4 hypermonomer. However, the low yield obtained was attributed due to the steric hindrance and inefficient reaction conditions. Gilat et al. synthesised high generation dye labelled dendrimers by utilising highly soluble and reactive monomers (Gilat, Adronov and Fréchet 1999).

6.5.3 Double Exponential Growth

The double exponential growth coupling method was first reported by Moore et al., who described the synthesis of a fourth-generation poly(phenylacetylene) dendron (G4.0). In this method, orthogonally protected focal and peripheral functionality monomers are required. Monomers can selectively be activated either by focal points or by peripherally (Sowinska and Urbanczyk-Lipkowska 2014). In this technique, the production of the growing dendron is multiplied with the repetition of a three-step sequence that involves two selective activation reactions and one coupling reaction (G1.0-G2.0-G4.0- ...), which reduces the reaction steps (Grayson and Frechet 2001). Ihre and co-workers described the use of this approach for the synthesis of polyester dendrimers. Orthogonal groups are used for the protection of end groups and focal groups. In this, 2,2 bis-hydroxymethyl propanoic acid was used as repeating units and protected with cyclic acetonide 105, and benzyl ester was used for protection at focal group 106. DCC and DPTS were used to catalyse the coupling reaction between these two molecules (105 and 106) to produce second generation dendron 107 (G2.0). Hydrogenolysis of 107 formed hypermonomer (second generation (G2.0)) 108 and 109. Again both 108 and 109 are coupled resulting in fourth generation dendron 110 (G4.0) with 91 % yield (Figure 6.28).

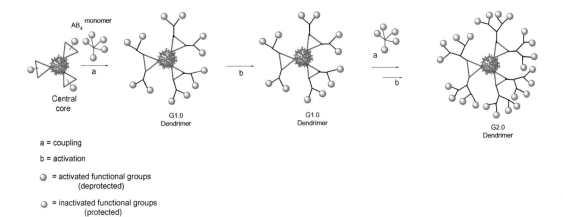

FIGURE 6.26 Synthesis of dendrimers using the hypermonomer approach.

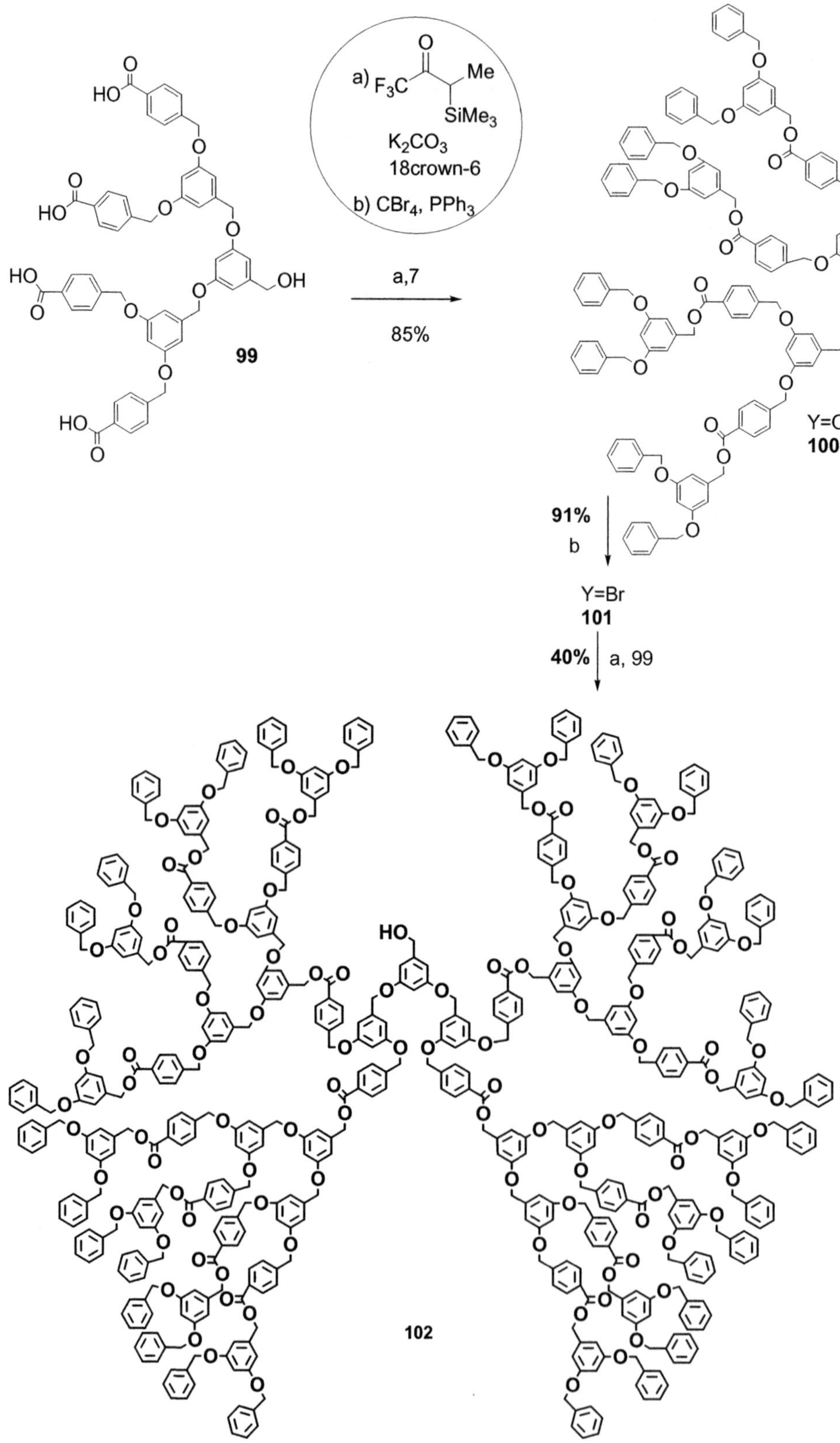

FIGURE 6.27 Synthesis of dendrimers through coupling of 103 and 104.

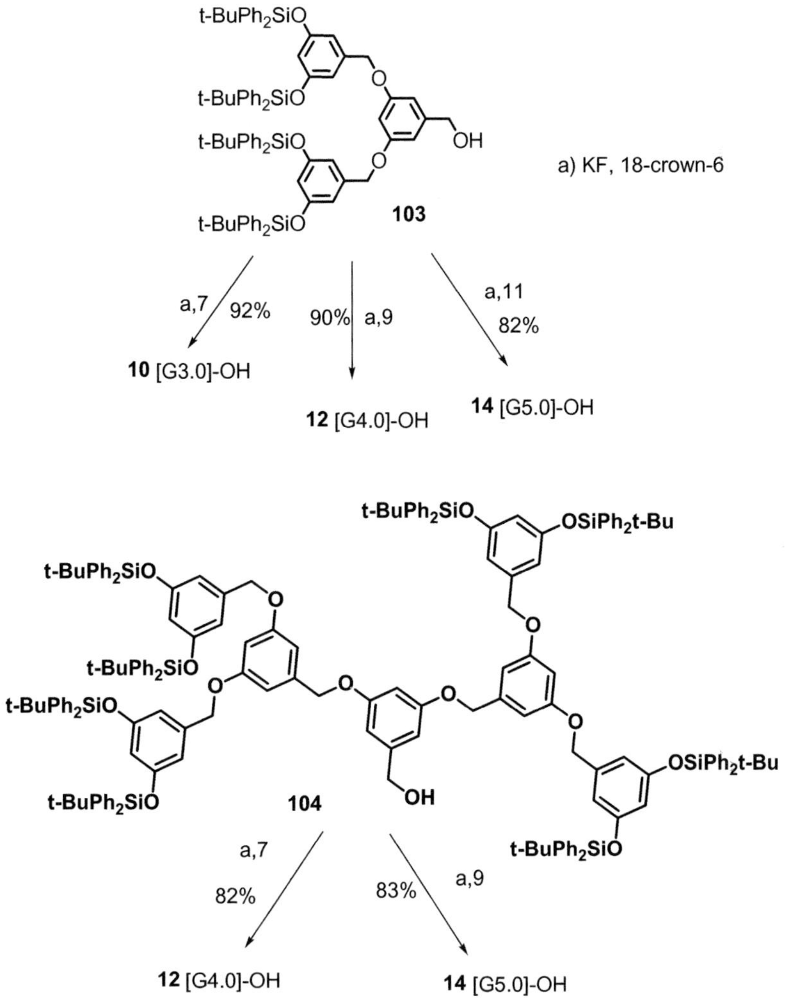

FIGURE 6.27 (Continued) Synthesis of dendrimers through coupling of 103 and 104.

6.5.4 Orthogonal Synthesis

Orthogonal synthesis is also famous by other names like the two-step approach, two monomer approach and AB_2–CD_2 approach. This process generally depends upon the divergent or convergent dendrimer growth by the use of two different monomers, specifically AB_2 and CD_2 that should have chemoselective functional groups (Grayson and Frechet 2001). The selection of AB_2 and CD_2 monomers is very critical because, in this process, the focal functionalities of every monomer will only react with the periphery of the other monomer (B couples only with C and D only with A), thus eliminating the need for activation reactions; consequently, there is a reduction in the number of steps during the dendrimer growth. The concept of orthogonal synthesis was introduced by Spindler and Frechet, who synthesised a G3.0 poly(etherurethane) dendron using 3,5-diisocyanatobenzyl chloride and 3,5-dihydroxybenzyl alcohol monomers (Zeng and Zimmerman 1996). Zimmerman and his co-worker used an orthogonal strategy for the synthesis of higher generation dendrimers. They described the synthesis route involving Mitsunobu esterification followed by Sonogashira coupling between aryliodide and alkyne groups present on surface groups as shown in Figure 6.29.

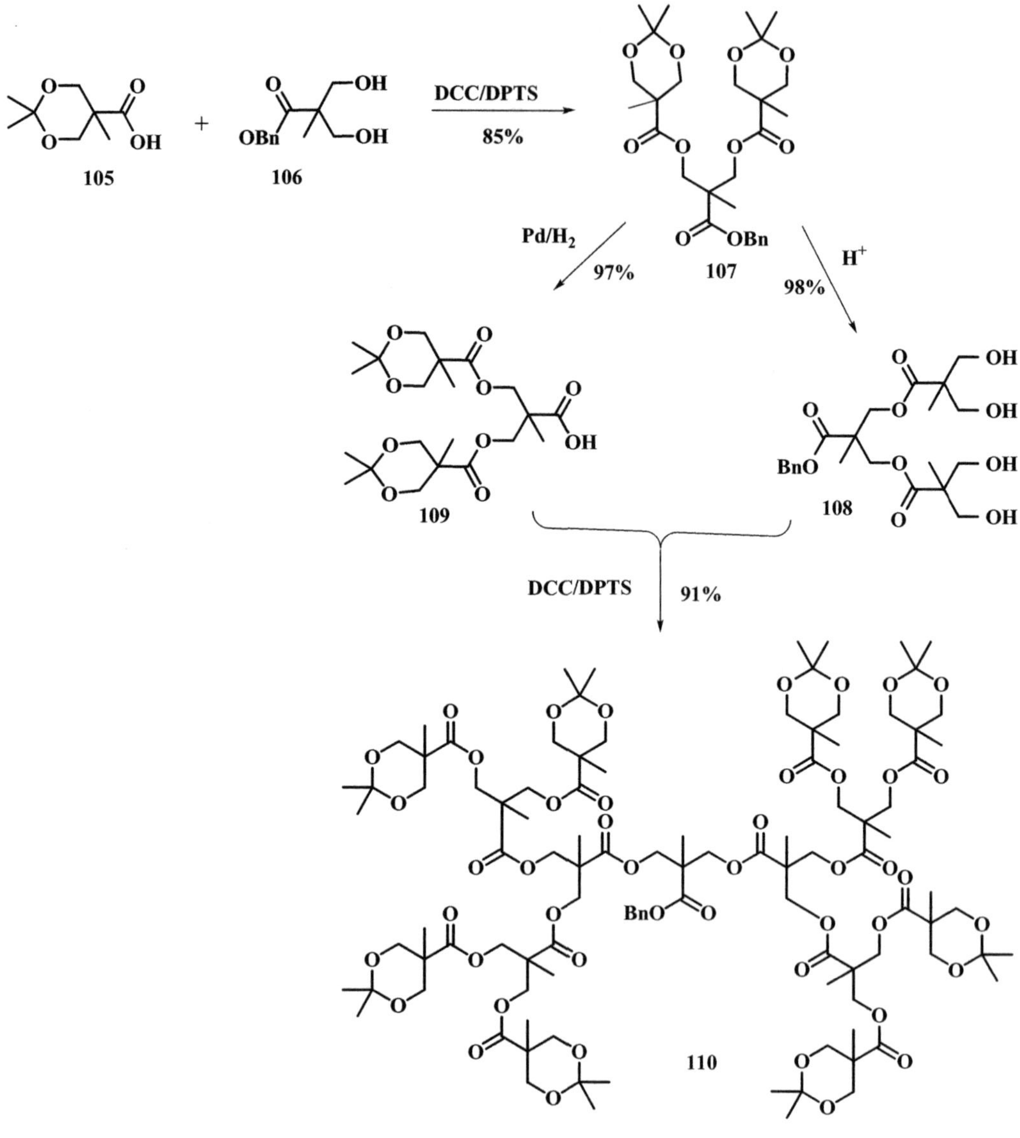

FIGURE 6.28   Synthesis of dendrimers via the double exponential growth method.

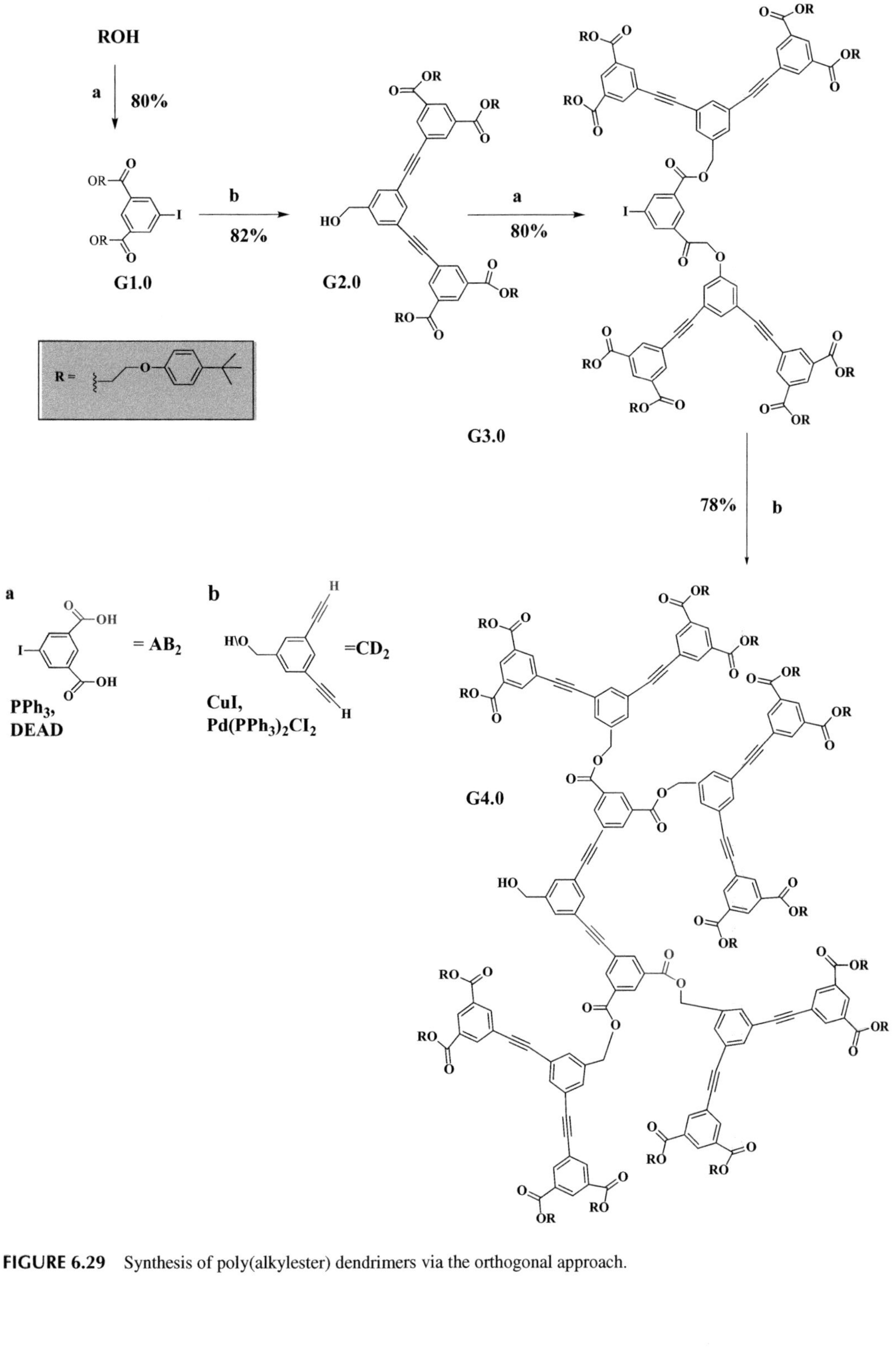

FIGURE 6.29 Synthesis of poly(alkylester) dendrimers via the orthogonal approach.

6.6 Application of Dendrimers in Drug Delivery

Dendrimers are the polymers of the twenty-first century and have gained huge attention in the development of targeted drug delivery systems. Dendrimers overcome many of the limitations of traditional drugs due to their wide properties. These classes of polymers are very effective in target delivery of drugs due to their unique property of organising their structure via controlling their size, shape, void spaces and functional groups over the peripheral area (Kannan, Nance, Kannan et al. 2014). They act as carriers by carrying the drug either into their void spaces or chemically linked/modify the functional groups present on the periphery of dendrimers. They are also able to enhance the properties of other drugs by increasing their effectiveness, site-specific delivery of drugs, decreasing their side effects, improving solubility and extending the onset of action of the drug. So they have shown a broad diversity of pharmaceutical and biological applications in the field of drug delivery systems, diagnosis, etc. (Dave and Krishna Venuganti 2017). Some of the dendrimer-based products that are commercially available in the market as well as in clinical trials are mentioned in Table 6.2.

TABLE 6.2

List of dendrimer-based products approved and under clinical trials

Product Name	Type	Role	Status	Inventor
Superfect® Reagent	Transfection agent	DNA transfection into a wide range of cell lines	Marketed	Qiagen
Stratus® CS 200 AcuteCare TM Troponin Analyser	Diagnostic agent	For cardiac assays in myocardial ischemia	Marketed	Siemens Healthcare Diagnostics
Vivagel® Condom (SPL7013 or astodrimer sodium)	Curative agents	Prevention of sexually transmitted infections (STIs)	Japan, Australia and Canada	Starpharma
Vivagel® BV	Curative agents	Prevention of recurrent bacterial vaginosis (BV) and treatment of BV	Phase-3 clinical trials	Starpharma
Vivagel® STI (vaginal microbicide)	Curative agents	Prevention of genital herpes (HSV-2), HIV and other STIs like human papillomavirus	Underway	Starpharma
DEP® Docetaxel	Targeted delivery agent	Anticancer agent	Phase 2 clinical trails	Starpharma
DEP® Cabazitaxel	Therapeutic agent (detergent free version of jevtana drug)	Anticancer agent to treat prostate cancer	Marketed and also under clinical development for other type of cancer (ovarian , breast, bladder)	Starpharma
DEP® Irinotecan	Targeted delivery agent	Anticancer agent	In phase ½ clinical trail	Starpharma
Partnered-DEP® AZD0466	Targeted delivery agent (nanomedicine formulation of Astra Zeneca)	Anticancer agent	Phase 1 clinical trial	Starpharma

6.7 Summary and Conclusion

Dendrimers are one of the important classes of polymers which are an emerging concept in the field of drug delivery systems. They offer a wide range of applications in the area of cosmetics, nanotechnology, biomedical and pharmaceuticals. Improvement in the biodistribution profile of the drug and their pharmacokinetic profile by restricting its size, shape, conformations and other parameters are the distinguishing features of these fascinating polymers as compared to other types of polymers. They act as a carrier to their target site by encapsulating the drugs or active molecules in their void spaces or attached to the functional groups present at the surface. Dendrimers are synthesised by both convergent and divergent routes. However, these existing approaches have some drawbacks like the synthesis of impure or defected dendrimers in large-scale production due to multistep synthesis. So the accelerated approach offers great advantages over the limitations of the existing approaches. This chapter covers the various synthetic strategies used for the synthesis of dendrimers of the desired generation. Advancement in their synthetic strategies will result in the synthesis of a potential dendrimer-based biocompatible molecule for the management of untreated diseases for improving the quality of life.

Abbreviations

18-crown-6	1,4,7,10,13,16-hexaoxacyclooctadecane
9-BBN	9-Borabicyclo(3.3.1)nonane
BBr$_3$	Boron tribromide
CBr$_4$	Carbontetrabromide
Cu	Copper
CuI	Copper iodide
CuSO$_4$.5H$_2$O	Copper sulfate pentahydrate
Dba	Dibenzylideneacetone
DCC	N, N′-Dicyclohexylcarbodiimide
DEAD	Diethylazodicarboxylate
DMAP	4-Dimethylaminopyrdine
DMF	Dimethylformamide
DMSO	Dimethylsulphoxide
DPTS	4-(dimethylamino)pyridinium 4-toulene sulfonate
EDA	Ethylenediamine
EDC	1-ethyl-3-(3-dimethylaminopropyl)carbodiimide
Et$_3$N	Triethylamine
G or Gen	Generation
H$_2$	Hydrogen
H$_2$O	Water
H$_2$O$_2$	Hydrogen peroxide
HCl	Hydrochloric acid
I	Iodo group
K$_2$CO$_3$	Potssium carbonate
KF	Potassium fluoride
KOC(CH$_3$)$_3$	Potassium tertiary butoxide
KOH	Potassium hydroxide
MeI	Methyl iodide
N$_2$	Nitrogen
N$_3$	Azide
Na$_2$CO$_3$	Sodium carbonate

NaH	Sodium hydride
NaN₃	Sodium azide
NaOH	Sodium hydroxide
NH	amine
OSiPh₂*t*-Bu	Tertiary butyldiphenyl silyl
OTs	Toulenesulfonyl
PAMAM	Polyamidoamine
Pd	Palladium
Pd(PPh₃)₂Cl₂	Bis(triphenylphosphine)palladium chloride
Pd(PPh₃)₄	Tetrakis(triphenylphosphine)palladium(0)
Pd/C	Palladium on carbon
Ph₃MeP⁺I⁻	Methyl triphenylphosphonium iodide
PPh₃	Triphenylposphine
S_N	Nucleophilic substitution
tert	Tertiary
THF	Tetrahydrofuran
TMS	Trimethylsilyl

ACKNOWLEDGEMENT

The authors are grateful to Dr. S.J.S Flora, Director, National Institute of Pharmaceutical Education and Research (NIPER-R) for his scientific input and constant support to compose this book chapter. NIPER-R/Communication/134.

CONFLICT OF INTEREST

The authors declare that there is no conflict of interest.

BIBLIOGRAPHY

Abasi, Elham, Sedigheh Fekri Aval, Abolfazl Akbarzadeh, et al. 2014. "Dendrimers: Synthesis, applications, and properties." *Nanoscale Research Letters* no. *9*:247. doi:10.1186/1556-276X-9-247.

Abd-El-Aziz, Alaa, and Christian Agatemor. 2017. "Emerging opportunities in the biomedical applications of dendrimers." *Journal of Inorganic and Organometallic Polymers and Materials* no. *28*. doi:10.1007/s10904-017-0768-5.

Arseneault, Mathieu, Isabelle Levesque, and Jean-François Morin. 2012. "Efficient and rapid divergent synthesis of ethylene oxide-containing dendrimers through catalyst-free click chemistry." *Macromolecules* no. *45* (9):3687–3694.

Astruc, Didier, Elodie Boisselier, and Catia Ornelas. 2010. "Dendrimers designed for functions: From physical, photophysical, and supramolecular properties to applications in sensing, catalysis, molecular electronics, photonics, and nanomedicine." *Chemical Reviews* no. *110* (4):1857–1959.

Augustus, Ebelegi Newton, Ekubo Tobin Allen, Ayawei Nimibofa, et al. 2017. "A review of synthesis, characterization and applications of functionalized dendrimers." *American Journal of Polymer Science* no. *7* (1):8–14.

Bharathi, P, Umesh Patel, Tohru Kawaguchi, et al. 1995. "Improvements in the synthesis of phenylacetylene monodendrons including a solid-phase convergent method." *Macromolecules* no. *28* (17):5955–5963.

Dave, K., and V. V. Krishna Venuganti. 2017. "Dendritic polymers for dermal drug delivery." *Therapeutic Delivery* no. *8* (12):1077–1096. doi:10.4155/tde-2017-0091.

Devadoss, Chelladurai, P Bharathi, and Jeffrey S Moore. 1996. "Energy transfer in dendritic macromolecules: Molecular size effects and the role of an energy gradient." *Journal of the American Chemical Society* no. *118* (40):9635–9644.

Dhaval, GG, MP Rinkesh, and MP Pravinkumar. 2014. "Dendrimers: Synthesis to applications: A review." *Macromolecules, MMAIJ* no. *10* (1):37–48.

Gilat, Sylvain L, Alex Adronov, and Jean MJ Fréchet. 1999. "Modular approach to the accelerated convergent growth of laser dye-labeled poly (aryl ether) dendrimers using a novel hypermonomer." *The Journal of Organic Chemistry* no. *64* (20):7474–7484.

Gonzaga, Ferdinand, Lukas P Sadowski, Talena Rambarran, et al. 2013. "Highly efficient divergent synthesis of dendrimers via metal-free "click" chemistry." *Journal of Polymer Science Part A: Polymer Chemistry* no. *51* (6):1272–1277.

Gorman, CB, and JC Smith. 2000. "Effect of repeat unit flexibility on dendrimer conformation as studied by atomistic molecular dynamics simulations." *Polymer* no. *41* (2):675–683.

Grayson, SM, and JM Frechet. 2001. "Convergent dendrons and dendrimers: From synthesis to applications." *Chemical Reviews* no. *101* (12):3819–3868. doi:10.1021/cr990116h.

Gupta, Vivek, and Surendra Nayak. 2015. "Dendrimers: A Review on Synthetic Approaches." *Journal of Applied Pharmaceutical Science* no. *5*:117–122. doi:10.7324/JAPS.2015.50321.

Han, Seung-Choul, Ji-Hyeon Kim, and Jae-Wook Lee. 2012. "Convergent synthesis of PAMAM dendrimers containing tetra (ethyleneoxide) at core using click chemistry." *Bulletin of the Korean Chemical Society* no. *33* (10):3501–3504.

Hawker, CJ, and JMJ Fréchet. 1990. *Journal of the Chemical Society, Chemical Communications* no. *15*:1010–1013.

Ihre, Henrik, Anders Hult, and Erik Söderlind. 1996. "Synthesis, characterization, and 1H NMR self-diffusion studies of dendritic aliphatic polyesters based on 2, 2-bis (hydroxymethyl) propionic acid and 1, 1, 1-tris (hydroxyphenyl) ethane." *Journal of the American Chemical Society* no. *118* (27):6388–6395.

Kannan, R. M., E. Nance, S. Kannan, et al. 2014. "Emerging concepts in dendrimer-based nanomedicine: From design principles to clinical applications." *Journal of Internal Medicine* no. *276* (6):579–617. doi:10.1111/joim.12280.

Kono, Kenji. 2012. "Dendrimer-based bionanomaterials produced by surface modification, assembly and hybrid formation." *Polymer Journal* no. *44*:531–540. doi:10.1038/pj.2012.39.

Lee, Jae Wook, Seung Choul Han, Seong-Hee Yun, et al. 2013. "Convergent synthesis of carbazole core PAMAM dendrimer via click chemistry." *Bulletin of the Korean Chemical Society* no. *34* (3):971–974.

Lehmann, Matthias, Bernhard Schartel, Manfred Hennecke, et al. 1999. "Dendrimers consisting of stilbene or distyrylbenzene building blocks synthesis and stability." *Tetrahedron* no. *55* (47):13377–13394.

Maraval, Valérie, Jaroslaw Pyzowski, Anne-Marie Caminade, et al. 2003. ""Lego" chemistry for the straight-forward synthesis of dendrimers." *The Journal of Organic Chemistry* no. *68* (15):6043–6046.

Meier, Herbert, and Matthias Lehmann. 1998. "Stilbenoid dendrimers." *Angewandte Chemie International Edition* no. *37* (5):643–645.

Mignani, Serge, Maria Bryszewska, Barbara Klajnert-Maculewicz, et al. 2015. "Advances in combination therapies based on nanoparticles for efficacious cancer treatment: An analytical report." *Biomacromolecules* no. *16* (1):1–27. doi:10.1021/bm501285t.

Montanez, Maria I, Luis M Campos, Per Antoni, et al. 2010. "Accelerated growth of dendrimers via thiol-ene and esterification reactions." *Macromolecules* no. *43* (14):6004–6013.

Moore, Jeffrey S, and Zhifu Xu. 1991. "Synthesis of rigid dendritic macromolecules: Enlarging the repeat unit size as a function of generation, permitting growth to continue." *Macromolecules* no. *24* (21):5893–5894.

Newkome, George R, Zhongqi Yao, Gregory R Baker, et al. 1985. "Micelles. Part 1. Cascade molecules: A new approach to micelles. A [27]-arborol." *The Journal of Organic Chemistry* no. *50* (11):2003–2004.

Pahovnik, David, Anja Čusak, Sebastjan Reven, et al. 2014. "Synthesis of poly (ester-amide) dendrimers based on 2, 2-Bis (hydroxymethyl) propanoic acid and glycine." *Journal of Polymer Science Part A: Polymer Chemistry* no. *52* (22):3292–3301.

Pillow, Jonathan NG, Mounir Halim, John M Lupton, et al. 1999. "A facile iterative procedure for the preparation of dendrimers containing luminescent cores and stilbene dendrons." *Macromolecules* no. *32* (19):5985–5993.

Samad, A, MI Alam, and K Saxena. 2009. "Dendrimers: A class of polymers in the nanotechnology for the delivery of active pharmaceuticals." *Current Pharmaceutical Design* no. *15* (25):2958–2969. doi:10.2174/138161209789058200.

Santos, Ana, Francisco Veiga, and Ana Figueiras. 2019. "Dendrimers as pharmaceutical excipients: Synthesis, properties, toxicity and biomedical applications." *Materials* no. *13*:65. doi:10.3390/ma13010065.

Sherje, A. P., M. Jadhav, B. R. Dravyakar, et al. 2018. "Dendrimers: A versatile nanocarrier for drug delivery and targeting." *International Journal of Pharmaceutics* no. *548* (1):707–720. doi:10.1016/j.ijpharm.2018.07.030.

Shifrina, Zinaida B, Marina S Rajadurai, Nina V Firsova, et al. 2005. "Poly (phenylene-pyridyl) dendrimers: Synthesis and templating of metal nanoparticles." *Macromolecules* no. *38* (24):9920–9932.

Sowinska, Marta, and Zofia Urbanczyk-Lipkowska. 2014. "Advances in the chemistry of dendrimers." *New Journal of Chemistry* no. *38* (6):2168–2203.

Tang, Zitao. 2017. "Research progress on synthesis and characteristic about dendrimers." *IOP Conference Series: Earth and Environmental Science* no. *100*:012024. doi:10.1088/1755-1315/100/1/012024.

Tiwari, Amit, Virendra Gajbhiye, Rajeev Sharma, et al. 2010. "Carrier mediated protein and peptide stabilization." *Drug Delivery* no. *17*:605–616. doi:10.3109/10717544.2010.509359.

Tomalia, Donald A, H Baker, J Dewald, et al. 1985. "A new class of polymers: Starburst-dendritic macromolecules." *Polymer Journal* no. *17* (1):117.

Tomalia, Donald A, Jørn B Christensen, and Ulrik Boas. 2012. *Dendrimers, Dendrons, and Dendritic Polymers: Discovery, Applications, and thes Future.* Cambridge University Press.

Villalonga-Barber, Carolina, Maria Micha-Screttas, Barry R Steele, et al. 2008. "Dendrimers as biopharmaceuticals: Synthesis and properties." *Current Topics in Medicinal Chemistry* no. *8* (14):1294–1309.

Walter, Marie V, and Michael Malkoch. 2012. "Simplifying the synthesis of dendrimers: Accelerated approaches." *Chemical Society Reviews* no. *41* (13):4593–4609.

Zeng, Fanwen, and Steven C Zimmerman. 1996. "Rapid synthesis of dendrimers by an orthogonal coupling strategy." *Journal of the American Chemical Society* no. *118* (22):5326–5327.

7

Functionalisation of Dendrimers

Divya Bharti Rai, Deep Pooja and Hitesh Kulhari

CONTENTS

7.1 Introduction

An ideal drug delivery carrier system should exhibit balance between aqueous solubility and permeability, high drug loading capacity, biodegradability, low toxicity and favourable retention with appropriate specificity and bioavailability. These characteristics are not fully covered by the currently available dendrimers.

Lower generation dendrimers exhibit low toxicity but cannot hold a sufficient amount of drug and show burst release of drug. On the other side, higher generation dendrimers suffer from the drawbacks of toxicity and high cost. Therefore, the most used generation of dendrimers is G3.5 to G5.0. Moreover, researchers are modifying the dendrimer surface as per the requirements. For biological application, the modification or functionalisation of dendrimers is for two basic objectives: first, to improve their biocompatibility and minimise toxicity associated with them and second, targeted dendrimers are synthesised for site-specific delivering of drugs. For other applications of dendrimers like in catalysis, supramolecular chemistry, etc., the internal structures of dendrimers such as core and branching units are also functionalised (Kulhari et al. 2015).

Functionalisation of dendrimers is the process of incorporating multiple active sites into dendrimers to create macromolecules with multifunctional architecture. Functionalised dendrimers are also known as structurally controlled dendrimers that have at least six well-defined nanoscale features known as critical nanoscale design parameters (CNDPs) such as size, shape, surface chemistry, flexibility, rigidity, architecture and elemental composition (Hawker and Frechet 1990). These CNDPs may be thoroughly manipulated to produce dendrimers with new emerging properties that may be desirable and critical for many industrial and commercial applications. Owing to systematic considerations, functionalisation is usually introduced into the dendrimer framework at either the core or periphery or even both (Hecht 2003).

7.2 Functionalisation of Dendrimers

Functionalisation of dendrimers is the modification of dendrimers by adding functional groups at sites within the dendrimer structure in order to increase their usefulness considering their applications (Hawker and Frechet 1990; Kono 2012). Functionalised dendrimers are tailored by incorporating functional groups at any of the constituent part, i.e., central core, inner branches and peripheral terminal groups (Figure 7.1) to index properties such as solubility, thermal stability and addition of compounds for particular applications (Tomalia 2016).

7.2.1 Functionalising Interior of Dendrimers

Depending on the nature of the synthesis, functionalisation is traditionally introduced at the core, the periphery or both. However, the specific incorporation of functional groups at the interior layers, i.e., generations, represents a considerable synthetic hurdle that must be overcome for the full potential of

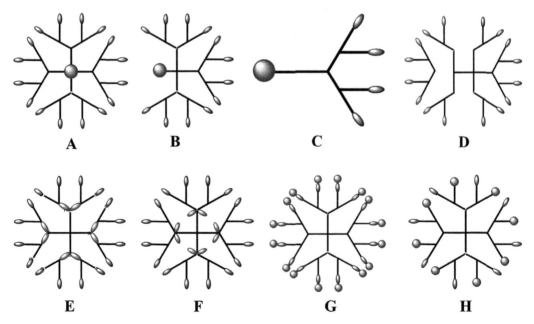

FIGURE 7.1 Types of bi-functional dendrimers showing core functionalisation different from the terminal functionalisation (A–C), branch functionalisation different from the terminal functionalisation (E and F) and two types of terminal functions (D, G and H).

dendrimers to be realised. Internal functionalisation dramatically increases the degree of complexity that can be implemented into a dendrimer macromolecule and, therefore, promises to lead to smart materials for future applications in bio- and nanotechnologies (Hecht 2003). Most examples concerning internally functionalised dendrimers are based on covalent modifications using organic building blocks, for example, the poly(benzyl ether) type which is typically synthesised by the classical convergent approach pioneered by Hawker and Frechet (Piotti et al. 1999).

The design of purely aliphatic dendrimers, although synthetically in some cases much more challenging, offers certain advantages, such as chemical robustness and ultraviolet (UV) transparency, necessary for specific applications, that is, catalysis and UV-photon harvesting, respectively. One important contribution to this area was made by Ford and co-workers, who investigated the catalytic activity of poly(propylene imine) (PPI) dendrimers that carried quaternised internal ammonium ions (Pan and Ford 2000). During their initial divergent synthesis of unimolecular hydrocarbon dendrimers with peripheral carboxylate groups, the authors used Wurtz couplings of branched acetylide fragments to a polyfunctional alkyl bromide core. Rather than removing the triple bond by complete hydrogenation as originally reported, the reaction with decaborane was used for the specific incorporation of *o*-carborane moieties within dendrimers. Furthermore, the molecule was rendered water-soluble by the introduction of sulphate groups at the periphery of dendrimers. These materials, allowing for high local concentrations of boron clusters in an aqueous environment, represent interesting candidates for disease treatment based on boron neutron capture therapy (Newkome et al. 1991).

7.2.2 Functionalising the Periphery of Dendrimers

Surface-functionalised dendrimers provide an opportunity to circumvent structural obstacles and obtain an effective delivery process. Dendrimers are functionalised with polyethylene glycol (PEG), acetyl anhydride, lipids, amino acids, vitamins, peptides, antibodies, aptamers and various pH, temperature, enzyme, redox and photosensitive polymers. These surface modified dendrimers show promising potential in drug and gene delivery (Jevprasesphant et al. 2003a). Lipid and amino acid-functionalised dendrimers exhibit high transfection efficiency due to the fusogenic property of the lipids and increasing buffering capacity of the histidine and imidazole groups (Guillot-Nieckowski et al. 2007). Peptide/protein-modified dendrimers are exploited for increased cellular uptake-mediated endocytosis into the specific site of tumour cells. Conjugation of peptide/protein, antibody and aptamers on the dendrimer surface as targeted ligands induces site-specific delivery of nanovectors to cancer cells that overexpress specific receptors. Stimuli-responsive dendrimers impart by conjugating various types of stimulus sensitive polymers on the dendrimer surface and exhibit controlled release based on the difference in biological conditions between cancer cells and normal cells (Calderón et al. 2010). Dendrimers are surface-functionalised for various purposes including dendrimer enhancement of transfection efficiency, avoiding uncontrollable release, inducing site-specific delivery of dendrimers based on various overexpressed receptors on cancer cells and improving biocompatibility of dendrimers in vivo (Ghaffari et al. 2018).

7.3 Various Types of Functionalisation Strategies

The commonly used strategies for surface functionalisation of dendrimers are as follows:

7.3.1 PEGylation

PEG is a US Food and Drug Administration approved and non-ionic polymer. PEG is being widely used in drug delivery because of its several advantages such as hydrophilicity, biodegradability, biocompatibility, low intrinsic toxicity and stealth behaviour. PEG can be tagged to the dendrimers to prevent the body's defence mechanisms from recognising and eliminating it. This allows the PEGylated delivery system to circulate for a prolonged period of time thereby increasing the opportunities for the drug to reach the required site of action (Knop et al. 2010).

7.3.2 Acetylation

Acetylation is another way for reducing dendrimer toxicity through conjugation to the terminal surface group which results in neutralisation of positive charges on the dendrimer surface (Cheng et al. 2011). Acetylation compared to PEGylation has several benefits including (i) high efficiency and simplicity of acetylation, (ii) low desirable amount of acetylation due to selection of an appropriate ratio of acetic anhydride and dendrimers and (iii) low steric hindrance of acetylation compared to the PEG chain. Besides, these acetylated dendrimers can cross the cell membrane and maintain permeability across the cell membrane but PEGylated one shows low cellular uptake (Wang et al. 2012).

7.3.3 Functionalisation with Vitamins

The glycosylphosphatidylinositol-anchored glycoprotein-folate generates a biomolecular signal, which is required in cell proliferation. They are overexpressed in various disease conditions like cancer and inflammation. For example, cancer cells and activated macrophages overexpress folate receptors and thus give an opportunity to design dendrimers conjugated with folic acid for targeted therapy (Yi 2016).

Essential micronutrients like biotin are required for certain functions of cells like gluconeogenesis, fatty acids, amino acid metabolism, growth and development. Biotin levels are extremely high in rapidly proliferating cells like cancer. Thus, researchers have used the strategy of conjugating the dendrimers with biotin for targeting them to cancerous cells via sodium-dependent multivitamin transporter uptake (Yellepeddi et al. 2011).

7.3.4 Amino acid/ Peptide/Protein Conjugation

Ongoing research in many fields is utilising proteins, peptides and even amino acids as a ligand for targeting a specific organ or tissue. Dendrimers have been conjugated with a wide range of amino acids, peptides and proteins. Therefore, dendrimer surfaces have been modified with these biomolecules for drug delivery purposes in cancer and human immunodeficiency virus (HIV) treatments. For example, Tuftsin, a tetrapeptide (Thr-Lys-Pro-Arg), increases natural killer activity of macrophages, monocytes and leukocytes by exhibiting specific binding to them. Tuftsin is an integral part of Fc portion of immunoglobulin IgG having dual advantages as a targeting ligand and phagocytosis inducer. Therefore, dendrimers with this peptide help in targeting and improving the therapeutic efficacy of the payload employed (Shukla et al. 2005).

Substituted amino acids are also being used for function of dendrimers. *N*-acetyl cysteine (NAC) is known to have clinical application by scavenging the reactive oxygen species through conversion to intracellular glutathione. NAC has wider applications in the treatment of cancer, HIV, heart disorders, etc. (Issels et al. 1988).

7.3.5 Carbohydrate Functionalisation

Receptors like lectin and asialoglycoprotein are overexpressed in many cancer cells, bacteria and virus cell membranes. Many carbohydrate molecules like lactobionic acid, fucose, mannose, galactose, lactose, sialic acid, etc., can be used as ligands to target various receptors that are overexpressed in various cancers and infectious conditions. Thus, they can be exploited for conjugation as ligands to target particular cancer or infection.

Hyaluronic acid (HA) is a linear mucopolysaccharide with inherent tumour-targeting potential through exhibiting binding efficiency to CD44, cell-surface glycoprotein known to overexpress in many tumours like ovarian, colon, lung, breast, etc, (Mattheolabakis et al. 2015). Kesharwani et al. synthesised HA-conjugated dendrimers for targeting of curcumin towards pancreatic cancer (MiaPaCa-2 cell line). They have used the derivative of curcumin, 3,4-difluorobenzylidene curcumin which was reported to have superior anti-cancer activity to curcumin. HA-conjugated dendritic formulations showed better cytotoxicity and cellular uptake. Receptor blockade assay showcased that endocytosis occurred because of binding of HA to CD44. It was concluded that even the cationic charge of G4.0 polyamidoamine (PAMAM) dendrimers is reduced because of conjugation with HA that will help in reduction of toxicity and thus the potential of dendrimer's usage into clinical translation (Kesharwani et al. 2015).

7.3.6 Lipid Functionalisation

According to hydrophobic nature of the cell membrane and intracellular vesicles, functionalisation of dendrimers with lipids results in both high cellular uptake and increased endosomal escape. Lipids, for example, cholesterol and fatty acid are fusogenic lipids and enhance cellular penetration of dendrimers in cancer cells (Mandeville et al. 2013).

The properties of dendrimers can be modified using ligands made of fatty acids like lauryl chains which help in decreasing the toxicity profile and improve the permeability of dendrimers across the biological membranes (Jevprasesphant et al. 2003b).

7.3.7 Antibody Grafting

Use of monoclonal antibodies (mAb) is a type of targeted immunotherapy with tumours overexpressing antigens or proteins known as tumour-associated antigens. By employing a mAb on the surface of the carrier, it can be directed towards tumours and avoiding their access to other sites. It is advantageous over other targeting ligands because other ligands are also inhibited competitively by the presence of endogenous moieties (Allen 2002).

7.3.8 Inorganic Ion Derivatisation

Research by Hayder et al. studied surface-modified G1.0 dendrimers with phosphorus-containing moieties. These modified dendrimers showed their ability in modulating innate immunity by exhibiting their effects on monocytes. These dendrimers are posed to have dual action in both inflammation and osteoclastogenesis which paves the opportunity for dendrimers to be used in rheumatoid arthritis (Hayder et al. 2011). Dendrimers with aldehyde on their surface are conjugated with phosphoric acid to give phosphonate modified dendrimers. The Poupot/Caminade group from France has adapted this approach and claimed that polyvalency of phosphorous G1.0 dendrimers with 16 phosphonate groups is important for biological activity with *in vitro* results, showing the decreased monocyte activation and increased number of natural killer cells (Griffe et al. 2007; Rolland et al. 2008).

The anionic surface created by sulphonation helps in increasing the binding efficiency of dendrimers to virus and also inhibit virus from invading the host cells thus avoiding infection (Javan et al. 1997; Price et al. 2011; O'Loughlin et al. 2009).

Fluorinated dendrimers are a new class of highly efficient gene vectors, and internalisation of fluorine on drug molecules increases their uptake and enhances metabolic stability and binding affinity towards the protein. Besides, fluorinated dendrimers exhibit improved transfection efficiency which was attributed to increased cellular uptake, serum stability and endosomal escape (Cai et al. 2016). These materials exhibit significant DNA condensation and protection ability with proper sizes and zeta-potentials. In addition, fluorination causes reducing positive charge density on dendrimers with excellent uptake and endosomal escape. Fluorination increases therapeutic efficacy, and pharmacokinetic behaviour of numerous drugs can be attributed to fluorination of dendrimers. One of the major advantages of fluorination of dendrimers is their low surface energy which leads to the formation of a complex with gene at low concentration and development of polyplex in an extremely low nitrogen to phosphorous (N/P) ratio. Therefore, gene transfection at a low N/P ratio results in low toxicity in the transfected cells (Johnson et al. 2016).

7.3.9 Miscellaneous

p-hydroxybenzoic acid (pHBA) is a small molecule with an affinity towards sigma receptors that overexpress in brain tumour. Swami et al. conjugated pHBA onto G4.0 PAMAM dendrimers loaded with docetaxel for targeting to brain. The prepared conjugates were characterised and preceded for evaluation of *in-vitro* studies on U87MG brain tumour cells and *in vivo* studies on mice for determining pharmacokinetic and biodistribution in comparison to marketed formulation Taxotere. Results demonstrated that pHBA-conjugated dendrimers showed superior activity in terms of cytotoxicity and cellular uptake. *In vivo* studies for conjugated formulations showed a two-fold increase in drug concentration in the brain when compared to unconjugated dendrimers (Swami et al. 2015).

7.4 Application of Functionalisation of Dendrimers

Dendrimers have been functionalised for various applications. These applications are presented in Figure 7.2 and listed in Table 7.1.

7.4.1 To Improve Cellular Penetration Across the Biological Membranes

Many ideas have been raised about the size of the dendrimers and the ability of terminal groups associated with lipid molecules, based on the assembly of dendrimer–lipid vesicles. The radius of the lipid heads (L) in contact with the surface of the dendrimers and the number of terminal groups of the dendrimers (P) appear to be the factors in the formation of the pore. For example, G7.0 PAMAM dendrimers support low ratios (L/P) allowing the formation of stable dendrimer–lipid vesicles (Mecke et al. 2005). In a study, it was showed that cationic G7.0 PAMAM dendrimers lead to the formation of a single pore in the membrane fluid phase, while the existence of a gel phase in the plasma membrane is not affected by the presence of these dendrimers, suggesting that the phase of the lipid bilayer could impact cellular uptake studies under experimental conditions such as chemical functionality and charge of the dendrimers (Wolinsky and Grinstaff 2008).

Teow et al. worked on G3.0 PAMAM dendrimers where they have conjugated with lauryl chains which are further grafted with paclitaxel (PTX) and fluorescein isothiocyanate (FITC) to check the increase

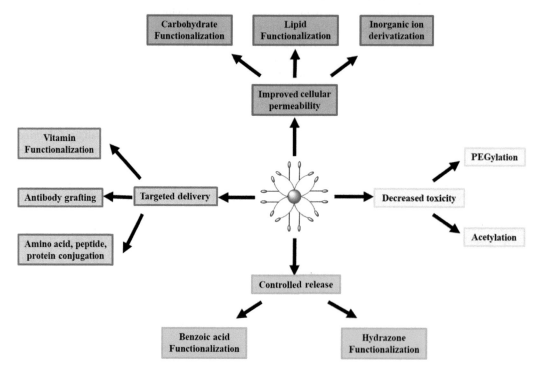

FIGURE 7.2   Functionalisation of the dendrimer surface for various applications.

TABLE 7.1

Functionalised dendrimer-based delivery of therapeutic molecules for various applications

Significance	Dendrimers	Drug Delivered	Targeting Ligand	Reference
Increased cellular permeability	G3.0 PAMAM	Paclitaxel	Lauroyl	Teow et al. (2013)
	G2.0 Polylysine	Camptothecin	PEG	Fox et al. (2009)
	G5.0 PAMAM	–	Galactose, lactose and NAG	Liu et al. (2012a); Medina et al. (2011)
	G5.0 PAMAM		NAG	Medina et al. (2011)
	G5.0 PAMAM	MTX	Folic acid	Thomas et al. (2011); O'Loughlin et al. (2009)
	Dendrimer	Pyrimethamine sulfadoxine	Transductive peptide	Moulton (2012)
Decreased non-specific toxicity	G4.0 PAMAM	Curcumin	HA	Kesharwani et al. (2015)
	G1.5–G9.5 PAMAM	–	Lysine, Arginine	Malik et al. (2000)
	G5.0 PPI		Amino acid	Agashe et al. (2006)
	G5.0 PAMAM	Adriamycin and MTX	PEG	Kojima et al. (2000)
	G3.0, G4.0 Polylysine	Doxorubicin	PEG	Kaminskas et al. (2012)
	G4.0 PAMAM	–	Sialic acid	Landers et al. (2002)
	G5.0 PPI	Amphotericin B	Muramyl dipeptide	Jain et al. (2015a)
	G5.0 PPI	primaquine phosphate	Galactose	Bhadra et al. (2005)
	G4.0 Polylysine	Chloroquine phosphate	Glycodendrimers	Agrawal et al. (2007)
	G1.0 ABP	–	Phosphonate	Fruchon et al. (2015)
	G4.0 PAMAM	ErbB antibody	PEG	Yoon et al. (2016)
	G5.0 PPI	Efavirenz, Lamivudine	Glycodendrimers	Dutta et al. (2007)
Controlled release	G4.0 PAMAM	Doxorubicin	PEG	Nie et al. (2016)
	G4.0 PAMAM	Adriamycin	PEG	Kono et al. (2008)
	G6.0 Polylysine	Doxorubicin	PEG	Niidome et al. (2014)
	G5.0 Polylysine	Doxorubicin	PEG	Kaminskas et al. (2011)
	G4.0 PAMAM	Doxorubicin	Hydrazone	Zhong et al. (2016)
Improved biodegradability	G1.0, G2.0, G3.0 PGLSA	–	Glycerol-succinic	Grinstaff (2002)
	G4.0, G5.0 Polyester dendrimers	–	PEG	Heyder et al. (2017)

(Continued)

TABLE 7.1 (*Continued*)

Functionalised dendrimer-based delivery of therapeutic molecules for various applications

Significance	Dendrimers	Drug Delivered	Targeting Ligand	Reference
Site/disease-specific targeting	G4.0 PAMAM	Docetaxel	pHBA	Swami et al. (2015)
	G4.0 PAMAM	5-fluorouracil	Folic acid	Singh et al. (2008)
	G4.0 PAMAM	Methotrexate	NT peptide	Falciani et al. (2007)
	G5.0 PAMAM	–	RGD	Shukla et al. (2005)
	PAMAM		Avidin	Dhanikula et al. (2008)
	G2.0PAMAM	Colchicine	β-glucose, α-galactose, N-acetyl-galactose, or lactose	Lagnoux et al. (2005)
	Glycodendrimers	Doxorubicin	Lactobionic acid	Fanfan et al. (2014)
	G4.0 PAMAM	Docetaxel	Trastuzumab	Kulhari et al. (2016)
	G3.5 PAMAM	Paclitaxel	Monoclonal antibody mAbK1	Jain et al. (2015b)
	PAMAM	–	Tryptophan	Buceta et al. (2015)
	G3.0, G4.0 Phosphorous dendrimers	–	HIV peptides	Ciepluch et al. (2014)
	G 3.0 PPI	Efavirenz, Lamivudine	Glycodendrimers	Dutta and Jain (2007)
	G5.0 PPI	Efavirenz	Tuftsin	Dutta et al. (2008)
	SPL7013	–	Sulphate	Javan et al. (1997)
	G4.0 PAMAM	Cy5.5.	Folic acid	Benchaala et al. (2014)
	G3.0 PAMAM	–	Carboxylic acid	Yamashita et al. (2017)
	G1.0 PAMAM	Naproxen	Poly(aspartic acid)	Ouyang et al. (2009)
	Janus-type dendrimer	5-fluorouracil	RGD	Jiang et al. (2013)
	G5.0 PAMAM	Fluorescein	c(RGDyK)	Hill et al. (2007)
	G4.0 PAMAM	Doxorubicin	Angiopep-2 (Ang2) peptide, EGFR-targeting peptide (EP-1)	Liu et al. (2019)
Combinational therapy	G4.0, G6.0 Cationic PAMAM	Rhodamine B isothiocyanate (RITC)	PEG	Santos et al. (2018)
	G4.0 PAMAM	Valproic Acid	N-acetyl-cysteine	Mishra et al. (2014)
	PAMAM	–	N-acetyl-cysteine	Bae and Park (2011)
	G4.0 PAMAM	siRNA, chemotherapeutic drug cis-diaminodichlorido platinum	Folic acid	Amreddy et al. (2018)
	G4.0 PAMAM	siRNA, doxorubicin	PEG	Biswas et al. (2013)
	G3.0 Polylysine	Doxorubicin, anti-tumour gene (pORF-hTRAIL)	T7	Liu et al. (2012b)
Biomimicking	G4.0 PAMAM	Hydroxyapatite	Dimethyl phosphate	Chen et al. (2014)
	G3.5 PAMAM	Glucosamine and glucosamine 6 sulfate	Anionic ligands	Shaunak et al. (2004)

in permeability using Caco-2 (human epithelial colorectal adenocarcinoma) cells. Results showed an increase of permeability by 12% of lauryl dendrimer conjugate in comparison to plain PTX alone (Teow et al. 2013).

Fox et al. showed that PEGylation of G2.0 polylysine dendrimer–camptothecin conjugate increased not only the circulation half-life but also the increased tumour uptake of the delivery system in comparison to free drug (Fox et al. 2009).

Liu Xiaopeng et al. demonstrated that glycans can be conjugated onto G5.0 PAMAM dendrimers and showed their binding activity *in vitro* on the HepG2 (human hepatic cancer cells) cell line. To study the glycodendrimer targeting efficiency, a small library of dendrimer surfaces modified with different glycans like galactose, lactose and *N*-acetyl galactosamine (NAG) were synthesised and tagged with fluorescein dye and checked for avidity on HepG2 cells. Results showed that dendrimers conjugated with NAG have significant specificity and binding efficacy (Liu et al. 2012b).

A similar study was performed by Medina et al. using G5.0 PAMAM dendrimers conjugated with NAG and checked for its targeting potential in HepG2 cells. They evaluated the effect of ligand selectivity, concentration, time of incubation and number of NAG molecules anchored on the uptake studies using HepG2. Results showed the role of asialoglycoprotein receptors (ASGPRs) in endocytosing the dendritic conjugated formulations, and it was also observed that conjugation of 12 NAG molecules is sufficient for efficient uptake by 100% cells within a short period (Medina et al. 2011).

Thomas et al. explored that G5.0 PAMAM dendrimers were loaded with methotrexate (MTX) and conjugated to folic acid for targeting to the macrophages which overexpress folate receptors under inflammatory conditions. Both *in vitro* and *in vivo* studies were carried out for establishing the targeting potential. Results revealed that folic acid conjugated dendrimers showed high internalisation in macrophages *in vitro* and enhanced anti-inflammatory by reducing arthritis-induced inflammatory parameters *in vivo* (Thomas et al. 2011).

Toxoplasma gondii is a pathogen that causes morbidity and mortality. Pyrimethamine and sulphadoxine are currently being used for the treatment of toxoplasmosis. The main problem with the drugs is that they do not eliminate the parasite because *Toxoplasma gondii* encysted bradyzoites. Therefore, dendrimer-based formulation was used that can effectively treat *Toxoplasma gondii* infection by crossing the host cell membrane, the parasitophorous vacuole and the tachyzoite membranes. Transductive peptide dendrimers are potential therapeutics because they can transport small bioactive molecules across multiple membranes through intracellular tachyzoites and encysted bradyzoites and they can also enhance the toxicity of the drugs (Moulton 2012).

7.4.2 To Decrease Toxicity of Dendrimers

Plain dendrimers show cytotoxicity depending on the polarity of the surface of the terminal groups. One can observe that the toxicity of dendrimers is dependent on the chemistry of the core but is more strongly influenced by the nature of the surface of the dendrimers. Dendrimers, with polar membranes non-invasive as PEG groups, have a non-toxic behaviour. In contrast, non-polar groups such as lipids interact with the cell membrane by hydrophobic interactions; this results in some cases cytotoxic dendrimers. Lipids (especially glycosphingolipids) may also have a positive influence because they can provide the dendrimers with immunostimulatory properties (Sebestik et al. 2011). Low generation dendrimers have a surface more accessible to end groups, which in the older generation is sterically hindered due to the high accumulation of terminal groups on the surface, the increase in branching (generation) and high surface coverage with biocompatible end groups like PEG are being widely used to create less toxic dendrimers (Hong et al. 2006).

For example, the cytotoxicity measured by the 3-(4,5-dimethylthiazolyl-2)-2, 5-diphenyltetrazolium bromide) assay (MTT assay) of a library of dendrimers based on melamine surfaces, including amines, guanidines, carboxylate, sulphonate, phosphonate, or PEGylated, showed that cationic dendrimers were much more cytotoxic than anionic dendrimers or PEGylated ones (Chen et al. 2004). Similarly, quaternaries PAMAM-OH showed lower levels of cytotoxicity that PAMAM-NH$_2$, given that the inner cationic charges were sealed by the surface of the hydroxyl groups (Lee et al. 2003). The surface modification of G4.0 PAMAM with lysine or arginine leads to the increase in toxicity compared to G4.0 PAMAM native, which is attributed to increased charge density and molecular size. It has also been found that anionic PAMAM dendrimers G1.5 – 9.5 and dendrimers DAB with surfaces –COOH are not cytotoxic in HepG2, CCRF and B16F10 cells, at a concentration of up to 5 mg/mL (72 h, MTT assay). Scanning electron microscopy (SEM) images of B16F10 cells exposed to these anionic dendrimers showed no morphological changes (Malik et al. 2000).

Many studies compared the cytotoxicity effect of plain G5.0 PPI dendrimers with amino acid and carbohydrate functionalised dendrimers on HePG2 and cos-7 cell lines. They reported a dose and time-dependent decrease in cell viability and concluded that it was attributed to positive charges on the dendrimer surface (Agashe et al. 2006). For example, Kojima et al. worked with G3.0 and G4.0 poly-lysine dendrimers with an equivalent ratio of surface conjugation with PEG1100 and doxorubicin (DOX). Dendritic formulations were compared with PEGylated liposomes in terms of anti-tumour efficacy and toxicity profile (rat and mouse tumour models). Results showed the potential of dendritic formulations in reducing the systemic toxicity of DOX, i.e., cardiotoxicity, extended plasma half-life (>30 h) and also reduced uptake of DOX to liver and spleen when compared to the liposomal formulations (Kojima et al. 2000; Kaminskas et al. 2012).

For prevention of influenza infection, sialic acids (SAs) are used to prevent viral attachment but they are prone to enzymatic cleavage which demands a higher concentration of SAs. At higher concentrations, SAs show toxicity. Therefore, as an alternative G4.0 PAMAM dendrimers have been used to conjugate sialic acid that can inhibit influenza haemagglutinin protein binding at non-toxic concentrations. Landers et al. adapted three approaches- *in vitro, in vivo* and computational modelling to show the inhibition of viral protein binding, and results have shown that the conjugate can prevent pulmonary infection caused by influenza A virus (Landers et al. 2002).

Functionalised dendrimers were also tried and tested in the treatment of parasitic infection such as leishmaniasis. Jain et al. encapsulated a common anti-parasitic drug, amphotericin B into G5.0 PPI dendrimers functionalised with muramyl dipeptide. The drug–dendrimer toxicity was determined by in vivo studies on macrophage cell lines and balb/c mice. Dendrimer–amphotericin B formulation showed less toxicity in comparison with the commercial formulation (Jain et al. 2015b).

Galactose, sugar-coated G5.0 PPI dendrimers have been synthesised for improving the delivery of primaquine phosphate. Results revealed that galactosylation has decreased not only the toxicity but also prolonged the drug release from 1–2 days of uncoated dendrimers to 5–6 days for galactose-coated dendrimers. Plasma studies showed that galactosylated formulations are targeted to liver with a prolonged period of circulation time and decreased haemolytic toxicity (Bhadra et al. 2005). In a separate study, galactose was conjugated to G4.0 polylysine dendrimers having PEG-1000 as the core for delivering chloroquine phosphate. The effect of galactose coating on safety, stability and efficacy was evaluated both *in vitro* and *in vivo*. There was a five-fold decrease in uptake of galactose-coated dendrimers by macrophages when compared to uncoated dendrimers, which is helpful to safeguard the dendrimers from macrophage action and target to liver. Further galactose coating decreased the toxicity of native dendrimers (Agrawal et al. 2007).

In the rheumatoid arthritis mouse model, the fully capped G1.0 phosphonate dendrimers with 24 phosphonate groups showed a dose-dependent decrease in pro-inflammatory cytokine levels from splenocytes, an increase in anti-inflammatory cytokine levels and the anti-osteoclastic activity which all will make phosphonate dendrimers a valuable tool in the treatment of rheumatoid arthritis. These fully capped bisphosphonate dendrimers were also found to be non-toxic in non-human primates up to 10 mg/kg for 4 weeks with intervals weekly when injected intravenously (Fruchon et al. 2015).

Yoon et al. addressed the limitation of systematically administered naked adenovirus, which causes hepatotoxicity and immunotoxicity that were overcome by use of epidermal growth factor receptor (EGFR) precise antibody (ErbB) and PEGylated conjugated G4.0 PAMAM dendrimer. It was demonstrated that exposure of adenovirus was avoided by the steric hindrance of PEGylated antibody conjugated dendrimers which specifically targeted the EGFR positive tumour cells and reduced immunotoxicity (Yoon et al. 2016).

Dutta et al. worked extensively on t-Boc–glycine and mannosylated G5.0 PPI dendrimers loaded with efavirenz for targeting the lectin receptors on macrophages, which act as main reservoirs for spreading of the virus in HIV infected persons. Toxicity of the PPI dendrimers was also decreased after conjugating with t-Boc–glycine and mannose. Cellular uptake studies on macrophages/monocytes showed increased uptake of conjugated dendrimers in comparison to plain dendrimers (Dutta et al. 2007).

7.4.3 To Achieve Controlled Drug Release

In general, dendrimers designed for controlled release should provide a release drug level in an optimum range over a long period compared to other methods of drug delivery. This increases the effectiveness of the drug and improves the possibility of using highly toxic drugs, poorly soluble or relatively unstable drugs. The first consideration in drug delivery is to achieve more effective therapies while eliminating the potential for overdose or inadequate dose. Another advantage of using controlled release systems includes the maintenance of drug levels within the desired range, the need for fewer administrations, optimal use of the drug in question and increased patient confidence. This is a particularly relevant disease which required chronic treatment like cancer, rheumatoid arthritis, etc (Alexis et al. 2008). Dendrimers released encapsulated molecules via a surface diffusion or swelling, in a time- or a condition-dependent manner. The release of the active agent may be constant over a long period, cyclical over a period, or can be triggered by environmental events or other external events, such as changes in pH and temperature, like folate-conjugated liposomes or Poly(N-isopropylacrylamide-co-propylacrylic acid) copolymers that respond to temperature and pH (Brown et al. 1996).

For example, hyperbranched PEGylated G4.0 PAMAM dendrimers have been grafted with DOX via acid-sensitive cis acotinyl bonds to release the drug in acidic pH near the tumour site. The selective targeting of designed formulation to gastric cancer cells was demonstrated by performing *in vitro* studies on mucinous gastric adenocarcinoma MGC803 cells and normal human foetal gastric epithelial cells. *In vivo* studies were performed on a mouse gastric cancer xenograft model. Results showed that conjugates are not cytotoxic to normal cells but showed cytotoxicity on tumour cells and *in vivo* studies showed anti-tumour efficacy in terms of tumour reduction in addition to reduced systemic toxicity (Nie et al. 2016). Kona et al. used G4.0 PAMAM dendrimers grafted with PEG and loaded with Adriamycin using glutamic acid as the linker through amide or hydrazone bonds (Kono et al. 2008).

Niidome et al. used G6.0 polylysine dendrimers and conjugated them with PEG pentapeptide to impart the hydrophobic core for enhancing the loading of DOX. The encapsulated DOX showed pH-dependent slow release. *In vivo* results showed tumour suppression with the conjugated dendrimers. Thus, the authors suggested that incorporating a hydrophobic core with pentapeptide is a valuable tool in delivering anticancer drugs to tumour sites (Niidome et al. 2014).

For the synthesis of stimuli-responsive formulations, PEGylated G5.0 polylysine dendrimers were grafted with DOX using acid labile 4-(hydrazinosulphonyl) benzoic acid (HSBA) linker which showed less than 10% release in pH 7.4, whereas nearly 100% drug release was observed at pH 5. *In vitro* cytotoxicity results of the DOX-loaded dendrimers and free DOX were similar. *In vivo* studies showed that DOX-loaded PEGylated dendrimers showed greater uptake in tumour tissue which indicates that acidic conditions in the tumour site provoked drug release and hence, more uptake of drug by tumours in comparison to other organs (Kaminskas et al. 2011).

In a study by Zhong et al., G4.0 PAMAM-DOX-loaded dendrimers were prepared for the treatment of lung metastasis. PAMAM and DOX were covalently conjugated using acid-labile hydrazone bonds. The PAMAM-DOX dendrimers showed particle sizes of 9.7 nm with a zeta potential of +13.8 mV, which is beneficial for improving cellular internalisation in pulmonary epithelia. The PAMAM–DOX dendrimers showed greater stability at extracellular physiological pH (pH 7.4) and sustained release at mild acidic pH (lysosomal pH 5). The acid-labile hydrazone linkage is responsible for the pH-dependent release of DOX. In acidic medium, more than 80% DOX was released over a period of 48 h, whereas an insignificant amount (4%) of DOX was released at pH 7.4 (Zhong et al. 2016).

7.4.4 To Enhance Biodegradability of Dendrimers

The need for biodegradable dendrimers gave rise to a strategy to produce the desired high molecular weight carrier to reach a high accumulation and retention in the diseased site which, in turn, allows the rapid and safe removal of the dendrimer fragments in the urine and avoids the non-specific toxicity

(Duncan and Izzo 2005). The biodegradable dendrimers are usually prepared by inclusion of ester groups in the polymer structure which will be chemically hydrolysed and/or enzymatically cleaved by esterases in physiological solutions. Consequently, the following four factors seem to control the rate of degradation of the dendrimers: (1) the chemical nature of the bridges connecting the monomer units with ester bonds, being the most susceptible to hydrolysis compared with the amide and ether bonds, (2) hydrophobicity of monomer units, where more hydrophilic polymeric units e.g., glycerol, lactic acid or succinic acid functionalisation result in rapid degradation compared to the hydrophobic monomers e.g., phenylalanine and alkyl amines, (3) the fact that dendrimers with large sizes and molecular weights are degraded more slowly compared to smaller given the tight packing of its surface, which effectively seals the hydrolysable surface functional groups and (4) the susceptibility of internal and external functional groups of the dendrimers to hydrolysis leads to faster degradation (Lee et al. 2006).

Grinstaff et al. compared the rate of degradation of dendrimers G1.0 polyester (poly (glycerol-succinic acid)) (PGLSA) in the presence of acid, base and esterases, with polyester-amide dendrimers (G2.0) and polyester-ether (G3.0) to identify factors that control its degradation kinetics under physiological conditions. The results of this study showed that the polyester–ether dendrimers were the fastest to degrade due to the increased hydrolytic susceptibility compared to polyester and polyester–amide functionalities (Grinstaff 2002).

For example, PEGylation modifies the degradation pattern of the dendrimers, exhibiting progression in a step-wise manner; first, the PEG layer is degraded, and afterwards, the dendrimers start shedding. Heyder et al. studied the effects of generation and surface PEGylation of polyester dendrimers on pulmonary cell transport. Hydroxyl-terminated polyester dendrimers were prepared using a divergent method and PEGylated. During an *in vitro* transport study using a polarised calu-3 epithelial monolayer, the PEGylated G-4 dendrimers showed poor internalisation in the epithelium but showed effective transport across cell monolayers compared to G3.0 polyester dendrimers. The PEGylated G4.0 dendrimers showed 4 and 2.3-fold improvements in cellular transport from the apical to basolateral side compared to non-PEGylated G4.0 and G3.0 dendrimers, respectively (Heyder et al. 2017).

7.4.5 To Target Therapeutic Molecules to the Diseased Site

In the biomedical field, dendrimers have been investigated for their applications as inherently active therapeutic agents and as vectors for targeted delivery of drugs, peptides and oligonucleotides or imaging agents. Targeted delivery means delivering of active therapeutic molecules to a specific site like infected organs, tissues or cells. Targeting is important to improve therapeutic efficacy of the therapy and to reduce the side effects of the payload. The polyvalent surface of dendrimers makes it an ideal structure for designing a system to overcome the biological barriers and deliver drugs to or near the target site. Many researchers have shown the potential of dendrimers for targeting various ailments like cancer, rheumatoid arthritis, bone disorders and many microbial infections (Bae and Park 2011; Teow and Valiyaveettil 2010).

Ideally after administration, an anticancer drug should be first able to achieve the desired tumour tissue through crossing of the barriers in the body with minimal loss of volume or activity in the bloodstream. Second, after reaching the tumour tissue, the drug should have the ability to selectively kill tumour cells without affecting normal cells, with a release mechanism controlled actively (Wolinsky and Grinstaff 2008). Drug targeting to the specific site occurs by two phenomena, passive and active targeting. In passive targeting, the drug or delivery system reaches and accumulates at a particular location or site because of the prevailing *in situ* abnormal physiological or pathological conditions like for example enhanced permeation and retention effect and acidic pH in the vicinity of tumour environment, and thus, these types of conditions help in targeting the medicament to treat that particular pathological condition existing in the body (Bazak et al. 2014).

While the passive distribution results in preferential accumulation of drug in tumours, non-specific delivery may also occur in healthy organs. Therefore, a large area of research involves the functionalisation of nanoparticles with specific fractions as monoclonal antibodies or ligands such as vitamins, carbohydrate residues or peptides, which identify and bind to receptors either overexpressed in tumours or in the endothelium associated with them, maximising the localisation and accumulation in the tumour

(Huynh et al. 2009). While designing a dendrimer-based system against tumours consisting of the drug and targeting molecules, certain factors must be considered to create efficient delivery systems. First, antigen or receptor should be expressed exclusively in the tumour cell and not expressed in normal cells. Second, antigen or receptor should be expressed uniformly in all tumour cells. Finally, the antigen or receptor should not be released into the bloodstream. The internalisation of the conjugate after binding to the tumour cell is an important criterion in selecting an appropriate targeting ligand. Dendrimers can be engineered for specific targeting by its surface modification.

For example, folate is a small organic molecule with high affinity to the folic acid receptors, highly overexpressed in certain tumours like brain, breast, kidney, lung and head and neck (100 to 300 times above endogenous levels). PEGylated G4.0 dendrimers functionalised with folate containing 5-fluoro-uracil (5-FU) were showed to have a high accumulation in tumour of 20.1% and 10% of the injected dose at 8 and 24 hours, respectively, resulting in a significant reduction of tumour growth approximately 40% compared with non-functionalised controls after two injections (days 0 and 7) over 20 days (Singh et al. 2008).

Other researchers have used different agents to reach the tumour including peptides to lead to the drug delivery systems based on dendrimers to specific tumour receptors. Falciani et al. used neurotensin (NT) peptides to develop a G4.0 dendrimer marked with NT carrying chlorine 6 (Cle6) and the chemotherapeutic agent MTX for various malignant tumours that express the receptor for NT, which includes carcinomas of colon, pancreatic, prostate and small cell lung. The treatment of mice carrying the tumour HT29 with conjugated dendrimers-MTX-NT for 20 days showed a reduction in tumour size in approximately one third in comparison to mice that received saline or conjugate without MTX, indicating the therapeutic benefit of the active approach (Falciani et al. 2007). Shukla et al. conjugated G5.0 PAMAM dendrimers with Arg-Gly-Asp (RGD) peptide and checked *in vitro* for their tumour targeting efficiency. They performed cellular uptake studies in cell lines like human umbilical vein endothelial cells and Jurkat cell lines which are known to have a high number of integrin receptors. Results showed that the conjugate was taken up by the cell lines indicating binding affinity of the conjugate to cancer cells (Shukla et al. 2005).

Other conjugated dendrimers used the tetrameric glycoprotein avidin to reach lectins differentially expressed on the surface of ovarian carcinoma cells (Dhanikula et al. 2008). Lagnoux et al. synthesised G2.0 glycodendrimers where dendrimers are surface decorated with four or more glycans like β-glucose, α-galactose, N-acetyl-galactose, or lactose and their branches were incorporated with amino acids like Ser, Thr, His, Asp, Glu, Leu, Val and Phe with cysteine moieties in the core which is used for attaching colchicine through disulphide linkage. These glycodendrimers were then evaluated for their cytotoxicity against HeLa cells (human cervical cancer cells) and non-transformed mouse embryonic fibroblasts. Results displayed that the glycoconjugated dendrimers showed more selectivity towards HeLa than mouse fibroblasts. This selectivity in terms of cell death was 163-fold more in HeLa cells than fibroblasts (Lagnoux et al. 2005). Lactobionic acid (LA), a diasaccharide consisted of gluconic acid and galactose, was conjugated to dendrimers for targeting hepatic cancer overexpressing ASGPRs. They have made surface conjugation sequentially using FITC and lactobionic acid or PEGylated LA and loaded with DOX. Remaining amine functional groups have been masked by acetylation. Authors concluded that use of PEG spacer has improved the encapsulation, release, targeting power and therapeutic efficacy of the DOX (Fanfan et al. 2014).

Kulhari et al. used G4.0 PAMAM dendrimers encapsulated with docetaxel (DTX) and conjugated using PEG linker with trastuzumab (TZ), which binds to human epidermal growth factor receptor type 2 (HER2), overexpressing in breast cancer. Two human breast cancer cells, i.e., MDA-MB-453 and MDA-MB-231 cells which are HER2 positive and negative respectively for checking the uptake of the unconjugated and conjugated formulations. The antibody-conjugated dendrimers showed enhanced uptake in comparison to unconjugated dendrimers in HER2 positive cells, on the contrary, no difference in uptake was observed in case of HER2 negative cells which explains the targeting efficiency of TZ. The cytotoxicity results also followed the same trend in which antibody conjugated formulations showed 3.6 folds higher than unconjugated dendrimers in HER2 positive cells (Kulhari et al. 2016). Jain et al. used carboxylic acid terminated G3.5 PAMAM dendrimers, loaded with paclitaxel and conjugated with monoclonal antibody mAbK1, mesothelin protein which is overexpressed in few cancers like ovarian. *In vitro* cytotoxicity studies were performed on OVCAR-3 (human ovarian cancer cell line) which showed an increase in cytotoxicity in comparison to plain dendrimers and free drug. *In vivo* studies showed a better survival rate and good tumour inhibition in case of antibody-conjugated dendrimers when compared to plain dendrimers and drug alone (Jain et al. 2015a).

Dendrimers have been functionalised for designing the formulations for the treatment of different virus infections. Buceta et al. used 9 to 18 tryptophan moieties to conjugate on PAMAM dendrimers and checked against HIV replication. Authors found that the conjugated dendrimers inhibited viral attachment to host surface by interacting with the glycoproteins gp120 and gp41of HIV envelop. They stated that 9 tryptophan residues are sufficient for efficient binding of dendrimers and thus anti-HIV activity (Buceta et al. 2015). In a study, HIV peptides Gp160, P24 and Nef were conjugated to two different types of phosphorus-containing dendrimers (G3.0 and G4.0) and evaluated for immunotherapy against HIV (Ciepluch et al. 2014). In a study, mannosylated G5.0 PPI dendrimers loaded with lamivudine were formulated and evaluated for anti-retroviral activity by estimating p24 antigen using MT2 cell lines along with cellular uptake studies where mannosylated dendrimers showed 21 fold higher uptake in comparison to free drug (Dutta and Jain 2007).

Likewise, tuftsin conjugated G5.0 PPI dendrimers showed the targeting potential of tuftsin in treatment of HIV using efavarinz, anti-HIV drug. It was found that, within 1 h, the cellular uptake of drug was increased by 34.5 times in the case of conjugate in comparison to pure drug. Accumulation of the conjugate was higher in infected macrophages in comparison to normal cells (Dutta et al. 2008).

The surface amine groups of dendrimers were sulphated to block the chemokine receptors CCR5 and CXCR4 which are crucial for viral way into CD4+ cells after binding of HIV-1 envelope gp120 to human cell surface CD4 (Javan et al. 1997). Starpharma developed a completely sulphated form of dendrimers SPL7013 which is formulated as aqueous gel into a mucoadhesive polymer Carbopol® that can be applied as a topical agent. The anionic surface charge of the SPL7013 is known to block the viral attachment to gp120 of human cells.

For the targeting of inflammatory arthritis caused by *Chlamydia trachomatis* known as Chlamydia-induced reactive arthritis, folate was conjugated with cyanine 5.5 (Cy 5.5) loaded G4.0 PAMAM dendrimers as the folate receptors are known to be overexpressed in inflammatory sites. Results showed the increased concentration of folate conjugated dendritic formulations by 3–4 fold in infected and inflammatory sites, i.e., paws and genital tracts compared to plain dendrimers (Benchaala et al. 2014).

Efficient bone targeting by a PAMAM dendrimer was successfully achieved by carboxylic acids. For example, G3.0 PAMAM dendrimers of third generation were conjugated with four different carboxylic acid molecules aspartic, glutamic, succinic and aconitic acids and were further PEGylated. *In vitro* studies were carried out to check their binding affinities to hydroxyapatite and calcium ions. Bone targeting and deposition were checked with dendrimers radiolabeled with ^{111}In *in vivo*. Results indicated that dendrimers conjugated with aspartic acid showed superior results among the others in bone targeting and bone deposition pattern associated with the pathogenesis of bone diseases such as rheumatoid arthritis and osteoporosis (Yamashita et al. 2017).

Similar bone targeting of naproxen prodrugs with poly(aspartic acid) moieties and with two and three poly(aspartic acid) sequences G1.0 peptide dendrimers was carried out. The modified naproxen conjugates were incubated with hydroxyapatite in PBS under physiological conditions over 16h. The study revealed the hydroxyapatite binding properties of poly(aspartic acid), and it was found that the peptide dendrimer prodrugs exhibited a faster initial binding and a greater total binding. The obtained binding data in vitro indicated that the peptide dendrimers with poly(aspartic acid) sequences were useful for the development of new bone targeting molecules for drug delivery to bone (Ouyang et al. 2009).

The RGD sequence has a potential dual-targeting property, with which the chemotherapeutic agents can be delivered to bone tissue and tumour. In this study, a Janus-type dendrimer which consisted of RGD dimer and 5-FU dimer was synthesised and characterised by ^{1}H-NMR and MS techniques. The hydroxyapatite binding assay and drug release study were performed. It was observed that Janus-type dendrimers with both an enhanced targeting property and optimised release property reduced the side effects in normal tissues and had a potential sustained-release effect. Thus, it may represent a novel opportunity to apply the versatile Janus-type dual-targeting delivery system to applications in bone tumour treatment (Jiang et al. 2013).

Dendrimers could be applied as a minimally invasive method of management of dental caries. Hill et al. demonstrated the use of cyclic RGD-G5.0 PAMAM for dental implants. They have synthesised the conjugate of c(RGDyK) labelled with fluorescein with PAMAM and evaluated for their binding and uptake properties in mouse dental papilla cell-23 (MDPC-23) cells which are odontoblast-like cells. Selective binding of the conjugate to integrin receptors was shown in *in vitro* cellular uptake studies conducted on human dermal micro-vessel endothelial cells, human vascular endothelial cells or MDPC-23 cells. It has been demonstrated that PAMAM dendrimers can be used to deliver RGD to tooth that can help in odontoblast regeneration (Hill et al. 2007).

Drug delivery to the central nervous system (CNS) is restricted by the blood–brain barrier (BBB). The BBB is mainly formed depending on the complex tight junctions between the adjacent endothelial cells and the other constituents, including extracellular matrix, astrocytes and pericytes, which can prevent diffusion of the toxic foreign substances into the brain parenchyma. However, it also prevents the penetration of therapeutic drugs such as the drugs against gliomas, Alzheimer's disease or Parkinson's disease into the CNS (Liu et al. 2019). However, with the onset of stroke, the BBB becomes leaky, providing a window of opportunity to passively target the brain. Cationic PAMAM of different generations was functionalised with PEG to reduce cytotoxicity and prolong blood circulation half-life, aiming for a safe *in vivo* drug delivery system in a stroke scenario. Rhodamine B isothiocyanate (RITC) was covalently tethered to the G4.0 and G6.0 dendrimers backbone and used as a small surrogate drug as well as for tracking purposes. The biocompatibility of PAMAM was markedly increased by PEGylation as a function of dendrimer generation and degree of functionalisation. The PEGylated RITC-modified dendrimers did not affect the integrity of an *in vitro* BBB model (Santos et al. 2018).

Another example is that of Liu et al., who developed an inclusion complex MTX-G4.0 dendrimer polyether co-polyester (PEPE) for the treatment of glioma. Glioblastoma multiform is a primary tumour type in the CNS with high malignancy, poor prognosis and high mortality. Hence, ligand D-glucosamine disposed in the conjugate to link the transporters GLUT-1 highly expressed on the luminal side of endothelial cells of the BBB and glioma cancer cells. In vitro studies showed that the conjugate MTX-PEPE exhibited 2 to 8 times greater accumulation in glioma cells, resulting in effectiveness of 2 to 4.5 times greater than non-conjugated dendrimers (Liu et al. 2019).

Mishra et al., evaluated for the first time the potential of OH-terminated G4.0 PAMAM dendrimers in reaching the damaged brain cells of dogs. Authors claim that the high dose of valproic acid and NAC required for neuroprotection can be reduced with the use of dendrimers. *In vivo* efficacy studies with reduced side effects showed the importance of drug dendrimer conjugates, i.e., NAC-dendrimer and dendrimer-PEG-valproic Acid conjugate in improving the neurological protection activity (Mishra et al. 2014).

Rett syndrome is a rare genetic brain disorder affecting girls. NAC conjugated with anionic PAMAM dendrimers was used in the treatment of this disorder. Bae et al. performed *in vitro* studies on glial cells from Methyl-CpG-binding protein 2 mouse gene (Mecp2)-null and wild type mice and *in vivo* on Mecp2-nullmice showed improved immune regulation and effective down phasing of Rett syndrome phenotype after treatment with the targeted dendritic formulations (Bae and Park 2011).

7.4.6 To Co-Deliver Multiple Therapeutic Agents

Polyvalency and availability of multiple binding sites in the dendrimer structure enable conjugation of multiple therapeutic agents in a single dendrimer molecule.

Amreddy et al. delivered two different therapeutic agents, i.e., a siRNA (small interfering RNA) and chemotherapeutic drug cis-diaminodichlorido platinum using folate conjugated G4.0 dendrimers targeting to lung cancer cells. RNA binding protein human antigen R (HuR) which is overexpressed in lung cancer is silenced using siRNA. Combining the drug with siRNA is proposed to give an additive effect. *In vitro* studies like cellular uptake, cytotoxicity and molecular studies were carried out. Results revealed that efficient targeting of the dendritic formulations was observed in case of folate-conjugated dendrimers and concluded that the delivery system has potential for *in vivo* studies on orthotopic induced lung cancer models (Amreddy et al. 2018).

Likewise, G4.0 PAMAM dendrimers have been modified with PEG and 1,2-dioleoyl-sn-glycero-3-phosphoethanolamine, i.e., G4.0-PEG-DOPE for co-delivery of a drug and a siRNA. The prepared formulations were evaluated for cytotoxicity, cellular internalisation, gel retardation, ethidium bromide exclusion assay and serum stability studies. The results indicate that incorporation of siRNA helps in preventing P-glycoprotein (Pgp) efflux of drug and is thus useful in multidrug resistance cancer. Use of long PEG chains helps in protecting siRNA from enzymatic degradation. Results show that G4.0-PEG-DOPE is more efficient in targeting siRNA to cancer cells than the micelles, whereas MDM nanomicelles delivered more DOX than G4.0-PEG-DOPE (Biswas et al. 2013).

Liu et al. worked on T7 peptide (His-Ala-Ile-Tyr-Pro-Arg-His), a ligand for transferrin receptor conjugated dendrimers loaded with anti-tumour gene (pORF-hTRAIL) and DOX. DOX was loaded onto the surface of G3.0 dendritic polylysine using an acid-sensitive spacer with glutamic acid. The designed polyfunctional dendrimer is subjected to both *in vivo* and *in vitro* studies. Results revealed the synergistic effect of gene and chemotherapy drugs in terms of apoptosis assay performed in both *in vitro* and *in vivo* conditions. Pharmacodynamic studies showed the potential of targeted co-delivery of gene and drug with pH-triggered drug release at the tumour site. The dose of DOX was reduced substantially with multifunctional dendrimers at which the side effects of DOX like weight loss and cardiotoxicity were negligible in mice (Liu et al. 2012a).

7.4.7 To Mimic the Biological System

Functionalised dendrimers are analogous to protein, enzymes and viruses. In a study, G4.0 PAMAM dendrimers are modified by dimethyl phosphate to obtain phosphate-terminated PAMAM dendrimers (PAMAM-PO$_3$H$_2$) to mimic the structure and surface components of amelogenin protein. Amelogenin is essential for amelogenesis, i.e., development of tooth enamel. The terminal phosphate group of PAMAM-PO$_3$H$_2$ tightly binds enamel and adsorbs calcium ions for deposition on the enamel similar to re-mineralisation of hydroxyapatite (HP), as illustrated by the adsorption test. Moreover, PAMAM-PO$_3$H$_2$ exhibited low toxicity in MTT assay. The enamel incubated with PAMAM-PO$_3$H$_2$ for 20 days was found to have 11.23 thick deposition of HP on the enamel as shown by SEM and X-ray diffraction analysis. The artificial regeneration of tooth was also clinically validated in vivo in the rat oral cavity. Thus, PAMAM-PO$_3$H$_2$ shows great potential as a biocompatible alternative of natural tooth enamel (Chen et al. 2014).

Shaunak et al. used anionic G3.5 PAMAM dendrimers to make water-soluble glucosamine and glucosamine 6-sulphate conjugates and showed that conjugates have immunomodulatory and anti-angiogenic properties, which are helpful in prevention of scar tissue formation after surgery. Dendrimer glucosamine inhibited toll-like receptor 4-mediated lipopolysaccharide-induced synthesis of pro-inflammatory chemokines (MIP-1 alpha, MIP-1 beta and IL-8) and cytokines (TNF-alpha, IL-1 beta and IL-6) from human dendritic cells and macrophages but allowed upregulation of the costimulatory molecules CD25, CD80, CD83 and CD86. Dendrimers glucosamine 6-sulphate blocked fibroblast growth factor-2 mediated endothelial cell proliferation and neoangiogenesis in human Matrigel and placental angiogenesis assays. In an animal model study, glucosamine and glucosamine 6-sulphate loaded dendrimers were administered post glaucoma filtration surgery in rabbits for evaluating the scar tissue formation. The results showed 30% to 80% rapid regeneration of scar tissue (Shaunak et al. 2004).

7.5 Conclusions

Plain dendrimers have limited therapeutic performance in drug delivery. Therefore, various strategies have been used for functionalisation of dendrimers in the core or at the surface. Modified dendrimers provide several advantages including improved cellular penetration, decreased non-specific toxicity, controlled drug release, better biodegradability, combinational therapy, bio-mimicking and disease-specific targeting of therapeutic drugs. Different classes of molecules can be used to functionalise cavity or surface moieties of dendrimers such as PEG, vitamins, amino acid, peptides, proteins, carbohydrate, lipid, antibody, inorganic ions, etc.

Abbreviations

5-FU	5-fluorouracil
ASGPR	Asialoglycoprotein receptors
BBB	Blood-brain barrier
Caco-2	Human epithelial colorectal adenocarcinoma cells
Cle6	Chlorine 6
CNDPs	Critical nanoscale design parameters
CNS	Central nervous system
Cy5.5	Cyanine 5.5
DOPE	1,2-dioleoyl-sn-glycero-3-phosphoethanolamine
DOX	Doxorubicin
DTX	Docetaxel
EGFR	Epidermal growth factor receptor
EPR	Enhanced permeation and retention
ErbB	Epidermal growth factor receptor (EGFR) precise antibody
FITC	Fluorescein isothiocyanate
GES-1	Human fetal gastric epithelial cells
GPI	Glycosylphosphatidylinositol
HA	Hyaluronic acid
HDMEC	Human dermal microvessel endothelial cells
HeLa cells	Human cervical cancer cells
HepG2 cells	Human hepatic cancer cells
HER2	Human epidermal growth factor receptor type 2
HP	Hydroxyapatite
HSBA	4-(hydrazinosulfonyl) benzoic acid
HuR	Human antigen R
HUVEC	Human umbilical vein endothelial cells
HVEC	Human vascular endothelial cells
LA	Lactobionic acid
mAb	Monoclonal antibodies
MDPC-23	Mouse dental papilla cell-23
Mecp2	Methyl-CpG-binding protein 2 mouse gene
MGC803 cells	Mucinous gastric adenocarcinoma cells
MTT assay	3-(4,5-dimethylthiazolyl-2)-2,5-diphenyltetrazolium bromide) assay
MTX	Methotrexate
NAC	*N*-acetyl cysteine
NAG	*N*-acetyl galactosamine
NT	Neurotensin
OVCAR-3	Human ovarian cancer cell line
PAMAM	Polyamidoamine
PAMAM-PO$_3$H$_2$	Phosphate-terminated dendrimer
PEG	Polyethylene glycol
PEPE	Polyether copolyester
PGLSA	Poly (glycerol-succinic acid)
Pgp	P-glycoprotein
pHBA	p-hydroxybenzoic acid
PPI	Poly (propyleneimine)
PTX	Paclitaxel
RITC	Rhodamine B isothiocyanate
ROS	Reactive oxygen species
SAs	Sialic acids

SEM	Scanning electron microscope
siRNA	Small interfering RNA
SMVT	Sodium-dependent multivitamin
TZ	Trastuzumab
USFDA	U.S. Food and Drug Administration

ACKNOWLEDGEMENT

The authors acknowledge the Central University of Gujarat, Gandhinagar for providing the necessary facilities and support. HK acknowledges the Department of Science and Technology, New Delhi for the INSPIRE Faculty Award. DBR acknowledges University Grant Commission, New Delhi, India for PhD Fellowship.

REFERENCES

Agashe, H.B., Dutta, T., Garg, M., Jain, N.K. 2006. Investigations on the Toxicological Profile of Functionalized Fifth-Generation Poly(Propylene Imine) Dendrimer. *Journal of Pharmacy and Pharmacology 58* (11): 1491–1498.

Agrawal, P., Gupta, U., Jain, N.K. 2007. Glycoconjugated Peptide Dendrimers-Based Nanoparticulate System for the Delivery of Chloroquine Phosphate. *Biomaterials 28* (22): 3349–3359.

Alexis, F., Rhee, J.-W., Richie, J.P., Radovic-Moreno, A.F., Langer, R., Farokhzad, O.C. 2008. New Frontiers in Nanotechnology for Cancer Treatment. *Urologic Oncology: Seminars and Original Investigations 26* (1): 74–85.

Allen, T.M. 2002. Ligand-Targeted Therapeutics in Anticancer Therapy. *Nature Reviews Cancer 2* (10): 750–763.

Amreddy, N., Babu, A., Panneerselvam, J., Srivastava, A., Muralidharan, R., Chen, A., Zhao, Y.D., Munshi, A., Ramesh, R. 2018. Chemo-Biologic Combinatorial Drug Delivery Using Folate Receptor-Targeted Dendrimer Nanoparticles for Lung Cancer Treatment. *Nanomedicine: Nanotechnology, Biology, and Medicine 14* (2): 373–384.

Bae, H.Y. Park, K. 2011. Targeted Drug Delivery to Tumours: Myths, Reality and Possibility. *Journal of Controlled Release: Official Journal of the Controlled Release Society 153* (3): 198–205.

Bazak, R., Houri, M., Achy, S. El, Hussein, W., Refaat, T. 2014. Passive Targeting of Nanoparticles to Cancer: A Comprehensive Review of the Literature. *Molecular and Clinical Oncology 2* (6): 904–908.

Benchaala, I., Mishra, M.K., Wykes, S.M., Hali, M., Kannan, R.M., Whittum-Hudson, J.A. 2014. Folate-Functionalized Dendrimers for Targeting Chlamydia-Infected Tissues in a Mouse Model of Reactive Arthritis. *International Journal of Pharmaceutics 466* (1): 258–265.

Bhadra, D., Yadav, A.K., Bhadra, S., Jain, N.K. 2005. Glycodendrimeric Nanoparticulate Carriers of Primaquine Phosphate for Liver Targeting. *International Journal of Pharmaceutics 295* (1): 221–233.

Biswas, S., Deshpande, P.P., Navarro, G., Dodwadkar, N.S., Torchilin, V.P. 2013. Lipid Modified Triblock PAMAM-Based Nanocarriers for SiRNA Drug Co-Delivery. *Biomaterials 34* (4): 1289–1301.

Brown, L.R., Edelman, E.R., Fischel-Ghodsian, F., Langer, R. 1996. Characterization of Glucose-Mediated Insulin Release from Implantable Polymers. *Journal of Pharmaceutical Sciences 85* (12): 1341–1345.

Buceta, R.E., Doyagüez, E.G., Colomer, I., Quesada, E., Mathys, L., Noppen, S., Liekens, S., Camarasa, M.J., Pérez, M.J., Balzarini, J., Félix, S.A. 2015. Tryptophan Dendrimers That Inhibit HIV Replication, Prevent Virus Entry and Bind to the HIV Envelope Glycoproteins Gp120 and Gp41. *European Journal of Medicinal Chemistry 106*: 34–43.

Cai, X., Jin, R., Wang, J., Yue, D., Jiang, Q., Wu, Y., Gu, Z. 2016. Bioreducible Fluorinated Peptide Dendrimers Capable of Circumventing Various Physiological Barriers for Highly Efficient and Safe Gene Delivery. *ACS Applied Materials & Interfaces 8* (9): 5821–5832.

Calderón, M., Quadir, M.A., Strumia, M., Haag, R. 2010. Functional Dendritic Polymer Architectures as Stimuli-Responsive Nanocarriers. *Biochimie 92* (9): 1242–1251.

Chen, H.T., Neerman, M.F., Parrish, A.R., Simanek, E.E. 2004. Cytotoxicity, Hemolysis, and Acute in Vivo Toxicity of Dendrimers Based on Melamine, Candidate Vehicles for Drug Delivery. *Journal of the American Chemical Society 126* (32): 10044–10048.

Chen, M., Yang, J., Li, Jiyao, Liang, K., He, L., Lin, Z., Chen, X., Ren, X., Li, Jianshu. 2014. Modulated Regeneration of Acid-Etched Human Tooth Enamel by a Functionalized Dendrimer That Is an Analog of Amelogenin. *Acta Biomaterialia 10* (10): 4437–4446.

Cheng, Y., Zhao, L., li, Y., Xu, T. 2011. Design of Biocompatible Dendrimers for Cancer Diagnosis and Therapy: Current Status and Future Perspectives. *Chemical Society Reviews 40* (5): 2673–2703.

Ciepluch, K., Ionov, M., Majoral, J.-P., Muñoz-Fernández, M.A., Bryszewska, M. 2014. Interaction of Phosphorus Dendrimers with HIV Peptides-Fluorescence Studies of Nano-Complexes Formation. *Journal of Luminescence 148*: 364–369.

Dhanikula, R.S., Argaw, A., Bouchard, J.-F., Hildgen, P. 2008. Methotrexate Loaded Polyether-Copolyester Dendrimers for the Treatment of Gliomas: Enhanced Efficacy and Intratumoural Transport Capability. *Molecular Pharmaceutics 5* (1): 105–116.

Duncan, R., Izzo, L. 2005. Dendrimer Biocompatibility and Toxicity. *Advanced Drug Delivery Reviews 57* (15): 2215–2237.

Dutta, T., Agashe, H.B., Garg, M., Balasubramanium, P., Kabra, M., Jain, N.K. 2007. Poly (Propyleneimine) Dendrimer Based Nanocontainers for Targeting of Efavirenz to Human Monocytes/Macrophages In Vitro. *Journal of Drug Targeting 15* (1): 89–98.

Dutta, T., Garg, M., Jain, N.K. 2008. Targeting of Efavirenz Loaded Tuftsin Conjugated Poly(Propyleneimine) Dendrimers to HIV Infected Macrophages In Vitro. *European Journal of Pharmaceutical Sciences 34* (2): 181–189.

Dutta, T., Jain, N.K. 2007. Targeting Potential and Anti-HIV Activity of Lamivudine Loaded Mannosylated Poly (Propyleneimine) Dendrimer. *Biochimica et Biophysica Acta (BBA) - General Subjects 1770* (4): 681–686.

Falciani, C., Fabbrini, M., Pini, A., Lozzi, L., Lelli, B., Pileri, S., Brunetti, J., Bindi, S., Scali, S., Bracci, L. 2007. Synthesis and Biological Activity of Stable Branched Neurotensin Peptides for Tumour Targeting. *Molecular Cancer Therapeutics 6* (9): 2441–2448.

Fanfan, F., Yilun, W., Zhu, J., Wen, S., Shen, M., Shi, X. 2014. Multifunctional Lactobionic Acid-Modified Dendrimers for Targeted Drug Delivery to Liver Cancer Cells: Investigating the Role Played by PEG Spacer. *ACS Applied Materials & Interfaces 6* (18): 16416–16425.

Fox, M.E., Guillaudeu, S., Fréchet, J.M.J., Jerger, K., Macaraeg, N., Szoka, F.C. 2009. Synthesis and In Vivo Antitumour Efficacy of PEGylated Poly(L-Lysine) Dendrimer–Camptothecin Conjugates. *Molecular Pharmaceutics 6* (5): 1562–1572.

Fruchon, S., Mouriot, S., Thiollier, T., Grandin, C., Caminade, A.-M., Turrin, C.O., Contamin, H., Poupot, R. 2015. Repeated Intravenous Injections in Non-Human Primates Demonstrate Preclinical Safety of an Anti-Inflammatory Phosphorus-Based Dendrimer. *Nanotoxicology 9* (4): 433–441.

Ghaffari, M., Dehghan, G., Abedi G.F., Kashanian, S., Baradaran, B., Dolatabadi, E.Z.J., Losic, D. 2018. Surface Functionalized Dendrimers as Controlled-Release Delivery Nanosystems for Tumour Targeting. *European Journal of Pharmaceutical Sciences 122*: 311–330.

Griffe, L., Poupot, M., Marchand, P., Maraval, A., Turrin, C.-O., Rolland, O., Métivier, P., Bacquet, G., Fournié, J.-J., Caminade, A.-M., Poupot, R., Majoral, J.-P. 2007. Multiplication of Human Natural Killer Cells by Nanosized Phosphonate-Capped Dendrimers. *Angewandte Chemie International Edition 46* (14): 2523–2526.

Grinstaff, M.W. 2002. Biodendrimers: New Polymeric Biomaterials for Tissue Engineering. *Chemistry – A European Journal 8* (13): 2838–2846.

Guillot-Nieckowski, M., Eisler, S., Diederich, F. 2007. Dendritic Vectors for Gene Transfection. *New Journal of Chemistry 31* (7): 1111–1127.

Hawker, C.J., Frechet, J.M.J. 1990. Preparation of Polymers with Controlled Molecular Architecture. A New Convergent Approach to Dendritic Macromolecules. *Journal of the American Chemical Society 112* (21): 7638–7647.

Hayder, M., Poupot, M., Baron, M., Nigon, D., Turrin, C.O., Caminade, A.M., Majoral, J.P., Eisenberg, R.A., Fournié, J.J., Cantagrel, A., Poupot, R., Davignon, J.L. 2011. A Phosphorus-Based Dendrimer Targets Inflammation and Osteoclastogenesis in Experimental Arthritis. *Science Translational Medicine 3* (81): 81ra35 LP-81ra35.

Hecht, S. 2003. Functionalizing the Interior of Dendrimers: Synthetic Challenges and Applications. *Journal of Polymer Science Part A: Polymer Chemistry 41* (8): 1047–1058.

Heyder, R.S., Zhong, Q., Bazito, R.C., da Rocha, S.R.P. 2017. Cellular Internalization and Transport of Biodegradable Polyester Dendrimers on a Model of the Pulmonary Epithelium and Their Formulation in Pressurized Metered-Dose Inhalers. *International Journal of Pharmaceutics 520* (1–2): 181–194.

Hill, E., Shukla, R., Park, S.S., Baker, J.R. 2007. Synthetic PAMAM–RGD Conjugates Target and Bind To Odontoblast-like MDPC 23 Cells and the Predentin in Tooth Organ Cultures. *Bioconjugate Chemistry 18* (6): 1756–1762.

Hong, S., Leroueil, P.R., Janus, E.K., Peters, J.L., Kober, M.M., Islam, M.T., Orr, B.G., Baker, J.R., Banaszak Holl, M.M. 2006. Interaction of Polycationic Polymers with Supported Lipid Bilayers and Cells: Nanoscale Hole Formation and Enhanced Membrane Permeability. *Bioconjugate Chemistry 17* (3): 728–734.

Huynh, N.T., Passirani, C., Saulnier, P., Benoit, J.P. 2009. Lipid Nanocapsules: A New Platform for Nanomedicine. *International Journal of Pharmaceutics 379* (2): 201–209.

Issels, R., Nagele, A., Eckert, K., Wilmanns, W. 1988. Promotion of Cystine Uptake and Its Utilization for Glutathione Biosynthesis Induced by and N-Acetyl-Cysteine. *Biochemical Pharmacology 37* (5): 881–888.

Jain, N., Tare, M., Mishra, V., Tripathi, P. 2015b. The Development, Characterization and in Vivo Anti-Ovarian Cancer Activity of Poly(Propylene Imine) (PPI)-Antibody Conjugates Containing Encapsulated Paclitaxel. *Nanomedicine: Nanotechnology, Biology and Medicine 11* (1): 207–218.

Jain, K., Verma, A., Mishra, P., Jain, N. 2015a. Characterization and Evaluation of Amphotericin B Loaded MDP Conjugated Poly(Propylene Imine) Dendrimers. *Nanomedicine: Nanotechnology, Biology, and Medicine 11* (3): 705–713.

Javan, C.M., Gooderham, N.J., Edwards, R.J., Davies, D.S., Shaunak, S. 1997. Anti-HIV Type 1 Activity of Sulfated Derivatives of Dextrin against Primary Viral Isolates of HIV Type 1 in Lymphocytes and Monocyte-Derived Macrophages. *AIDS Research and Human Retroviruses 13* (10): 875–880.

Jevprasesphant, R., Penny, J., Attwood, D., McKeown, N., D'Emanuele, A. 2003a. The Influence of Surface Modification on the Cytotoxicity of PAMAM Dendrimers. *International Journal of Pharmaceutics 252* (1): 263–266.

Jevprasesphant, R., Penny, J., Jalal, R., Attwood, D., McKeown, N.B., D'Emanuele, A. 2003b. Engineering of Dendrimer Surfaces to Enhance Transepithelial Transport and Reduce Cytotoxicity. *Pharmaceutical Research 20* (10): 1543–1550.

Jiang, B., Zhao, J., Li, Y., He, D., Pan, J., Guo, J.C. 2013. Dual-Targeting Janus Dendrimer Based Peptides for Bone Cancer: Synthesis and Preliminary Biological Evaluation. *Letters in Organic Chemistry 10* (8): 594–601.

Johnson, M.E., Shon, J., Guan, B.M., Patterson, J.P., Oldenhuis, N.J., Eldredge, A.C., Gianneschi, N.C., Guan, Z. 2016. Fluorocarbon Modified Low-Molecular-Weight Polyethylenimine for SiRNA Delivery. *Bioconjugate Chemistry 27* (8): 1784–1788.

Kaminskas, L.M., Kelly, B.D., McLeod, V.M., Sberna, G., Owen, D.J., Boyd, B.J., Porter, C.J.H. 2011. Characterisation and Tumour Targeting of PEGylated Polylysine Dendrimers Bearing Doxorubicin via a PH Labile Linker. *Journal of Controlled Release: Official Journal of the Controlled Release Society 152* (2): 241–248.

Kaminskas, L.M., McLeod, V.M., Kelly, B.D., Cullinane, C., Sberna, G., Williamson, M., Boyd, B.J., Owen, D.J., Porter, C.J.H. 2012. Doxorubicin-Conjugated PEGylated Dendrimers Show Similar Tumouricidal Activity but Lower Systemic Toxicity When Compared to PEGylated Liposome and Solution Formulations in Mouse and Rat Tumour Models. *Molecular Pharmaceutics 9* (3): 422–432.

Kesharwani, P., Xie, L., Banerjee, S., Mao, G., Padhye, S., Sarkar, F.H., Iyer, A.K. 2015. Hyaluronic Acid-Conjugated Polyamidoamine Dendrimers for Targeted Delivery of 3,4-Difluorobenzylidene Curcumin to CD44 Overexpressing Pancreatic Cancer Cells. *Colloids and Surfaces B: Biointerfaces 136*: 413–423.

Knop, K., Hoogenboom, R., Fischer, D., Schubert, U.S. 2010. Poly(Ethylene Glycol) in Drug Delivery: Pros and Cons as Well as Potential Alternatives. *Angewandte Chemie International Edition 49* (36): 6288–6308.

Kojima, C., Kono, K., Maruyama, K., Takagishi, T. 2000. Synthesis of Polyamidoamine Dendrimers Having Poly(Ethylene Glycol) Grafts and Their Ability To Encapsulate Anticancer Drugs. *Bioconjugate Chemistry 11* (6): 910–917.

Kono, K. 2012. Dendrimer-Based Bionanomaterials Produced by Surface Modification, Assembly and Hybrid Formation. *Polymer Journal 44* (6): 531–540.

Kono, K., Kojima, C., Hayashi, N., Nishisaka, E., Kiura, K., Watarai, S., Harada, A. 2008. Preparation and Cytotoxic Activity of Poly(Ethylene Glycol)-Modified Poly(Amidoamine) Dendrimers Bearing Adriamycin. *Biomaterials 29* (11): 1664–1675.

Kulhari, H., Pooja, D., Shrivastava, S., Kuncha, M., Naidu, V.G.M., Bansal, V., Sistla, R., Adams, D.J. 2016. Trastuzumab-Grafted PAMAM Dendrimers for the Selective Delivery of Anticancer Drugs to HER2-Positive Breast Cancer. *Scientific Reports 6* (1): 1–13.

Kulhari, H., Pooja, D., Singh, M.K., Chauhan, A.S. 2015. Optimization of Carboxylate-Terminated Poly(Amidoamine) Dendrimer-Mediated Cisplatin Formulation. *Drug Development and Industrial Pharmacy 41* (2): 232–238.

Lagnoux, D., Darbre, T., Schmitz, M., Reymond, J.-L. 2005. Inhibition of Mitosis by Glycopeptide Dendrimer Conjugates of Colchicine. *Chemistry (Weinheim an Der Bergstrasse, Germany) 11* (13): 3941–3950.

Landers, J.J., Cao, Z., Lee, I., Piehler, L.T., Myc, P.P., Myc, A., Hamouda, T., Galecki, A.T., Baker Jr., J.R. 2002. "Prevention of Influenza Pneumonitis by Sialic Acid–Conjugated Dendritic Polymers." *The Journal of Infectious Diseases 186* (9): 1222–1230.

Lee, C.C., Cramer, A.T., Szoka Francis C., Fréchet, J.M.J. 2006. An Intramolecular Cyclization Reaction Is Responsible for the in Vivo Inefficacy and Apparent PH Insensitive Hydrolysis Kinetics of Hydrazone Carboxylate Derivatives of Doxorubicin. *Bioconjugate Chemistry 17* (5): 1364–1368.

Lee, J.H., Lim, Y., Choi, J.S., Lee, Y., Kim, T., Kim, H.J., Yoon, J.K., Kim, K., Park, J. 2003. Polyplexes Assembled with Internally Quaternized PAMAM-OH Dendrimer and Plasmid DNA Have a Neutral Surface and Gene Delivery Potency. *Bioconjugate Chemistry 14* (6): 1214–1221.

Liu, C., Zhao, Z., Gao, H., Rostami, I., You, Q., Jia, X., Wang, C., Zhu, L., Yang, Y. 2019. Enhanced Blood-Brain-Barrier Penetrability and Tumour-Targeting Efficiency by Peptide-Functionalized Poly(Amidoamine) Dendrimer for the Therapy of Gliomas. *Nanotheranostics 3* (4): 311–330.

Liu, S., Guo, Y., Huang, R., Li, J., Huang, S., Kuang, Y., Han, L., Jiang, C. 2012a. Gene and Doxorubicin Co-Delivery System for Targeting Therapy of Glioma. *Biomaterials 33* (19): 4907–4916.

Liu, X., Liu, J., Luo, Y. 2012b. Facile Glycosylation of Dendrimers for Eliciting Specific Cell–Material Interactions. *Polymer Chemistry 3* (2): 310–313.

Malik, N., Wiwattanapatapee, R., Klopsch, R., Lorenz, K., Frey, H., Weener, J.W., Meijer, E.W., Paulus, W., Duncan, R. 2000. Dendrimers: Relationship between Structure and Biocompatibility in Vitro, and Preliminary Studies on the Biodistribution of 125I-Labelled Polyamidoamine Dendrimers In Vivo. *Journal of Controlled Release 65* (1): 133–148.

Mandeville, J.S., Bourassa, P., Tajmir-Riahi, H.A. 2013. Probing the Binding of Cationic Lipids with Dendrimers. *Biomacromolecules 14* (1): 142–152.

Mattheolabakis, G., Milane, L., Singh, A., Amiji, M.M. 2015. Hyaluronic Acid Targeting of CD44 for Cancer Therapy: From Receptor Biology to Nanomedicine. *Journal of Drug Targeting 23* (7–8): 605–618.

Mecke, A., Lee, D.-K., Ramamoorthy, A., Orr, B.G., Banaszak Holl, M.M. 2005. Synthetic and Natural Polycationic Polymer Nanoparticles Interact Selectively with Fluid-Phase Domains of DMPC Lipid Bilayers. *Langmuir 21* (19): 8588–8590.

Medina, S.H., Tekumalla, V., Chevliakov, M. V., Shewach, D.S., Ensminger, W.D., El-Sayed, M.E.H. 2011. N-Acetylgalactosamine-Functionalized Dendrimers as Hepatic Cancer Cell-Targeted Carriers. *Biomaterials 32* (17): 4118–4129.

Mishra, M.K., Beaty, C.A., Lesniak, W.G., Kambhampati, S.P., Zhang, F., Wilson, M.A., Blue, M.E., Troncoso, J.C., Kannan, S., Johnston, M. V., Baumgartner, W.A., Kannan, R.M. 2014. Dendrimer Brain Uptake and Targeted Therapy for Brain Injury in a Large Animal Model of Hypothermic Circulatory Arrest. *ACS Nano 8* (3): 2134–2147.

Moulton, H.M. 2012. Cell-Penetrating Peptides Enhance Systemic Delivery of Antisense Morpholino Oligomers. *Methods in Molecular Biology (Clifton, N.J.) 867*: 407–414.

Newkome, G.R., Moorefield, C.N., Baker, G.R., Johnson, A.L., Behera, R.K. 1991. Alkane Cascade Polymers Possessing Micellar Topology: Micellanoic Acid Derivatives. *Angewandte Chemie International Edition in English 30* (9): 1176–1178.

Nie, J., Wang, Y., Wang, W. 2016. In Vitro and In Vivo Evaluation of Stimuli-Responsive Vesicle from PEGylated Hyperbranched PAMAM-Doxorubicin Conjugate for Gastric Cancer Therapy. *International Journal of Pharmaceutics 509* (1–2): 168–177.

Niidome, T., Yamauchi, H., Takahashi, K., Naoyama, K., Watanabe, K., Mori, T., Katayama, Y. 2014. Hydrophobic Cavity Formed by Oligopeptide for Doxorubicin Delivery Based on Dendritic Poly(L-Lysine). *Journal of Biomaterials Science. Polymer Edition 25* (13): 1362–1373.

O'Loughlin, J., Millwood, I., McDonald, H., Price, C., Kaldor, J., Paull, J. 2009. Safety, Tolerability, and Pharmacokinetics of SPL7013 Gel (VivaGel®): A Dose Ranging, Phase I Study. *Sexually Transmitted Diseases 37* (2): 100–104.

Ouyang, L., Huang, W., Guo, G.H. 2009. Bone Targeting Prodrugs Based on Peptide Dendrimers, Synthesis and Hydroxyapatite Binding In Vitro. *Letters in Organic Chemistry. 6* (4): 272–277

Pan, Y., Ford, W.T. 2000. Amphiphilic Dendrimers with Both Octyl and Triethylenoxy Methyl Ether Chain Ends. *Macromolecules 33* (10): 3731–3738.

Piotti, M.E., Rivera Felix, Bond, R., Hawker, C.J., Fréchet, J.M.J. 1999. Synthesis and Catalytic Activity of Unimolecular Dendritic Reverse Micelles with 'Internal' Functional Groups. *Journal of the American Chemical Society 121* (40): 9471–9472.

Price, C., Tyssen, D., Sonza, S., Davie, A., Evans, S., Lewis, G., Xia, S., Spelman, T., Hodsman, P., Moench, T., Humberstone, A., Paull, J., Tachedjian, G. 2011. SPL7013 Gel (VivaGel®) Retains Potent HIV-1 and HSV-2 Inhibitory Activity Following Vaginal Administration in Humans. *PloS One 6* (9): e24095.

Rolland, O., Griffe, L., Poupot, M., Maraval, A., Ouali, A., Coppel, Y., Fournié, J.J., Bacquet, G., Turrin, C.O., Caminade, A.-M., Majoral, J.-P., Poupot, R. 2008. Tailored Control and Optimisation of the Number of Phosphonic Acid Termini on Phosphorus-Containing Dendrimers for the Ex-Vivo Activation of Human Monocytes. *Chemistry – A European Journal 14* (16): 4836–4850.

Santos, S.D., Xavier, M., Leite, D.M., Moreira, D.A., Custódio, B., Torrado, M., Castro, R., Leiro, V., Rodrigues, J., Tomás, H., Pêgo, A.P. 2018. PAMAM Dendrimers: Blood-Brain Barrier Transport and Neuronal Uptake after Focal Brain Ischemia. *Journal of Controlled Release 291*: 65–79.

Sebestik, J., Niederhafner, P., Jezek, J. 2011. Peptide and Glycopeptide Dendrimers and Analogous Dendrimeric Structures and Their Biomedical Applications. *Amino Acids 40* (2): 301–370.

Shaunak, S., Thomas, S., Gianasi, E., Godwin, A., Jones, E., Teo, I., Mireskandari, K., Luthert, P., Duncan, R., Patterson, S., Khaw, P., Brocchini, S. 2004. Polyvalent Dendrimer Glucosamine Conjugates Prevent Scar Tissue Formation. *Nature Biotechnology 22* (8): 977–984.

Shukla, R., Thomas, T.P., Peters, J., Kotlyar, A., Myc, A., Baker James R. J. 2005. Tumour Angiogenic Vasculature Targeting with PAMAM Dendrimer–RGD Conjugates. *Chemical Communications 14* (46): 5739–5741.

Singh, P., Gupta, U., Asthana, A., Jain, N.K. 2008. Folate and Folate–PEG–PAMAM Dendrimers: Synthesis, Characterization, and Targeted Anticancer Drug Delivery Potential in Tumour Bearing Mice. *Bioconjugate Chemistry 19* (11): 2239–2252.

Swami, R., Singh, I., Kulhari, H., Jeengar, M.K., Khan, W., Sistla, R. 2015. P-Hydroxy Benzoic Acid Conjugated Dendrimer Nanotherapeutics As Potential Carriers For Targeted Drug Delivery To Brain: An In Vitro And In Vivo Evaluation. *Journal of Nanoparticle Research 17* (6).

Teow, H.M., Zhou, Z., Najlah, M., Yusof, S.R., Abbott, N.J., D'Emanuele, A. 2013. Delivery of Paclitaxel across Cellular Barriers Using a Dendrimer-Based Nanocarrier. *International Journal of Pharmaceutics 441* (1): 701–711.

Teow, Y., Valiyaveettil, S. 2010. Active Targeting of Cancer Cells Using Folic Acid-Conjugated Platinum Nanoparticles. *Nanoscale 2* (10): 2607–2613.

Thomas, T.P., Goonewardena, S.N., Majoros, I.J., Kotlyar, A., Cao, Z., Leroueil, P.R., Baker Jr, J.R. 2011. Folate-Targeted Nanoparticles Show Efficacy in the Treatment of Inflammatory Arthritis. *Arthritis and Rheumatism 63* (9): 2671–2680.

Tomalia, D.A. 2016. Special Issue: 'Functional Dendrimers.' *Molecules (Basel, Switzerland) 21* (8): 1035.

Wang, F., Cai, X., Su, Y., Hu, J., Wu, Q., Zhang, H., Xiao, J., Cheng, Y. 2012. Reducing Cytotoxicity While Improving Anti-Cancer Drug Loading Capacity of Polypropylenimine Dendrimers by Surface Acetylation. *Acta Biomaterialia 8* (12): 4304–4313.

Wolinsky, J.B., Grinstaff, M.W. 2008. Therapeutic and Diagnostic Applications of Dendrimers for Cancer Treatment. *Advanced Drug Delivery Reviews 60* (9): 1037–1055.

Yamashita, S., Katsumi, H., Hibino, N., Isobe, Y., Yagi, Y., Kusamori, K., Sakane, T., Yamamoto, A. 2017. Development of PEGylated Carboxylic Acid-Modified Polyamidoamine Dendrimers as Bone-Targeting Carriers for the Treatment of Bone Diseases. *Journal of Controlled Release 262*: 10–17.

Yellepeddi, V.K., Kumar, A., Maher, D.M., Chauhan, S.C., Vangara, K.K., Palakurthi, S. 2011. Biotinylated PAMAM Dendrimers for Intracellular Delivery of Cisplatin to Ovarian Cancer: Role of SMVT. *Anticancer Research 31* (3): 897–906.

Yi, Y.S. 2016. Folate Receptor-Targeted Diagnostics and Therapeutics for Inflammatory Diseases. *Immune Network 16* (6): 337–343.

Yoon, A.R., Kasala, D., Li, Y., Hong, J., Lee, W., Jung, S.J., Yun, C.O. 2016. Antitumour Effect and Safety Profile of Systemically Delivered Oncolytic Adenovirus Complexed with EGFR-Targeted PAMAM-Based Dendrimer in Orthotopic Lung Tumour Model. *Journal of Controlled Release 231*: 2–16.

Zhong, Q., Bielski, E.R., Rodrigues, L.S., Brown, M.R., Reineke, J.J., da Rocha, S.R.P. 2016. Conjugation to Poly(Amidoamine) Dendrimers and Pulmonary Delivery Reduce Cardiac Accumulation and Enhance Antitumour Activity of Doxorubicin in Lung Metastasis. *Molecular Pharmaceutics 13* (7): 2363–2375.

8

Characterisation of Dendrimers: Methods and Tools

Sreejan Manna and Sougata Jana

CONTENTS

8.1 Introduction

Dendrimers were initially discovered as cascade molecules in 1978 by Buhleier and his co-workers (Buhleier et al. 1978). The synthesis of cascade molecules implies repetition of reaction sequences in such an order that a certain functional group is present twice in the molecule. The word dendrimer originated from two Greek words: dendron and meros, which means tree-like. Dendrimers are mono-dispersed macromolecules containing symmetric branches surrounding a core or small molecules (Hawker and Fréchet 1990). Dendrimers can be defined as a highly branched and functionalised macromolecule having a carefully designed architecture. Donald Tomalia and his co-workers have also synthesised dendrimers in early 1980s and reported some discriminate structural features (Tomalia et al. 1986) of dendrimers such as an initiator core consisting of an atom or a cluster of atoms, interior layers consisting of radially attached repeating units and the surface or exterior layer composed of terminal groups as shown in Figure 8.1 (Dias et al. 2020).

Like other nanocomposites, dendrimers have also received significant attention from researchers due to their application in delivering poorly water-soluble compounds as soluble complexes and decreasing overall toxicity. Dendritic polymers represent nanosized, spherical, uniform and highly branched polymers with excellent complexation abilities and an easily modifiable structure. The tailored architectural structure (Bosman and Meijer 1999) and functionalised end groups impart the ability to modify the physicochemical and biological behaviour of dendrimers (Smith and Diederich 1998). The molecular architecture of dendrimers is responsible for their improved characteristics compared to linear polymers. This plays a significant role in determining rheological behaviour by lowering the viscosity of dendrimers in the solution phase than linear polymers (Fréchet 1994). The presence of numerous terminal groups is responsible for binding ability (Pourianazar et al. 2014) solubility and reactivity (Peetla et al. 2009). Depending on their property, the terminal groups can accommodate ligands, which ensures protection from degradation and target specific delivery (Biswas and Torchilin 2014).

Dendrimers were extensively studied as a drug carrier based on their structural architecture and precise control over numerous surface groups with negligible variations from one batch to another (Fox et al. 2009). A high degree of structural deformability allows flexibility to the system resulting in an engineered biological interaction. The structural flexibility offers powerful multivalent interactions between the receptor and the ligands present on the surface, which increases the targeting ability (Hong et al. 2007; Patton et al. 2006). One of the major advantages, which separates dendrimers from other conventional nanocarriers, is a low polydispersity index (PDI) achievable at large-scale production (Esfand and Tomalia 2001). In the polymeric blocks of dendrimers, different functional features can be incorporated such as biodegradability, imaging, targeting, etc. Various stimuli responsive functions (temperature, pH, UV light and enzymes) can also be incorporated into the designs of dendrimeric blocks (Jin et al. 2011).

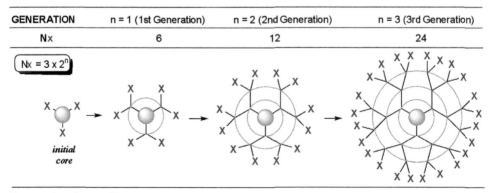

GENERATION	n = 1 (1st Generation)	n = 2 (2nd Generation)	n = 3 (3rd Generation)
Nx	6	12	24

$Nx = 3 \times 2^n$

initial core

Nx = number of groups X

FIGURE 8.1 Dendrimeric structure including core, generations and end groups. Adapted and reproduced from Dias, A. P., Santos, S. S., Vitor J. S., et al. 2020. Dendrimers in the context of nanomedicine. *International Journal of Pharmaceutics* 573:118814.

Due to their advantages of having a defined size and shape along with their versatility, dendrimers have drawn the attention of researchers in numerous biomedical fields. Selectivity towards a specific bio-target can be attained with dendrimer-based delivery of therapeutics. Stability (Hsu et al. 2017) and solubility (Kesharwani et al. 2014) of certain drugs can be enhanced with the help of dendrimers. They also facilitate other biological applications, such as delivery of vaccines, as a gene carrier and also in diagnostic purposes (Svenson and Tomalia 2012). Dendrimers have been extensively studied for targeting cancer tissues and hence to minimise off-target toxicity. Dendrimers were also investigated as a carrier of imaging agents, such as folic acid for targeting cancer tissues (Low and Kularatne 2009). Dendrimers were also studied in different central nervous system-related diseases such as Alzheimer's disease, different neuro-degenerative disorders, etc. Dendrimer-based delivery of therapeutics was investigated for various infectious diseases including viral and tropical infections. Different studies revealed the ability of dendrimers to conjugate with polymers and different targeting moieties such as carbohydrates, antibodies and aptamers (Dias et al. 2020).

8.2 Characterisation of Dendrimers

The repetitive structure along with numerous terminal groups has subjected dendrimers to the demanding need of different characterisation techniques. The size, shape, homogeneity and morphology are equally important aspects as the chemical composition of dendrimers. The conceptual structure of dendrimers synthesised by different methods is always in question and hence needs to be ensured. The characterisation of dendrimers plays a significant role in knowing their interaction with other moieties. It also helps to predict the biocompatibility of dendrimer conjugates (Biricova and Laznickova 2009). Different tools for dendrimer characterisation are depicted in Table 8.1.

8.2.1 Spectroscopic and Spectrometric Methods

These characterisation techniques help to understand the chemical compositions as well as identify the presence of structural defects.

8.2.1.1 Nuclear Magnetic Resonance (NMR)

NMR is the most commonly used characterisation technique for dendrimers. For studying the transformation of the terminal groups during high generation dendrimer synthesis, the routine NMR analysis is very useful. In characterisation of organic dendrimers, ^{1}H and ^{13}C NMR methods are mostly used. Structural defects can be detected for poly(propyleneimine) (PPI), which is an organic dendrimer (Tomalia et al. 1985). In the ^{1}H NMR spectra, the characteristic presence of a triplet corresponding to NCH_2CH_2CN is found for NH_2-terminated PPI dendrimers. In the case of poly(phosphorhydrazone) (PPH) dendrimer synthesis, the dissipation of the aldehyde signal in ^{1}H NMR indicates the condensation step completion (Launay et al. 1995). Water-soluble poly(amidoamine)(PAMAM) dendrimers are characterised by ^{13}C NMR and ^{1}H NMR. It also ensures the purity of the dendrimers (Ertürk et al. 2016a).^{13}C NMR spectra can detect PPI dendrimers, which are incompletely cyanoethylated as well as the occurrence of retro-Michael addition, by identifying their by-products (de Brabander-van den Berg and Meijer 1993). Similarly, retro-Michael addition along with other structural defects during the synthesis of PAMAM dendrimers is shown in Figure 8.2 (Lyu et al. 2019). Characterisation of the folic acid–PPI dendrimer conjugate was performed by ^{1}H NMR spectroscopy for confirming the conjugation. The spectrum showed peaks for folic acid (Kaur et al. 2017). The purity of the aminosugar and aminopolyol G 0 dendrimers was confirmed using NMR spectra.^{1}H and ^{13}C NMR techniques were applied for thorough investigation of the chemical shift of carbon and protons (Pyziak et al. 2016).

Dendrimers containing heteroatoms are usually characterised by resonance, which affords important information. In particular, ^{31}P NMR is very sensitive to minute changes that allow differentiating the layers up to the fourth generation (G4.0) (Launay et al. 1994) and a minimum of three outermost layers for substantially large dendrimers up to G12.0 (Lartigue et al. 1997) Rotaxane dendrimers (up to G3.0) were

TABLE 8.1

Different tools for dendrimer characterisation

Sl. No.	Characterisation Methods	Tools	Dendrimer Type	References
1.	NMR	TitroLine® 7000autotitrator and PG TG70 UV-Vis spectrophotometer (Software v5.0.5)	Water-soluble PAMAM dendrimers	Strašák et al. (2016)
		Varian VNMRS 700 apparatus employing DOSY(VnmrJ_3.x software)	Biomedical-grade G3.0 and G5.0 PAMAM	vanDongen et al. (2014)
		Bruker DRX 500 NMR spectrometer	DAB-Am8, DAB-Am16, DAB-Am32 and DAB-Am62 PPIdendrimers	Appelhans et al. (2010)
2.	XRD	Ultima IV multipurpose XRD system	Glass nanoparticles coated with PAMAM dendrimers	Bae et al. (2019)
3.	MS	A commercial Finnigan MATTSQ 7000 triple quadrupole mass spectrometer	smaller Starburst PAMAM dendrimers	Schwartz et al. (1995)
4.	IR spectroscopy	FT-IR DIGILAB, Scimitar Series Spectrometer	Liquid crystalline poly(amidoamine) codendrimer (PAMAM (L1)16-(L2)16) generation 3.0	Popescu et al. (2006)
5.	Raman spectroscopy	Lab Ram HR Evolution Raman spectrometer	PAMAM dendrimer, decorated with silver nanoparticles	Saleh et al. (2016)
6.	UV-vis spectroscopy	Hewlett-Packard HP8453 spectrometer	G4-OH, G6-OH, G4-NH2 and G6-NH2 PAMAM dendrimers	Pande and Crooks (2011)
7.	Florescence spectroscopy	Perkin-Elmer LS-55 spectrofluorometer	G2.0, G3.0 and G4PAMAM dendrimers	Konopka et al. (2018)
8.	Electrochemical Impedance Spectroscopy	IM6 Zahner-Elektrik potentiostat-galvanostat (Thales software)	Prussian Blue/PAMAM Dendrimer Films	Mendez et al. (2017)
9.	EPR	EMX-Bruker spectrometer	DAB-Am8, DAB-Am16, DAB-Am32,DAB-Am62 PPI dendrimers	Appelhans et al. (2010)
10.	AFM	a Nanoscope III Multimode scanning probe microscope from Digital Instruments (Veeco Metrology Group; Santa Barbara, CA) using an "E" scanner	PAMAM dendrimers having an ethylenediamine core	Muller et al. (2002)
11.	Intrinsic viscosity	Rheometric Scientific Advanced Rheometric Expansion System (ARES) controlled-strain rheometer	G4 PPI and G5 PPI	Tande et al. (2003)
12.	DSC	Mettler Toledo DSC822e heat-flux DSC	PAMAM G4.0 and G3,5	Gardikis et al. (2006)
13.	Dielectric spectroscopy	Novocontrol α-analyzer and HewlettPackard 4291B RF impedance analyzer	Generations 0-5 of PAMAM dendrimers	Mijovic et al. (2007)

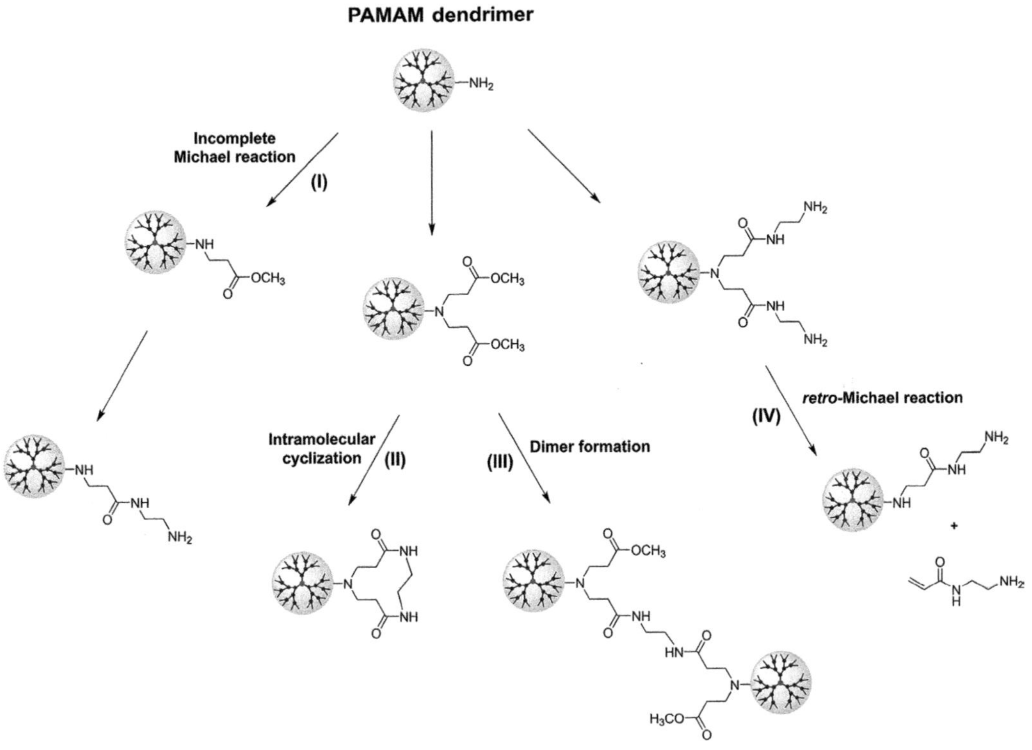

PAMAM dendrimer

FIGURE 8.2 Different structural defects during the synthesis of PAMAM dendrimers caused by various side-reactions. . Adapted and reproduced from Lyu et al. 2019.

characterised by ^{31}P NMR, which showed a consistent and symmetrical signal peak. With the increase in generations, a broad effect was observed for the NMR spectra of rotaxane dendrimers (Wang et al. 2018). Through the resonance of the heteroatom, coexistence of different branches (Galliot et al. 1997) and occurrence of coupling can be detected in the complex dendrimer structure (Brauge et al. 2001a). Carbosilane dendrimers of sodium montmorillonite were characterised by solid-state ^{29}Si MAS NMR. The spectra indicated the impact of modification on the micro-level structural aspects of the dendrimeric nanocomposite (Strašák et al. 2016). Dendrimers containing silicon are usually characterised by ^{29}Si NMR. Six different silyl groups can be distinguished through ^{29}Si NMR in phenylsilane-containing polysiloxane dendrimers (Morokawa et al. 1991). A similar technique was applied to characterise carbosilane dendrimers. A two-dimensional ^{29}Si, ^{29}Si incredible natural abundance double quantum transfer experiment was utilised for confirming the first generation polysilane dendrimer structure (Lambert et al. 1998). ^{15}N NMR was used to characterise selective protonation of the surface and core for PPI dendrimers (Koper et al. 1997). A special NMR method in solution was used for characterising the morphology and size of dendrimers. Based on relaxation time, azobenzene dendrimers were found to have a rigid surface and a non-constrained inner part (Jiang and Aida 1997).

Non-identical dendrimers containing other NMR detectable nuclei were used as structural substituents. Perfluorinated polyphenylene dendrimers were characterised by ^{19}F NMR where fluorine is used as a supplementary marker (Caminade et al. 2003). A similar technique was also incorporated in the characterisation of fluorinated substituents attached to carbosilane and PPH dendrimers (Turrin et al. 2003). Covalently modified PAMAM dendrimers were used to obtain non-invasive images in the systemic circulation of mice with magnetic resonance image technology of ^{19}F (Criscione et al. 2009). Generation 5 (G5.0) PAMAM dendrimers were characterised by using ^{19}F NMR (Dougherty et al. 2014). Boron clusters attached with C=C bonds of organic dendrimers can be characterised by ^{11}B NMR (Escamilla and Newkome 1994).

Determination of the molecular diffusion coefficient is performed by pulse field-gradient spin echo ^{1}H NMR with the help of the Stejskal–Tanner equation, which can be used to estimate hydrodynamic radii by following the Stokes–Einstein equation, considering the spherical structure of dendrimers. This technique is useful in characterising PPI dendrimers (Rietveld and Bedeaux 2000), aliphatic polyester (Ihre et al. 1996), PAMAM dendrimers with hydrophobic chains (Menger et al. 2001) and carbosilane dendrimers (Sagidullin et al. 2002). The pH and concentration influenced radius change can be determined for PPI and poly(allylcarbosilane) dendrimers, respectively, by this technique (Rietveld et al. 2001). Similarly, the free diffusion coefficient for partially fluorinated PAMAM dendrimers can be compared by the ^{19}F NMR technique in water and bicontinuous cubic phases comprising hydrated lipids (Jeong et al. 2002). The nuclear magnetisation recovery towards the equilibrium value can be described by T_1 (spin-lattice relaxation time). This dependence of frequency can be investigated by the field cycling technique. Stronger pronounced dispersion was evidenced at low temperature where the determination of relaxations was reported by local dynamics (Mohamed et al. 2015). Apart from the solution phase characterisation, magic angle spinning NMR was used to characterise dendrimers in solid phase. Poly(phenylene) dendrimers were characterised by this technique (Wind et al. 2002). Some dendrimers were reported to be characterised in solid state by ^{2}H NMR using deuterium (Lehmann et al. 2004). HCl can be replaced with DCl in the case of PAMAM dendrimers, where quaternarisation of branching points up to G 9.0 allowed for developing a dendrimeric structure, which employed echo line shapes of deuteron quadrupole (Malyarenko et al. 2000). In characterisation of denatured PAMAM dendrimers and rational-echo double resonance poly(benzyl ether), quadrupole echo NMR was applied which showed penetration of intermolecular chains (Kao et al. 2000). In the case of mesoporous silica, phosphorus dendrons were found by solid-state characterisation using ^{31}P high-power decoupling NMR(Turrin et al. 2000). Solid-state characterisation was performed for deuterated dendrimers by using ^{2}H NMR. The angular frequency of the NMR was dependent on the spherical angles between the direction of the magnetic field and EFG tensor (Mohamed et al. 2015).

8.2.1.2 X-ray Diffraction (XRD)

The XRD technique allows specific determination of the components apart from characterising the dendrimeric size and shape. Most of the dendrimers being solid in nature are present in amorphous state, which in condensed phase lacks the regularity in structural arrangement. Apart from the first generation dendrimers, the exact molecular structures of other dendrimers are difficult to confirm by this technique (Ropponen et al. 2004). Glass nanoparticles coated with PAMAM dendrimers were characterised by XRD, which showed amorphous nature along with calcite peaks (Bae et al. 2019). Some of the second generation dendrimer structures can also be determined (PCSi dendrimers), but thermal motions were observed in peripheral atoms (Seyferth et al. 1994). The existence of aromatic groups can help to obtain crystals of dendrimers. Different first generation dendrimers which contain phosphorus can also be characterised by the XRD technique. First generation PPH dendrimers which were synthesised from first generation dendrons (Maraval et al. 2000) a trifunctional core P(S)Cl$_3$ and a large dendrimers belonging to first generation have a core comprising hydroalkyl chains functionalised by allyl or methyl groups (Larre et al. 1998). SO$_3$H dendrimer functionalised nanoparticles were characterised by the XRD technique. The XRD pattern confirmed that the functionalisation process does not alter the crystalline structure of the Fe$_3$O$_4$ magnetic nanoparticles (Maleki et al. 2019).

8.2.1.3 Mass Spectrometry (MS)

The MS method is used to acquire information regarding the molecular mass along with any structural defects present in dendrimers. Small dendrimers having molecular mass less than 3000 D can be characterised by a classical approach of MS like rapid atomic bombardment and chemical ionisation (Hawker and Fréchet 1990). Dendrimers which can form multicharge containing stable species can be characterised by applying electrospray ionisation. The fifth generation dendrimers showed an approximately 20% dendrimeric purity with a poly-dispersity of 1.002. A similar technique is applied for characterising PAMAM dendrimers for determining polydispersity (Tolic et al. 1997). Matrix-assisted laser desorption ionisation time of flight (MALDI-ToF) is an extensively used technique for determining dendrimeric purity. To increase the sensitivity, a superconducting cryodetector was used with ultrahigh MALDI-ToF to characterise polyphenylene dendrimers (Maraval et al. 2003). Peripherally terminated PPI dendrimers were characterised by MALDI-ToF MS, and the result confirmed interaction between generation 2.0 PPI dendrimers and propionyl chloride (Zhou and Peng 2019). The MALDI-ToF technique is used for characterising different varieties of dendrimers, such as PAMAM (Zhou et al. 2001),polybenzylacetylenes (Kawaguchi et al. 1995), PBzE (Pollak et al. 1998) and dendrimers containing phosphorus (Maraval et al. 2003). The obtained spectra depend on the matrix type; the acidic matrices present in persulphonylated dendrimers showed negative results for MALDI MS, whereas the electrospray ionisation technique did not produce similar results (Lukin et al. 2006). For PAMAM dendrimers, the MALDI-ToF technique was used to monitor and analyse the dendrimer synthesis. A positive MALDI-ToF spectrum is shown in Figure 8.3 for metastable ions of G1.0 PAMAM dendrimers (Leriche et al. 2014). The molecular mass of amino-sugar and aminopolyol was characterised by the MALDI-TOF technique. The existence of side products can also be detected by this method (Pyziak et al. 2016). Apart from matrix absorption, dendrimeric absorption of UV light was evidenced by PPH dendrimers, which induces fragmentation and rearrangement of the fragments (Blais et al. 2000). Electrostatic spray coupling with high-resolution mass spectroscopy has elucidated the structural configuration of carbosilane dendrimer–sodium montmorillonite clay nanocomposites (Strašák et al. 2016).

8.2.1.4 Infra-red and Raman Spectroscopy

The infra-red spectroscopic method is used as a regular analytical technique to detect the chemical transformation at the dendrimeric surface. This technique is useful to detect removal of aldehyde groups while synthesising PMMH dendrimer synthesis (Galliot et al. 1995) or H-bonding in PPI glycine dendrimers (Bosman et al. 1998). The Fourier-transform infrared spectroscopy (FT-IR) technique was applied to characterise G0 PAMAM dendrimers, where the infrared (IR) peaks confirmed conversion of amino groups to amide groups (Sohail et al. 2020). For characterising delocalised π-π interaction between the terminal groups of naphthalene diimide modified PAMAM dendrimers, near IR spectroscopy was applied (Miller et al. 1997).

Raman spectroscopy is useful for providing useful information regarding PPI dendrimers (Perrot et al. 1999) and polyphenylene cyclodehydrogenation (Simpson et al. 2004). Raman spectra with lower frequency were applied for the investigation of end group vibrations. For PPH dendrimers, Fourier transform

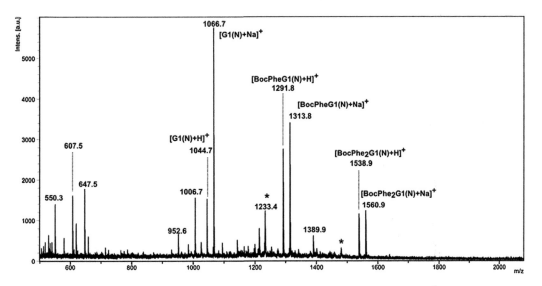

FIGURE 8.3 Positive spectra of MALDI-TOF for N-Boc Phenylalanine PAMAM G1(N). Adapted and reproduced from Leriche et al. 2014.

Raman spectra of 0–10 generations were analysed. It was evidenced that Raman spectra higher than generation four are the same. In the case of phosphorus starburst dendrimers containing deuterophenoxy end groups, the Raman spectra of initial six generations were recorded and compared (Furera et al. 2015). A hybrid graphene oxide-PAMAM dendrimer was characterised using a Raman spectrometer with the help of an argon ion laser. The obtained data have confirmed the graphene oxide functionalisation with PAMAM dendrimers. The Raman spectra higher than sixth generation have revealed the fact regarding the terminal group conformation, which was explained by steric congestion (Furer et al. 2004).

8.2.1.5 Ultraviolet-Visible (UV-vis) Spectroscopy

UV-vis spectroscopy is applied for monitoring dendrimer synthesis. With the increase in the generation number, the chromophoric group also increases by following the Beer–Lambert law. Hence, the absorption band intensity changes proportionally with chromophoric group's number. The azobenzene-containing dendrimers such as PPI dendrimers consisting of azobenzene end groups or phosphorus dendrimers comprising azobenzenes in their branches can be tested for purity by using the UV-vis spectroscopic method (Deloncle and Caminade 2010). This method is also applied to understand morphological aspects. When a solvatochromic probe was attached with the polybenzylether dendrimeric core up to the sixth generation, the first six generations showed a shift in absorption maxima from generation 3 to generation 4 in order to transform into a globular shape (Hawker et al. 1993). Amine-terminated fourth generation PAMAM dendrimers were characterised by UV-vis spectroscopy, which showed the presence of *d-d* transition bands (Ertürk et al. 2016b).

- Various study findings suggested that deviation in the Beer–Lambert law not always symbolises the existence of defects. In the case of PBZE dendrimers containing a core of zinc, tetraporphyrin with CO_2^- end groups experiences hypochromicity in aqueous medium due to the decrease in pH or improving the ionic strength. This symbolises the shrinkage of the lipophilic dendrimeric framework (Sadamoto et al. 1996). Phthalocyanine was used as the phosphorous dendrimer core, which can be used as a sensor as well as a probe to analyse internal structural aspects. A hyperchromic as well as bathochromic effect was observed for the UV-vis spectra of Q bands of phthalocyanine. With an increase in generation, phthalocyanine indicates an isolated chromophore along with a dendrimeric shell mimicking an extremely polar solvent. In the case of phenylazomethine dendrimers, the step-by-step complexation was monitored by UV-vis spectroscopy. Four major changes were monitored in the isosbestic point while titrating G 4.0 dendrimers, which indicated occurrence of stepwise complexation (Yamamoto et al. 2002). G3.0 and G4.0 PAMAM dendrimers were monitored and characterised by UV-vis spectroscopy in different stages of preparation. The spectroscopic study confirmed coordination of amine groups in PAMAM dendrimers containing ethanol amine groups (Erturk et al. 2016b).

8.2.1.6 Chirality, Optical Rotation and Circular Dichroism

The relation of dendrimers and chirality was highlighted by many scientists (Peerlings and Meijer 1997; Gibson and Rendell 2008). The synthesis and evaluation of triamine-coordinated chiral dendrimers comprising polyol terminal groups were reported (Miyake et al. 2015). In the case of completely chiral dendrimers, it has been reported that 0–3rd generation dendrimers contain a core along with double and triple branched building blocks. In the case of triply branched dendrimers, reports suggested a reversed rotation from first to second generation (Murer and Seebach 1995). PPI dendrimers containing a chiral amino acid on their surface induce a rapid optical rotation decrease from first to fifth generation (Jansen et al. 1995). The paracyclophanes present on the PPI dendrimer surface have an almost constant activity with the increase in generations. In contrast to PPI dendrimers, ferrocene-linked PMMH dendrimers produce additive values for optical rotation with increasing generations (Turrin et al. 2001). The dendrimeric structure of PBzE containing a binaphthyl group at the core showed a decrease in optical rotation with the increase in the generation number. A drastic change in CD spectra was reported for the first three generations, for the chiral polyarylether dendrimers and dihydroxypyrrolidine dendrimers. The chiroptical nature of these two dendrimers symbolises the presence of conformational substructures in dendrimeric branches (Cicchi et al. 1998). Phenylalanine dendrimers, which are fully chiral in nature, show minimum steric interaction up to generation 2. Polyether dendrimers were studied for this parameter, and it was reported that up to generation 4, the molar rotation is roughly proportional to the chiral unit number (Chang et al. 1996).

8.2.1.7 Fluorescence Spectroscopy

The highly sensitive fluorescence was employed to detect structural defects during dendrimer synthesis. Fluorescence spectroscopy consists of a covalently attached photo-chemical probe, which is used for characterising the dendrimeric structure. Pyrene molecules linked with PPI dendrimers showed an increasing aggregation with the increase in generations (Baker and Crooks 2000). The selective attachment of pyrene molecules to different dendrimers, such as poly(aryl esters) dendrimers, can be applied for investigating the molecular dynamics of the exterior region (Wilken and Adams 1997). A spectrofluorometer was utilised for studying the intrinsic fluorescence of PAMAM dendrimers. The results revealed information regarding the molecular environment (Konopka et al. 2018). In characterisation of polyphenyl (Devadoss et al. 1997) or PBzE dendrimers, the fluorescent part attached with the dendrimeric core, after a specific generation, changes were observed in size and shape (Sadamoto et al. 1996). In pyrene-linked PPH dendrimers, the available free space in the dendrimeric structure was exhibited by the development of excimers (Brauge et al. 2001b). Internal quenching of chromophore-containing dendrimers was investigated by many scientists. Polyamide dendrimers generated from the porphyrin core exhibit significant quenching. This porphyrin fluorescence quenching was reported to decrease with the increase in generation 1 to generation 3 (Capitosti et al. 2001). On the other hand, when the maleimide group was attached to phosphorhydrazone dendrimers, the quenching was reported to be increased with the increase in generations (Franc et al. 2007). Dendrimers with a fluorescent core can be inspected by applying confocal microscopy. Metallodendrimers derived from rhodamine B were observed by the atomic force microscopy technique containing a probe with an opening which helps in localising the molecules (Veerman et al. 1999).

8.2.1.8 Electrochemical Impedance Spectroscopy

The electrochemical impedance spectroscopy method was applied to determine resistance values of electronic charge transfer for redox reactions. In the case of PAMAM dendrimers, it uses the equivalent circuit of the interfacial structures, for determining induced effects. It also helps to understand about interphases for designing electrochromic films. This technique is also useful to determine electronic transport, which can be employed in constructing electrochromic films (Mendez et al. 2017). Another major application of this technique was reported for ferrocene-linked PPI dendrimers (Cuadrado et al. 1996) and PPH dendrimers (Turrin et al. 2001). This method is very useful for determining the internal burying of the electroactive part (Cardona et al. 2000). It has been reported that the electroactive group present in the dendrimeric core shows a simultaneous decrease in the rate constant for electron transfer in the case of a wide range of dendrimers apart from tiny PBzE dendrimers (Ceroni et al. 2001). Apart from molecular recognition of dendrimers, this technique also helps to predict the possible interaction for electroactive terminal groups (Cuadrado et al. 1997).

8.2.1.9 Electron Paramagnetic Resonance (EPR)

This technique is also known as electron spin resonance, which is applied to investigate compounds containing unpaired electrons. In the case of PAMAM dendrimers, it is applied for quantitative estimation of substitution efficiency with the help of nitronyl nitroxide radicals (Francese et al. 2003). By using nitroxide radicals, the carboxylate group terminated PAMAM dendrimer binding sites can be assessed (Ottaviani et al. 1995). The EPR technique is also used to characterise PPI dendrimers for detecting interactions between nitroxyl terminal groups. The large size PPI dendrimers showed that the increase in relaxivity depends on the generation of dendrimers (Maliakal et al. 2003). The EPR technique is also used to determine the complexation capacity of Cu(II) through a number of PPI dendrimers (Appelhans et al. 2010). The complexation nature of PAMAM dendrimers of second to sixth generation was investigated by varying the molar ratio. The computer aided EPR technique was employed for characterising PAMAM dendrimers, which proved the effectiveness of this technique for studying interactions of Cu(II) in nano-range dendrimers (Tang et al. 2017). The paramagnetic probe is also reported to be used in determining the dendrimeric structure.

8.2.1.10 Magnetometry

The inclusion of magnetic atoms in dendrimeric structures can help in characterisation of dendrimers. The gadolinium-capped PPH dendrimers can be characterised by this method (Franc et al. 2009). The structure of dendritic polyaryl methane was characterised by superconducting quantum interface device magnetometry after generating polyradicals inside the structure. The study reported a reduced spin value, which might be the result of ferromagnetic coupling disruptions (Rajca and Utamapanya 1993). The Cr(III)- and Fe(III)-capped melamine dendrimers can also be characterised by this method (Uysal and Koc 2010). Magnetised PEGylated Fe_3O_4 dendrimeric nanoparticles were characterised by vibrating sample magnetometry. A decrease in magnetism was reported, which was probably caused by polymer backbone insertion and dendrimeric branching on the surface (Hooshyar et al. 2019).

8.2.1.11 X-ray Spectroscopy

When X-ray strikes a sample with adequate energy, the inner electrons can be ejected causing the phenomenon of X-ray fluorescence. In the course of de-excitation, the atoms emit typical radiation, which affords information regarding the dendrimeric composition. A specific application of this technique is observed in PPI dendrimers containing cobaltocenium and ferrocene end groups. The ratio of these two end groups can be determined by X-ray spectroscopy (Casado et al. 2000). The X-ray photoelectron spectroscopy technique was employed to determine the chemical composition of different dendrimers (Miksa et al. 1999). A precise analysis of structural features of platinum-PAMAM dendrimers was performed by X-ray absorption fine structure spectroscopy. The study findings helped in evaluating the local surroundings during preparation steps of PAMAM dendrimers (Alexeev et al. 2006). A recent study on carboxyl terminated PAMAM dendrimers was characterised by energy dispersive X-ray spectroscopy (Lin et al. 2017).

8.2.1.12 Mossbauer Spectroscopy

This technique was applied to characterise iron-containing dendrimers, such as dendrimers with a ferrocenium group or complexed benzyl ether dendrimers containing carboxylic acid (Zhao et al. 2008). It is also used to detect ferrocenuim hexafluorophosphate entrapment in the dendrimeric core of polyphenyl azomethine (Ochi et al. 2010). Dendrimers with different amine cores were characterised by Mossbauer spectroscopy by using a transmission spectrometer (Nariaki et al. 2014). Dendrimer-based iron complexes were characterised by using Mossbauer spectroscopy with the help of ^{57}Co. The metastable FeII and FeIII intermediates were characterised by the Mossbauer method.

8.2.2 Size Exclusion Chromatography (SEC)

SEC, a separation technique based on the size of molecules, is widely used for the characterisation of different types of dendrimers including self-assembled dendrimers (Zeng et al. 2002). SEC apparatus coupled with a differential refractive index or LLS detector determines polydispersity of molecules. However, SEC measured the change of ARB dendrimer's size with respect to pH (Newkome et al. 1993). This method can also be used for determining molecular weight distribution, polydispersity and dendrimeric purity (Tomalia et al. 2012). Low weight impurities can be eliminated by this method (Mullen et al. 2012). G5.0 PAMAM dendrimers were characterised by SEC (van Dongen et al. 2013).

8.2.3 Scattering Methods

Scattering methods are applied to analyse the size and shape of the dendrimeric structure based on the deflection of electromagnetic waves after interacting with the structure.

8.2.3.1 Small Angle X-ray Scattering (SAXS)

An analytical technique, SAXS, measures the intensities of X-rays scattered by a testing sample as a function of scattering angle 2θ: close to $0°$, is applied for the characterisation of hyperbranched polymers (Chu and Hsiao 2001). From the SAXS study, several data regarding dendrimers such as the average radius of gyration (Rg) in solution and architecture of the polymer networked structure are obtained. The SAXS technique provides the Rg values of fluorinated carbosilane dendrimers (Omotowa et al. 1999) and PAMAM dendrimers (Prosa et al. 1997). SAXS provides structural information of DNA-based enzyme linked dendrimers. For DNA dendrimers, the internal heterogeneous structure can be obtained from the Kratky plot of the data (Walsh et al. 2015). A relatively open boundary was indicated for third generation PAMAM dendrimers, whereas an intensely sharp outer boundary was reported for tenth generation dendrimers. A gradual transition was observed from the third generation (star-like) to tenth generation (sphere-like) PAMAM dendrimers (Prosa et al. 2001).

8.2.3.2 Wide-Angle X-ray Scattering (WAXS)

The intensity of X-rays scattered by a sample as a function of scattering angle larger than $5°$ is measured by the wide-angle X-ray scattering technique. Several studies revealed that the WAXS technique detects the thermotropic phases and layered mesophase formed by the perfluoroalkyl group of carbosilane dendrimers (Lorenz et al. 1997). PAMAM dendrimers were characterised by using the wide angle X-ray scattering technique (Berényi et al. 2014).

8.2.3.3 Small Angle Neutron Scattering (SANS)

The SANS technique indicates the internal architecture and radius of gyration of dendrimers more accurately than SAXS and laser light scattering. Additionally, SANS may determine the molecular weight of PPI (Potschke et al. 1999), PAMAM (Topp et al. 1999b) and PBzE dendrimers (Evmenenko et al. 2001). Carboxylic acid-terminated dendrimeric analysis in solution phase by the SANS technique showed that an increased dendrimer concentration can broaden a peak obtained from inter-dendrimer interaction (Huang et al. 2005). Several studies revealed the peripheral (Topp et al. 1999a) and overall distribution (Rosenfeldt et al. 2002) of the labelled or unlabelled end groups in the networked structure of PAMAM dendrimers and PPI dendrimers, respectively. The neutron spin-echo technique, a recent advancement of SANS, was also adapted for the characterisation of PAMAM dendrimers (Funayama et al. 2003).

8.2.3.4 Dynamic Light Scattering (DLS)

The DLS method is widely used for investigating the dendrimeric structure. This technique is helpful in detecting the hydrodynamic radius (RH) of dendrimers. The value of the diffusion coefficient, D, can be calculated, which can be utilised to derive the RH of dendrimers (Wang et al. 2012). The RH can also be calculated by laser light scattering. PBzE dendrimers with a phthalocyanine core can be characterised by the dynamic LLS technique (Li et al. 2000a).

8.2.4 Microscopy

8.2.4.1 Atomic Force Microscopy (AFM)

AFM has been widely used for high-resolution imaging of dendrimers. An atomic force microscope was used to characterise metallodendrimers having the focal point rhodamine B (Veerman et al. 1999) and dendritic macromolecular films. It can also be used for estimating significant molecular parameters, such as molecular weight and volume. In the case of PAMAM dendrimers of generation 5.0 to generation 10.0, the spin coating technique was used for visualising isolated dendrimeric molecules. The dendrimeric diameters were found to be larger than heights indicating a hemispherical structure of surface deposited dendrimers (Li et al. 2000b). This technique is also used for studying morphology, molecular dynamics and molecular packing of dendrimers (Schlüterand Rabe 2000).

8.2.4.2 Transmission Electron Microscopy (TEM)

In this technique, a beam of electron is used that produces the image of high resolution regulated by the wavelength. Moreover, the visualisation of individual molecules of covalently attached gold PMMH dendrimers (G3.0–G10.0) has been performed by TEM (Slany et al. 1995). The size distribution of PAMAM dendrimers was measured by TEM (Oddone et al. 2016). TEM revealed a strict linear correlation between the size and the logarithm of the molecular weight of G5.0–G10.0 PAMAM dendrimers (Jackson et al. 1998). Optical microscopy has been applied to visualise the single molecule of dendrimers having a fluorescent core. Observation of fluorescence of a dihydropyrrolopyrroledione core containing PBzE dendrimers (third generation) (Hofkens et al. 1998) and polyphenylene dendrimers (having peryleneimide end groups) (Gensch et al. 1999) has been successfully carried out by confocal microscopy.

8.2.4.3 Polarising Optical Microscopy

The self-assembled liquid crystalline dendrons or dendrimers are characterised by this technique. Liquid crystals of PPI and PAMAM dendrimers showed columnar mesomorphism and lamellar mesomorphism (Martin-Rapun et al. 2005). The nematic phase of PAMAM- and PPI-based liquid crystal dendrimers was detected by the polarising optical microscopy technique (Martin-Rapun et al. 2015). The characterisation of a diethanolamine-based dendrimer by POM showed droplet texture at 66°C during the cooling process (Didehban et al. 2009).

8.2.5 Physical and Rheological Characterisation

8.2.5.1 Intrinsic Viscosity

The main objective of the rheological study is to determine the morphological structure of dendrimers as an analytical probe. This is an important property for characterising the rheological character of dendrimeric solutions. Several studies on G2.0 or G3.0 for PbzE dendrimers (Matos et al. 2000), G3.0 phosphorus dendrimers with two different types of end groups (Merino et al. 2001), G4.0 for PAMAM (Dvornic and Uppuluri 2001) and G5.0 for PPI (Scherrenberg et al. 1998) revealed that the maxima of the intrinsic viscosity varied with density of the branches. On the contrary, an exceptional result was observed for PLy dendrimers (Aharoni et al. 1982) due to close segmentation along with asymmetrical geometry (Tomalia et al. 1987). In the case of polyether dendrimers, the viscosity increases with generations. Branch density and molecular characteristics also had significant impact on intrinsic viscosity (Mourey et al. 1992).

8.2.5.2 Differential Scanning Calorimetry (DSC)

The DSC technique, a thermal analysis, detects glass transition temperature (T_g) of polymers. It depends on the degree of branching, molecular weight and end group functionality of dendrimers. A longer chain on the dendrimeric surface lowers the T_g for polyamidoamine-g-poly(N,N-dimethylaminoethyl methacrylate) (Gill et al. 2010). The T_g of PBzE dendrimers (Farrington et al. 1998) and phosphorus dendrimers (Merino et al. 2001) was found to be varied with the end group substitutions and molecular mass. Similar observation was found for PPI dendrimers (Tande et al. 2003) to a lesser extent. However, generation has no impact on the T_g values of liquid crystals of poly(phenyl acetylene) (Pesak and Moore 1997) or carbosilazane dendrimers (Elsasser et al. 2001). Besides DSC, Temperature modulated calorimetry has been successfully employed to detect physical aging of PMMH dendrimers (Dantras et al. 2002).

8.2.5.3 Dielectric Spectroscopy

This method affords information regarding the molecular dynamics for dendrimeric polymers. This method was applied in characterising different dendrimers, where the obtained α relaxation value supports the data obtained by DSC. For instance, the α relaxation value for PPI dendrimers reflectsa similar result as obtained by the DSC method (Mohamed et al. 2015).

8.2.6 Miscellaneous

Elemental analysis is also applied to characterise dendrimers except for the repetitive structure of high generation dendrimers. The sedimentation rate and dipole moments were observed on lactosylated PAMAM dendrimers (Pavlov et al. 2001),PBzE dendrimers (Wooley et al. 1993) and PMMH dendrimers (Lartigue et al. 1997), respectively. A titrimetric study was performed for the number of NH2 end groups of PAMAM dendrimers (Zhuo et al. 1999). Quartz crystal microbalance with dissipation and multi-parametric surface plasmon resonance has also been successfully used for the immobilisation controlling parameters of PAMAM dendrimers on the Au surface (Jachimska 2019). Purity and homogeneity of water-soluble dendrimers such as PAMAM dendrimers (Brothers II et al. 1998), PPI dendrimers (Welch and Hoagland 2003), nucleic acid dendrimers (Hudson and Damha 1993) or phenylacetylene dendrimers (Pessac et al. 1997) have been successfully analysed by gel electrophoresis. Studies reported that gel electrophoresis was used to determine an interaction between positively charged dendrimers and DNA for PAMAM dendrimers (Kukowska-Latallo et al. 1996) and poly(ethylene glycol)-block-poly(l-lysine) dendrimers (Choi et al. 1999). The scanning electron microscopy (SEM) method was also performed to study the morphological characteristics of dendrimers. Liquid crystalline PPI dendrimers were characterised by SEM (Figure 8.4), where the image showed

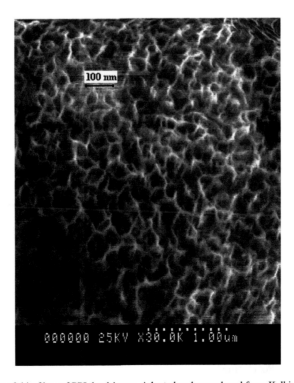

FIGURE 8.4   SEM image of thin films of PPI dendrimers. Adapted and reproduced from Kulbickas et al. 2007.

a complicated self-assembly of dendrimers (Kulbickas et al. 2007). The polymer molecular weight and PDI of charge-neutral carboxybetaine and sulphobetaine zwitterionic dendrimers were determined using gel permeation chromatography (Roeven et al. 2019).

8.3 Conclusions

Specific characterisation of dendrimers plays a significant role for establishing them as an attractive carrier system in targeted delivery of drugs. With high branching ability, spherical architecture and multivalency, dendrimers have become a very interesting carrier system to characterise. The step-wise synthesis of dendrimers has even made it more challenging for the scientists. The most widely applied methods for characterising dendrimers were discussed in this chapter. Different spectroscopic and spectrometric methods were described briefly in context with dendrimeric characterisation. Different methods for determining structural integrity were explored. Many dendrimer-based encapsulated materials were recently investigated which include nanomaterials, model drugs, biological entities and dyes. Depending on the encapsulated materials, the characterisation techniques differ. For instance, the nano-material-encapsulated dendrimeric core can be commonly characterised by using FT-IR, UV-vis spectroscopy and TEM techniques. Similarly, electrochemistry along with MALDI-ToF is very useful in characterising encapsulated materials. Fluorescence and UV-vis spectroscopy along with ^{1}H NMR were applied in characterisation of dendrimers containing fluorescent compounds. ^{1}H NMR spectroscopy and TEM are widely used to characterise biologically relevant smaller entities. These techniques are not only applied for characterising dendrimers, but also for characterising the interactions with the microenvironment surrounded by dendrimers. Apart from these, different separation techniques and scattering methods along with rheological characterisation are frequently used as characterisation techniques for macromolecule-based dendrimers.

Abbreviations

AFM	Atomic force microscopy
Boc	Butoxycarbonyl
DLS	Dynamic light scattering
DNA	Deoxyribonucleic acid
DSC	Differential scanning calorimetry
EFG	Electric field gradient
EPR	Electron paramagnetic resonance
ESI	Electrospray ionization
ESR	Electron spin resonance
HPDEC	High-power decoupling
HRMS	High resolution mass spectroscopy
INADEQUATE	Incredible natural abundance double-quantum transfer experiment
IR	Infrared spectroscopy
MALDI-ToF	Matrix-assisted laser desorption ionization-time of flight
MAS	Magic-angle spinning
MS	Mass spectrometry
NMR	Nuclear magnetic resonance
PAMAM	Polyamidoamine
PBzE	Poly(benzyl ether)
PDI	Polydispersity index
PEG	Polyethylene glycol
PFGSE	Pulse field-gradient spin echo
PLy	Polylysine
PMMH	Phenoxymethyl(methylhydrazono)

PPH	Poly(phosphorhydrazone)
PPI	Poly(propyleneimine)
RH	Hydrodynamic radius
SAXS	Small angle X-ray scattering
SEC	Size exclusion chromatography
SEM	Scanning electron microscope
SQUID	Superconducting quantum interface device
TEM	Transmission electron microscopy
TMC	Temperature modulated calorimetry
UV-vis spectroscopy	*Ultraviolet–visible spectroscopy*
WAXS	Wide-angle x-ray scattering
XRD	X-ray diffraction

REFERENCES

Aharoni, S. M., Crosby, C. R., Walsh, E. K. 1982. Size and solution properties of globular tert-butyloxycarbonyl-poly (a, q-llysine). *Macromolecules 15*:1093–1098.

Alexeev, O. S., Siani, A., Lafaye, G., Williams, C. T., Ploehn, H. J., Amiridis, M. D. 2006. EXAFS Characterization of dendrimer-Pt nanocomposites used for the preparation of Pt/ç-Al$_2$O$_3$ catalysts. *Journal of Physical Chemistry B 110*:24903–24914.

Appelhans, D., Oertel, U., Mazzeo, R., et al. 2010. Dense-shell glycodendrimers: UV/Vis and electron paramagnetic resonance study of metal ion complexation. *Proceedings of the Royal Society A 466*:1489–1513.

Bae, J., Son, W. S., Yoo, K. H., et al. 2019. Effects of poly(amidoamine) dendrimer-coated mesoporous bioactive glass nanoparticles on dentin remineralization. *Nanomaterials 9*:591. doi:10.3390/nano9040591.

Baker, L. A., Crooks, R. M. 2000. Photophysical properties of pyrene—functionalized poly(propylene imine) dendrimers. *Macromolecules 33*:9034–9039.

Berényi, S., Mihály, J., Wacha, A., Tőke, O., Bóta, A. 2014. A mechanistic view of lipid membrane disrupting effect of PAMAM dendrimers. *Colloids and Surfaces B: Biointerfaces 118*:164–171.

Biricova, V., Laznickova, A. 2009. Dendrimers: analytical characterization and applications. *Bioorganic Chemistry 37*:185–192.

Biswas, S., Torchilin, V. P. 2014. Nanopreparations for organelle-specific delivery in cancer. *Advanced Drug Delivery Reviews 66*:26–41.

Blais, J. C., Turrin, C. O., Caminade, A. M., Majoral, J. P. 2000. MALDI ToF mass spectrometry for the characterization of phosphorus—Containing dendrimers. Scope and limitations. *Analytical Chemistry 72*:5097–5105.

Bosman, A. W., Bruining, M. J., Kooijman, H., Spek, A. L., Janssen, R. A. J., Meijer, E. W. 1998. Concerning the localization of end groups in dendrimers. *Journal of the AmericanChemical Society 120*: 8547–8548.

Bosman, A. W., Meijer, E. W. 1999. About dendrimers: structure, physical properties, and applications. *Chemical Reviews 99*:1665–1688.

Brauge, L., Caminade, A-M., Majoral, J-P., Slomkowski, S., Wolszczak, M. 2001a. Segmental mobility in phosphorus—Containing dendrimers. Studies by fluorescent spectroscopy. *Macromolecules 34*:5599–5606.

Brauge, L., Magro, G., Caminade, A-M., Majoral, J-P. 2001b. First divergent strategy using two AB$_2$ unprotected monomers for the rapid synthesis of dendrimers. *Journal of the American Chemical Society 123*:6698–6699.

BrothersII, H. M., Piehler, L. T., Tomalia, D. A. 1998. Slab-gel and capillary electrophoretic characterization of polyamidoamine dendrimers. *Journal of Chromatography A 814*:233–246.

Buhleier, E., Wehner, W., Vögtle, F. 1978. "Cascade"- and "nonskid-chain-like" synthesis of molecular cavity topologies. *Synthesis 9*:155–158.

Caminade, A. M., Turrin, C. O., Sutra, P., Majoral, J. P. 2003. Fluorinated dendrimers. *Current Opinion in Colloid & Interface Science 8*:282–295.

Capitosti, G. J., Cramer, S. J., Rajesh, C. S., Modarelli, D. A. 2001. Photoinduced electron transfer within porphyrin—Containing poly(amide) dendrimers. *Organic Letters 3*:1645–1648.

Cardona, C. M., Mendoza, S., Kaifer, A. E. 2000. Electrochemistry of encapsulated redoxcenters. *Chemical Society Reviews 29*:37–42.

Casado, C. M., Gonzalez, B., Cuadrado, I., Alonso, B., Moran, M., J. Losada. 2000. Mixed ferrocene- cobalto-cenium dendrimers: the most stable organometallic redox systems combined in a dendritic molecule. *Angewandte Chemie 39*:2135–2138.

Ceroni, P., Vicinelli, V., Maestri, M., et al. 2001. Dendrimers with a 4, 4′ - bipyridinium core and electron – donor branches. Electrochemical and spectroscopic properties. *New Journal of Chemistry 25*:989–993.

Chang, H. T., Chen, C. T., Kondo, T., Siuzdak, G., Sharpless, K. B. 1996. Asymmetric dihydroxylation enables rapid construction of chiral dendrimers based on 1, 2-diols. *Angewandte Chemie 35*:182–186.

Choi, J. S., Lee, E. J., Choi, Y. H., Jeong, Y. J., Park, J. S. 1999. Poly(ethylene glycol)-block-poly(l-lysine) dendrimer: novel linear polymer/dendrimer block copolymer forming a spherical water-soluble polyionic complex with DNA. *Bioconjugate Chemistry 10*:62–65.

Chu, B., Hsiao, B. S.2001. Small-angle X-ray scattering of polymers. *Chemical Reviews 101*:1727–1762.

Cicchi, S., Goti, A., Rosini, C., Brandi, A. 1998. Enantiomerically pure dendrimers based on a trans-3, 4-dihy-droxypyrrolidine. *European Journal of Organic Chemistry* 2591–2597. doi:10.1002/(SICI)1099-0690(199811)1998:11<2591::AID-EJOC2591>3.0.CO;2-Q

Criscione, J. M., Le, B. L., Stern, E., et al. 2009. Self-assembly of pH – responsive fluorinated dendrimer – based particulates for drug delivery and noninvasive imaging. *Biomaterials 30*:3946–3955.

Cuadrado, I., Casado, C. M., Alonso, B., Moran, M., Losada, J., Belsky, V. 1997. Dendrimers containing organo-metallic moieties electronically communicated. *Journal of the American Chemical Society 119*:7613–7614.

Cuadrado, I., Moran, M., Casado,C. M., et al.1996. Ferrocenyl – functionalized poly(propylenimine) den-drimers. *Organometallics 15*:5278–5280.

Dantras, E., Dandurand, J., Lacabanne, C., Caminade, A. M., Majoral, J. P. 2002. Enthalpy relaxation in phos-phorus-containing dendrimers. *Macromolecules 35*:2090–2094.

de Brabander-van den Berg, E. M. M. and Meijer, E. W. 1993. Poly(propyleneimine) dendrimers – large – scale synthesis by hetereogeneously catalyzed hydrogenations. *Angewandte Chemie 32*:1308–1311.

Deloncle, R. and Caminade, A. M. 2010. Stimuli-responsive dendritic structures: the case of light-driven azo-benzene-containing dendrimers and dendrons. *Journal of Photochemistry and Photobiology C: Photo-chemistry Reviews 11*:25–45.

Devadoss, C., Bharathi, P., Moore, J. S. 1997. Anomalous shift in the fluorescence spectra of a high-generation dendrimer in nonpolar solvents. *Angewandte Chemie 36*:1633–1635.

Dias, A. P., Santos, S. S., Vitor, J. S., et al. 2020. Dendrimers in the context of nanomedicine. *International Journal of Pharmaceutics 573*:118814.

Didehban, K., Namazia, H., Entezami, A. A. 2009. Synthesis and characterization of liquid crystalline dietha-nolamine-based dendrimers. *Polymers for Advanced Technology 20*:1127–1135.

Dougherty, C. A., Furgal, J. C., vanDongen, M. A., et al. 2014. Isolation and Characterization of Precise Dye/Dendrimer Ratios. *Chemistry: A European Journal 20*:4638–4645.

Dvornic, P. R. and Uppuluri, S. 2001. Rheology and solution properties of dendrimers. In *Dendrimers and Other Dendritic Polymers*, ed. J. M. J. Frechet, and D. A. Tomalia, 331–358. John Wiley & Sons.

Elsasser, R., Mehl, G.H., Goodby, J.W., Veith, M. 2001. Nematic dendrimers based on carbosilazane cores. *Angewandte Chemie 40*:2688–2690.

Ertürk, A. S., Gürbüz, M. U., Tülü, M., Bozdoğan, A. E. 2016a. Preparation of Cu Nanocomposites from EDA, DETA, and Jeffamine Cored PAMAM Dendrimers with TRIS and Carboxyl Surface Functional Groups. *Acta Chimica Slovenica 63*:763–771.

Ertürk, A. S., Gürbüz, M. U., Tülü, M., Bozdoğan, A. E. 2016b. Characterization of co(ii)-pamam dendrimer complexes with polypropylene oxide core by using UV-vis spectroscopy. *Macedonian Journal of Chem-istry and Chemical Engineering 35*:263–270.

Escamilla, G. H. and Newkome, G. R. 1994. Bolaamphiphiles – from golf balls to fibers. *Angewandte Chemie 33*:1937–1940.

Esfand, R., Tomalia, D. A. 2001. Poly(amidoamine) (PAMAM) dendrimers: from biomimicry to drug delivery and biomedical applications. *Drug Discovery Today 6*:427–436.

Evmenenko, G., Bauer, B. J., Kleppinger, R., et al. 2001. The influence of molecular architecture and solvent type on the size and structure of poly(benzyl ether) dendrimers by SANS. *Macromolecular Chemistry and Physics 202*:891–899.

Farrington, P. J., Hawker, C. J., Frechet, J. M. J., Mackay, M. M. 1998. The melt viscosity of dendritic poly(benzyl ether) macromolecules. *Macromolecules 31*:5043–5050.

Fox, M. E., Szoka, F. C., Frechet, J. M. J. 2009. Soluble polymer carriers for the treatment of cancer: the importance of molecular architecture. *Accounts of Chemical Research 42*:1141–1151.

Franc, G., Mazères, S., Turrin, C. O., et al. 2007. Synthesis and properties of dendrimers possessing the same fluorophore(s) located either peripherally or off-center. *Journal of Organic Chemistry 72*:8707–8715.

Franc, G., Turrin, C. O., Cavero, E., et al. 2009. gem-bisphosphonate-ended group dendrimers: design and gadolinium complexingproperties. *European Journal of Organic Chemistry* 4290–4299.

Francese, G., Dunand, F. A., Loosli, C., Merbach, A. E., Decurtins, S. 2003. Functionalization of PAMAM dendrimers with nitronyl nitroxide radicals as models for the outer-sphere relaxation in dentritic potential MRI contrast agents. *Magnetic Resonance in Chemistry 41*:81–83.

Fréchet, J. M. J. 1994. Functional polymers and dendrimers: Reactivity, molecular architecture,and interfacial energy. *Science 263*:1710–1715.

Funayama, K., Imae, T., Seto, H., et al. 2003. Fast and slow dynamics of water-soluble dendrimers consisting of amido-amine repeating units by neutron spin-echo. *Journal of Physical Chemistry B 107*:1353–1359.

Furer, V. L., Majoral, J. P., Caminade, A. M., Kovalenko, V. I. 2004. Elementoorganicdendrimer characterization by Raman spectroscopy. *Polymer 45*:5889–5895.

Furera, V. L., Vandyukov, A. E., Padie, C., Majoral, J. P., Caminade, A. M., Kovalenko, V.I. 2015. Raman spectroscopy studies of phosphorus dendrimers with phenoxy and deuterophenoxy terminal groups. *Vibrational Spectroscopy 80*:17–23.

Galliot, C., Larré, C., Caminade, A. M., Majoral, J. P. 1997. Regioselective step wise growth of dendrimer units in the internal voids of a main dendrimer. *Science 277*:1981–1984.

Galliot, C., Prevote, D., Caminade, A. M., Majoral, J. P. 1995. Polyaminophosphines containing dendrimers. Syntheses and characterizations. *Journal of the American Chemical Society* 117: 5470–76.

Gardikis, K., Hatziantoniou, S., Viras, K., Wagner, M., Demetzos, C. 2006. A DSC and Raman spectroscopy study on the effect of PAMAM dendrimer on DPPC model lipid membranes. *International Journal of Pharmaceutics 318*:118–123.

Gensch. T., Hofkens, J., Heirmann. A., et al. 1999. Fluorescence detection from single dendrimers with multiple chromophores. *Angewandte Chemie 38*:3752–3756.

Gibson, S. E. and Rendell, J. T. 2008. The quest for secondary structure in chiral dendrimers. *Chemical Communications*:922–941.

Gill, P., Moghadam, T. T., Ranjbar, B. 2010. Differential scanning calorimetry techniques: applications in biology and nanoscience. *Journal of Biomolecular Techniques 21*:167–193.

Hawker, C. J. and Fréchet, J. M. J. 1990. Preparation of polymers with controlled molecular architecture – a new convergent approach to dendritic macromolecules. *Journal of the American Chemical Society 112*:7638–7647.

Hawker, C. J., Wooley, K. L., Fréchet, J. M. J. 1993. Solvatochromism as a probe of the microenvironment in dendritic polyethers: transition from an extended to a globular structure. *Journal of the American Chemical Society 115*:4375–4376.

Hofkens, J., Verheijen, W., Shukla, R., Dehaen, W., DeSchryver, F. C. 1998. Detection of a single dendrimer macromolecule with a fluorescent dihydropyrrolopyrroledione (DPP) core embedded in a thin polystyrene polymer film. *Macromolecules 31*:4493–4497.

Hong, S., Leroueil, P. R., Majoros, I. J., Orr, B. G., Baker, J. R., Holl, M. M. B. 2007. The binding avidity of a nanoparticle-based multivalent targeted drug delivery platform. *Chemistry & Biology 14*:107–115.

Hooshyar, S. P., Mehrabian, R. Z., Panahi, H. A., Jouybari, M. H., Jalilian, H. 2019. Synthesis and characterization of magnetized-PEGylated dendrimer anchored to thermosensitive polymer for letrozole drug delivery. *Colloids and Surfaces B: Biointerfaces 176*:404–411.

Hsu, H. J., Bugno, J., Lee, S. R., Hong, S. 2017. Dendrimer-based nanocarriers: a versatile platform for drug delivery. *WIREs Nanomedicine and Nanobiotechnol 9*: 1–21.

Huang, Q. R., Dubin, P. L., Lal, J., Moorefield, C. N., Newkome, G. R. 2005. Small-angle neutron scattering studies of charged carboxyl-terminated dendrimers in solutions. *Langmuir 21*:2737–2742.

Hudson, R. H. E. and Damha, M. J. 1993. Nucleic acid dendrimers: novel biopolymer structures. *Journal of American Chemical Society 115*:2119–2124.

Ihre, H., Hult, A., Söderlind, E. 1996. Synthesis, characterization, and ¹H NMR self-diffusion studies of dendritic aliphatic polyesters based on 2,2-bis(hydroxymethyl)propionic acid and 1,1,1-tris(hydroxyphenyl) ethane. *Journal of the American Chemical Society 118*:6388–6395.

Jachimska, B. 2019. Physicochemical characterization of PAMAM dendrimer as a multifunctional nanocarriers. In *Nanoparticles in Pharmacotherapy*, ed. A. M. Grumezescu, 251–271. Poland: Polish Academy of Sciences.

Jackson, C. L., Chanzy, H. D., Booy, F. P., et al. 1998. Visualization of dendrimer molecules by transmission electron microscopy (TEM): staining methods and cryo-TEM of vitrified solutions. *Macromolecules 31*:6259–6265.

Jansen, J. F. G. A., Peerlings, H. W. I., de Brabander-van den Berg, E. M. M., Meijer, E. W. 1995. Optical activity of chiral dendritic surfaces. *Angewandte Chemie 34*:1206–1209.

Jeong, S. W., O'Brien, D. F., Oradd, G., Lindblom, G. 2002. Encapsulation and diffusion of water-soluble dendrimers in a bicontinuous cubic phase. *Langmuir 18*:1073–1076.

Jiang, D. L. and Aida, T. 1997. Photoisomerization in dendrimers by harvesting of low energy photons, *Nature 388*:454–456.

Jin, Y., Ren, X., Wang, W.et al. 2011. A 5-fluorouracil-loaded pH-responsive dendrimer nanocarrier for tumor targeting. *International Journal of Pharmeceutics 420*:378–384.

Kao, H. M., Stefanescu, A. D., Wooley, K. L., Schaefer, J. 2000. Location of terminal groups of dendrimers in the solid state by rotational-echo double-resonance NMR, *Macromolecules 33*:6214–6216.

Kaur, A., Jain, K., Mehra, N. K., Jain, N. K. 2017. Development and characterization of surface engineered PPI dendrimers for targeted drug delivery. *Artificial Cells, Nanomedicine, and Biotechnology 45*:414–425.

Kawaguchi, T., Walker, K. L., Wilkins, C. L., Moore, J. S. 1995. Double exponential dendrimer growth. *Journal of the American Chemical Society 117*:2159–2165.

Kesharwani, P., Jain, K., Jain, N. K. 2014. Dendrimer as nanocarrier for drug delivery. *Progress in Polymer Science 39*:268–307.

Konopka, M., Janaszewska, A., Klajnert-Maculewicz, B. 2018. Intrinsic fluorescence of PAMAM dendrimers-quenching studies. *Polymers 10*:540. doi:10.3390/polym10050540

Koper, G. J. M., Van Genderen, M. H. P., Elissen-Roman, C., Baars, M. W. P. L., Meijer, E. W., Borkovec, M. 1997. Protonation mechanism of poly(propylene imine) dendrimers and some associated oligo amines. *Journal of the American Chemical Society 119*:6512–6521.

Kukowska-Latallo, J. F., Bielinska, A. U., Johnson, J., Spindler, R., Tomalia, D. A., Baker Jr, J. R. 1996. Efficient transfer of genetic material into mammalian cells using starburst polyamidoamine dendrimers. *Proceedings of the National Academy of Sciences of the United States of America 93*:4897–4902.

Kulbickas, A., Tamuliene, J., Rasteniene, L., et al. 2007. Optical study and structure modelling of PPI liquid crystalline dendrimer derivatives. *Photonics and Nanostructures – Fundamentals and Applications 5*:178–183.

Lambert, J. B., Basso, E., Qing, N., Lim, S. H., Pflug, J. L. 1998. Two-dimensional silicon-29 inadequate as a structural tool for branched and dendritic polysilanes. *Journal of Organometallic Chemistry 554*:113–116.

Larré, C., Donnadieu, B., Caminade, A-M., Majoral, J-P. 1998. Phosphorus-containing dendrimers: chemoselective functionalization of internal layers. *Journal of the American Chemical Society 120*:4029–4030.

Lartigue, M. L., Donnadieu, B., Galliot, C., Caminade, A. M., Majoral, J. P., Fayet, J. P. 1997. Unexpected very large dipole moments of phosphorus containing dendrimers. *Macromolecules 30*:7335–7337.

Launay, N., Caminade, A. M., Lahana, R., Majoral, J. P. 1994. A general synthetic strategy for neutral phosphorus-containing dendrimers. *Angewandte Chemie 33*:1589–1592.

Launay, N., Caminade, A. M., Majoral, J. P. 1995. Synthesis and reactivity of unusual phosphorus dendrimers – a useful divergent growth approach up to the 7th generation. *Journal of the American Chemical Society 117*:3282–3283.

Lehmann, M., Fischbach, I., Spiess, H. W., Meier, H. 2004. Photochemistry and mobility of stilbenoid dendrimers in their neat phases. *Journal of the American Chemical Society 126*:772–784.

Leriche, E.-D., Marie, H.-R., Grossel, M. C., Lange, C. M., Afonso, C., CorinneL.-B. 2014. Direct TLC/MALDI–MS coupling for modified polyamidoamine dendrimers analyses. *Analytica Chimica Acta 808*: 144–150.

Li, J., Piehler, L. T., Qin, D., Baker, J. R., Tomalia, D. A., Meier, D. J. 2000a. Visualization and characterization of poly(amidoamine) dendrimers by atomic force microscopy. *Langmuir 16*:5613–5616.

Li, X., He, X., Ng, A. C. H., Wu. C., Ng. K. P. D. 2000b. Influence of surfactants on the aggregation behaviour of water-soluble dendritic phthalocyanine. *Macromolecules 33*:2119–2123.

Lin, X., Xie, F., Ma, X., Hao, Y., Qin, H., Long, J. 2017. Fabrication and Characterization of Dendrimer-functionalized Nanohydroxyapatite and Its Application in Dentin Tubule Occlusion. *Journal of Biomaterials Science, Polymer Edition 28*:846–863.

Lorenz, K., Frey, H., Stuhn, B., Mulhaupt, R. 1997. Carbosilane dendrimers with perfluoroalkyl end groups core-shell macromolecules with generation-dependent order. *Macromolecules 30*:6860–6868.

Low, P. S., Kularatne, S.A. 2009. Folate-targeted therapeutic and imaging agents for cancer. *Current Opinion in Chemical Biology 13*: 256–262.

Lukin, O., Gramlich, V., Kandre, R., et al. 2006. Designer dendrimers: branched oligosulfonimides with controllable molecular architectures. *Journal of the American Chemical Society 128*:8964–8974.

Lyu, Z., Ding, L., Huang, A.Y.-T., Kao, C.-L., Peng, L. 2019. Poly(amidoamine) dendrimers: covalent and supramolecular synthesis. *Materials Today Chemistry 13*:34–48.

Maleki, B., Reiser, O., Esmaeilnezhad, E., Choi, H. G. 2019. SO_3H-dendrimer functionalized magnetic nanoparticles (Fe_3O_4@D-NH-$(CH_2)_4$-SO_3H): synthesis, characterization and its application as a novel and heterogeneous catalyst for the one-pot synthesis of polyfunctionalized pyrans and polyhydroquinolines. *Polyhedron 162*:129–141.

Maliakal, A. J., Turro, N. J., Bosman, A. W., Cornel, J., Meijer, E. W. 2003. Relaxivity studies on dinitroxide and polynitroxyl functionalized dendrimers: effect of electron exchange and structure on paramagnetic relaxation enhancement. *Journal of Physical Chemistry A 107*:8467–8475.

Malyarenko, D. I., Vold, R. L., Hoatson, G. L. 2000. Solid state deuteron NMR studies of polyamidoamine dendrimer salts. 1. Structure and hydrogen bonding. *Macromolecules 33*:1268–1279.

Maraval, V., Caminade, A. M., Majoral, J. P., Blais, J. C. 2003. Dendrimer design: how to circumvent the dilemma of a reduction of steps or an increase of function multiplicity? *Angewandte Chemie 42*:1822–1826.

Maraval, V., Laurent, R., Donnadieu, B., Mauzac, M., Caminade, A. M., Majoral, J. P. 2000. Rapid synthesis of phosphorus-containing dendrimers with controlled molecular architectures: first example of surface-block, layer-block, and segment-block dendrimers issued from the same dendron. *Journal of the American Chemical Society 122*:2499–2511.

Martín-Rapún, R., Cano, M., McKenna, M., Serrano, J. L., Marcos, M. 2015. Side-on nematic liquid crystal dendrimers based on PAMAM and PPI as dendritic scaffolds: synthesis and characterization. *Macromolecular Chemistry and Physics 216*:950–957.

Martin-Rapun, R., Marcos, M., Omenat, A., Barbera, J., Romero, P., Serrano, J. L. 2005. Ionic thermotropic liquid crystal dendrimers. *Journal of American Chemical Society 127*:7397–7403.

Matos, M. S., Hofkens, J., Verheijen, W., et al.2000. Effect of core structure on photophysical and hydrodynamic properties of porphyrin dendrimers. *Macromolecules 33*:2967–2973.

Mendez, P. F., Lopez, J. R., Lopez-Garc, U., et al. 2017. Voltammetric and electrochemical impedance spectroscopy study of prussian blue/polyamidoamine dendrimer films on optically transparent electrodes. *Journal of the Electrochemical Society 164*:H85–H90.

Menger, F. M., Peresypkin, A. V., Wu, S. X. 2001. Do dendritic amphiphiles self – assemble in water? A Fourier transform pulse-gradient spin-echo NMR study. *Journal of Physical Organic Chemistry 14*:392–399.

Merino, S., Brauge, L., Caminade, A. M., Majoral, J. P., Taton, D., Gnanou, Y. 2001. Synthesis and characterisation of linear, hyperbranched and dendrimer-like polymers constituted of the same repeating unit. *Chemistry: A European Journal 7*:3095–3105.

Mijovic, J., Ristic, S., Kenny, J. 2007. Dynamics of six generations of PAMAM dendrimers as studied by dielectric relaxation spectroscopy. *Macromolecules 40*:5212–5221.

Miksa, B., Slomkowski, S., Chehimi, M. M., Delamar, M., Majoral, J. P., Caminade, A. M. 1999. Tailored modification of quartz surfaces by covalent immobilization of small molecules (gamma-aminopropyltriethoxysilane), monodisperse macromolecules (dendrimers), and poly(styrene/acrolein/divinylbenzene) microspheres with narrow diameter distribution. *Colloid and Polymer Science 277*:58–65.

Miller, L. L., Duan, R. G., Tully, D. C., Tomalia, D. A. 1997. Electrically conducting dendrimers. *Journal of the American Chemical Society 119*:1005–1010.

Miyake, Y., Kimura, Y., Orito, N., et al. 2015. Synthesis and functional evaluation of chiral dendrimer-triamine-coordinated Gd complexes with polyaminoalcohol end-groups as highly sensitive MRI contrast agents. *Tetrahedron 71*:4438–4444.

Mohamed, F., Hofmann, M., Pötzschner, B., Fatkullin, N., Rössler, E. A. 2015. Dynamics of PPI dendrimers: a study by dielectric and ^{2}H NMR spectroscopy and by field-cycling ^{1}H NMR relaxometry. *Macromolecules 48*:3294–3302.

Morokawa, A., Kakimoto, M. A., Imai, Y.1991. Synthesis and characterization of new polysiloxane starburst polymers. *Macromolecules 24*:3469–3474.

Mourey, T. H., Turner, S. R., Rubinstein, M., Frechet, J. M. J., Hawker, C. J., Wooley, K. L. 1992. Unique behavior of dendritic macromolecules: intrinsic viscosity of polyether dendrimers. *Macromolecules 25*:2401–2406.

Mullen, D. M., Desai, A., vanDongen, M. A., Barash, M., Baker, J.R., Holl, M. M. B. 2012. Best practices for purification of PAMAM dendrimer. *Macromolecules 45*:5316–5320.

Muller, T., Yablon, D. G., Karchner, R., et al. 2002. AFM studies of high-generation PAMAM dendrimers at the liquid/solid interface. *Langmuir 18*:7452–7455.

Murer, P. and Seebach. D.1995. Synthesis and Properties of First to Third Generation Dendrimers with Doubly and Triply Branched Chiral Building Blocks. *Angewandte Chemie 34*:2116–2119.

Nariaki, D., Lekovic, F., Homenya, P., et al. 2014. Iron(III) complexes on a dendrimeric basis and various amine core investigated by Mössbauer spectroscopy. *Journal of Physics: Conference Series 534*:012003. doi:10.1088/1742-6596/534/1/012003

Newkome, G. R., Young, J. K., Baker, G. R., et al. 1993. Cascade polymers pH dependence of hydrodynamic radii of acid terminated dendrimers. *Macromolecules 26*: 2394–2396.

Ochi, Y., Suzuki, M., Imaoka, T., et al.2010. Controlled storage of ferrocene derivatives as redox-active molecules in dendrimers. *Journal of the American Chemical Society 132*:5061–69.

Oddone, N., Lecot, N., Fernández, M., et al. 2016. In vitro and in vivo uptake studies of PAMAM G4.5 dendrimers in breast cancer. *Journal of Nanobiotechnology 14*:45. doi:10.1186/s12951-016-0197-6.

Omotowa, B. A., Keefer, K. D., Kirchmeier, R. L., Shreeve, J. M. 1999. Preparation and characterization of nonpolar fluorinated carbosilane dendrimers by APcI mass spectrometry and small-angle X-ray scattering. *Journal of the American Chemical Society 121*:11130–11138.

Ottaviani, M. F., Cossu, E., Turro, N. J., Tomalia, D. A. 1995. Characterization of Starburst dendrimers by electron-paramagnetic-resonance. 2. Positively charged nitroxide radicals of variable chain-length used as spin probes. *Journal of the American Chemical Society 117*:4387–4398.

Pande, S. and Crooks, R. M.2011. Analysis of poly(amidoamine) dendrimer structure by UV-vis spectroscopy. *Langmuir 27*:9609–9613.

Patton, D. L., Cosgrove, S. Y. T., McCarthy, T. D., Hillier, S. L. 2006. Preclinical safety and efficacy assessments of dendrimer-based (SPL7013) microbicide gel formulations in a nonhuman primate model. *Antimicrobial Agents and Chemotherapy 50*:1696–1700.

Pavlov, G. M., Errington, N., Harding, S. E., Korneeva, E. V., Roy, R. 2001. Dilute solution properties of lactosylated polyamidoamine dendrimers and their structural characteristic. *Polymer 42*:3671–3678.

Peerlings, H. W. I. and Meijer, E. W. 1997. Chirality in dendritic architectures. *Chemistry: A European Journal 3*:1563–1570.

Peetla, C., Stine, A., Labhasetwar, V. 2009. Biophysical interactions with model lipid membranes: applications in drug discovery and drug delivery. *Molecular Pharmaceutics 6*:1264–1276.

Perrot, M., Rothschild, W. G., Cavagnat, R. M. 1999. Picosecond mobilities of neat diamino poly(propylene imine) dendrimer $DAB(CN)_{64}$ from a Raman study of its CN stretching fundamental. *Journal of Chemical Physics 110*:9230–9234.

Pesak, D. J. and Moore, J. S. 1997. Columnar liquid crystals from shape persistent dendritic molecules. *Angewandte Chemie 36*:1636–1639.

Pessac, D. J., Moore, J. S., Wheat, T. E. 1997. Synthesis and characterization of water-soluble dendritic macromolecules with a stiff, hydrocarbon interior. *Macromolecules 30*:6467–6482.

Pollak, K. W., Sanford, E. M., Fréchet, J. M. J. 1998. Comparison of two convergent routes for the preparation of metalloporphyrin-core dendrimers: direct condensation vs. chemical modification. *Journal of Materials Chemistry 8*:519–527.

Popescu, M., Filip, D., Vasile, C., et al. 2006. Characterization by Fourier transform infrared spectroscopy (FT-IR) and 2D IR correlation spectroscopy of PAMAM dendrimer. *Journal of Physical Chemistry B 110*:14198–14211.

Potschke, D., Ballauff, M., Lindner, P., Fischer, M., Vogtle, F. 1999. Analysis of the structure of dendrimers in solution by small angle neutron scattering including contrast variation. *Macromolecules 32*: 4079–4087.

Pourianazar, N. T., Mutlu, P., Gunduz, U. 2014. Bioapplications of poly(amidoamine)(PAMAM) dendrimers in nanomedicine. *Journal of Nanoparticle Research 16*:2342.

Prosa, T. J., Bauer, B. J., Amis, E. J. 2001. From stars to spheres: a SAXS analysis of dilute dendrimer solutions. *Macromolecules 34*:4897–4906.

Prosa, T. J., Bauer, B. J., AmisE. J., Tomalia, D. A., Scherrenberg, R. 1997. A SAXS study of the internal structure of dendritic polymer systems. *Journal of Polymer Science, Part B: Polymer Physics 35*:2913–2924.

Pyziak, M. A., Bartkowiak, G., Popenda, L., Jurga, S., Schroeder, G. 2016. Synthesis of G0 aminopolyol and aminosugar dendrimers, controlled by NMR and MALDI TOF mass spectrometry. *Designed Monomers and Polymers 20*:144–156.

Rajca, A. and Utamapanya, S. 1993. Toward organic synthesis of a magnetic particle: dendritic polyradicals with 15 and 31centers for unpaired electrons. *Journal of the American Chemical Society 115*: 10688–10694.

Rietveld, I. B. and Bedeaux, D. 2000. Self-diffusion of poly(propylene imine) dendrimers in methanol. *Macromolecules 33*:7912–7917.

Rietveld, I. B., Bouwman, W. G., Baars, M. W. P. L., Heenan, R. K. 2001. Location of the outer shell and influence of pH on carboxylic acid-functionalized poly(propyleneimine) dendrimers. *Macromolecules 34*:8380–8383.

Roeven, E., Scheres, L., Smulders, M. M. J., Zuilhof, H. 2019. Design, synthesis, and characterization of fully zwitterionic, functionalized dendrimers. *ACS Omega 4*:3000–3011.

Ropponen, J., Nattinen, K., Lahtinen, M., Rissanen, K. 2004. Synthesis, thermal properties and X-ray structural study of weak C – H … O = C hydrogen bonding in aliphatic polyester dendrimers. *CrystEngComm 91*:559–566.

Rosenfeldt, S., Dingenouts, N., Ballauff, M., Werner, N., Vogtle, F., Lindner, P. 2002. Distribution of end groups within a dendritic structure: a SANS study including contrast variation. *Macromolecules 35*:8098–8105.

Sadamoto, R., Tomioka, N., Aida, T. 1996. Photoinduced electron transfer reactions through dendrimer architecture. *Journal of the American Chemical Society 118*:3978–3979.

Sagidullin, A. I., Muzafarov, A. M., Krykin, M. A., Ozerin, A. N., Skirda, V. D., Ignat'eva, G. M. 2002. Generalized concentration dependence of self-diffusion coefficients in poly(allylcarbosilane) dendrimer solutions. *Macromolecules 35*:9472–9479.

Saleh, T. A., Al-Shalalfeh, M. M., Al-Saadi, A. A. 2016. Graphene dendrimer-stabilized silver nanoparticles for detection of methimazole using surface-enhanced Raman scattering with computational assignment. *Scientific Reports*. doi:10.1038/srep32185.

Scherrenberg, R., Coussens, B., vanVliet, P., et al. 1998. The molecular characteristics of poly(propyleneimine) dendrimers as studied with small-angle neutron scattering, viscosimetry, and molecular dynamics. *Macromolecules 31*:456–461.

Schlüter, A. D. and Rabe, J. P. 2000. Dendronized polymers: synthesis, characterization, assembly at interfaces, and manipulation. *Angewandte Chemie 39*:864–883.

Schwartz, B. L., Rockwood, A. L., Smith, R. D. 1995. Detection of high molecular weight starburst dendrimers by electrospray ionization mass spectrometry. *Rapid Communication in Mass Spectroscopy 9*:1552–1555.

Seyferth, D., Son, D. Y., Rheingold, A. L., Ostrander, R. L. 1994. Synthesis of an organosilicondendrimer containing 324 Si – H bonds. *Organometallics 13*:2682–2690.

Simpson, C. D., Mattersteig, G., Martin, K., et al. 2004. Nanosized molecular propellers by cyclodehydrogenation of polyphenylene dendrimers. *Journal of the American Chemical Society 126*:3139–3147.

Slany, M., Bardajı, M., Casanove, M. J., Caminade, A. M., Majoral, J. P., Chaudret, B. 1995. Dendrimer surface chemistry. An easy access to polyphosphines and their gold complexes. *Journal of American Chemical Society 117*:9764–9765.

Smith, D. K. and Diederich, F. 1998. Functional dendrimers: unique biological mimics. *Chemistry: A European Journal 4*:1353–1361.

Sohail, I., Bhatti, I. A., Ashar, A., et al. 2020. Polyamidoamine (PAMAM) dendrimers synthesis, characterization and adsorptive removal of nickel ions from aqueous solution. *Journal of Materials Research and Technology 9*:498–506.

Strašák, T., Maly, M., Mullerova, M., et al. 2016. Synthesis and characterization of carbosilane dendrimer–sodium montmorillonite clay nanocomposites. Experimental and theoretical Studies. *RSC Advances 49*:43356–43366.

Svenson, S. and Tomalia, D. A. 2012. Dendrimers in biomedical applications-reflections on the field. *Advanced Drug Delivery Reviews 64*:102–115.

Tande, B. M., Wagner, N. J., Kim, Y. H. 2003. Influence of end groups on dendrimer rheology and conformation. *Macromolecules 36*:4619–4623.

Tang, Y. H., Cangiotti, M., Kao, C. L., Ottaviani, M. F. 2017. EPR Characterization of Copper (II) Complexes of PAMAM-Py Dendrimers for Biocatalysis in the Absence and Presence of Reducing Agents and a Spin Trap. *Journal of Physical Chemistry B 121*:10498–10507.

Tolic, L. P., Anderson, G. A., Smith, R. D., Brothers, H. M., Spindler, R., Tomalia, D. A. 1997. Electrospray ionization Fourier transform ion cyclotron resonance mass spectrometric characterization of high molecular mass Starburst (TM) dendrimers. *International Journal of Mass Spectrometry 165*:405–441.

Tomalia, D. A., Baker, H., Dewald, J.et al. 1985. A new class of polymers: starburst- dendritic macromolecules. *Polymer Journal 17*:117–132.

Tomalia, D. A., BakerH, Dewald J., et al. 1986. Dendritic macromolecules: synthesis of starburst dendrimers. *Macromolecules 19*:2466–2468.

Tomalia, D. A., Hall, M., Hedstrand, D. M. 1987. Starburst dendrimers. III. The importance of branch junction symmetry in the development of topological shell molecules. *Journal of American Chemical Society 109*:1601–1603.

Topp, A., Bauer, B. J., Klimash, J. W., Spindler, R., Tomalia, D. A., Amis, E. J. 1999a. Probing the location of the terminal groups of dendrimers in dilute solution. *Macromolecules 32*:7226–7231.

Topp, A., Bauer, B. J., Tomalia, D. A., Amis, E. J. 1999b. Effect of solvent quality on the molecular dimensions of PAMAM dendrimers. *Macromolecules 32*:7232–7237.

Turrin, C. O., Chiffre, J., Daran, J. C., et al. 2001. New chiral phosphorus-containing dendrimers with ferrocenes on the periphery. *Tetrahedron 57*:2521–2536.

Turrin, C. O., Maraval, V., Caminade, A. M., Majoral, J. P., Mehdi, A., Reye, C. 2000. Organic – inorganic hybrid materials incorporating phosphorus-containing dendrimers. *Chemistry of Material 12*:3848–3856.

Turrin, C. O., Maraval, V., Leclaire, J., et al. 2003. Surface, core, and structure modifications of phosphorus-containing dendrimers. Influence on the thermal stability. *Tetrahedron 59*:3965–3973.

Uysal, S. and Koc, Z. E. 2010. Synthesis and characterization of dendrimeric melamine cored [salen/salophFe(III)] and [sale n/salophCr(III)] capped complexes and their magnetic behaviors. *Journal of Hazardous Materials 175*:532–539.

vanDongen, M. A., Desai, A., Orr, B. G., Baker, J. J. R., Holl, M. M. B. 2013. Quantitative analysis of generation and branch defects in G5 poly(amidoamine) dendrimer. *Polymer 54*:4126–4133.

vanDongen, M. A., Orr, B. G., Banaszak, H. M. M. 2014. Diffusion NMR Study of generation-five PAMAM dendrimer materials. *Journal of Physical Chemistry B 118*:7195–7202.

Veerman, J. A., Levi, S. A., vanVeggel, F. C. J. M., Reinhoudt, D. N., vanHulst, N. F. 1999. Near-field scanning optical microscopy of single fluorescent dendritic molecules. *Journal of Physical Chemistry A 103*:11264–11270.

Walsh, R., Morales, J. M., Skipwith, C. G., Ruckh, T. T., Clark, H. A. 2015. Enzyme linked DNA dendrimer nanosensors for acetylcholine. *Scientific Reports 5*:14832. doi:10.1038/srep14832.

Wang, X., Guerrand, L., Wu, B., et al. 2012. Characterizations of polyamidoamine dendrimers with scattering techniques. *Polymers 4*:600–616.

Wang, X. Q., Wang, W., Li, W. J., et al. 2018. Dual stimuli-responsive rotaxane-branched dendrimers with reversible dimension modulation. *Nature Communications 9*:3190.doi:10.1038/s41467-018-05670-y

Welch, C. F. and Hoagland, D. A. 2003. The electrophoretic mobility of PPI dendrimers: do charged dendrimers behave as linear polyelectrolytes or charged spheres? *Langmuir 19*:1082–1088.

Wilken, R. and Adams, J. 1997. End group dynamics of fluorescently labeled dendrimers. *Macromolecular Rapid Communications 18*:659–665.

Wind, M., Saalwachter, K., Wiesler, U. M., Mullen, K., Spiess, H. W. 2002. s. *Macromolecules 35*:10071–10086.

Wooley, K. L., Hawker, C. J., Frechet, J. M. J. 1993. Unsymmetrical three-dimensional macromolecules: preparation and characterization of strongly dipolar dendritic macromolecules. *Journal of American Chemical Society 115*:11496–11505.

Yamamoto, K., Higuchi, M., Shiki, S., Tsuruta, M., Chiba, H. 2002. Stepwise radial complexation of imine groups in phenylazomethine dendrimers. *Nature 415*:509–511.

Zeng, F., Zimmerman, S. C., Kolotuchin, S. V., Reichert, D. E. C., Ma, Y. 2002. Supramolecular polymer chemistry: design, synthesis, characterization, and kinetics, thermodynamics, and fidelity of formation of self-assembled dendrimers. *Tetrahedron 58*:825–843.

Zhao, M., Helms, B., Slonkina, E., et al. 2008. Iron complexes of dendrimer-appended carboxylates for activating dioxygen and oxidizing hydrocarbons. *Journal of the American Chemical Society 130*:4352–4363.

Zhou, L., Russell, D. H., Zhao, M. Q., Crooks, R. M. 2001. Characterization of poly(amidoamine) dendrimers and their complexes with Cu by matrix-assisted laser desorption ionization mass spectrometry. *Macromolecules 34*:3567–3573.

Zhou, W. and Peng, X. 2019. Enhanced separation capability of rhodium ionic catalyst encapsulated by propionation-terminated poly(propylene imine) dendrimer. *Macromolecular Research 27*:238–242.

Zhuo, R. X., Du, B., Lu, Z. R. 1999. In vitro release of 5-fluorouracil with cyclic core dendritic polymer. *Journal of Controlled Release 57*:249–257.

9

Dendrimer–Guest Interaction Chemistry and Mechanism

Valamla Bhavana, Thakor Pradip, Keerti Jain and Neelesh Kumar Mehra

CONTENTS

9.1 Introduction

Hyperbranched polymers have been playing a pivotal role in the scientific community with more emphasis on drug delivery and targeting. Dendrimers are mono-dispersed, globular, interior cavities with large end-terminal functional groups at the peripheral group. Dendrimers encompass the host–guest interaction phenomenon with three-dimensional polymeric nanoarchitectures. In dendrimers, all bonds emerge radially from a central core focal point along with repeating units; each contributes a branching point in dendrimers. The US Food and Drug Administration approved Vivagel™ i.e. SPL 7013 Gel in human immunodeficiency virus/herpes simplex virus prevention, and it is a water-based vaginal product using Carbopol® gel. It is available in market and physiologically compatible with the normal human vagina (Sharma et al. 2020; Jain et al. 2015; Mehra et al. 2015a, 2015b, Gupta et al. 2014; Agrawal et al. 2013). These branching points and core give rise to different kinds of generations with multivalent surface functional groups. Three architectural components of dendrimers are (i) core, (ii) interior layers or generations comprising repeating units and (iii) outer functional chemical groups attached at the outermost generation shown in Figure 9.1 (Jain et al. 2020; Carvalho et al. 2020; Santos et al. 2020; Mishra and Kesharwani 2016; Jain et al. 2015).

Synonymous terms for dendrimers include 'arborols' and cascade molecules. Dendrimers were first introduced in 1985, and the term was coined by Prof. Donald A. Tomalia (Tomalia et al. 1990;

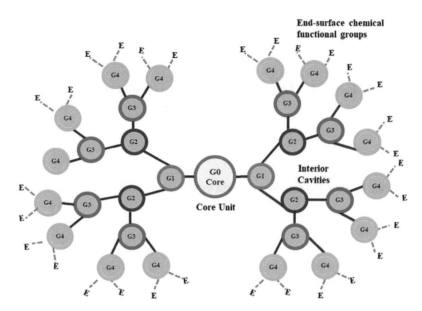

FIGURE 9.1 Structure of dendrimers (G0, initial generation, G1: first generation, G2: second generation, G3: third generation and G4: fourth generation).

Tomalia et al. 1985). The numerous bioactive(s) including drug, nucleic acid, imaging, targeting, contrast agents, proteins, peptide, antibodies and aptamers could be easily conjugated with the available end surface functional groups of dendrimers and make a perfect platform or scaffold in supramolecular chemistry (Jain et al. 2020; Wei et al. 2020; Mehra et al., 2013; Esfand and Tomalia 2001). In this book chapter, we exhaustively discuss and summarise the probable bonding interaction among biomolecules and other bioactive(s) with dendrimers.

9.1.1 Surface Charge of Dendrimers

Surface charge of the existing dendrimers plays an important role in theranostic application along with the negative aspects in drug delivery and targeting. Surface charge may sometimes indicate the toxicity of dendrimers due to the presence of end-terminal-NH_2 chemical functional groups. Multiple cationic charges of dendrimers restrict their use in drug delivery and targeting due to cytotoxicity issues. The cationic nature of dendrimers including poly-(amido) amine (PAMAM), poly-propylene-imine (PPI) and poly-l-lysine dendrimers shows the *in vitro* cytotoxicity. The safety of dendrimers is conflicting because several research groups suggest that the concentration and generation-dependent toxicities are caused by the presence of free amine functional groups at the peripheral end of the dendrimers (Zhong et al. 2017; Jain et al. 2020; Xiao et al. 2020; Kesharwani et al. 2015; Kesharwani et al. 2014; Mishra et al. 2015; Mehra et al. 2016; Jain et al. 2015). Therefore, exploration is required to better understand the toxicity issues of dendrimers. The salient characteristics of dendrimers with respect to the surface charge are discussed below (Jain et al. 2020; Mishra et al. 2019):

1. DNA has negative charge, and cationic dendrimers such as PPI and PAMAM formed a DNA–dendrimer complex due to their ionic interaction.
2. The cationic charge of the DNA–dendrimer complex may allow them to interact with a negatively charged biological membrane. It leads to membrane destabilisation and endocytosis after internalisation of the complex.
3. Dendrimers are soluble in aprotic solvents and polar solvents owing to their surface charge.
4. The presence of amine (cationic surface charge) is associated with haemolysis, drug leakage, immunogenicity and cytotoxicity.

5. The surface charge, molecular weight and molecular geometry may affect the microvascular extravasation of dendrimers.

6. In the acidic pH microenvironment, cationic dendrimers show extended conformation due to electrostatic repulsion. However, at neutral pH, back folding may happen due to hydrogen bonding between the cationic amine and uncharged tertiary amine in the interior.

9.2 Dendrimer–Guest Interaction Chemistry

The dendrimer–guest interaction chemistry plays a promising and significant role in the design of complexes of guest moieties in drug delivery and targeting. The guest–host interaction chemistry is divided into endo- or exo-complexation, which is determined by whether the guest molecule is bound in the interior or to the end-surface chemical functional groups of the dendrimers (Ficker et al. 2015a).

1-(4-Carbomethoxy) pyrrolidone-coated PAMAM dendrimers show promising potential in complexation and release of drug molecules with unique and favourable properties in organic and aqueous solutions with a benign toxicity profile (Ficker et al. 2015b). The biomolecule(s) are conjugated with the end-terminal chemical functional groups and the interior cavities of the dendrimers. We reviewed the various mechanisms of dendrimer–drug interactions such as hydrogen bonding, hydrophobic linkage, electrostatic interaction, cleavable hydrazine bond interaction, host–guest interaction and physical encapsulation (Singh et al. 2016; Madaan et al. 2014; Menjoge et al. 2010; Fox et al. 2018; Shcharbin et al. 2017; Emanuele and Attwood 2005). Readers may refer the review article entitled "Interaction of dendrimers with biological drug targets: reality or mystery-a gap in drug delivery and development research" to understand the impact of mechanistic analysis of dendrimer interaction with therapeutic biological targets at a molecular level (Ahmed et al. 2016).

The available interaction and conjugation mechanisms involved in conjugation are shown in Figure 9.2. Drug–dendrimer conjugates have been synthesised by conjugating free functional groups of drugs with chemical functional groups of dendrimers using various covalent linkages (amide, ester, hydrazine, disulphide and azo bonds). Anticancer drugs such as paclitaxel (PTX) (Cline et al. 2013), doxorubicin (DOX) (Almuqbil et al. 2020; Zhang et al. 2018), gemcitabine (Oztruck et al. 2017), methotrexate (MTX) (Thomas et al. 2012), cisplatin (Pt) (Nguyen et al. 2015), camptothecin (CPT) (Laskar et al. 2019; Zolotarskaya et al. 2015), 5-fluorouracil (5-FU) (Bhadra et al. 2003) and other classes of drug molecules

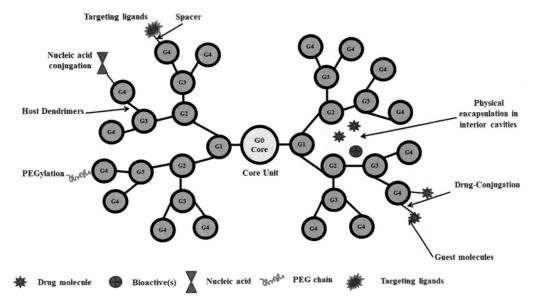

FIGURE 9.2 Host–guest interaction mechanism of dendrimers.

such as ibuprofen, piroxicam and indomethacin have been shown to form stable drug–dendrimer complexes (Pooja et al. 2018; Jain et al. 2020).

Hansen and co-workers reported the study of endo-complexation of oxacillin to three generation of G4.0 1,4-diaminobutane-core 1-(4-carbomethoxy) pyrrolidone functionalised PAMAM dendrimers using nuclear magnetic resonance (NMR) in CD_3OD and $CDCl_3$. Authors found that the stoichiometry of the guest–host complexes showed solvent- and generation-dependency. This oxacillin and other penicillin are sold as alkali metal salts owing to their low stability of the free carboxylic acid. The weaker binding of oxacillin upon increasing the size i.e. generation of dendrimers (Hansen et al. 2013) was also observed.

9.2.1 Dendrimer-Drug Loaded Nanovehicles

Drug(s) have been directly loaded into the interior cavities of dendrimers using physical encapsulation or non-covalent interaction. The hydrophobic, electrostatic and hydrogen bonding interactions are associated in non-covalent interaction of drug(s) molecules and dendrimers. The interior cavities of the dendrimers work as a container to load the drug(s) molecules and release them in a controlled manner as well reduce the toxicities associated with the drug molecules (Kulhari et al. 2011; He et al. 2015: Tekade et al. 2009). Until now, huge research reports have been available on dendrimer-drug loaded nanoformulations. In this section, we discussed the drug loaded-dendrimer formulation for biomedical and pharmaceutical applications.

In 2014, sulfasalazine (NF-κB inhibitor drug) was loaded into the fucose-conjugated G5.0 PAMAM dendrimers for kupffer cell targeting in efficient management of cytokine-induced liver damage (Gupta et al. 2014). DOX was encapsulated within the amine-terminated fifth generation PAMAM dendrimers after modification with the chelator/gadolinium (Gd) complexes with folic acid (FA) through polyethylene glycol (PEG) as spacer yielding G5.0.NHAc-DOTA(Gd)-PEG-FA complexes. These complexes showed better stability under different pH conditions. The formed dendrimer nanocomplexes also exhibit greater therapeutic efficacy of DOX by FAR-overexpressing cancer cells and can be a promising theranostic nanoplatform for magnetic resonance imaging and targeted chemotherapy (Zhu et al. 2015).

9.2.2 Dendrimer–Drug Conjugation

In the last few decades, several researchers have been published on dendrimer–drug conjugation and interaction for their role in drug targeting and delivery. Several biomolecule(s) have been conjugated with the end surface chemical functional groups for temporal and spatial drug release patterns. These dendrimer–drug conjugations could open new opportunities in targeted and controlled drug delivery. The dendritic systems have a un-customised drug release pattern (entrapped drug is not being well retained and escapes out rapidly) and haemolytic toxicities. The conjugation of drug molecules with the multi-valent dendrimeric structure *via* covalent conjugation chemistry was also formed. The therapeutic efficacy of dendrimer–drug conjugates depends upon the generation and the chemical linkage of dendrimers (Pooja et al. 2018; Tekade et al. 2009). The dendrimer–drug conjugation is an important approach and shows promising benefits in drug delivery as discussed below:

1. Enhances the drug loading capacity.
2. Increases the aqueous solubility of poorly water-soluble drugs (e.g. BCS class II and IV).
3. Enhances the pharmacokinetic parameters of the conjugated drug (such as volume of distribution, plasma half-life, clearance, etc.)
4. Releases the drug at desired/specific site(s) due to break down of the sensitive linkage at the tumour microenvironment.
5. Reduces the premature drug release from dendrimers.
6. Increases the targeting and enhanced permeability and retention of conjugates in cancer therapy.

In 2008, Navath and co-workers reported the G4.0-NH_2 and anionic $G_{3.5}$-COOH PAMAM dendrimer-N-acetyl-L-cysteine (NAC) conjugates with cleavable disulphide linkages for intracellular delivery based on glutathione (GSH) levels and with 16 and 18 NAC per dendrimer. The payload of NAC was confirmed by

^{1}H-NMR and the matrix-assisted laser desorption/ionisation-time of flight analytical technique. The NAC is an antioxidant and anti-inflammatory agent with stroke and neuro-inflammation application. The dendrimer–NAC conjugates with ~70% payload were released within 1 h at intracellular GSH concentration (~10 μM) and negligible release at extracellular GSH levels (2 μM). These conjugates showed improved efficacy in activated microglial cells and better nitrite inhibition as compared with free NAC (Navath et al. 2008).

Taxol derivatives (PTX, docetaxel and cabazitaxel) are the anticancer drugs useful in the enhancement of drug targeting and delivery. Cline and co-workers reported hydrophilic PTX-conjugated G5.0 PAMAM dendrimers ($(NH_2)_{114}$-G5) in targeting of cultured cancerous cells. Authors demonstrated that the dendrimer–PTX conjugate adversely affects microtubules through protonation of tertiary amines present in the interior cavity of dendrimers in PTX independent manner (Cline et al. 2013).

Multivalent drug-dendrimers are a highly explored pathway for improvement of the therapeutic index. The MTX conjugate is formed with G5.0 PAMAM dendrimers using G5.0-$(COG-MTX)_n$ and G5.0-$(MFCO-MTX)_n$ small linkers with the G5.0-G5.0(D) dimer, D-$(COG-MTX)_n$ and D-$(MFCO-MTX)_n$ (Dongen et al. 2014). Dendrimer–DOX conjugates were reported to improve the anticancer activity in a murine hepatocellular carcinoma model and reduce the cardio-toxicity caused by DOX. The free functional group of DOX is conjugated with the end-functional groups of G5.0 PAMAM dendrimers using L3 and L4 enzyme-sensitive linkages for controlled DOX release. In this, the G5.0-PAMAM-DOX conjugate is formed *via* aromatic azo-linkages, which selectively recognise and are cleavable by azo-reductase enzymes. The N-acetyl galactosamine ($NAcGal_\beta$) sugar molecules conjugate with the free tip of the PEG brush, which are anchored onto the surface of G5.0 PAMAM through acid-labile cis-aconityl C linkages, and formulations are able to avoid P-gp-mediated efflux of receptor-mediated internalisation instead of the passive diffusion phenomenon (Kuruvilla et al. 2017a, 2017b).

The multifunctional DOX and FA-conjugated G5.0 PAMAM dendrimers have been designed and developed as a unique platform for pH-responsive drug release and targeted chemotherapy. In this, DOX is covalently conjugated onto the periphery of partially acetylated FA-modified G5.0 PAMAM dendrimers using pH sensitive *cis*-aconityl linkage (G5.0.NHAc-FA-DOX conjugate) (Zhang et al. 2018).

A dual drug nanodelivery system is an attractive strategy approach in drug delivery and targeting. The hyaluronic acid (HA)-coated G4.0-PAMAM dendrimers with Pt and DOX have been synthesised through covalent conjugation and esterification reactions (HA@PAMAM-Pt-DOX). Both Pt and DOX were conjugated with PAMAM dendrimers as Pt and DOX possess the CH_3 group, indicating that the existence of CH_3 infers to conjugation with PAMAM. The obtained results from findings suggested that the conjugate has great potential to improve the chemotherapeutic efficacy of both drugs in breast cancer (Guo et al. 2019).

CPT is a naturally occurring plant alkaloid isolated from *Camptotheca acuminate*. It is extremely hydrophobic in nature, which presents major hurdles during drug product development. The closed lactone ring and the C-20-OH groups of CPT are mainly responsible for cell damage by binding with DNA topoisomerase-I by the lactone ring. CPT has poor aqueous solubility, rapid blood clearance, severe toxicity and inactivation of its lactone ring at physiological pH. The aqueous solubility and lactone ring stability at physiological pH increase due to the presence of 20-OH in the CPT conjugate. The hydrophobic CPT is easily conjugated with any type of dendrimer, which led to the formation of varied nanostructures using hydrophobic–hydrophilic balance (Cheng et al. 2008; Thiagarajan et al. 2010; Laskar et al. 2019).

9.2.3 Dendrimer–Aptamer Conjugation and Interaction Mechanism

The virus encoded small structured ribonucleic acids (RNAs) have the specificity and affinity to bind with viral or various ranges of biomolecules including drugs, peptides, proteins and so on. These small structured RNAs, so-called aptamers are short, single stranded oligonucleotides DNA and RNA biomaterials comprising 20–80 nucleotides selected from the systematic evolution of ligands by exponential enrichment process. These aptamers have been considered as molecular agents for drug delivery and targeting purpose due to the distinct, ease of modification, high affinity, sensitivity, specificity, low immunogenicity, high stability, rapid tissue penetration and minimal toxicities as compared with antibodies. These aptamers are unable to cross the biological barriers (cell membranes) for drug targeting purpose for internalisation into the infected cells (Wang et al. 2015; Mehra et al. 2014).

In 2015, G5.0 PAMAM dendrimer-PEG-aptamer connection (PAM-Ap) using S6 (an aptamer) and the aptamer-PAMAM for non-small cell lung cancer cells (NSCLCs) were developed. The bi-functional maleimide PEG succinimidyl ester (Mal-PEG-NHS) is attached to the amino group of PAMAM and 5′-thiolated S6 aptamer. These novel and potential dendrimer–aptamer conjugates show a potential therapeutic solution to deliver microRNA to NSCLCs (Wang et al. 2015).

A controllable step-by-step self-assembly, programmable, sgc8 aptamer-based self-assembled DNA G3 dendrimer was synthesised for high affinity targeting, bio-imaging, DOX delivery and selective recognition of target CCRF-CEM (CCL-119, T-cell line, human acute lymphoblastic leukaemia), HeLa (Human cervical cancer cell line) and Ramos (CRL-1596, B-cell line, human Burkitt's lymphoma) cancer cells. For cellular uptake organic dye, FITC was labelled on Y_3 to stain G3.0 PAMAM dendrimer-sgc8 aptamer for specific binding and green fluorescence. The DNA–PAMAM dendrimers have several remarkable features (**1**) facile design and preparation, (**2**) multi-functionality, (**3**) good biocompatibility and (**4**) excellent stability for biomedical applications (Zhang et al. 2015).

Nucleolin (NCL) is a nuclear and cytoplasmic nucleolus ribosomal phosphor-protein that plays a crucial role in polymerase I transcription for biological and cellular function (Behrooz et al. 2017). Colorectal cancer is a vital cause of cancer associated death worldwide, and CPT drugs and their derivatives have been considered as first-line chemotherapeutic regimens. However, their limited application in cancer therapy is reported owing to their poor aqueous solubility, low bioavailability, systemic toxicity and short half-life. The CPT-loaded PEGylated G5.0 PAMAM dendrimers were synthesised, and then functionalised with AS1411 antinucleolin aptamers for targeting against colon adenocarcinoma by over-expressing nucleolin receptors. AS1411, a 26-mer single strand DNA (ssDNA) aptamer, has high and specific binding affinity to nucleolin (Alibolandi et al. 2017). The synthesis of PEGylated dendrimers and conjugation of AS1411 aptamers to the maleimide groups of MAL-PEG-PAMAM-CPT and preparation of Apt-PEG-PAMAM-CPT are shown in Figure 9.3. Similarly, smart bomb AS1411

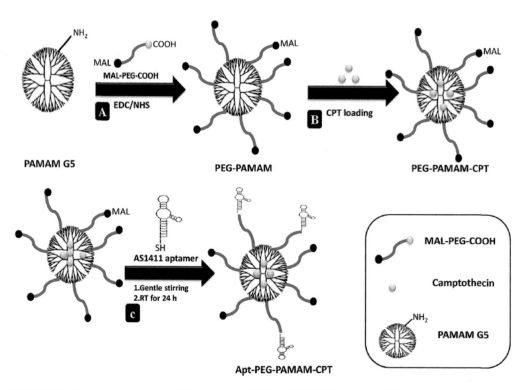

FIGURE 9.3 Schematic representation of (A) synthesis of PEGylated PAMAM dendrimer (PEG-PAMAM); (B) CPT loading in the cavities of PEG-PAMAM; (C) conjugation of thiolated AS1411 aptamers to the maleimide groups of MAL-PEG-PAMAM-CPT and preparation of Apt-PEG-PAMAM-CPT (Reproduced with copyright permission from Alibolandi et al., 2017).

aptamers (APTAS1411)-functionalised with PAMAM-PEG complex for delivery of 5-FU were synthesised. The nanoparticle-PAMAM dendrimer-aptamer-5-FU complex aids in effective delivery of 5-FU to cancer cells (Behrooz et al. 2017).

9.2.4 Dendrimer–Liposome Interaction Mechanism and Chemistry

The liposome '*locked in*' dendrimer concept came into existence for fulfilling the requirement of cancer targeting therapy and drug delivery. The liposomal locked-in dendrimer (LLD), a combination of dendrimers and liposomes, is a new term in the drug delivery technology and has gained benefits of liposomal and dendrimeric nanoarchitectures. This LLD technology viewed as a dendrimer-based class of modulatory liposomal controlled release systems depicts the high loading and alteration of the drug release profile from liposomes. The dendrimers are few nanometres (1–100 nm) in particle size, and the highly cationic dendrimers interact with oppositely charged liposomes and result in the encapsulation of the dendrimers inside the aqueous phase of the liposomes or in the lipid bilayer (Purohit et al. 2001; Tekade et al. 2009; Gardikis et al. 2010).

The milestone for the creation of the liposome locked in dendrimer concept was the work by Khopade and co-workers. The milestone for utilisation of the 'lock in' dendrimer concept in drug delivery and targeting was reported by Khopade and co-workers under collaboration with the Max Planck Institute of Colloids and Interfaces, Potsdam, Germany. Khopade and co-workers studied the effect of cationic charged PAMAM dendrimers with the aqueous layers of the liposomes for encapsulation and release of the acidic anticancer agent (MTX) from the conjugates. The dendrimers have a very few nanometre size range, almost equivalent to the thickness of aqueous space between two liposomal bilayers. Thus, it is easier to get encapsulated in between the aqueous space of liposomes. Cationic charged dendrimers interact with the oppositely charged liposomes as shown in Figure 9.4 (Khopade et al. 2002).

Later in 2010, G1.0 and G2.0 hydroxy-terminated dendrimers were 'locked in' in liposomes consisting of DOPC/DPPG; then DOX was loaded into pure liposomes and LLDs followed by lyophilisation. The interaction of lipids and dendrimers is of entropic in nature, derived by thermal analysis. It suggests that there is no bond formation among liposomes and dendrimers but the interaction is steric in nature also favoured in the interior of the acyl chains (Gardikis et al. 2010).

The interaction of different generations G2.0, G4.0 and G6.0 PAMAM dendrimers and liposomes of different compositions was performed using combined turbidity, dynamic light scattering and atomic force microscopy studies. The interaction between PAMAM dendrimers and liposomes is governed by hydrogen bonding and electrostatic interaction in the following order G6.0>G4.0>G2.0 generation. Authors reported the formation of liposome–dendrimer complexes by the term '*dendriosomes*', and they were formed by the adhesion of the dendrimers with individual liposomes, and dendrimers act as 'glue' (Roy et al. 2014). In 2018, the same research group also studied the interaction of cationic G3.0, G4.0 and G5.0 PAMAM dendrimers with the negatively charged liposomes using different physicochemical techniques. The interaction mainly depends upon the dendrimers, generation, surface charge, head group and hydrocarbon chain length of the lipids. The dendrimer–lipid bilayer interaction is formed by electrostatic interaction observed through charge neutralisation. The dendrimer–liposome aggregates are exothermic in nature and non-toxic to healthy human blood cells (Roy et al. 2018). The above LLD technology has not yet been completely elucidated, and many things need to be investigated using various novel physicochemical and analytical techniques to further clarify many issues in drug delivery and targeting.

9.2.5 PEGylation (Dendrimer–PEG interaction)

The conjugation of the PEG chain with dendrimers is considered as a successful technique to enhance the therapeutic potential of bioactive(s) including biomolecules and known as PEGylation. It increases the particle size of dendrimers and molecular weight and improves the pharmacokinetics and aqueous solubility. In addition, it protects bioactive(s) from enzymatic degradation and importantly reduces the toxicities associated with dendrimers. PEGylation of dendrimers has the following

(a)

(b)

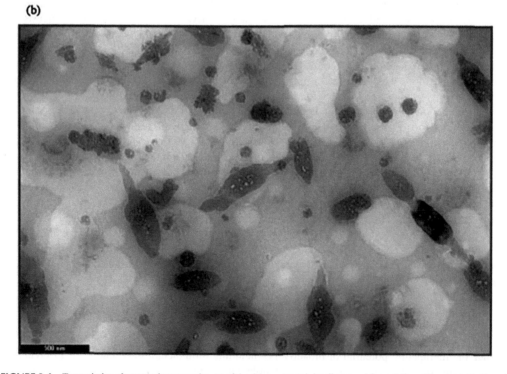

FIGURE 9.4   Transmission electron microscopy image of dendrimers containing liposomal formulation; (b) a single lipid parti-cle showing encapsulated dendrimer grains (Reproduced with copyright permission form Elsevier Pvt Ltd. Khopade et al., 2002).

benefits (Barraza et al. 2017; Hutnick et al. 2017; Jain et al. 2010; Freitas et al. 2015; Gajbhiye et al. 2007, 2009; Kesharwani et al. 2014; Jiang et al. 2016; Tekade et al. 2009; Bhadra et al. 2003):

1. Increases the drug loading capacity and aqueous solubility.
2. Protecting from enzymatic degradation.
3. Increases the size and molecular weight of dendrimers.
4. Reduces haemolytic and haematological toxicities.
5. Improves the stability and drug therapy.
6. Increases the bio-distribution and pharmacokinetics.
7. Assists in shielding of peripheral cationic groups for a better delivery system.
8. Controlled and sustained drug delivery.
9. Improves transfection efficiency and decreases immunogenicity.

The milestone study by Bhadra and co-workers in 2003 explored PEGylation strategies for improvement of therapeutic potential of G4.0 PAMAM dendrimers. The resulting PEGylation reduced the haematological toxicity of PEG-coated-PAMAM dendrimers with improvement in drug loading capacity and reduction in drug leakage (Bhadra et al. 2003). The PEG chain arms, numbers of PEG molecules and dendrimer generation are also responsible for increasing the encapsulation of the drug. Until now, numerous research studies have been available for comparison of PEGylated and non-PEGylated dendrimers for drug delivery and drug targeting.

Glioblastoma is a most common and aggressive primary central nervous system tumour with short survival time. It can be classified based on the cells, which give rise to them: oligodendrocytes give rise to oligodendrogliomas, ependymal cells generate ependymomas and astrocytes produce astrocytomas. The World Health Organization defined the astrocytomas into grade I to IV based on the degree of malignancy and categorised glioblastoma as the frequent and fatal form into grade IV. Chemotherapy fails due to the low transport of chemotherapeutic agents across the blood brain tumour barrier and poor penetration (Bark et al. 2020; Jiang et al. 2016).

The PEGylated PAMAM dendrimer nanoparticles conjugated with pep-1 (Pep-PEG-PAMAM) were developed to evaluate the glioma tumour targeting efficiency. The peptide was conjugated with PEGylated PAMAM through the maleimide–thiol reaction. The zeta potential and particle size of the conjugate were found to be decreased. This small size of PAMAM and pep-1 modified dendrimers improved the active targeting efficiency to glioma cells, which were considered as safe, effective theranostic nanocarriers (Jiang et al. 2016).

Somani and co-workers studied the impact of PEG molecular weight and diaminobutyric polypropylenimine (DAB) dendrimer generation (in this G3.0 and G4.0 generation was used) for non-viral gene delivery systems and enhancement of transfection efficacy. The G4.0 dendrimers PEGylated with PEG_{2K} at a dendrimer:DNA ratio of 20:1 show a higher level of β-galactosidase gene expression ($10.07 \times 10^{-3} \pm 0.09 \times 10^{-3}$ U/mL) after treatment of B16f10-Luc cells (Somani et al. 2018).

9.2.6 Dendrimer–Lipid Biological Membrane Interaction

Recently, a new class of carbosilane dendrimers formed by Si-C or C-C bonds in the interior has been increasingly investigated in biomedical application. It depicts the thermal and hydrolytic stability with hydrophobic core properties. Tailoring of this carbosilane dendrimer by control of synthesis, branching degree, density and distribution of chemical functional groups is easier as well introducing the numerous functional groups on its peripheral surface. The influence of the dendrimer formulation on cell membrane properties and integrity is also required to investigate the interaction and possible toxic effects. The interaction with the biological membrane is difficult as the biological membrane is a multicomponent and complicated system (Wrobel et al. 2018; Astruc and Boisselier 2010; Hsu et al. 2016; Strasak et al. 2017).

The interaction of PAMAM dendrimers and cell membranes remains poorly understood and needs to be well explored with more focus on the cell membrane like supported lipid bilayers, liposomes and Langmuir monolayers. Readers may refer the review article "*PAMAM dendrimers-cell membrane interactions*" published in Advances in Colloid and Interface Sciences to understand the interactions of PAMAM dendrimers and cell membranes (Fox et al. 2018). Additionally, for well understanding about the nature, physicochemical properties of the dendrimer–protein interaction, comparison of dendrimer-based nanomedicine and dendrimer–protein interaction refer the review articles entitled "Dendrimers-protein interaction versus dendrimers-based nanomedicine" by Scharbin & co-workers (Scharbin et al. 2017).

The interaction of the carbosilane dendrimers with the lipid membrane was reported and studied. Three generations of five different phosphonium carbosilane dendrimers and one ammonium carbosilane dendrimer as a reference (PMe$_3$, PBu$_3$, P(Et)$_2$(CH$_2$)$_3$OH, PPh$_3$, P(MeOPh)$_3$ and NMe$_3$ and peripheral functional groups) on dimyristoylphosphatidylcholine (DMPC) or a lipid mixture DMPC/dimyristoylphosphatidylglycerol of liposomes was reported by Wrobel and co-workers (2018). Authors also reported that the dendrimers with PPh$_3$ and P(MeOPh)$_3$ peripheral functional groups interact much more strongly and enhance the rigidity of liposomes. They also reported that the electrostatic interaction, the hydrophobicity of substituents and charge shielding on the peripheral phosphonium group are important factors in the interaction (Wrobel et al. 2018).

9.2.7 Domino Dendrimers as a Potential Platform for Bio-sensing

Domino dendrimers have been designed in such a pattern, to release their tail units through domino-like chain disintegration initiated by single cleavage, which could be catalysed by a suitable enzyme to work as a single triggered multiprodrug unit. In this domino dendrimer, incorporation of drug molecules is considered as the tail unit and an enzyme substrate as a trigger. Domino dendrimers may be a potential platform for biosensor molecules, used to detect enzymatic activity (Amir et al. 2006). The domino dendrimers could be designed and formulated as promising drug delivery systems in the treatment of cancer and biosensors for diagnosis purpose. The self-immolative polymers with a polyurethane backbone were designed by Sagi and co-workers in 2008, which could be disassembled sequentially into its building block upon initiation by triggering event at the head of polymers into developing a highly sensitive molecular sensor with large signal-to-noise ratios (Sagi et al. 2008).

9.3 Molecular Modelling (MM) of Dendrimers

MM is a promising tool to predict the properties of dendrimers with defined and repetitive branching architecture, terminal functional groups which are present on the end of dendrimers and their possible interactions at the molecular level in designing of the drug delivery systems. The possible interaction with bioactive(s), drugs, proteins, amino acids, nucleic acid and so on can be easily performed using the MM technique. In MM, the particle size, poly-dispersity index and surface charge have been considered as critical process parameters, so there is ease modification for enhancing the performance of dendrimers. Computational and MM studies could support the experimental outcomes through valuable inputs about the structure of dendrimers and interaction in prediction and understanding of the chemistry of several interactions (Jain et al. 2020; Martinho et al. 2014; Lombardo 2014).

Computer simulation techniques are an important and valuable tool help to predict the properties of dendrimers at the molecular level. Upon alteration of the end functional groups of dendrimers, the biological properties and toxicities are changed. The various physicochemical properties such as conformational analysis, molecular interaction, and validation of the experimental data could be determined by MM techniques, which will reduce the laborious and expensive laboratory trials. Thus, a thorough understanding of the interaction between the macromolecules (dendrimers) with the biological systems is required for the development of safe and effective therapeutics (Martinho et al. 2014; Lombardo 2014).

9.4 Conclusion and Future Perspectives

Surface-functionalised multifunctional dendrimers have been attracting and promising a great deal of attention for drug delivery and targeting. Dendrimers are more effective and efficient nanocarriers owing to their nanometric size, free functional groups, stability, high payload and so on. The various biomolecules can be easily loaded and conjugated onto the periphery of the dendrimers using various interaction mechanisms such as covalent conjugation, hydrogen bonding, cleavable linkage and so on. The degree of PEGylation (PEG chain and molecular weight) and dendrimer generation (chemical functional groups) are critical for designing of biocompatible and safe dendrimeric formulations. In our opinion, multifunctional dendrimers may open a new door in upcoming scenarios for pharmaceutical and biomedical applications.

Abbreviations

5-FU	5-fluorouracil
BBTYB	Blood brain tumor barrier
CPT	Camptothecin
DMPC	Dimyristoylphosphatidylcholine
DOX	Doxorubicin
EPR	Enhanced permeability and retention
FA	Folic acid
GEM	Gemcitabine
GSH	Glutathione
HA	Hyaluornic acid
HIV	Human immunodeficiney virus
HSV	Herpes simplex virus
LLD	Liposomal locked-in dendrimers
MALDI	Matrix assisted laser desorption/ionization
MRI	Magnetic resonance imaging
MTX	Methotrexate
NAC	N-acetyl-L-cysteine
NAcGal	N-acetyl galactosamine
NCL	Nucleolin
NMR	Nuclear magnetic resonance
NSCLC	Non-small cell lung cancer cells
PAMAM	Poly (amido) amine
PDI	Poly-dispersity index
PEG	Poly-ethylene glycol
PLL	Poly-l-lysine
PPI	Poly-propylene-imine
PTX	Paclitaxel
RNA	Ribo-nucleic acids
SELEX	Systematic evolution of ligands by exponential enrichment
SLBs	Supported lipid bilayers
USFDA	US Food and Drug Administration

ACKNOWLEDGEMENT

The authors would like to thank National Institute of Pharmaceutical Education and Research (NIPER), Raebareli and National Institute of Pharmaceutical Education and Research (NIPER), Hyderabad for extending the facilities to write this chapter.
The NIPER, Raebareli communication number for this manuscript is NIPER-R/Communication/141.

REFERENCES

Agrawal, U., Mehra, N.K., Gupta, U., Jain, N.K. 2013. Hyperbranched dendritic nano-carriers for topical delivery of dithranol. *Journal of Drug Targeting* no. *21*(5): 497–506.

Ahmed, S., Vepuri, S.B., Kalhapure, R.S., Govender, T. 2016. Interactions of dendrimers with biological drug targets: reality or mystery–a gap in drug delivery and development research. *Biomaterials science* no. *4*(7): 1032–1050.

Alibolandi, M., Taghdisi, S.M., Ramezani, P. et al. 2017. PAMAM dendrimer for the superior delivery of camptothecin to colon adenocarcinoma in vitro and in vivo. *International Journal of Pharmaceutics* no. *519*(1–2): 352–364.

Almuqbil, R.M., Heyder, R.S., Bielski, E.R. et al. 2020. Dendrimer conjugation enhances tumour penetration and efficacy of doxorubicin in extracellular matrix-expressing 3D lung cancer models. *Molecular Pharmaceutics.* doi:10.1021/acs.molpharmaceut.0c00083.

Amir, R.J., Danieli, E., Shabat, D. 2006. Domino dendrimers. *Advanced Polymer Science* no. *192*: 59–94.

Amir, R.J., Danieli, E., Shabat, D. 2007. Receiver-amplifier, self-immolative dendritic device. *Chemistry* no. *13*(3): 812–821.

Astruc, D., Boisselier, E. 2010. Dendrimers designed for functions: From physical, photophysical, and supramolecular properties to applications in sensing, catalysis, molecular electronics, photonics, and nanomedicine. *Chemical Review* no. *110*: 1857–1959.

Bark, J.M., Kulasinghe, A., Chua, B., Day, B.W., Punyadeera, C. 2020. Circulating biomarkers in patients with glioblastoma. *British Journal Cancer* no. *122*: 295–305.

Barraza, L.F., Jimenez, V.A., Alderete, J.B. 2017. Association of methotrexate with native and PEGylated PAMAM-G4 dendrimers: effect of the PEGylation degree on the drug-loading capacity and preferential binding sites. *Journal of Physics Chemistry B* no. *121*(1): 4–12.

Behrooz, A.B., Nabavizadeh, F., Adiban, J. et al. 2017. Smart bomb AS1411 aptamer-functionalized/PAMAM dendrimer nanocarriers for targeted drug delivery in the treatment of gastric cancer. *Clinical Expert Pharmacology Physiology* no. *44*(1): 41–51.

Bhadra, D., Bhadra, S., Jain, S., Jain, N.K. 2003. A PEGylated dendritic nanoparticulate carrier of fluorouracil. *International Journal of Pharmaceutics* no. *257*: 111–124.

Carvalho, M.R., Reis, R.L., Oliveira, J.M. 2020. Dendrimer nanoparticles for colorectal cancer applications. *Journal Material Chemistry B* no. *8*: 1128–1138.

Cheng, Y., Li, M., Xu, T. 2008. Potential of poly(amidoamine) dendrimers as drug carriers of camptothecin based on encapsulation studies. *European Journal Medicinal Chemistry* no. *43*(8): 1791–1795.

Cline, E.N., Li, M.H., Choi, S.K. et al. 2013. Paclitaxel-conjugated PAMAM dendrimers adversely affect microtubules structure through two independent modes of action. *Biomacromolecules* no. *14*: 654–664.

Dongen, M.A., Rattan, R., Slipe, J. et al. 2014. Poly(amidoamine) dendrimer-methotrexate conjugates: the mechanism of interaction with folate binding protein. *Molecular Pharmaceutics* no. *11*(11): 4049–4058.

Emanuele, A.D., Attwood, D. 2005. Dendrimer-drug interactions. *Advanced Drug Delivery Reviews* no. *57*(15): 2147–2162.

Esfand, R., Tomalia, D.A. 2001. Poly(amidoamine) (PAMAM) dendrimers: from biomimicry to drug delivery and biomedical applications. *Drug Discovery Today* no. *6*(8): 427–436.

Ficker, M., Pereson, J.F., Hansen, J.S., Christensen, J.B., 2015a. Guest-host chemistry with dendrimers-binding of carboxylates in aqueous solution. *PLoS ONE* no. *10*(10): e0138706.

Ficker, M., Petersen, J.F., Gschneidtner, T. et al. 2015b. Being two is better than one-catalytic reductions with dendrimer encapsulated copper- and copper-cobalt-subnanoparticles. *Chemical Communications* no. *51*: 9957–9960.

Fox, L.J., Richardson, R.M., Briscoe, W.H. 2018. PAMAM dendrimer-cell membrane interactions. *Advanced Colloid Interface Sciences* no. *257*: 1–18.

Freitas, D., Boldrini-Franca, J., Arantes, E.C. 2015. PEGylation: a successful approach to improve the biopharmaceutical potential of snake venom thrombin-like serine protease. *Protein Peptide Letters* no. *22*(12): 1133–1139.

Gajbhiye, V., Kumar, P.V., Tekade, R.K., Jain, N.K., 2007. Pharmaceutical and biomedical potential of PEGylated dendrimers. *Current Pharmaceutical Design* no. *13*; 415–429.

Gajbhiye, V., Kumar, P.V., Tekade, R.K., Jain, N.K., 2009. PEGylated PPI dendritic architectures for sustained delivery of H2 receptor antagonist. *European Journal of Medicinal Chemistry* no. *44*(3): 1155–1166.

Gardikis, K., Hatziantoniou, S., Bucos, M., Fessas, D., Signorelli, M., Felekis, T., Zervou, M., Screttas, C.G., Steele, B.R., Ionov, M., Micha-Screttas, M. 2010. New drug delivery nanosystem combining liposomal and dendrimeric technology (liposomal locked-in dendrimers) for cancer therapy. *Journal of Pharmaceutical Sciences* no. *99*(8): 3561–3571.

Guo, X.L., Kang, X.X., Wang, Y.Q. et al. 2019. Co-delivery of cisplatin and doxorubicin by covalently conjugating with polyamidoamine dendrimer for enhanced synergistic cancer therapy. *Acta Biomaterial* no. *84*: 367–377.

Gupta, R., Mehra, N.K., Jain, N.K. 2014. Development and characterization of sulfasalazine loaded fucosylated PPI dendrimer for the treatment of cytokine-induced liver damage. *European Journal of Pharmaceutics Biopharmaceutics* no. *86*(3): 449–458.

Hansen, J.S., Ficker, M., Petersen, J.F., Nielsen, B.E., Gohar, S., Christensen, J.B. 2013. Study of the complexation of Oxacillin in 1-(4-Carbomethoxypyrrolidone)-Terminated PAMAM dendrimers. *The Journal of Physical Chemistry B* no. *117*: 14865–14874.

He, X.C., Lin, M., Lu, T.J., Qu, Z.G., Xu, F., 2015. Molecular analysis of interactions between a PAMAM dendrimer paclitaxel conjugate and a biomembrane. *Physics Chemistry* no. *17*: 29507–29517.

Hsu, H., Bugno, J., Lee, S., Hong, S. 2016. A versatile platform for drug delivery. *Wiley Interdisciplinary Reviews: Nanomedicine and Nanobiotechnology* no. *9*: 1–21.

Hutnick, M.A., Ahsanuddin, S., Guan, L., Lam, M., Baron, E.D. Pokorski, J.K. 2017. PEGylated dendrimers as drug delivery vehicles for the photosensitizer silicon phthalocyanine Pc 4 for candida infections. *Biomacromolecules* no. *18*(2): 379–385.

Jain, K., Kesharwani, P., Gupta, U., Jain, N.K. 2010. Dendrimer toxicity: let's meet the challenge. *International Journal of Pharmaceutics* no. *394*: 122–142.

Jain, K., Mehra, N.K., Jain, N.K. 2015. Nanotechnology in drug delivery: safety and toxicity issues. *Current Pharmaceutical Design* no. *21*(29): 4252–4261.

Jain, K., Mehra, N.K., Jain, V., and Jain, N.K. 2020. IPN dendrimers in drug delivery. In *Interpenetrating polymer network: Biomedical application*, ed. Jan, S., Jana, S., 143–182. Springer.

Jiang, Y., Lv, L., Shi, H., Hua, Y., Lv, W., Wang, X., Xin, H., Xu, Q. 2016. PEGylated Polyamidoamine dendrimer conjugated with tumour homing peptide as a potential targeted delivery system for glioma. *Colloids Surface B: Biointerfaces* no. *147*: 242–249.

Kesharwani, P., Jain, K., Jain, N.K., 2014. Dendrimer as nanocarriers for drug delivery. *Progress in Polymer Science* no. *39*(2): 268–307.

Kesharwani, P., Mishra, V., Jain, N.K. 2015. Generation dependent hemolytic profile of folate engineered poly(propyleneimine) dendrimer. *Journal of Drug Delivery Science Technology* no. *28*: 1–6.

Khopade, A.J., Caruso, F., Tripathi, P., Nagaich, S., Jain, N.K. 2002. Effect of dendrimer on entrapment and release of bioactive from liposomes. *International Journal of Pharmaceutics* no. *232*: 157–162.

Kulhari, H., Pooja, D., Prajapati, S.K., Chauhan, A.S. 2011. Performance evaluation of PAMAM dendrimer based simvastatin formulations. *International Journal of Pharmaceutics* no. *405*(1–2): 203–209.

Kuruvilla, S.P., Tiruchinapally, G., Crouch, A.C., ElSayed, M.E.H., Greve, J.M. 2017a. Dendrimer-doxorubicin conjugates exhibit improved anticancer activity and reduce doxorubicin-induced cardiotoxicity in a murine hepatocellular carcinoma model. *Plos ONE* no. *12*(8): e0181944.

Kuruvilla, S.P., Tiruchinapally, G., ElAzzouny, M., ElSayed, M.E.H. 2017b. N-Acetylgalactosamine-targeted delivery of dendrimer-doxorubicin conjugates influences doxorubicin cytotoxicity and metabolic profile in Hepatic cancer cells. *Advanced Health Materials* no. *6*(5): 1601046.

Laskar, P., Somani, S., Campbell, S.J. et al. 2019. Camptothecin-based dendrimersomes for gene delivery and redox-responsive drug delivery to cancer cell. *Nanoscale* no. *11*: 20058–20071.

Lombardo, D. 2014. Modeling dendrimers charge interaction in solution relevance in Biosystems. *Biochem Research International*. Article ID 837651 (1–14).

Madaan, K., Kumar, S., Poonia, N., Lather, V., Pandita, D. 2014. Dendrimers in drug delivery and targeting: drug-dendrimer interactions and toxicity issues. *Journal of Pharmaceutical Bioallied Science* no. *6*: 139.

Martinho, N., Florindo, H., Silva, L., Brocchini, S., Zloh, M., Barata, T. 2014. Molecular modeling to study dendrimers for biomedical applications. *Molecules* no. *19*: 20424–20467.

Mehra, N.K., Jain, K., Jain, N.K. 2015a. Design of multifunctional nanocarriers for delivery of anti-cancer therapy. *Current Pharmaceutical Design* no. *21*(42): 6157–6164.

Mehra, N.K., Jain, K., Jain, N.K. 2015b. *Novel triazine dendrimer: Encyclopedia of biomedical polymers and polymeric biomaterials*. CRC Press. doi:10.1081/E-EBPP-120049287.

Mehra, N.K., Mishra, V., Jain. N.K. 2013. Receptor based therapeutic targeting. *Therapeutics Delivery* no. *4*(3): 1–26.

Mehra, N.L., Cai, D., Kuo, L., Hein, T., Palakurthi, S. 2016. Safety and toxicity of nanomaterials for ocular drug delivery applications. *Nanotoxicology* no. *10*(7): 836–860.

Menjoge, A.R., Kannan, R.M., Tomalia, D.A. 2010. Dendrimer-based drug and imaging conjugates: design considerations for nanomedical applications. *Drug Discovery Today* no. *15*: 171–185.

Mishra, V., Kesharwani, P. 2016. Dendrimer technologies for brain tumour. *Drug Discovery Today* no. *21*(5): 766–778.

Mishra, V., Yadav, N., Saraogi, G.K., Tambuwala, M.M., Giri. N. 2019. Dendrimer based nanoarchitectures in diabetes management: An overview. *Current Pharmaceutical Design* no. *25*(23): 2569–2583.

Navath, R.S., Kurtoglu, Y.E., Wang, B., Kannan, S., Romero, R., Kannan, R.M. 2008. Dendrimers-drug conjugates for tailored intracellular drug release based on glutathione levels. *Bioconjugate Chemistry* no. *19*(12): 2246–2455.

Nguyen, H., Nguyen, N.H. Tran, N.Q., Nguyen, C.K. 2015. Improved method for preparing cisplatin-dendrimer nanocomplex and its behavior against NCl-H460 lung cancer cell. *Journal of Nanoscience Nanotechnology* no. *15*(6): 4106–4110.

Oztruck, K., Esendagli, G., Gurbuz, M.U., Tulu, M., Calis, S. 2017. Effective targeting of gemcitabine to pancreatic cancer through PEG-cored Flt-1 antibody-conjugated dendrimers. *International Journal of Pharmaceutics* no. *517*(1–2): 157–167.

Pooja, D., Sistla, R., Kulhari, H. 2018. *Dendrimers-drug conjugates: Synthesis strategies, stability and application in anticancer drug delivery*. Elsevier. 273–303.

Purohit, G., Sakthivel, T., Florence, A.T. 2001. Interaction of cationic partial dendrimers with charged and neutral liposomes. *International Journal of Pharmaceutics* no. *214*: 71–76.

Roy, B., Guha, P., Nahak, P., et al. 2018. Biophysical correlates on the composition, functionality, and structure of dendrimer-liposome aggregates. *CACS Omega* no. *3*(9): 12235–12245.

Roy, B., Panda, A.K., Parimi, S., Ametov, I., Barnes, T., Prestidge, C.A. 2014. Physico-chemical studies on the interaction of dendrimers with lipid bilayers. 1. Effect of dendrimers generation and liposomes surface charge. *Journal Oleo Science* no. *63*: 1185–1193.

Sagi, A., Weinstain, R., Karton, N., Shabat, D. 2008. Self-immolative polymers. *Journal of American Chemical Society* no. *130*(16): 5434–5435.

Santos, A., Veiga, F., Figueiras A. 2020. Dendrimers as pharmaceutical Excipients: synthesis, properties, toxicity and Biomedical applications. *Materials* no. *13*: 65: 1–31.

Scharbin, D., Shcharbina, N., Dzmitruk, V. et al. 2017. Dendrimers-protein interaction versus dendrimer-based nanomedicine. *Colloid Surface B: Biointerfaces* no. *152*: 414–422.

Sharma, A., Sharma, R., Zhang, Z. et al. 2020. Dense hydroxyl polyethylene glycol dendrimer targets activated glia in multiple CNS disorders. *Science Advanced* no. *6*: eaay8514 (1–14).

Shcharbin, D., Shcharbina, N., Dzmitruk, V. et al., 2017. Dendrimer-protein interaction versus dendrimer-based nanomedicine. *Colloid Surface B: Biointerface* no. *152*: 414–422.

Singh, J., Jain, K., Mehra, N.K., Jain, N.K. 2016. Dendrimers in anticancer drug delivery: mechanism of interaction of drugs and dendrimers. *Artificial Cell Nanomedicine Biotechnology* no. *44*(7): 1626–1634.

Somani, S., Laskar, P., Altwaijry, N. et al. 2018. PEGylation of polypropylenimine dendrimers: effects on cyto-toxicity, DNA condensation, gene delivery and expression in cancer cells. *Scientific Reports* no. *8*(1–13): 9410.

Strasak, T., Maly, J., Wrobel, D. et al. 2017. Phosphonium carbosilane dendrimers for biomedical applications synthesis, characterization and cytotoxicity evaluation. *RSC Advances* no. *7*: 18724–18744.

Tekade, R., Kumar, P.V., Jain, N.K. 2009. Dendrimers in oncology: an expanding horizon. *Chemical Review* no. *109*(1): 49–87.

Thiagarajan, G., Ray, A., Malugin, A., Ghandehari, H. 2010. PAMAM-camptothecin conjugate inhibits prolif-eration and induces nuclear fragmentation in colorectal carcinoma cells. *Pharmaceutical Research* no. *27*(11): 2307–2316.

Thomas, T.P., Huang, B., Choi, S.K. et al. 2012. Polyvalent dendrimer-methotrexate as a folate receptor-targeted cancer therapeutic. *Molecular Pharmaceutics* no. *9*(9): 2669–2676.

Tomalia, D.A., Baker, H., Dewald, J.R. et al. 1985. A new class of polymers: starburst dendritic molecules. *Polymer Journal* no. *17*, 117–132.

Tomalia, D. A., Naylor, A. M., Goddard, W. A. 1990. Starbust dendrimers: molecular-level control of size, shape, surface chemistry, topology, and flexibility from atoms to macroscopic matter. *Angewandte Chemie International Edition* no. *29*(2): 138–175.

Wang, H., Zhao, X., Guo, C., et al. 2015. Aptamer-dendrimer bioconjugates for targeted delivery of miR-34a expressing plasmid and antitumour effects in non-small cell lung cancer cells. *PLoS ONE* no. *0*(9): e0139136.

Wei, C., Lin, L., Zhao, Y. et al. 2020. Fabrication of pH-sensitive superhydrophilic/underwater superoleophobic poly(vinylidene fluoride)-graft-(SiO2 nanoparticles and PAMAM dendrimers) membranes for oil-water separation. *ACS Applied Material Interfaces* doi:10.1021/acsami.9b22881.

Wrobel, D., Kubikova, R., Mullerova, M. et al. 2018. Phosphonium carbosilane dendrimers-interaction with a simple biological membrane model. *Physical Chemistry Chemical Physics* no. *20*(21): 14753–14764.

Xiao, T., Li, D., Shi, X., Shen, M. 2020. PAMAM dendrimer-based nanodevices for nuclear medicine applications. *Macromolecular Biosciences* no. *20*(2): doi:10.1002/mabi.201900282.

Zhang, H., Ma, Y., Xie, Y. et al. 2015. A controllable aptamer-based self-assembled DNA dendrimer for high affinity targeting, bioimaging and drug delivery. *Science Reports* no. *5*(10099): 1–8.

Zhang, M., Zhu, J., Zheng, Y. et al. 2018. Doxorubicin-conjugated PAMAM dendrimers for pH-responsive drug release and folic acid-targeted cancer therapy. *Pharmaceutics* no. *10*(1–13): 162.

Zhong, Q., Humia, B.V., Punjabi, A.R., Padilha, F.F., da Rocha, S.R.P. 2017. The interaction of dendrimer-doxorubicin conjugates with a model pulmonary epithelium and their cosolvent-free, pseudo-solution formulations in pressurized metered-dose inhalers. *European Journal Pharmaceutical Sciences* no. *109*: 86–95.

Zhu, J., Xiong, Z., Shen, M., Shi, X. 2015. Encapsulation of doxorubicin within multifunctional gadolinium-loaded dendrimer nanocomplexes for targeted theranostics of cancer cells. *RSC Advances* no. *5*: 30286–30296.

Zolotarskaya, O.Y., Xu, L., Valerie, K., Yang, H. 2015. Click synthesis of a polyamidoamine dendrimer-based camptothecin prodrug. *RSC Advances* no. *72*(5): 58600–58608.

10

Dendrimers in Gene Delivery

Dnyaneshwar Baswar, Ankita Devi and Awanish Mishra

CONTENTS

10.1 Introduction

Gene therapy is a promising technique that focuses on the utilisation of genes to treat or prevent diseases ranging from single gene disorder to multi-gene disorder. The experimental approaches of gene therapy may authorise doctors to treat any diseases by introducing a gene into patient's cells in place of surgery or drugs (Sung and Kim 2019). Gene therapy looks to alter genes to ameliorate genetic defects and cure genetic diseases. It is one of the most advanced strategies for therapeutic prospects searching for stopping genetic diseases to fight cancer. The basic process of gene therapy is substituting a faulty gene that creates disease with a normal gene and knocking out/deactivating a mutant gene that is working inappropriately (Maguire et al. 2014). Gene therapy in multiple forms will be giving clinical improvements in patients with neuromuscular disease, blindness, haemophilia, cancer and immune deficiencies (Dunbar et al. 2018). Currently, researchers are focusing on the safety of gene therapy, for future studies whether therapy is a safe and effective treatment option for the prevention of diseases such as viral infection, cancer and inherited disorders (Yang et al. 2015).

10.2 Gene Delivery and Its Application

A gene delivery system is essential to treat genetic diseases. Gene delivery is a process of delivering genetic materials such as nucleic acid into the patient's cells to generate the therapeutic effect. Genomic molecules grasp into the nuclei of defective cells to induce gene expression. For gene delivery, the foreign genetic molecule (DNA or RNA) is needed, which remains stable within the host cells (Sung and

Kim 2019). Victorious gene delivery systems can depend on relevant vehicles (dendrimers) for carrying the genomic molecules (DNA or RNA) safely to the target cells and the capacity to target both tissues and cells including high specificity (Wang et al. 2014b).

Somatic and germline therapy are types of gene therapy. Somatic gene therapy is considered a much safer and most common therapy, which contains gene vectors, such as viral vectors and non-viral vectors (Nayerossadat et al. 2012). Vectors are basically transporters that may deliver the therapeutic gene to the infected cells. The general concept of gene therapy is inserting the exogenous gene into somatic cells that develop organs to produce a therapeutic effect. Germline gene therapy has the potential to manipulate reproductive sperm cells and eggs to create heritable changes (Misra 2013). Some therapeutic agents contain nucleic acid materials, which upon administration results in cellular adaptation, repairing, insertion or deletion of a gene sequence. Genetic materials such as DNA, mRNA, miRNA, siRNA and antisense oligonucleotides are commonly used in deficient target cells or tissue to re-establish the specific gene function for disease management (Chen et al. 2018). On this basis, gene therapy may appear as a promising pharmacotherapy tool (Klug et al. 2012; Ingusci et al. 2019).

Gene delivery might become a unique tool for the treatment of central nervous system disorders (Ingusci et al. 2019). To overcome neurological disorders such as Alzheimer's disease, Parkinson's disease and epilepsy, new genetic tools are developed, i.e., optogenetic and chemogenetic tools such as designer receptors exclusively activated by designer drugs (Seeger-Armbruster et al. 2015; Ingusci et al. 2019). Gene therapy allows us the opportunity to eliminate and treat untreatable diseases. It is the potential medical option for the prevention of hereditary diseases such as cystic fibrosis and haemophilia (Kaufmann et al. 2013). Polycation-condensed DNA (LPD) and the lipid-entrapped method have been produced for brain gene delivery, utilising an adeno-associated viral (AAV) vector. Liver-directed gene delivery for hemophilia has associated with the use of AAV vectors for targeting hepatocytes. Complete clinical practice in liver-directed gene delivery appears from experimental trials for the treatment of hemophilia B using AAV vectors (Anguela and High 2019).

10.3 Vectors for Gene Delivery

Various types of vectors are available for delivering therapeutic genomic molecules (such as DNA, siRNA and oligonucleotides) for treating various genetic diseases. Generally, both viral and non-viral systems play a role as vectors/vehicles for transferring genes/nucleic acids into the target cells. However, cationic polymer-based vector systems and their modified multi-functional forms, nanocomposites, are more relevant for the safe and efficient transfer of therapeutic genes (Zhang and Wagner 2017). These vectors appear as most promising agents in achieving personalised gene therapy against various genetic diseases. However, due to the highly negatively charged surface of plasmid DNA, gene delivery systems having high transfection efficiency are required for successful nucleic acid delivery (Shim et al. 2018). Gene delivery vectors are generally divided into viral and non-viral vector systems.

10.4 Dendrimers as Gene Delivery Vectors

Dendrimers have ability to form a complex with genomic materials such as RNA, plasmid DNA, antisense oligonucleotides, etc. They are artificial macromolecules, which are designed by a grouping of various functional groups and a compact molecular structure. Structural modification of dendrimers and multivalent external moieties appears significant for the development of vectors for gene delivery. Dendrimers possess great potential for gene delivery, due to their nanoscale size, shape and high density of functional groups. They have been used widely for gene delivery to specific body parts, and they act as transfection agents (Chaplot and Rupenthal 2014). The transfection potency of dendrimers mostly depends on their generation (G1.0, G2.0, G3.0, etc.); higher generations are more compressed and orbicular than the lower generation and generate a surface with a high density of essential amines and, thus, form stable dendriplexes with higher effectiveness.

Dendrimers have capability to form polycations under various physiological conditions with the ability to bind genetic molecules containing negative charge, and they also interact with anionic groups of the nucleic acid. Dendrimers are highly efficient novel gene carriers for gene delivery (Wang et al. 2011). Chemical modification of dendrimers facilitates the fixing of various functional groups in high density and improves the *in vivo* performance of gene delivery (Lee and Larson 2011). Structurally optimised dendrimers are used to improve delivery efficiency and lowering cytotoxicity.

10.4.1 Biological Properties of Dendrimers

A family of dendrimers has different biological properties such as electrostatic interactions, poly-valency, self-assembling, chemical stability, solubility and less cytotoxicity. Biological properties are major important features that require to be studied for the effective biomedical treatment with dendrimers, for example, gene delivery, photodynamic therapy (PDT), imaging and neutron capture therapy (Abbasi et al. 2014). Antibody-conjugated dendrimers, dendritic boxes and peptide-conjugated dendrimers can be prepared through modification of surface groups present on dendrimers that enhances the guest encapsulation property of dendrimers (Szymański et al. 2011).

These are the following surface properties of dendrimers, which appears suitable for their diverse biological properties. Molecular identification results at dendrimer surfaces are characterised through a vast number of usually similar terminal groups present on the dendritic organism. The surface of dendrimers contains polyelectrolyte groups, which are attracted electrostatically toward oppositely charged molecules. For example, electrostatic interactions within charged species affect the accumulation of methylene blue on the dendrimer surface and polyelectrolyte dendrimers and the coupling of copper complexes and nitroxide cation radicals (Abbasi et al. 2014; Ahmed et al. 2016). Polyvalency is a biological property of dendrimers, which is required to provide various interactions with biological receptors. Self-assembly is an intrinsic property of molecules that results in the formation of defined arrangement of molecules through various inter and intramolecular forces. Recently, it has been found that dendritic structures also possess a self-assembling property due to their specific structure that comprises terminal groups, core unit and branched units. Through this property, dendrimers convert into the dendrons containing a polytopic or ditopic core structure (Abbasi et al. 2014).

10.4.2 Types of Dendrimers

Dendrimers are novel gene delivery vectors belonging to the class of non-viral polymeric vectors. Types of dendrimers including poly(amidoamine) (PAMAM) dendrimers, poly(propyleneimine) (PPI) dendrimers, poly-L-lysine (PLL) dendrimers, carbosilane dendrimers, triazine dendrimers, poly(etherimine) dendrimers, polyglycerol-based dendrimers such as polyglycerolamine (PG-NH$_2$) and polyglyceryl pentaethylene hexamine carbamate are used as carriers for siRNA delivery (Wu et al. 2013). Cationic dendrimers have capability to bound the siRNA with electrostatic interaction. These dendrimers improve the efficiency and safety of siRNA (Biswas and Torchilin 2013).

PAMAM dendrimers contain an alkyl-diamine core and a tertiary amine branch. These are mostly composed of an ethylenediamine centre; their spreading units comprise amine bunches, which can be utilised to stack drugs, antibodies, chemicals and other bioactive chemicals (Shukla et al. 2016). These branching units, different cores and functional groups have been developed and created for a wide range of applications (Lyu et al. 2019). Most suitable dendrimers for siRNA delivery are PAMAM dendrimers. They were joined to cell-penetrating TAT peptides for improving intracellular delivery. These types of dendrimers may be interacting and compacting with DNA plasmid by electrostatic interaction to form nanocomposites facilitating the cell uptake process and branching of tertiary amine groups, responsible for liberating DNA in the cell through "sponge effect" (Wu et al. 2013). The cationic dendrimers, including PAMAM, carbosilane and phosphorus dendrimers, were effective and safe carriers of the gene for cancer therapy (Ionov et al. 2015). PLL dendrimers are essentially used as gene transporters; they comprise two essential amines, which are regularly changed to improve their beneficial effects (Shukla et al. 2016). The core of PPI is normally based on 1,4-diaminobutane or ethylenediamine, and the branching units comprise propylene imine monomers. They are most commonly used for diagnosis purpose (Mhlwatika and Aderibigbe 2018). The different types of dendrimers which are most commonly used for gene delivery are illustrated in Table 10.1.

TABLE 10.1

Types of dendrimers for the treatment of diseases

Types of Dendrimers	Combination with Genetic Molecules	Diseases	References
PAMAM	siRNA	HIV	Zhou et al. (2011)
	DNA	Schistosomiasis	Wang et al. (2014a)
PAA	DNA	Influenza	Lazniewska et al. (2012); Mhlwatika and Aderibigbe (2018)
Peptide dendrimers	siRNA	Cervical cancer	Dutta et al. (2010)
Carbosilane dendrimers	siRNA	HIV	Jiménez et al. 2010; Yang et al. (2015)
PETIM	siRNA	Hepatitis	Lakshminarayanan et al. (2015)

10.4.3 Classification of Dendrimers

Different classes of dendrimers exist based on their mode of generation, terminal groups, core material, architecture and side chain conjugation on the core of dendrimers. Currently, PPI and PAMAM dendrimers are extensively used for gene delivery due to their higher transfection efficiency and better commercial availability. The classification of dendrimers is represented in Table 10.2.

a) **Based on terminal groups**

Cationic dendrimers: These dendrimers contain amine groups as terminal groups. Example: PAMAM, PPI and polyetherimine (PETIM). The commonly used core in these dendrimers is triethanolamine.

b) **Based on dendrimer architecture**

Symmetrical dendrimers: These dendrimers include barbell shaped triblocks. Example: PLL dendrimer block-methoxy (ethylene glycol)-block PLL dendrimer.

Asymmetrical dendrimers: These dendrimers include water-soluble amphiphilic polylysine and polyornithine peptide with hydrophobic α-aminomyristic acid as the core and methoxy (ethylene glycol)-block PLL dendrimer. These dendrimers possess good transfection efficiency.

Intact dendrimer: These dendrimers possess a rigid, globular shaped structure with high molecular weight than fractured dendrimers. Example: PAMAM dendrimer.

Fractured dendrimer: It contains partially degraded PAMAM dendrimers with 50 fold increased transfection efficacy compared to intact dendrimers.

c) **Based on dendrimer generation**

Generations of dendrimers (G1–G10) are classified based on the number of layers added to the core. Lower generation dendrimers have low efficiency to condense the foreign DNA compared to high generation dendrimers. However, dendrimers of G2 generation possess a more flexible structure, which leads to their comparatively good DNA binding, high transfection efficiency and low cytotoxicity. Examples: PAMAM and PPI dendrimers.

d) **Based on surface functionality**

Surface-engineered dendrimers were improved by adding lipidmoieties, fluorous compounds, amino acid units, saccharides, cationic moieties, proteins and peptides, polymers and nanoparticles to enhance their transfection efficiency and biocompatibility. Such dendrimers are commonly used in *in vivo* and *in vitro* gene delivery (Yang et al. 2015). Therapeutic nucleic acids bind to the surface groups especially amine groups of cationic dendrimer (Florendo et al. 2018). For achieving target delivery and improved delivery of genes with dendrimers, their surface can be modified by adding more moieties to the functional groups already present on the surface of dendrimers (Palmerstan Mendes et al. 2017).

Lipid modified dendrimers: Lipids have strong fusogenic activity because they consist of fatty acids and cholesterol. Therefore, lipid-modified vectors (like lipofectin, lipofectam and lipofectamine) have higher transfection efficacy in different cell lines. The conjugation process between

TABLE 10.2

Various classification systems of dendrimers

Modification of Dendrimers	Dendrimer Class	Examples	Properties	References
Terminal groups	Cationic	Amine groups containing PAMAM, PPI and PETIM	Good transfection efficacy, cell membrane destabilisation, more cytotoxicity/apoptosis and more genome binding affinity	Vidal and Guzman (2015) Ciolkowski et al. (2012)
	Anionic	Sodium carboxylate group containing PAMAM and PPI.	Less genome binding affinity and less cytotoxicity	Posadas et al. (2017)
	Neutral	High generation phosphorous dendrimers	Voluminous internal cavities and good nucleic acid entrapment efficiency	Dzmitruk et al. (2018) Shcharbin et al. (2018)
Architecture	Symmetrical	Triblock dendrimer such as PLL dendrimer block-methoxy (ethylene glycol)-block PLL dendrimerand PLL-PEG-PLL	Good transfection efficacy and biodegradability	Fu et al. (2011)
	Asymmetrical	Methoxy(ethylene glycol)-block PLL dendrimer, polylysine and polyornithine peptide containing an aminomyristic acid core	More water solubility, low toxicity, forming stable toroids by condensing DNA and moderate transfection efficiency	Shah et al. (2011)
	Intact	PAMAM dendrimer and higher generation triazine dendrimers	Less transfection efficiency compared to fractured dendrimers	Wang et al. (2018)
	Fractured	Partially degraded PAMAM dendrimers	50 fold more transfection efficiency than intact dendrimers	Wang et al. (2018)
Generation (number of layers added to core)	G1–G10	PAMAM and PPI dendrimers	G2 generation dendrimers possess more nucleic acid binding affinity	Palmerston Mendes et al. (2017)

lipids and nucleic acids such as siRNA allows an effective gene delivery process. Adjusting the lipid content and cationic charge of dendrimers (polymeric gene vector) increases both cellular uptake and endosomal release of the polyplexes. These dendrimers have a beneficial effect on serum stability, polyplex resistance and intracellular DNA separation from the cationic polymers (Yang et al. 2015). Fatty acid (palmitic acid, lauric acid and myristic acid) conjugated-dendrimers have increased the gene transfection efficacy on stem cells (Willibald et al. 2012).

Dendrimers conjugated with lipids efficiently escape from the intracellular acidic vesicles like endosomes/lysosomes due to the proton sponge effect, and they also show a good cell membrane fusion property due to the presence of lipids. Modification of the surface of dendrimers with lipids such as cholesterol or fatty acids increases the DNA binding affinity and enhances the transfection efficiency of dendrimers. Lipid-bearing dendrons containing two octadecenyl chains are more flexible and hence show a more DNA condensing property.

Fluorinated dendrimers: Fluorination enhances the plasma resistance, cellular uptake, endosomal release and intracellular DNA relief profiles of dendrimers. Fluorination energetically elevates the transfection potency of PAMAM dendrimers. These dendrimers are more efficient vectors as they show transfection efficiency even at the N/P ratio below 1:1. For example, heptafluorobutyric acid modified $G5-F7_{68}$ fluorinated dendrimers have more serum resistance properties and high transfection efficiency than commercially available superfect and lipofectamine 2000. Fluoroaromatics such as 2,3,5,6-tetrafluoro-p-toluic acid increase the transfection efficacy many fold (Yang et al. 2015).

Modification of dendrimers by amino acids: Amino acids such as arginine, histidine, guanidine and imidazole show considerable functions in gene delivery progressions. Lysine and arginine amino acids have been ameliorating the transfection efficacy of PAMAM dendrimers. These amino acids consist of two positive charges in their structure, and their adjustment increases the charge density on the dendrimer surface, which promotes DNA absorption and their usefulness for polyplex stability (Wang et al. 2014a). Hydrophobic amino acids (phenylalanine and leucine) can be used for improving the cellular uptake of dendrimers.

Amino acids are added to modify the surface of dendrimers to increase the nucleic acid condensing property of dendrimers; 6-48 amino acid-modified 2,2-bis(methylol)propionic acid (bis-MPA) based cationic polymers of generation G2 and G4 can be used for transferring siRNAs as they protect siRNA complexes from RNase degradation (Stenström et al. 2018). PAMAM-apoptin gene polyplex seems to be a promising candidate against primary glioma cells in glioblastoma multiform (brain tumour) (Bae et al. 2016).

Cyclodextrin cored dendrimers: These dendrimers possess a hydrophobic cavity for entrapment of guest molecules. They form more stable dendriplexes that can be used in photoresponsive therapy.

Folic acid modified dendrimers: PAMAM dendrimers functionalised with folic acid can be used for more successful targeted delivery of genes to head and neck cancer cells in future (Abedi-Gaballu et al. 2018). Folic acid and folate binding protein-modified dendrimers bind to folate receptors with 17,000 fold more affinity, which suggests that the dendritic structure plays a critical role in their target specificity and cellular interactions (Hsu et al. 2017). The formulation containing peptide dendrimer–lipid (DOTMA/DOPE)-oligonucleotide ternary nanocomplexes with sucrose shows 40-time higher transfection efficiency and more liver targeting (90% of injected dose) in animal models with lowered cytotoxicity. So, this type of vector can be used to deliver splice-switching oligonucleotides (Saher et al. 2018). The transfection efficiency of surface-engineered dendrimers depends on the groups of amino acids as well as on the conjugation ratio (Wang et al. 2018). Transferrin conjugated to PAMAM dendriplexes increases gene expression in the brain by two-fold. Dendrimer-conjugated triamcinolone acetonide gives relief for peripheral neuropathic pain (Kim et al. 2017).

Biomembrane similar modified peptide dendrons: Recently, peptide dendron-based triblock copolymers modified to contain structural arrangements either similar or reverse to biomembranes are relevant for safe and effective gene delivery through intramuscular administration (Pu et al. 2016).

Porphyrin cored dendrimers: These contain photoresponsive porphyrin as their core. Porphyrin is used to achieve increased cytoplasmic delivery of DNA as after entering into the cell these dendrimers get localised at the endosomal membrane and cause photochemical disruption of the endosomal membrane (Wang et al. 2014a).

Polyphenylene vinylene cored dendrimers: These dendrimers can be used for monitoring intracellular trafficking of dendriplexes as these contain fluorescent cores made up of polyphenylene vinylene.

PEGylated dendrimers: Polyethylene glycol is used to modify the core of dendrimers, which has been reported to increase the solubility and biocompatibility and mitigate the toxic effects of dendrimers. The properties of the PEGylated core like inert nature and its negligible non-specific interaction with the serum proteins or lipids make this compound an effective material to modify the dendrimer surface and reduce the toxicity. Multiarmed PEG cored PAMAM dendrimers are more flexible and hence possess good DNA binding affinity. Lactoferrin conjugated to PEGylated PAMAM dendrimers increases the gene expression in brain by 4.6 fold (Florendo et al. 2018).

Hybrid graphene-oleate-PAMAM dendrimers: Oleic acid modified graphene oxide-PAMAM dendrimers exhibit more efficient gene transfection in comparison to unmodified graphene oxide-PAMAM dendrimers (Liu et al. 2014).

Fullerene cored dendrimers: In these types of dendrimers, fullerenes such as C_{60} and C_{70} are used as the core. These are used as light responsive gene delivery systems and in PDT or gene combinational therapy. Surface engineering of dendrimers also affects the toxicity profile of dendrimers used in gene delivery. For example, amine-terminated dendrimers are more cytotoxic than dendrimers containing hydroxyl (–OH) and carboxyl (–COOH) as surface functional groups. However, PEGylated or 4-carbomethoxypyrrolidone-modified PAMAM dendrimers show less cytotoxicity (Vidal and Guzman 2015). Thus, mixed-surface dendrimers with minimal amine groups and surface engineered with the hydroxyl groups which renders them more biocompatible and hence less cytotoxic are relevant for successful gene delivery. Dendrimers decorated with hydroxyl groups do not stick to all cells or biomolecules inside body and are non-immunogenic and highly water-soluble. So, these can be used for targeting brain diseases (Hu et al. 2016).

10.4.4 Dendrimer-Mediated Gene Delivery

Delivery of foreign gene or gene of interest into the nucleus of target cell using dendrimers as a vector involves six major steps: formation of dendriplexes, serum-stable formulation, endocytosis, endosomal escape, intracellular release and entry into the nucleus of target cells as shown in Figure 10.1. Then these genes get incorporated into host cell DNA and express specific proteins in the transfected cells (Hu et al. 2016).

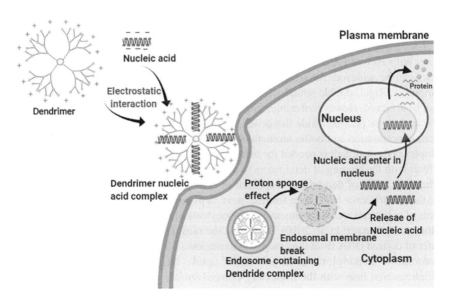

FIGURE 10.1 Schematic representation of dendrimer-mediated gene delivery.

(a) **Dendrimers and nucleic acid interaction**

Dendrimers act as a vector for introducing foreign nucleic acid (DNA or RNA) into the target cell. Dendrimers form dendriplexes with nucleic acid to provide protection to desired nucleic acid from the enzymatic degradation by nucleases. Dendriplexes are the complexes formed by the electrostatic interaction within the multivalent positively charged dendrimers and negatively charged nucleic acid resulting in the condensation of nucleic acid into the small molecular structures (Florendo et al. 2018). For example, the size of dendriplexes formed by most of the cationic dendrimers is around 50 nm. Lower generation dendrimers bind to the major groove of DNA, but higher generation dendrimers have the ability to bind across the entire helical turn of DNA.

These higher generation dendrimers contain multiple attachment sites leading to a higher surface to volume ratio leading to good binding affinity of these vectors. In addition to this, high generation vectors also show a '*starburst effect*'. The starburst effect refers to the presence of a barrier in form of a high-density outer shell which creates a microenvironment inside the core of dendrimers to facilitate the encapsulation of guest molecules. Small sized dendriplexes are more suitable for undergoing the next step that is endocytosis. Generally, all higher generation dendrimers bind more efficiently to DNA than lower generation because of the presence of more amine groups and wrapping around DNA. $PAMAM_{G2}$ is an exception to this which possesses higher binding affinity than $PAMAM_{G7}$ owing to its more fluid structure. Thus, more the dendrimer is flexible such as TEA cored dendrimers more is its binding affinity to DNA. Nature of dendriplex influences the transfection efficacy. More stable dendriplexes fuse to the cell membrane more efficiently.

Morphology and stability of dendriplexes depend on various factors including stoichiometry or dendrimer to DNA phosphate (N/P) ratio, types of dendrimers, types of DNA and solvent properties. To develop suitable dendriplexes for successful gene delivery, the distinctive ratio of dendrimers to DNA is required for each type of dendrimer. If the dendrimer to DNA ratio is greater than 1, it forms a stable dendriplex, irrespective of types of dendrimers and is suitable for gene delivery. But for PAMAM this ratio is 1:1 for making stable dendriplexes. At a ratio of greater than 1, PPI generations$_{G1-G2}$ result in the formation of electroneutral complexes. Around G3 and G4 generations of PPI dendrimers, optimal binding to DNA can be achieved. On increasing this ratio, the size of dendriplexes formed by fractured PAMAM goes on decreasing. 'Salting in' causes the resolubilisation of dendriplexes. Salting in occurs at a ratio greater than 100 and 200 for PAMAMG2 and PAMAMG6, respectively.

PPIG4 and PPIG5 form hexagonal mesophase dendriplexes, which finally appear as extended fibrils. On the other hand, PAMAM, polylysine and cationic dendrimers form toroid shaped dendriplexes. Most of the dendriplexes exist in cluster form with the exception of fractured PAMAM and PEI, which exists as discrete units. With higher generation dendrimers, electroneutral dendriplexes cannot be formed due to the presence of uncomplexed positively charged groups on the surface. With 20% of PAMAMG7 dendriplexes of low density and solubility, 90% of transfection efficacy can be achieved. An adequate amount of PPI results in the formation of water-insoluble dendriplexes. Rigid, double strand DNA interacts only with surface amines. So, electroneutrality cannot be achieved. However, flexible, single strand DNA interacts with all amines whether present on the surface or present inside the dendrimers. Different properties of solvent systems such as pH, salt concentration and buffer strength of solvent influence the shape, stability and solubility of dendriplexes formed. Salt is needed for balancing equilibrium but an excess concentration of NaCl interferes with the binding of dendrimers to nucleic acids.

(b) **Mechanistic effect of dendrimer transfection**

Once the dendriplexes have been formed, next steps required for successful gene delivery to the nucleus of target cell are endocytosis/micropinocytosis of dendriplexes and proton sponge effect of dendrimers (Albertazzi et al. 2010). After this, the most critical step of transfection occurs that is transfer of desired DNA from the cytoplasm to nucleus.

Endocytosis: Dendriplexes undergo cellular uptake by the endocytosis mechanism. Cationic dendriplexes first fuse with the negatively charged surface of the cell membrane, and then these dendrimers create transient small pores (15–40 nm) in the cell membrane. Thus, these dendriplexes get entrapped inside the endosomes. Lipid-bearing dendrimers show more interaction with the cell membrane than other dendrimes. Once endosomes are formed, next thing required is the release of DNA from the endosomes into the cytosol. For achieving this, the endosomal escape property is prerequisite in the dendrimer.

Endosomal escape: Buffering capacity of dendrimers leads to the decelerated acidification of endosomes, increases the osmotic accumulation of Cl⁻ inside the endosomal membrane and results in 140% increase in the endosomal volume. This is called the '*proton sponge effect*' exhibited by dendrimers inside the endosomes (pH= 5.5) that results in the endosomal escape by dendriplexes (Abedi-Gaballu et al. 2018). Thus, dendrimers provide the protection to nucleic acid from enzymatic degradation such as nucleases and lysosomal enzymes. Thus, this effect facilitates the release of desired DNA into the cytosol by disrupting the endosomal membrane. In addition to it, some dendrimers possess a photo-responsive porphyrin core, which causes photochemical disruption by localisation at the endosomal membrane, thus, facilitating the more release of desired DNA into the cytosol of cells.

10.4.5 Dendrimers as Cellular Transfection Agents

Dendrimers possess the ability to fuse with the cell membrane due to positively charged surface groups and possess buffering capacity, which makes them efficient cell transfecting agents, for example, PAMAM$_{G6}$ dendrimers. These can be used for delivering the antisense oligonucleotide and desired plasmid DNA into the target cell. Some dendrimers such as PLL do not possess the endosomal escape property so lysomotropic agents are added to them to prevent enzymatic degradation of nucleic acid. Mostly, amine-containing dendrimers have endosome escape ability due to their pH buffering action resulting in swelling of endosomes. High generation dendrimers possess more ability to bind DNA to make stable dendriplexes and hence increase transfection efficiency with the increase in generation from G5 to G10. Trimesyl cored PAMAM$_{G6}$ dendrimers have more transfection efficacy. However, dendrimers lead to cytotoxicity, which is less with the use of triblock dendrimers PAMAM-PEG-PAMAM. This triblock dendrimer also increases the colloidal stability of dendriplexes. Fractured PAMAM possesses 50 times more transfection efficiency compared to intact PAMAM. More the surface group density more is the transfection efficiency of dendrimers. Large DNA segments (60 Mb artificial chromosome) can be delivered using PAMAM dendrimers. Decreasing order of transfection efficacy of various dendrimers is linear PEI > activated PAMAM dendrimers > branched PEI >pluronic PEI graft block copolymer. DAB cored PPI dendrimers have least transfection efficacy and more toxicity at a high N/P ratio. Thus, a good transfecting agent is that dendrimers, which have high DNA binding affinity and low cytotoxicity (Abedi-Gaballu et al. 2018).

10.4.6 Mode of Administration

Dendrimers can be administered in various ways. However, for circumventing the vascular system to avoid unknown side effects, a local or *ex-vivo* route of administration is preferred. 6–10% cells transfected by local application of activated PAMAM dendriplex formulation on cornea of eye. Subcutaneous transplants of PAMAM–EGFP hBMP2 plasmid DNA complex-coated titanium are available for orthopaedic application (Chen et al. 2018). To treat tumours, intratumoral injection of dendriplexes can be given. Intratracheal instillation can also be performed for gene delivery into lungs. Metered-dose inhalers of aerosol formulation of TPP-conjugated siRNA dendriplexes are also available for local delivery to lungs through inhalational route (Bielski et al. 2017). Intravascular administration of PAMAM$_{G9}$ dendriplexes results in the transfection of lung parenchyma only, while systemic administration of PAMAM$_{G3}$ and cyclodextrin conjugated PAMAM$_{G3}$ results mostly in the transfection of spleen and liver. Specific delivery of genes into liver rather than lungs can be efficiently achieved by using PPI$_{G3}$ or quaternised PPI$_{G2}$ through the systemic route. Through systemic administration, intratumoral transgene expression can be achieved by using dendrimers PPI$_{G3}$. PPI$_{G3}$ dendrimers possess no apparent toxicity and have intrinsic anti-proliferative property. Systemic administration of folic acid-modified dendrimers is suitable for taking into consideration in clinical trials (Cheng et al. 2008). Triblock copolymers can be administered intramuscularly for delivering genes (Pu et al. 2016).

10.4.7 Conjugate Approach

Dendrimers can be conjugated with DNA/RNA or drugs. Coumarin-attached low generation dendrimers possess high DNA binding affinity and targeted gene delivery. So, these are used as light responsive gene delivery vectors(Wang et al. 2018). Fullerene (C60/C70)-cored dendrimers are used for PDT. TPP-conjugated dendrimers can be used as a vector for siRNA delivery to epithelial cells of lungs and their aerosol formulation for local siRNA delivery to lungs are also available (Bielski et al. 2017). 29-amino acid rabies virus glycoprotein (RVG29) conjugated to PAMAM dendriplexes crosses the blood–brain barrier efficiently (Florendo et al. 2018). Small siRNA conjugated to carbosilane dendrimers also enters the brain (Serramía et al. 2015). Thus, the conjugate approach is also used for targeting various brain diseases by gene- or drug-loaded dendrimers.

10.4.8 Advantages with Dendrimers

Large DNA construct such as 60 mb artificial chromosome can be delivered easily with the dendrimers. Moreover, both drugs and genes can be delivered using dendrimers. These can be used to achieve higher transfection efficiency and biocompatibility as they allow easy surface modification with multi-functional ligands. These increase the stability of dendriplexes and provide protection to DNA from intracellular enzymatic degradation. These vectors increase the solubility and bioavailability of hydrophobic molecules (Hsu et al. 2017). Their manufacturing process is facile and easy to reproduce the preparation of a mono-dispersed, uniform formulation (Hsu et al. 2017). The drug conjugated to PAMAM dendrimers results in the controlled release of the drug (Abedi-Gaballu et al. 2018). Experimentally sustained delivery of insulin has been achieved by using PEGylated PPI dendrimers (Parashar et al. 2019).

10.4.9 Application of Dendrimers

Dendrimers have a few therapeutic and practical applications. In future, they can be utilised for gene delivery, tissue building and diagnosis. Dendrimers are used as non-viral gene delivery vectors due to their consistency and multi-valency provided by functional groups present on their surface. Now dendrimers are also being explored for various biomedical applications. PAMAM and PPI dendrimers are used as gene carriers due to their relative low toxicity and good affinity to bind negatively charged genetic materials. PAMAM has been utilised widely for gene delivery and tissue designing because it has shown a non-immunogenic effect, biocompatibility and hydrophilic nature (Mhlwatika and Aderibigbe 2018). Through electrostatic interaction, amine residues on the surface of PAMAM dendrimers bind to the phosphate group of nucleic acids. Currently, G6–7 generation of PAMAM dendrimers is used for gene transfection. Few types of dendrimers are also applicable for photodynamic gene delivery, for example, fullerene C_{60} and C_{70} containing dendrimers. Significant enhancement in gene expression efficiency has been achieved by exposing the DNA-polycation dendrimer ternary complex to light. This type of ternary complex has been formed by mixing cationic peptide molecules containing a nuclear localisation signal sequence and anionic phthalocyanine dendrimers.

REFERENCES

Abbasi, E., Aval, S.F., Akbarzadeh, A. et al. 2014. Dendrimers: synthesis, applications, and properties. *Nanoscale Research Letters* 9:247.

Abedi-Gaballu, F., Dehghan, G., Ghaffari, M. et al. 2018. PAMAM dendrimers as efficient drug and gene delivery nanosystems for cancer therapy. *Applied Materials Today* 12:177–190.

Ahmed, S., Vepuri, S.B., Kalhapure, R.S. and Govender, T. 2016. Interactions of dendrimers with biological drug targets: reality or mystery–a gap in drug delivery and development research. *Biomaterials Science* 4:1032–1050.

Albertazzi, L., Serresi, M., Albanese, A. and Beltram, F. 2010. Dendrimer internalization and intracellular trafficking in living cells. *Molecular Pharmaceutics* 7:680–688.

Anguela, X.M. and High, K.A. 2019. Entering the modern era of gene therapy. *Annual Review of Medicine* 70:273–288.

Bae, Y., Green, E.S., Kim, G.Y. et al. 2016. Dipeptide-functionalized polyamidoamine dendrimer-mediated apoptin gene delivery facilitates apoptosis of human primary glioma cells. *International Journal of Pharmaceutics* 515:186–200.

Bielski, E., Zhong, Q., Mirza, H. et al. 2017. TPP-dendrimer nanocarriers for siRNA delivery to the pulmonary epithelium and their dry powder and metered-dose inhaler formulations. *International Journal of Pharmaceutics* 527:171–183.

Biswas, S. and Torchilin, V.P. 2013. Dendrimers for siRNA delivery. *Pharmaceuticals* 6:161–183.

Chaplot, S.P. and Rupenthal, I.D. 2014. Dendrimers for gene delivery—a potential approach for ocular therapy? *Journal of Pharmacy and Pharmacology* 66:542–556.

Chen, C., Yang, Z. and Tang, X. 2018. Chemical modifications of nucleic acid drugs and their delivery systems for gene-based therapy. *Medicinal Research Reviews* 38:829–869.

Cheng, Y., Xu, Z., Ma, M. and Xu, T. 2008. Dendrimers as drug carriers: applications in different routes of drug administration. *Journal of Pharmaceutical Sciences* 97:123–143.

Ciolkowski, M., Petersen, J.F., Ficker, M. et al. 2012. Surface modification of PAMAM dendrimer improves its biocompatibility. *Nanomedicine: Nanotechnology, Biology and Medicine* 8:815–817.

Dunbar, C.E., High, K.A., Joung, J.K., Kohn, D.B., Ozawa, K. and Sadelain, M. 2018. Gene therapy comes of age. *Science 359*:e4672.

Dutta, T., Burgess, M., McMillan, N.A. and Parekh, H.S. 2010. Dendrosome-based delivery of siRNA against E6 and E7 oncogenes in cervical cancer. *Nanomedicine: Nanotechnology, Biology and Medicine* 6:463–470.

Dzmitruk, V., Apartsin, E., Ihnatsyeu-Kachan, A., Abashkin, V., Shcharbin, D. and Bryszewska, M. 2018. Dendrimers show promise for siRNA and microRNA therapeutics. *Pharmaceutics* 10:126.

Florendo, M., Figacz, A., Srinageshwar, B. et al. 2018. Use of polyamidoamine dendrimers in brain diseases. *Molecules 23*:2238.

Fu, C., Sun, X., Liu, D., Chen, Z., Lu, Z. and Zhang, N. 2011. Biodegradable tri-block copolymer poly (lactic acid)-poly (ethylene glycol)-poly (L-lysine)(PLA-PEG-PLL) as a non-viral vector to enhance gene transfection. *International Journal of Molecular Sciences* 12:1371–1388.

Hsu, H.J., Bugno, J., Lee, S.R. and Hong, S. 2017. Dendrimer-based nanocarriers: a versatile platform for drug delivery. *Wiley Interdisciplinary Reviews: Nanomedicine and Nanobiotechnology* 9:e1409.

Hu, J., Hu, K. and Cheng, Y. 2016. Tailoring the dendrimer core for efficient gene delivery. *Acta Biomaterialia* 35:1–11.

Ingusci, S., Verlengia, G., Soukupova, M., Zucchini, S. and Simonato, M. 2019. Gene therapy tools for brain diseases. *Frontiers in Pharmacology* 10:724.

Ionov, M., Lazniewska, J., Dzmitruk, V. et al. 2015. Anticancer siRNA cocktails as a novel tool to treat cancer cells. Part (A). Mechanisms of interaction. *International Journal of Pharmaceutics* 485:261–269.

Jiménez, J.L., Clemente, M.I., Weber, N.D. et al. 2010. Carbosilane dendrimers to transfect human astrocytes with small interfering RNA targeting human immunodeficiency virus. *BioDrugs* 24:331–343.

Kaufmann, K.B., Büning, H., Galy, A., Schambach, A. and Grez, M. 2013. Gene therapy on the move. *EMBO Molecular Medicine* 5:1642–1661.

Kim, H., Choi, B., Lim, H. et al. 2017. Polyamidoamine dendrimer-conjugated triamcinolone acetonide attenuates nerve injury-induced spinal cord microglia activation and mechanical allodynia. *Molecular Pain 13*:1–11.

Klug, B., Celis, P., Carr, M. and Reinhardt, J. 2012. Regulatory structures for gene therapy medicinal products in the European Union. *Methods in Enzymology* 507:337–354.

Lakshminarayanan, A., Reddy, B.U. and Raghav, N. 2015. A galactose-functionalized dendritic siRNA-nanovector to potentiate hepatitis C inhibition in liver cells. *Nanoscale* 7:16921–16931.

Lazniewska, J., Milowska, K. and Gabryelak, T. 2012. Dendrimers—revolutionary drugs for infectious diseases. *Wiley Interdisciplinary Reviews: Nanomedicine and Nanobiotechnology* 4:469–491.

Lee, H. and Larson, R.G. 2011. Membrane pore formation induced by acetylated and polyethylene glycol-conjugated polyamidoamine dendrimers. *The Journal of Physical Chemistry C 115*:5316–5322.

Liu, X., Ma, D., Tang, H. et al. 2014. Polyamidoamine dendrimer and oleic acid-functionalized graphene as biocompatible and efficient gene delivery vectors. *ACS Applied Materials & Interfaces* 6:8173–8183.

Lyu, Z., Ding, L., Huang, A.Y.T., Kao, C.L. and Peng, L. 2019. Poly (amidoamine) dendrimers: covalent and supramolecular synthesis. *Materials Today Chemistry* 13:34–48.

Maguire, C.A., Ramirez, S.H., Merkel, S.F., Sena-Esteves, M. and Breakefield, X.O. 2014. Gene therapy for the nervous system: challenges and new strategies. *Neurotherapeutics 11*:817–839.

Mhlwatika, Z. and Aderibigbe, B.A. 2018. Application of dendrimers for the treatment of infectious diseases. *Molecules 23*:2205.

Misra, S. 2013. Human gene therapy: a brief overview of the genetic revolution. *The Journal of the Association of Physicians of India 61*:127–133.

Nayerossadat, N., Maedeh, T. and Ali, P.A. 2012. Viral and nonviral delivery systems for gene delivery. *Advanced Biomedical Research 1*:27.

Palmerston Mendes, L., Pan, J. and Torchilin, V.P. 2017. Dendrimers as nanocarriers for nucleic acid and drug delivery in cancer therapy. *Molecules 22*:1401.

Parashar, A.K., Patel, P., Gupta, A.K., Jain, N.K. and Kurmi, B.D. 2019. Synthesis, characterization and in vivo evaluation of PEGylated PPI Dendrimer for safe and prolonged delivery of insulin. *Drug Delivery Letters 9*:248–263.

Posadas, I., Romero-Castillo, L., El Brahmi, N. et al. 2017. Neutral high-generation phosphorus dendrimers inhibit macrophage-mediated inflammatory response in vitro and in vivo. *Proceedings of the National Academy of Sciences 114*:E7660–E7669.

Pu, L., Wang, J., Li, N. et al. 2016. Synthesis of electroneutralized amphiphilic copolymers with peptide dendrons for intramuscular gene delivery. *ACS Applied Materials & Interfaces 8*:13724–13734.

Saher, O., Rocha, C.S., Zaghloul, E.M. et al. 2018. Novel peptide-dendrimer/lipid/oligonucleotide ternary complexes for efficient cellular uptake and improved splice-switching activity. *European Journal of Pharmaceutics and Biopharmaceutics 132*:29–40.

Seeger-Armbruster, S., Bosch-Bouju, C., Little, S.T. et al. 2015. Patterned, but not tonic, optogenetic stimulation in motor thalamus improves reaching in acute drug-induced Parkinsonian rats. *Journal of Neuroscience 35*:1211–1216.

Serramía, M.J., Álvarez, S., Fuentes-Paniagua, E. et al. 2015. In vivo delivery of siRNA to the brain by carbosilane dendrimer. *Journal of Controlled Release 200*:60–70.

Shah, N., Steptoe, R.J. and Parekh, H.S. 2011. Low-generation asymmetric dendrimers exhibit minimal toxicity and effectively complex DNA. *Journal of Peptide Science 17*:470–478.

Shcharbin, D., Shcharbina, N., Pedziwiatr-Werbicka, E. et al. 2018. Phosphorus Dendrimers as vectors for gene therapy in cancer, *Phosphorous Dendrimers in Biology and Nanomedicine: Syntheses, Characterization, and Properties 227*(ebook).

Shim, G., Kim, D., Le, Q.V., Park, G.T., Kwon, T. and Oh, Y.K. 2018. Nonviral delivery systems for cancer gene therapy: strategies and challenges. *Current Gene Therapy 18*:3–20.

Shukla, S.K., Govender, P.P. and Tiwari, A. 2016. Polymeric micellarstructures for biosensor technology. *Advances in Biomembranes and Lipid Self-Assembly 24*:143–161.

Stenström, P., Manzanares, D., Zhang, Y., Ceña, V. and Malkoch, M. 2018. Evaluation of amino-functional polyester dendrimers based on Bis-MPA as nonviral vectors for siRNA delivery. *Molecules 23*:2028.

Sung, Y.K. and Kim, S.W. 2019. Recent advances in the development of gene delivery systems. *Biomaterials Research 23*:8.

Szymański, P., Markowicz, M. and Mikiciuk-Olasik, E. 2011. Nanotechnology in pharmaceutical and biomedical applications: Dendrimers. *Nano 6*:509–539.

Vidal, F. and Guzman, L. 2015. Dendrimer nanocarriers drug action: perspective for neuronal pharmacology. *Neural Regeneration Research 10*:1029–1031.

Wang, H., Miao, W., Wang, F. and Cheng, Y. 2018. A self-assembled coumarin-anchored dendrimer for efficient gene delivery and light-responsive drug delivery. *Biomacromolecules 19*:2194–2201.

Wang, H., Shi, H.B. and Yin, S.K. 2011. Polyamidoamine dendrimers as gene delivery carriers in the inner ear: How to improve transfection efficiency. *Experimental and Therapeutic Medicine 2*:777–781.

Wang, X., Dai, Y., Zhao, S. et al. 2014a. PAMAM-Lys, a novel vaccine delivery vector, enhances the protective effects of the SjC23 DNA vaccine against *Schistosoma japonicum* infection. *PLoS One 9*:e86578.

Wang, Z., Liu, G., Zheng, H. and Chen, X. 2014b. Rigid nanoparticle-based delivery of anti-cancer siRNA: challenges and opportunities. *Biotechnology Advances 32*:831–843.

Willibald, J., Harder, J., Sparrer, K., Conzelmann, K.K. and Carell, T. 2012. Click-modified anandamide siRNA enables delivery and gene silencing in neuronal and immune cells. *Journal of the American Chemical Society 134*:12330–12333.

Wu, J., Huang, W. and He, Z. 2013. Dendrimers as carriers for siRNA delivery and gene silencing: a review. *The Scientific World Journal* Article ID 630654.

Yang, J., Zhang, Q., Chang, H. and Cheng, Y. 2015. Surface-engineered dendrimers in gene delivery. *Chemical Reviews 115*:5274–5300.

Zhang, P. and Wagner, E. 2017. History of polymeric gene delivery systems. *Topics in Current Chemistry 375*:26.

Zhou, J., Neff, C.P., Liu, X. et al. 2011. Systemic administration of combinatorial dsiRNAs via nanoparticles efficiently suppresses HIV-1 infection in humanized mice. *Molecular Therapy 19*:2228–2238.

11

Dendrimers in Diagnostics Application

Biplab Sikdar, Gagandeep Maan and Awanish Mishra

CONTENTS

11.1 Introduction

Diagnosis of different medical conditions is an important part of the health care management system. Early prediction of diseases with the help of advanced diagnostic tools is helpful for effective management of diseases and reduces chances of mortality associated with life-threatening diseases. Especially for cancer, early diagnosis emerged as a promising tool for providing effective chemotherapy as well as patient recovery. In addition, accurate diagnosis results in proper therapy, which will be most efficacious.

Initial diagnosis of patients is done by observing patient's sign and symptoms. Diagnosis also includes physical examination and patient's history along with a specific diagnostic test. For the diagnosis, different kinds of diagnostic and imaging agents are used to confirm the disease stage and condition. At present, development of molecular diagnostics and *in-vivo* imaging agents appears as one of the major focuses in medical research. These diagnostic and imaging techniques can be used in pre-clinical studies and also to monitor the efficacy of therapeutic agents in patients. Magnetic resonance imaging (MRI), computed tomography (CT), positron emission tomography (PET) and X-ray spectroscopy appear as highly used imaging techniques in the diagnosis of diseases (Akbarzadeh et al. 2018; Noriega-Luna et al. 2014; Kunjachan et al. 2012).

Since the advent of nanotechnology, nanoparticles as nanoformulations have surprised us with their diverse and useful physicochemical properties. Nanomaterials have proven their efficacy as theranostics by virtue of their use in diagnosis and mitigation of various pathological conditions (Key and Leary 2014; George et al. 2019). Though the use of nanoparticles in medicine appears promising, their utility in diagnosis has not been explored well (Rizzoet al. 2013).

Dendrimers as a nanomolecular structure involved in designing of various molecules have various biomedical utilities. They are greatly explored for different applications such as in drug delivery, bio-imaging, diagnosis, tissue engineering and cancer diagnosis as well as in therapy, in gene delivery and other areas (Chen et al. 2019; Gan et al. 2019). Dendrimers are greatly used due to their advantages over other materials. Their utility is more due their shape, size and presence of interior void space, and also they can be synthesised simply using a single polymeric unit. They provide a wide range of flexibility and architecture, which is helpful in designing of molecular nanostructures (Menjoge et al. 2010; Kesharwani et al. 2014; Lyu et al. 2019).

The limitations of carrier-free bio-diagnostics are poor solubility, poor systemic availability, high clearance rate, non-specific accumulation and distribution and toxic and immunological reaction (Noriega-Luna et al. 2014; Pooja et al. 2018; Dias et al. 2020). In this chapter, we have described the benefits of dendrimers as a diagnostic agent and also their biomedical application, such as cancer diagnosis, as a MRI contrast agent and as a biosensor.

11.2 Dendrimers

The word dendrimer basically originated from the Greek word *Dendron* that means 'tree-like' due to its shape and *Meros* means 'part' for its chemical structure. It was first discovered by Donal Tomalia in 1985 (Sherjeet al. 2018; Dias et al. 2020) but the polypropyleneimine, which is a dendrimer-like compound, was first synthesised by Vogtle et al in 1978 (Mintzer and Grinstaff 2011). A dendrimer is a nanosized structure, monodispersed macromolecule with a radially symmetrical chemical structure, consisting of well-defined tree shape arms or branches with different terminal functional groups extended from a central core. The size of the dendrimer ranges from 1 to 100nm (Abbasi et al. 2014; Kesharwani et al. 2014; Akbarzadeh et al. 2018).

Dendrimers comprise three basic parts: a central core with multifunctionality, radially symmetrical branched extended from the core and the external covering groups. The dendrimer can be made larger in size by increasing the size of branching with repetitive addition of monomer units, which is also known as generation growth (DaSilva Santos et al. 2016; Araújo et al. 2018). By using such larger branching, the dendrimers can develop into a hyperbranched structure radiating from the central core. Depending upon the structure, the dendritic structure can be classified as hyper-grafted polymers, star-like polymers, dendronised or dendrigrafts, dendritic linear block polymers or dendrimers (Ma 2013; Ma et al. 2016; Dias et al. 2020).

Dendrimers can also be categorised depending upon the functional groups as polythene dendrimers, polyester dendrimers, tecto or core–shell dendrimers, triazine dendrimers, citric acid dendrimers, phosphate dendrimers, melamine dendrimers, polyether imine dendrimers and polyether–polyester dendrimers. These dendrimers are mostly used in the field of pharmaceuticals and in medical diagnosis. They can be synthesised from a simple monomer unit with branches to connect the core with the outer capping groups. Dendrimers are synthesised mainly through convergent and divergent approaches (Dayyani et al. 2018; Ray and Khan 2018; Lyu et al. 2019). Apart from this, they can be also synthesised by covalent conjugation strategies, by using the polyvalency concept, by self-assembling of dendrimers and by electrostatic interactions. The convergent approach involves the addition of the monomer unit from the core to the outer direction in a stepwise manner, whereas in the divergent approach, the synthesis starts from a terminal functional group and finally forming the core of the dendrimer. The branching monomer units of dendrimers are organised in layers known as generation (Abbasi et al. 2014; Sowinska and Urbanczyk-Lipkowska 2014; Chen et al. 2016).

Dendrimers are the synthesised hyper-grafted polymer systems that can be combined to different chemical agents for various biomedical applications such as diagnostic agents, imaging agents, targeting therapeutic agents, etc. Other applications of dendrimers include as a carrier for drug targeting, for gene delivery, in cancer therapy, as RNA interference transporters, in targeting the central nervous system, in bacterial and viral infections, in other neglected tropical diseases, etc. (Mintzer and Grinstaff 2011; Ray and Khan 2018; Lyu et al. 2019).

11.3 Properties and Benefits of Dendrimers as a Diagnostic Agent

The polymeric structure of dendrimers shows a variety of specific properties than other diagnostic materials. Dendrimers have some of the unique properties such as nanomolecular globule shape, multifunctional covering groups at the outer surface, both hydrophobic or hydrophilic cavities, with a very low polydispersity (Li et al. 2018). Owing to these properties, dendrimers possess a wide range of applications in the pharmaceutical field as well as in the medical field (Figure 11.1). The nanosized dendrimers can be easily taken up by cells via endocytosis than the microsized particles. Dendrimers have a very high degree of uniformity within molecules, low molecular weight and large distribution with specific characteristics according to shape and size containing highly functional covering groups at the outer surface. As dendrimers have cellular uptake through endocytosis, they can bound the drug to the cell (DaSilva Santos et al. 2016; Vieira Gonzaga et al. 2018; Akbarzadeh et al. 2018). Dendrimers have several benefits over other molecules in diagnosis and imaging due to the following properties as dendrimers have good structural control on the of size and shape of diagnostic or imaging agents.

Dendrimers are highly compatible with biological systems as compared to other molecular agents. They have very low toxicity depending upon the extent and the type of monomeric unit used in the formulation of the contrasting or diagnostic agent. Due to the nanosize structure, the imaging capacity of the agent also increases. Dendrimers have a very low immunogenic reaction in the biological system. The uptake of dendrimers by endocytosis allows the diagnostic agent to access the nucleus and cytoplasm of the cell. Dendrimers have a very wide range of acceptable bioelimination and biodegradation with a great range of biodistribution. Dendrimers have very limited non-specific cellular and protein binding with high retention inside the body. Dendrimers are compatible with the magnetic field, suitable with ultrasound and also with X-rays that supply good contrast. The synthesis of dendrimers can be performed

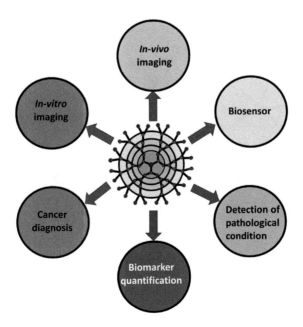

FIGURE 11.1 Application of dendrimers in diagnosis.

with consistency and reproducibility with clinically acceptable quality (Mintzer and Grinstaff 2011; Kesharwani et al. 2014; Chen et al. 2016; Li et al. 2018; Liu et al. 2019).

11.4 Types of Diagnostics Based on Dendrimers

Dendrimers can be used in live imaging systems by conjugating them with various contrast agents. The contrasting agent either can be captured inside the core or otherwise fixed or attached on the terminal groups of the dendrimers. Incorporation of contrasting agents inside the void spaces of dendrimers provides gives the controlled release along with reduced toxicity. The schematic representation of various possible modifications of dendrimers for their diagnostic purposes is illustrated in Figure 11.2.

This incorporation is done by the physical method or non-covalent interaction of the contrasting agent with the functional groups of dendrimers (Kulhari et al. 2011; Pooja et al. 2018). Depending upon the chemical structure and nature of some imaging agents and radioligands, they are attached on the surface of the branched dendrimers. There are several diagnostic techniques (Figure 11.3), which are based on dendrimers and are discussed below:

11.4.1 Dendrimer-Based MRI for Diagnosis

MRI is an important non-invasive technique, which is widely used to diagnose the disease condition and treatment strategy. By using this tool, the comprehensive images of the different tissues, various body organs and bones can be obtained. This technique uses radio waves along with the strong magnetic field and to get the images within the living system. Examples of encapsulating MRI agents in the void space of dendrimers include G3.0–G6.0-PAMAM dendrimers, G8.0-PAMAM dendrimer, G5.0–G7.0 PMPA dendrimers, A 4.0 PAMAM, G2.0-PAMAM-Cysteamine dendrimer, G1.0-5.0 Polypropylene amine dendrimer, DAB-Am64 (G6.0) and DAB-Am64 (G4.0) dendrimers. Dendrimers on which conjugation of contrasting agents on the surface is done are PEG-G4.5PAMAM and G6.0-PAMAM-Cystamine (Noriega-Luna et al. 2014).

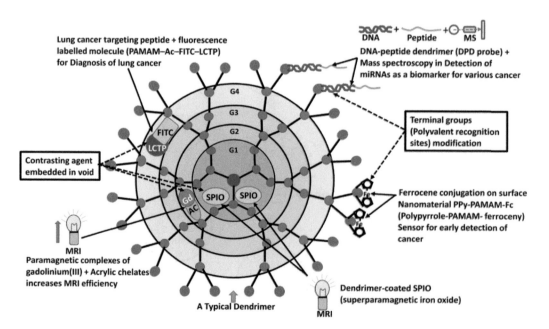

FIGURE 11.2 Modification of dendrimer for different diagnostic purpose.

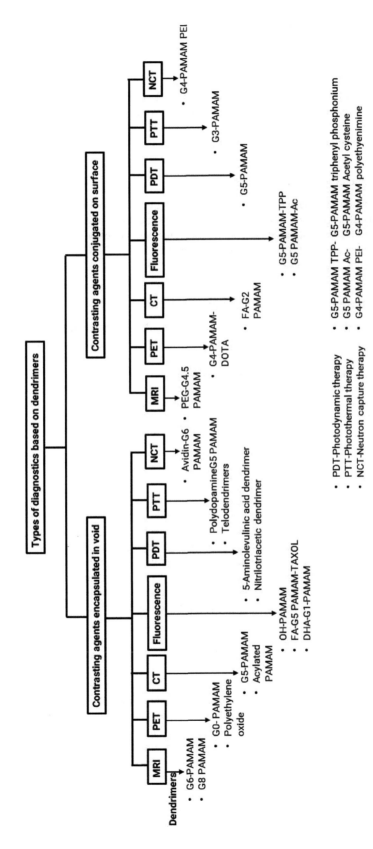

FIGURE 11.3 Dendrimers used in different diagnostic tools.

11.4.2 Dendrimer-Based PET for Diagnosis

PET uses gamma rays to produce the three-dimensional (3D) image of the different metabolic processes within the body. It is used to understand the exact functioning of the specific tissue and organ. In this technique, a radioactive tracer is used to detect the functioning of the tissue, which is incorporated into the core of the dendrimer. Following are the dendrimers used to encapsulate the tracer inside the void spaces of dendrimer: (LyP-1) 4-dendrimer, trifluoroboroaryl- PAMAM- biotin dendrimer, polyethylene oxide-dendrimer and G0-PAMAM dendrimer. G4-PAMAM-DOTA and maleimide-dendrimers are used to form a conjugate at the outer surface with the contrasting agents (Zhao et al. 2017).

11.4.3 Dendrimer-Based CT for Diagnosis

To obtain a high resolution 3D image of the organ of any living body, this tool is used extensively. This technique works on the absorption phenomenon of X-rays. The structure of dendrimers stabilises the contrasting agent and gives a good compatible system for the living cell. For encapsulation of the CT agent, the following dendrimers are used: G5.0-PAMAM, acylated PAMAM, FA-G2.0 PAMAM, lactobonic acid-PAMAM and ^{131}I-PAMAM (Liu et al. 2013; Sharma et al. 2017).

11.4.4 Dendrimer-Based Fluorescence Imaging for Diagnosis

In this technique, excitation of fluorescent molecules is done initially by an optimum wavelength after that lower energy light is used to obtain an image. The different dendrimers which are used for incorporation of contrasting agents are OH-PAMAM dendrimers, G5.0-PAMAM dendrimers, lauroyl/propranolol-G3.0 PAMAM dendrimers, PAMAM dendrimers, FA-G5.0 PAMAM-Taxol dendrimers, DHA-G1.0 PAMAM and Gly-Lys-G6.0 dendrimers. G5.0-PAMAM-Ac dendrimers and G5.0-PAMAM-TPP dendrimers are used for conjugation with contrasting agents at the outer surface (Biswas et al. 2012; Dougherty et al. 2015; Gan et al. 2019).

11.5 Biomedical Applications of Dendrimers

By virtue of the structural framework of dendrimers, they have shown wide potential in theranostics. Dendrimers are used in drug delivery as a carrier of drug; some dendrimers also act as a drug (anti-microbialand anti-viral). In drug delivery, they are used to target the therapeutics in specific tissue or different organs of the body. Dendrimers are used extensively in gene delivery, tissue engineering, targeted delivery of orthopaedic drugs, etc. The dendrimers are also used in cancer diagnosis and its management. Clinically, these are used in the diagnosis of various disease conditions. Conjugating various contrast media with dendrimers enhances the efficacy of contrast agents by enhancing their solubility, tissue penetration and targeted delivery. Dendrimers are also used in the preparation of various biosensors to detect the physiochemical changes in the body.

11.5.1 Dendrimer in Cancer Diagnosis

Nanotechnology is developing widely nowadays and hence also plays an important role in cancer diagnosis and its treatments. Dendrimers can be designed and regulated in terms of their structural aspects such as their size, shape, surface and interior void space as per required. Their nanoscale scaffolding and nanocontainer properties play a beneficial role in loading medicine or diagnostic agents. Various biocompatible dendrimers are used to fabricate and stabilise the diagnostic agents. Even sometimes they are used for the functionalisation of magnetite nanoparticles for diagnosis as well for the treatment. Generally, polyamidoamine (PAMAM) dendrimers are preferred, but peptide dendrimers can also be used, as they are biocompatible, and their metabolic product has very low or negligible toxicity. Example the doxorubicin loaded peptide dendrimer-based nanoformulation an alternative to PAMAM dendrimer shows an efficient result in combinatorial therapy (Nigam and Bahadur 2017). The novel biomarker for various cancer diagnosis like breast cancer is microRNAs (miRNAs). However, the quantification of the biomarker which is crucial for diagnosis is not so easy. Dendrimers are used to increase the efficiency of the quantification assay, for example peptide dendrimers combined with a specific DNA sequence to form a novel DNA–peptide dendrimer (DPD), which is combined with mass spectrometry to detect the miR-NAs (Liu et al. 2019). The combination of the DPD probe along with mass spectrometry provides a novel technique for determination of miRNAs as a biomarkers of various cancer cells (Liu et al. 2019). Various generation of PAMAM dendrimer were tested with combination of polydopamine to coat iron oxide nanoparticles for imaging. G 5.0 PAMAM was found suitable for conjugating iron oxide nanoparticles as over the (G) 4.0 and 6.0 generation of PAMAM for cancer diagnosis (Jędrzak et al. 2019). Different PAMAM dendrimers are used in functionalisation of LFC131 (d-Tyr-Arg-Arg-2-Nal-Gly) peptide for determination of the CXCR4 (a chemokine receptor), which is a biomarker of the various cancer such as lung cancer, breast cancer, etc. (Wang et al. 2015; Chittasupho et al. 2017).

Dendrimer-coated superparamagnetic iron oxide nanoparticles are the best candidate for the simultaneous detection and treatment particularly in case of malignant tumours (Ray and Khan 2018). Their adjustable size, compatibility in humans and low toxicity have made them more effective. Different diagnostic tools such as CT, PET and fluorescence imaging can be combined with MRI to make it more advantageous by overcoming the limitations of one imaging technique with the other.

11.5.2 Dendrimer-Based MRI Contrast Agents

MRI, which is also known as nuclear magnetic resonance imaging (NMRI), is used for diagnosis of the disease condition. This is a non-invasive tool, which uses the radio waves along with a strong magnetic field, to provide an image of the anatomy or physiological processes and, helps the physician for diagnosis as well as the treatment of the disease condition.

This tool mostly consists of large tube-shaped magnets, a contrasting agent like paramagnetic substance such as metal ions, which reduces the relaxation time of in vivo protons of water molecules. These paramagnetic agents include different groups of compounds such as gadolinium chelates and particles of iron oxides, which are super magnetic. However, these contrasting agents have several pitfalls, such as low diffusivity in body or a high excretion rate, a short period of bioavailability, inefficiency to distinguish diseased tissue from normal healthy tissues, etc, which makes the researcher to advance the tools or to search the agents, which can be used to increase the efficiency of MRI agents, which can improve the performance of the imaging technique (Noriega-Luna et al. 2014).

Dendrimer-based contrasting agents are tested in animals and are found to be efficacious in improving the performance of the contrasting agents. The conjugation of the paramagnetic agents with the dendrimers increases the sensitivity of the contrasting agents and also can be targeted to specific parts or organs of the body i.e., increases specificity. Examples include the fourth-generation (G4.0) PAMAM dendrimers, used as carriers for gadolinium complexes of indole acetic acid derivatives. Due to the use of dendrimers, the signal intensity in liver increased (Markowicz-Piasecka et al. 2015). The use of PAMAM dendrimers as a carrier of paramagnetic complexes of gadolinium(III) with acyclic or macrocyclic chelates increases the contrast and specificity of the MRI contrasting agents (Gündüz et al. 2016).

11.5.3 Dendrimers as Biosensors

Dendrimers have wide applications in the field of biosensors due to their properties, which meets the specification for the material suitable for construction of biosensors. It has been seen that the PAMAM dendrimer is being used to prepare the biosensor for any biotinylated bio-receptors. Examples include streptavidin supramolecular immobilisation on the surface of PAMAM dendrimers. The biotin-streptavidin supramolecular attraction is the strong non-covalent biospecific interactions known ever (having affinity constant Ka $\approx$ 1013 M^{-1}). Thus, dendrimers streptavidin can be used as a versatile platform for immobilising the large number of biotinylated bioreceptors such as DNA, aptamers, enzymes and others with an improved detection limit (Soda and Arotiba 2017). Other dendrimers such as polypyrrole (PPy), which is one of the conducting polymers, are extensively used for the preparation of biosensors. The fabrication based on polypyrrole (PPy) film and PAMAM dendrimers of fourth-generation (PAMAM G4.0) by electrochemical patterning method involves the attachment of ferrocenyl groups on the surface as a redox marker, which gives modified nanomaterial (PPy-PAMAM-Fc). This modified material shows good sensitivity and selectivity in DNA sensing with a detection limit of 0.4 fM (Miodek et al. 2016). DNA/gold nanoparticles (GNP)(AuNPs)/dendrimers are also used for sensing the microRNA specific to some disease such as cancer, which is not easily detected in early stages. After detection, obtained signals are amplified by coupling the signals with an enzymatic amplification process. Thus the modified sensor can be used as good tool for early detection of the cancer (Guo et al. 2016).

DNA dendrimers are also used because these are superior to other nanocarriers due to its larger surface, having very good *in-vivo* stability, highly branched structure and very good monodispersity. There are specially designed DNA dendrimers, which carry a luminescent system and DNA nanomachine. This helps in ultrasensitive detection of laminin (LN), which is an important glycoprotein and is involved in many physiological processes such as cellular adhesion, its migration, in its differentiation, and as well in its growth. LN is one of the efficient biomarkers used for liver fibrosis. Thus, this biosensor may be used in detection of liver damage very efficiently (Li et al. 2018).

Third generation PAMAM (PAMAM G3.0) was also modified and used as a sensor to detect H_2O_2. In this detector, AuNPs were encaptured inside of the dendrimer structure and haemoglobin (Hb). This whole material is immobilised on a glass carbon electrode (GCE). The resultant modified electrode (Hb/PAMAM-AuNPs/GCE) shows remarkable results for H_2O_2 determination within the range of 20–950.22 µM (Elancheziyan and Senthilkumar, 2019). Moreover, PPy-PAMAM G2.0 dendrimers were modified with a monosaccharide inhibitor (N-(5-phosphate-D-arabinoyl)-2-amino ethanolamine) of autocrine motility factor (AMF) on its surface as a bioreceptor. This sensor has ability to detect the AMF, which is the biomarker for a variety of the cancers. Phosphoglucose isomerase (PGI), which is an intracellular glycolytic enzyme, and AMF are strictly identical in 3D structure having the same sequence of amino acids. When PGI is secreted by tumours, it acts similarly to AMF. The modified sensor displays high sensitivity to AMF-PGI detection (LOD 43 fM) in HEPES buffer. It also detects the AMF-PGI in human plasma within 10 minutes of short time period (Ahmad et al. 2019).

For the determination of the glucose, pyranose oxidase enzyme is fixed on the surface of the modified material of PVA/PAMAM-Mt. This modified dendrimer material successfully detects the glucose in samples due to the presence of the pyranose oxidase enzyme (Unal et al. 2018). It has also been reported that the sensor system also provides the platform for the enzymes, which helps to evaluate their corresponding substrate in the samples. For examples, the use of carboxymethyl cellulose-modified graphene oxide with a platinum nanoparticle-decorated PAMAM-MNP hybrid nanomaterial shows sensitivity for the assessment of the xanthine in the samples up to the nanomolar range with a sensitivity of 140 mA/M cm^2 and a detection limit of 13 nM (Borisova et al. 2016). In the same way, urease enzyme fixed on PAMAM grafted multiwalled carbon nanotubes (MWCNTs) (MWCNT-PAMAM G5) on a gold electrode are used to determine the urine in human blood plasma samples. This modified sensor system is very sensitive and gives results in very short time (3s) with 0.4 mM detection limit and 6.6 nA/mM sensitivity. Results do not fluctuate in the presence of variable factors such as different acids (uric acids, ascorbic acid and lactic acid), glucose or cholesterol in sample, which produce possible interference during analysis of urea in human blood. Hence, this dendrimer-based sensor is an efficient tool to determine the uric acid in human blood (Dervisevic et al. 2018).

11.5.4 Dendrimers Along with Other Nanomaterials as a Diagnostic Tool

Nanotechnology is extensively used in the biomedical field for diagnosis as well as for the treatment of various diseases. Quantum dots (QDs) are synthetic nanocrystals ranges from 2 to 10 nm in diameter. QDs are the nanostructured semiconductor having the ability to accept and carry the electrons within it. Both dendrimers and QDs are nanosized compounds which are combined to enhance the various imaging techniques. QDs have excellent luminescence capacity with a broader range of excitation, having longer lifetime and resistance to photobleaching and are suitable to be used in fluorescence spectroscopy over traditional fluorescent dyes (Kairdolf et al. 2013; Bajwa et al. 2016; Horozić et al. 2018). One of the examples is the PAMAM encapsulated gold QDs for the quantification of the human IgG immunoglobu-lins. Triulzi and co-workers have developed this biosensor complex based on PAMAM dendrimers, which have the high efficiency of the assay due to QDs (AuNCs) and greater cell permeability due to dendrimers (Triulzi et al. 2008). Cadmium and zinc QDs were crosslinked to the second generation PAMAM den-drimer and subsequently subjected to PEGylation. These PAMAM complexed QD-PEG conjugates were used to visualise the brain vasculature utilising fluorescence microscopy, and also for the visualisation of tumours using fluorescence imaging and MRI (Bakalova et al. 2011).

Among all kinds of QDs, carbon quantum dots (CDs) are one of the important QDs that can be obtained from organic compounds. CDs are the carbon nanoparticles that are stable in aqueous media and provide a greater opportunity to be used for detection and bioimaging (Gayen et al. 2019). The CDs were conju-gated with PAMAM dendrimers to form the CD-PAMAM complex (CDP). This CDP was then further combined with arginylglycylaspartic acid (RGDS) peptide for v 3 integrin (a vitronectin receptor) whose expression increases in triple-negative breast cancer (TNBC). This CDP3 results in effective detection of copper ions due to quenching of fluorescence in the presence of Cu(II), which is upregulated in TNBC. This CD in combination with PAMAM dendrimers becomes a theranostic tool for the diagnosis of the TNBC (Ghosh et al. 2019).

The aptamer is an oligonucleotide nanosized particle, having very high affinity and specificity towards its ligand. Both nucleic acid and peptide-based aptamers can be conjugated with dendrimers for the development of diagnostic or imaging tools. The aptamer–dendrimer bioconjugates are the novel system used in the preparation of various biosensors and are also used in *in vivo* imaging (Pednekar et al. 2012). The dendrimer-conjugated QD nanocomplexed structure can also be combined with DNA aptamer to improve its imaging efficiency. Example- The PAMAM dendrimer was modified using QDs, which is then complexed with a DNA aptamer, tenascin-C (GBI-10). This aptamer-conjugated nanoprobe was able to identify the protein tenascin-C on the surface of human U251 glioblastoma cells specifically (Li et al. 2010). Similar kinds of aptamer-conjugated dendrimer-modified QDs can be prepared, which will have great potential in molecular imaging for the diagnosis of cancer and other diseases.

One of the most captivating nanomaterials is carbon nanotubes (CNTs) in the biomedical field. These are the tube-like nanostructure obtained from graphite and appear as a rolled layer of hexagonal carbon sheet. These can be fabricated into single-walled CNTs, double-walled CNTs and MWCNTs by using the respective layer of carbon. These are used in the preparation of diagnostic tools, in biosensors as well as in bioimaging due to their high solubility in the aqueous environment and good compatibility with a bio-logical system, after functionalisation of their surface molecule (Anzar et al. 2020; Sharma et al. 2016).

Nowadays, promising nanostructures like CNTss and dendrimers are combined to reduce the demerits of each other and to provide a better diagnostic tool (Jain 2019; Anzar et al. 2020). A platinum nano-cluster and CNTs are combined with G6.0 PAMAM dendrimers for the detection of H_2O_2. A simple electrochemical sensor is developed based on amine surfaced PAMAM (G6.0-NH_2 PAMAM dendrimers) dendrimer encapsulation of the Pt nanocluster and covalently bonded to the carbonylated CNT. This mod-ified GCE acts as a sensor in the quantification of H_2O_2 in MCF-7 cells (Liu and Ding, 2017).

An electrochemical biosensor for the identification of the serum marker of prostate cancer, Prostate-specific antigen (PSA) was developed by using PSA-antibody, and AuNP-conjugated PAMAM den-drimers and MWCNTs/ionic liquid/chitosan nanocomposite (MWCNTs/IL/Chit). This immunosensor is prepared by covalently attaching the PSA-antibody and thionine with AuNP-PAMAM dendrimers, in which MWCNTs/IL/Chit nanocluster is used to provide a suitable platform for transduction. The pre-pared immunosensor is used for the detection of PSA in biological samples, which provide a promising tool for the diagnosis of prostate cancer in the patient (Kavosi et al. 2014).

11.6 Safety Aspects of Dendrimers

Dendrimers are polymeric architectures, having characteristics similar to biomolecules. Dendrimers offer an outstanding carrier in the field of medicine for different kinds of drugs including anti-cancer, anti-viral, anti-bacterial, anti-tubercular, etc., due to their derivable branched architecture and their ability to modify it in numerous ways (Jain et al. 2010). Because of wide use in the biomedical field, it is important that they must comply with the living system and should be free from toxicity.

PAMAM dendrimers are observed to be considerably safe as they have no significant effect on feeding behaviour, body weight, haematological profile and vital organ toxicity. However, a high dose of PAMAM-G4.0 has shown some toxicity, which is recovered on discontinuation of administration of dendrimers (Chauhan et al. 2010). It was observed that PAMAM higher generation dendrimers show more toxicity than lower generation dendrimers (0–4th generation), and there is a workable safe window for use of these dendrimers as carriers of drugs (Sadekar and Ghandehari 2012). Generally, cationic PAMAM dendrimers are relatively more toxic than anionic dendrimers. The anionic dendrimers are safe nearly ten times than cationic ones. Anionic dendrimers up to G6.5 and smaller molecules with less molecular weight along with terminal groups like carboxyl, amine or hydroxyl were safe at much higher doses up to 500mg/kg without any toxic effects (Thiagarajan et al. 2013). However, surface modification of PAMAM dendrimers tends to reduces the toxicity profile. *In-vitro* studies indicates that SN-38 complexed G4.0-PAMAM dendrimer, lauroyl, and PEGylated PAMAM dendrimers have a better safety profile than unmodified dendrimers (Jain et al. 2010).

Among polypropylene imine (PPI) dendrimers, it has been seen that unmodified and modified dendrimers are not really toxic. It has been seen that at high dose (4mg/kg/day) of unmodified PPI-G4, after 4–9 days there is decrease in body weight and food intake was observed. These toxic effects are due to the presence of cationic amino group on the outer surface of the dendrimer. These toxic effects are less observed in the partially modified cationic glycodendrimer (PPI-G4-25%) in which, 25% or more of the amino acid groups are coated with carbohydrate such as maltotriose. Moreover, it was seen that prolonged treatment of the unmodified dendrimers results in reduction of undesirable side effects possibly due to counteract mechanism. However, on discontinuation of the treatment, these side effects return to normal (Ziemba et al. 2011; Ciepluch et al. 2012). Toxicity of the higher generation dendrimers was more as compared to lower generation which has less or nearly no toxicity (Bodewein et al. 2016). It has also been seen that glycine coated, phenylalanine coated, mannose coated, lactose coated and galactose coated 5.0 G PPI dendrimers are seen less toxic with less cytotoxicity and haematotoxicity profile which favours the use of these dendrimers in various applications (Jain et al. 2010).

Phosphorus (or phosphorus-containing) dendrimers also seen safe to use due to no available defined toxicity data but some *in-vitro* studies indicate that G2.0 and G3.0 dendrimers are toxic to neuroblastoma cells and murine embryonic hippocampal cells (Lazniewska et al. 2013) but further studies indicate that phosphorus-based dendrimers which consist of anionic Aza Bis Phosphonate groups (ABP dendrimer) as terminal groups exert no adverse response when injected repeatedly. All physiological parameters remain in the normal range after continuous injections (Fruchon et al. 2015).

Apart from it, there are a number of dendrimers with a biodegradable core and branching units. These dendrimers are synthesised by those monomer units which forms the biodegradable dendrimers and are relatively safe to use. Polyether dendrimers, polyester dendritic system, polyether imine dendrimers, polyether–copolyester (PEPE) dendrimers, phosphate dendrimer, citric acid dendrimers, peptide dendrimers and triazine dendrimers are some examples of biodegradable dendrimers.

11.7 Conclusions and Future Perspectives

Owing to their unique physicochemical properties, dendrimers serve as a useful tool in theranostics. Dendrimers alone or in combination of other nanomaterials may be useful in differential diagnosis of diseases. Use of dendrimers as MRI contrast media has gained wide attention in diagnosis of cancers. These dendrimers may be combined with other nanostructures to enhance their diagnostic potentials. In diagnostics, acute exposure of these dendrimers is expected, through the safety aspects appear important to establish. Therefore more emphasis should be given on biodegradable and biocompatible dendrimers should be explored. Further combination of dendrimers with other nanostructures like QDs, CNTs, etc. may be explored for wider application in diagnostics.

Abbreviations

ABP dendrimer	Aza Bis Phosphonate groups
AMF	Autocrine motility factor
AuNPs	Gold nanoparticles
CD	Carbon quantum dots
CD-PAMAM	Complex (CDP)
CdSe/ZnS	Cadmium and zinc quantum dots
CNTs	Carbon nanotubes
CT	Computed tomography
CXCR4	C-X-C chemokine receptor type 4
DAB-Am64	Polypropylenimine tetra hexaconta amine Dendrimer
DNA	Deoxyribose nucleic acid
DWCNTs	Double-walled carbon nanotubes
GCE	Glass carbon electrode
GCE	Glass carbon electrode
GNP	Gold nanoparticle
H_2O_2	Hydrogen peroxide
HEPES	4-(2-hydroxyethyl)-1-piperazineethanesulfonic acid
LFC131	d-Tyr-Arg-Arg-2-Nal-Gly peptide
LN	laminin
miRNAs	MicroRNAs
MNP	Magnetic nanoparticles
MRI	Magnetic resonance imaging
MWCNTs	Multi-walled carbon nanotubes
MWCNTs/IL/Chit	Multiwalled carbon nanotubes/ionic liquid/chitosan nanocomposite
NCT	Neutron capture therapy
NMRI	Nuclear magnetic resonance imaging
PAMAM	Polyamidoamine
PAMAM-Ac	Polyamidoamine-acetylcysteine
PAMAM-PEI	Polyamidoamine-polyethyleneimine
PAMAM-TPP	Polyamidoamine-triphenyl phosphonium
PDT	Photodynamic therapy
PEG	Polyethylene glycol
PEI	Polyether imine
PEPE	Polyether–copolyester
PET	Positron emission tomography
PGI	Phosphoglucose isomerase
PMPA	Poly 2,2-bis(hydroxymethyl)-propanoic acid
PPE	Polyether-polyester
PPI	Polypropylene imine
PPy	Polypyrrole
PPy-PAMAM-Fc	Polypyrrole-polyamidoamine-ferrocenyl
PSA	Prostate-specific antigen
PTT	Photothermal therapy
PVA/PAMAM-Mt	Polyvinyl alcohol PAMAM montmorillonite clay
RGDS	Arg-Gly-Asp-Ser, arginylglycylaspartic acid
RNA	Ribonucleic acid
SPIO	Superparamagnetic iron oxide
SWCNTs	Single-walled carbon nanotubes
TNBC	Triple-negative breast cancer

REFERENCES

Abbasi, E., Aval, S.F., Akbarzadeh, A. et al. 2014. Dendrimers: synthesis, applications, and properties. *Nanoscale Research Letters 9*: 247.

Ahmad, L., Salmon, L., Korri-Youssoufi, H. 2019. Electrochemical detection of the human cancer biomarker 'autocrine motility factor-phosphoglucoseisomerase' based on a biosensor formed with a monosaccharidic inhibitor. *Sensors and Actuators B: Chemical 299*: 126933.

Akbarzadeh, A., Khalilov, R., Mostafavi, E. et al. 2018. Role of dendrimers in advanced drug delivery and biomedical applications: a review. *Experimental Oncology 40*: 178–183.

Anzar, N., Hasan, R., Tyagi, M. et al. 2020. Carbon nanotube—A review on synthesis, properties and plethora of applications in the field of biomedical science. *Sensors International 1*: 100003.

Araújo, R.V.D., Da Silva Santos, S., Igne Ferreira, E., Giarolla, J. 2018. New advances in general biomedical applications of PAMAM dendrimers. *Molecules 23*: 2849.

Bajwa, N., Mehra, N.K., Jain, K., Jain, N.K. 2016. Pharmaceutical and biomedical applications of quantum dots. *Artificial cells Nanomedicine and Biotechnology 44* (3): 758–768.

Bakalova, R., Zhelev, Z., Kokuryo, D., Spasov, L., Aoki, I., Saga, T. 2011. Chemical nature and structure of organic coating of quantum dots is crucial for their application in imaging diagnostics. *International Journal of Nanomedicine 6*: 1719–1732.

Biswas, S., Dodwadkar, N.S., Piroyan, A., Torchilin, V.P. 2012. Surface conjugation of triphenylphosphonium to target poly (amidoamine) dendrimers to mitochondria. *Biomaterials 33*: 4773–4782.

Bodewein, L., Schmelter, F., Di Fiore, S., Hollert, H., Fischer, R., Fenske, M. 2016. Differences in toxicity of anionic and cationic PAMAM and PPI dendrimers in zebrafish embryos and cancer cell lines. *Toxicology and Applied Pharmacology 305*: 83–92.

Borisova, B., Sánchez, A., Jiménez-Falcao, S. 2016. Reduced graphene oxide-carboxymethylcellulose layered with platinum nanoparticles/PAMAM dendrimer/magnetic nanoparticles hybrids. Application to the preparation of enzyme electrochemical biosensors. *Sensors And Actuators B: Chemical 232*: 84–90.

Chauhan, A.S., Jain, N.K., Diwan, P.V., 2010. Pre-clinical and behavioural toxicity profile of PAMAM dendrimers in mice. *Proceedings of the Royal Society A: Mathematical, Physical and Engineering Sciences 466*: 1535–1550.

Chen, G., Roy, I., Yang, C., Prasad, P.N. 2016. Nanochemistry and nanomedicine for nanoparticle-based diagnostics and therapy. *Chemical Reviews 116*: 2826–2885.

Chen, K., Xin, X., Qiu, L. 2019. Co-delivery of p53 and MDM2 inhibitor RG7388 using a hydroxyl terminal PAMAM dendrimer derivative for synergistic cancer therapy. *Acta Biomaterialia 100*: 118–131.

Chittasupho, C., Anuchapreeda, S., Sarisuta, N. 2017. CXCR4 targeted dendrimer for anti-cancer drug delivery and breast cancer cell migration inhibition. *European Journal of Pharmaceutics and Biopharmaceutics 11*: 310–321.

Ciepluch, K., Ziemba, B., Janaszewska, A. 2012. Modulation of biogenic amines content by poly (propylene imine) dendrimers in rats. *Journal of Physiology and Biochemistry 68*: 447–454.

Da Silva Santos, S., Igne Ferreira, E., Giarolla, J. 2016. Dendrimer prodrugs. *Molecules 21*: 686.

Dayyani, N., Ramazani, A., Khoee, S., Shafiee, A. 2018. Synthesis and characterization of the first generation of polyamino-ester dendrimer-grafted magnetite nanoparticles from 3-aminopropyltriethoxysilane (APTES) via the convergent approach. *Silicon 10*: 595–601.

Dervisevic, M., Dervisevic, E., Şenel, M. 2018. Design of amperometric urea biosensor based on self-assembled monolayer of cystamine/PAMAM-grafted MWCNT/Urease. *Sensors and Actuators B: Chemical 254*: 93–101.

Dias, A.P., daSilva Santos, S., daSilva, J.V. et al. 2020. Dendrimers in the context of nanomedicine. *International Journal of Pharmaceutics 573*: 118814.

Dougherty, C.A., Vaidyanathan, S., Orr, B.G., Banaszak Holl, M.M. 2015. Fluorophore: dendrimer ratio impacts cellular uptake and intracellular fluorescence lifetime. *Bioconjugate Chemistry 26*: 304–315.

Elancheziyan, M., Senthilkumar, S. 2019. Covalent immobilization and enhanced electrical wiring of hemoglobin using gold nanoparticles encapsulated PAMAM dendrimer for electrochemical sensing of hydrogen peroxide. *Applied Surface Science 495*: 143540.

Fruchon, S., Mouriot, S., Thiollier, T. 2015. Repeated intravenous injections in non-human primates demonstrate preclinical safety of an anti-inflammatory phosphorus-based dendrimer. *Nanotoxicology 9*: 433–441.

Gan, B.H., Siriwardena, T.N., Javor, S., Darbre, T., Reymond, J.L. 2019. Fluorescence Imaging of Bacterial Killing by Antimicrobial Peptide Dendrimer G3KL. *ACS Infectious Diseases 5*: 2164–2173.

Gayen, B., Palchoudhury, S., Chowdhury, J. 2019. Carbon dots: a mystic star in the world of nanoscience. *Journal of Nanomaterials 2019*: 3451307.

George, S., Shebitha, A.M., Kannan, V., Mathew, S., Asha, K.K., Sreekumar, K. 2019. Amphiphilic Dendrimer as reverse micelle: synthesis, characterization and application as homogeneous organocatalyst. *Tetrahedron 75*: 130676.

Ghosh, S., Ghosal, K., Mohammad, S.A., Sarkar, K. 2019. Dendrimer functionalized carbon quantum dot for selective detection of breast cancer and gene therapy. *Chemical Engineering Journal 373*: 468–484.

Gündüz, S., Savić, T., Toljić, D., Angelovski, G. 2016. Preparation and in vitro characterization of dendrimer-based contrast agents for magnetic resonance imaging. *Journal of Visualized Experiments 2016*: e54776.

Guo, Y., Wang, Y., Yang, G., Xu, J.J., Chen, H.Y. 2016. MicroRNA-mediated signal amplification coupled with GNP/dendrimers on a mass-sensitive biosensor and its applications in intracellular microRNA quantification. *Biosensors and Bioelectronics 85*: 897–902.

Horozić, E., Begić, S., Cipurkovic, A. et al. 2018. Application of dendrimers and quantum dots in cancer diagnosis and therapy. *Acta Medica Saliniana 48* (1–2): 23–31.

Jain, K. 2019. Nanohybrids of Dendrimers and carbon nanotubes: a benefaction or forfeit in drug delivery? *Nanoscience& Nanotechnology-Asia 9* (1): 21–29.

Jain, K., Kesharwani, P., Gupta, U., Jain, N.K. 2010. Dendrimer toxicity: let's meet the challenge. *International Journal of Pharmaceutics 394*: 122–142.

Jędrzak, A., Grześkowiak, B.F., Coy, E. 2019. Dendrimer based theranostic nanostructures for combined chemo-and photothermal therapy of liver cancer cells in vitro. *Colloids and Surfaces B: Biointerfaces 173*: 698–708.

Kairdolf, B.A., Smith, A.M., Stokes, T.H. et al. 2013. Semiconductor quantum dots for bioimaging and biodiagnostic applications. *Annual Review of Analytical Chemistry 6*: 143–162.

Kavosi, B., Salimi, A., Hallaj, R., Amani, K. 2014. A highly sensitive prostate-specific antigen immunosensor based on gold nanoparticles/PAMAM dendrimer loaded on MWCNTS/chitosan/ionic liquid nanocomposite. *Biosensors and Bioelectronics 52*: 20–28.

Kesharwani, P., Jain, K., Jain, N.K. 2014. Dendrimer as nanocarrier for drug delivery. *Progress in Polymer Science 39*: 268–307.

Key, J., Leary, J.F. 2014. Nanoparticles for multimodal in vivo imaging in nanomedicine. *International Journal of Nanomedicine 9*: 711.

Kulhari, H., Pooja, D., Prajapati, S.K., Chauhan, A.S. 2011. Performance evaluation of PAMAM dendrimer based simvastatin formulations. *International Journal of Pharmaceutics 405*: 203–209.

Kunjachan, S., Jayapaul, J., Mertens, M.E., Storm, G., Kiessling, F., Lammers, T. 2012. Theranostic systems and strategies for monitoring nanomedicine-mediated drug targeting. *Current Pharmaceutical Biotechnology 13*: 609–622.

Lazniewska, J., Milowska, K., Zablocka, M.et al. 2013. Mechanism of cationic phosphorus dendrimer toxicity against murine neural cell lines. *Molecular Pharmaceutics 10*: 3484–3496.

Li, J., Liang, H., Liu, J., Wang, Z. 2018. Poly (amidoamine)(PAMAM) dendrimer mediated delivery of drug and pDNA/siRNA for cancer therapy. *International Journal of Pharmaceutics 546*: 215–225.

Li, Z., Huang, P., He, R. et al. 2010. Aptamer-conjugated dendrimer-modified quantum dots for cancer cell targeting and imaging. *Materials Letters 64* (3): 375–378.

Liu, H., Xu, Y., Wen, S. et al. 2013. Targeted tumor computed tomography imaging using low-generation dendrimer-stabilized gold nanoparticles. *Chemistry–A European Journal 19*: 6409–6416.

Liu, J.X., Ding, S.N. 2017. Non-enzymatic amperometric determination of cellular hydrogen peroxide using dendrimer-encapsulated Pt nanoclusters/carbon nanotubes hybrid composites modified glassy carbon electrode. *Sensors and Actuators B: Chemical 251*: 200–207.

Liu, L., Kuang, Y., Yang, H., Chen, Y. 2019. An amplification strategy using DNA-Peptide dendrimer probe and mass spectrometry for sensitive MicroRNA detection in breast cancer. *Analytica Chimica Acta 1069*: 73–81.

Lyu, Z., Ding, L., Huang, A.Y.T., Kao, L., Peng, L. 2019. Poly (amidoamine) dendrimers: covalent and supramolecular synthesis. *Materials Today Chemistry 13*: 34–48.

Ma, Y., Mou, Q., Wang, D., Zhu, X., Yan, D. 2016. Dendritic polymers for theranostics. *Theranostics 6*: 930.

Ma, Y.Q. 2013. Theoretical and computational studies of dendrimers as delivery vectors. *Chemical Society Reviews 42*: 705–727.

Markowicz-Piasecka, M., Sikora, J., Szymański, P., Kozak, O., Studniarek, M., Mikiciuk-Olasik, E. 2015. PAMAM dendrimers as potential carriers of gadolinium complexes of iminodiacetic acid derivatives for magnetic resonance imaging. *Journal of Nanomaterials 2015*: 394827.

Menjoge, A.R., Kannan, R.M., Tomalia, D.A. 2010. Dendrimer-based drug and imaging conjugates: design considerations for nanomedical applications. *Drug Discovery Today 15*: 171–185.

Mintzer, M.A., Grinstaff, M.W. 2011. Biomedical applications of dendrimers: a tutorial. *Chemical Society Reviews 40*: 173–190.

Miodek, A., Mejri-Omrani, N., Khoder, R., Korri-Youssoufi, H., 2016. Electrochemical functionalization of polypyrrole through amine oxidation of poly (amidoamine) dendrimers: application to DNA biosensor. *Talanta 154*: 446–454.

Nigam, S., Bahadur, D. 2017. Dendrimer-conjugated iron oxide nanoparticles as stimuli-responsive drug carriers for thermally-activated chemotherapy of cancer. *Colloids and Surfaces B: Biointerfaces 155*: 182–192.

Noriega-Luna, B., Godínez, L.A., Rodríguez, F.J. et al. 2014. Applications of dendrimers in drug delivery agents, diagnosis, therapy, and detection. *Journal of Nanomaterials 2014*: 507273.

Pednekar, P.P., Jadhav, K.R., Kadam, V.J. 2012. Aptamer-dendrimer bioconjugate: a nanotool for therapeutics, diagnosis, and imaging. *Expert Opinion on Drug Delivery 9* (10): 1273–1288.

Pooja, D., Sistla, R., Kulhari, H., 2018. Dendrimer-drug conjugates: synthesis strategies, stability and application in anticancer drug delivery. *Design of Nanostructures for Theranostics Applications*, 277–303. doi: https://doi.org/10.1016/B978-0-12-813669-0.00007-5.

Ray, A., Khan, S. 2018. Convergent synthesis of novel mono-and di-substituted 1, 2-isopropylideneglucofuranose appended Dendrimers with a Ferrocenecore and their electrochemical studies. *Synlett 14*: 1367–1372.

Rizzo, L.Y., Theek, B., Storm, G., Kiessling, F., Lammers, T. 2013. Recent progress in nanomedicine: therapeutic, diagnostic and theranostic applications. *Current Opinion in Biotechnology 24*: 159–1166.

Sadekar, S., Ghandehari, H., 2012. Transepithelial transport and toxicity of PAMAM dendrimers: implications for oral drug delivery. *Advanced Drug Delivery Reviews 64*: 571–588.

Sharma, A.K., Gothwal, A., Kesharwani, P., Alsaab, H., Iyer, A.K., Gupta, U. 2017. Dendrimer nanoarchitectures for cancer diagnosis and anticancer drug delivery. *Drug Discovery Today 22*: 314–326.

Sharma, P., Mehra, N.K., Jain, K., Jain, N.K. 2016. Biomedical applications of carbon nanotubes: a critical review. *Current Drug Delivery 13* (6): 796–817.

Sherje, A.P., Jadhav, M., Dravyakar, B.R., Kadam, D. 2018. Dendrimers: a versatile nanocarrier for drug delivery and targeting. *International Journal of Pharmaceutics 548*: 707–720.

Soda, N., Arotiba, O.A. 2017. A polyamidoamine dendrimer-streptavidin supramolecular architecture for biosensor development. *Bioelectrochemistry 118*: 14–18.

Sowinska, M., Urbanczyk-Lipkowska, Z. 2014. Advances in the chemistry of dendrimers. *New Journal of Chemistry 38*: 2168–2203.

Thiagarajan, G., Greish, K., Ghandehari, H. 2013. Charge affects the oral toxicity of poly (amidoamine) dendrimers. *European Journal of Pharmaceutics and Biopharmaceutics 84*: 330–334.

Triulzi, R.C., Micic, M., Orbulescu, J. et al. 2008. Antibody–gold quantum dot–PAMAM dendrimer complex as an immunoglobulin immunoassay. *Analyst 133* (5): 667–672.

Unal, B., Yalcinkaya, E.E., Demirkol, D.O., Timur, S. 2018. An electrospunnanofiber matrix based on organoclay for biosensors: PVA/PAMAM-Montmorillonite. *Applied Surface Science 44*: 542–551.

Vieira Gonzaga, R., daSilva Santos, S., Da Silva, J.V. et al. 2018. Targeting groups employed in selective dendrons and dendrimers. *Pharmaceutics 10*: 219.

Wang, R.T., Zhi, X.Y., Yao, S.Y., Zhang, Y. 2015. LFC131 peptide-conjugated polymeric nanoparticles for the effective delivery of docetaxel in CXCR4 overexpressed lung cancer cells. *Colloids and Surfaces B: Biointerfaces 133*: 43–50.

Zhao, L., Zhu, M., Li, Y., Xing, Y., Zhao, J. 2017. Radiolabeled dendrimers for nuclear medicine applications. *Molecules 22*: 1350.

Ziemba, B., Janaszewska, A., Ciepluch, K. et al. 2011. In vivo toxicity of poly (propyleneimine) dendrimers. *Journal of Biomedical Materials Research Part A 99*: 261–268.

12

Dendrimer–Nanomaterial Conjugation: Concept, Chemistry and Applications

Sonali Batra, Samridhi Thakral, Amit Singh and Sumit Sharma

CONTENTS

12.1 Introduction

The Lycurgus Cup is the oldest known example of nanotechnology date back to the fourth century when Roman dispersed colloidal gold and silver nanoparticles in glass (Freestone et al. 2007). This experimental mixing gave rise to a dichroic glass material that is sensitive to the direction of light and responds with a different colour. However, a long time was taken to understand that nanostructured gold under certain lighting conditions produces different-coloured solutions. Finally, in 1857 the discovery of 'Ruby' gold by Michael Faraday explained this phenomenon, and his explanation laid down the foundation of modern nanotechnology (Thompson 2007), whereas in 1956, Arthur von Hippel coined the term molecular engineering that looks more promising and revealed a more comprehensive scope of nanotechnology. In the 1980s, research in the field of nanotechnology started blooming, and interestingly at the same time, Vogtle (Buhleier, Wehner and VOGtle 1978) defined a cascade synthesis procedure for well-defined branched structures followed by Denkewalter (Denkewalter, Kolc and Lukasavage 1979), who patented the synthesis of amino acid-based highly branched macromolecules. However, structural characterisation of these so-called well-defined, branched macromolecules remained unclear until 1985. Then Tomalia et al. synthesised a new class of topological macromolecules referred to as starburst polymers of polyamidoamine (PAMAM) (Tomalia et al. 1985) followed by Newkome's unimolecular micelles (Newkome, He and Moorefield 1999). These two detailed investigations mentioned divergent highly branched macromolecules, but authors referred fundamental building blocks of these new polymer classes as 'dendrimers'. Eventually, a dendrimer is a result of an iterative sequence of reaction steps that started with a fundamental core where each additional iteration leads to an additional branch to the boundary.

Now, the field of dendrimers is very vast and summarised by various authors (Bosman, Janssen and Meijer (1999), Bhyrappa et al. (1996), Trollsås and Hedrick (1998), Astruc, Boisselier and Ornelas (2010), Klajnert and Bryszewska (2001), Abbasi et al. (2014), Vögtle, Richardt and Werner (2009), Caminade, Yan and Smith (2015). In contrast, molecular engineering techniques enable customised dendritic polymers according to desired physicochemical properties such as size, shape, solubility, toxicity, etc. These diverse combinations yield dendrimers of different shapes and sizes with shielded interior cores for promising candidates for applications in medicines and pharmaceutical sciences (Liu and Fréchet 1999, Yoo, Sazani and Juliano 1999, Tang, Redemann and Szoka 1996, Twyman et al. 1999). The formation or the coupling of dendrimers with drug molecules or genetic materials can be achieved in three different ways, i.e., encapsulation, complexation or conjugation (Liu and Fréchet 1999, Tang, Redemann and Szoka 1996, Twyman et al. 1999). Out of these, dendrimer-nanoconjugation is a promising technique for targeting drug delivery systems.

In view of narrow polydispersity and various terminal amino groups of PAMAM dendrimers, they can be used for conjugation with various therapeutic, targeting and imaging agents for *in vitro* imaging, drug delivery systems (Otis et al. 2016, Choi et al. 2005) and for encapsulating the guest molecules, where the cavities of dendrimers act as the carriers (He et al. 2015). To overcome the issues pertaining to biodistribution and solubility, the PAMAM–drug conjugate can be utilised, which provides better pharmacodynamic effects (Dichwalkar et al. 2017). For example, PEGylation (partial or full) of dendrimers can efficiently improve their biocompatibility and also prolongs their blood stream circulation, while carboxylation and acetylation of dendrimers with amine groups at the terminal position can certainly enhance their biocompatibility (He et al. 2015). Due to lower toxicity of PAMAM, a conjugated dendrimer is most widely used for on-site delivery of methotrexate in uterine sarcoma cell lines (Zong et al. 2012, Choi et al. 2012, van Dongen et al. 2014). Dendrimer-conjugated magnetic nanoparticles have been widely employed for effective removal and recovery of various heavy metals from aqueous solutions (Zhu et al. 2015, Chou and Lien 2011, Vunain, Mishra and Mamba 2016, Hayati et al. 2017). Dendrimer-based nanodrugs were found to be a potential candidate against the transmission of sexually transmitted diseases, human immunodeficiency virus and other viral infections (McCarthy et al. 2005, Chonco et al. 2012, Tyssen et al. 2010). With the advances in technology, dendrimers are extensively studied to be employed in immunotherapy and hormone therapy, where dendrimers are engineered in a way to mimic viral outer cover proteins to train macrophages or to deliver the antigenic peptide for an effective antigen cross-presentation and cancer immunotherapy (Swager and Concellón 2019, Bonam et al. 2020, Canas-Arranz, Forner and Defaus 2020, Xu et al. 2019). This chapter further covers in detail the synthesis of dendrimer–nanomaterial conjugation systems and its improved efficacy using various examples with emphasis on medicine and pharmaceutical sciences.

12.2 Chemistry and Applications

Dendrimers are nanosized architectural structures, possessing an extremely branched polyvalent organic surface, which can be conjugated with various drug molecules to overcome associated issues of bio-distribution and their solubility, which have influence on the pharmacodynamic effect of drug molecules.

12.2.1 Conjugation with Peptides

Conjugation of dendrimers with peptides inculcates in them various amazing properties. This is mainly due to the terminal groups, which have high functional properties and well-defined architecture. The chain of amino acids attached offers better water solubility, biocompatibility and reduced digestion by the proteolytic enzymes (Zhang et al. 2014a). Conjugation of dendrimers with polyethylene glycol (PEG) is one of the safest methods to be used as a promising drug delivery vehicle as it is a neutral molecule, and it helps the dendrimer conjugate to assemble into small size particles (Longmire et al. 2011). Large generations of dendrimers are difficult to choose as they can cause toxicity due to low clearance and higher residence time in blood. Therefore, low generation dendrimers are being used for conjugation with the peptides as they can be easily formulated into nanodrug delivery systems. An enzyme-sensitive vehicle was prepared by conjugation of the mPEGylated dendron with doxorubicin GFLG-DOX (Gly-Phe-Leu-Gly and doxorubicin) for treatment of breast tumours by Li et al. by first linking the PEGylated ligands to the dendron, yielding the G 2.0 and G 3.0 mPEGylated Dendron. In the next step, this prepared dendron was reacted with the cathepsin B sensitive N_3-GFLG-DOX moiety via click reactions as presented in Figure 12.1. The GFLG tetra-peptide spacer was used as an enzyme-sensitive linker. This conjugate also self-assembled as nanoparticles, which is demonstrated microscopically by transmission electron microscopy (TEM) and light scattering studies. The combination of enzyme-sensitive linker with the peptide and dendrimers demonstrated effective in vitro anti-tumour activity against breast cells (Li et al. 2014).

Many of the studies have been performed that conjugated dendrimers with peptides; one such study carried out by He et al. synthesised G 5.0 PAMAM conjugated cyclic arginine-glycine-aspartic acid (RGD) peptide dendrimers for encapsulation of doxorubicin (DOX), a prominent anticancer drug. G 5.0 dendrimer was modified by using thiourea linkage and fluorescein isothiocyanate (FI) to yield PEGylated RGD. The remaining amine terminal groups of dendrimers were then acetylated to obtain multifunctional G 5.0 conjugated dendrimers (NHAc-FI-PEG-RGD). Then, these prepared dendrimers were used to encapsulate DOX as given in Figure 12.2. About six DOX molecules were found to be encapsulated in each dendrimer. This water soluble and stable complex with a sustained release profile was found to target $\alpha_v\beta_3$ integrin cells, which are believed to be over expressed in cancer and were shown to be therapeutically efficient towards the cancer cells. Also, under physiological pH conditions, a high release rate of DOX was observed as compared to the normal acidic pH conditions (He et al. 2015).

In another study conducted by Zhang et al., nanoparticles were synthesised using the mPEGylated dendrimer-GFLG-DOX conjugate, as an enzyme-sensitive drug delivery vehicle for ovarian cancer. First, the dendrimer was conjugated with azido mPEG chains (methoxy poly(ethylene glycol), and then further the dendrimer was conjugated with GFLG-modified DOX via a copper-catalysed alkyne-azide click cycloaddition reaction (Figure 12.3). The DOX was attached at the peripheral groups of the mPEGylated dendrimer using an enzyme-responsive linker, i.e., peptide sequence of GFLG. TEM and dynamic light scattering microscopy studies also proved the self-assembly potential of conjugate as a nanoparticle.

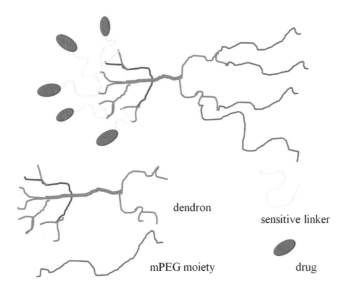

dendron

sensitive linker

mPEG moiety

drug

FIGURE 12.1 The schematic presentation of the amphiphilic mPEGylate Dendron-GFLG-DOX conjugate (Adapted with permission from Li et al., 2014).

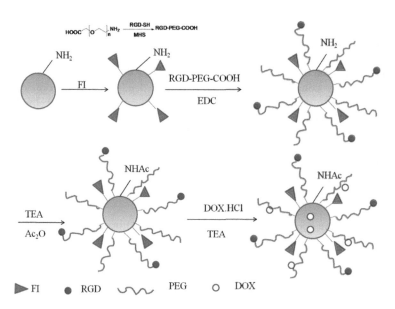

FIGURE 12.2   HOOC-PEG-RGD and G5.NHAc-FI-PEG-RGD/DOX complex preparation scheme (Adapted with permission from He et al., 2015).

1) TFA/CH$_2$Cl$_2$
2)N$_3$-mPEG
CuSO$_4$/Sodium ascorbate
Alkyne-Boc-dendrimer ————————————————→ mPEGylated Boc-dendrimers

1) TFA/CH$_2$Cl$_2$
2) 5-hexynoic acid/ HBTU/HOBT/
DIPEA   1) N$_3$-GFGL-DOX
                                                                       2) CuSO$_4$/ Sodium ascorbate
————————————————→ mPEGylated Alkyne-dendrimer ————————————————→

mPEGylated dendrimer-GFLG-DOX Conjugate

TFA– trifluoroacetic acid, CuSO$_4$– copper sulphate, mPEG– methoxy poly(ethylene glycol), HBTU–
N,N,N',N'-tetramethyl-(1H-benzotriazol-1-yl) uranium hexafluorophosphate, HOBT– 1-hydroxy-
benzotriazole, DIPEA– N,N'-diisopropylethylamine,

FIGURE 12.3   The scheme for the synthesis of the mPEGylated dendrimer-GFLG-DOX conjugate (Adapted with permission from Zhang et al., 2014b).

This study revealed the efficiency of the prepared conjugate as a specific enzyme responsive system which delivers DOX effectively for the treatment of ovarian cancer in comparison of DOX when used alone (Zhang et al. 2014b).

12.2.2 Conjugation with Hormones

Delivering of hormones has always been complicated especially in case of cancer-related malignancies. Amongst these, ovarian cancer is mainly fatal. Although chemotherapy has been one of the tradition-ally followed treatment methods, surgery being the next, recurrence of the disease is quite prominent (Gilks 2010). Highly effective treatment for ovarian cancer is therefore achieved by targeting the follicle

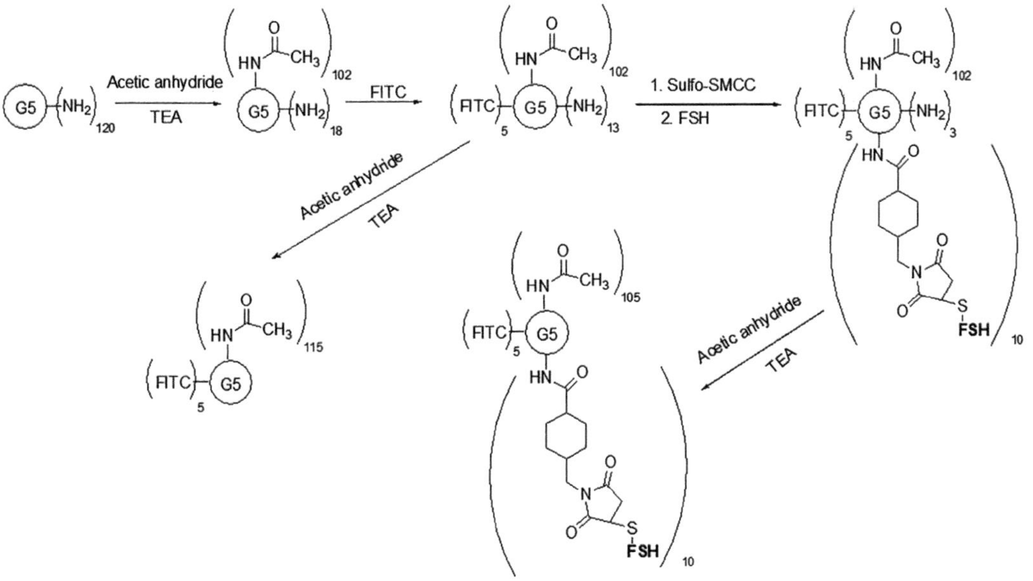

FIGURE 12.4 The schematic synthesis of FSH3-targeted G5 PAMAM dendrimer conjugates (Adapted with permission from Modi et al., 2014).

stimulating hormone receptor (FSHR), which is assumed to show elevated expression in the disease (Bose 2008). In order to target FSHR, Modi et al. synthesised G 5.0 PAMAM-Ac-FI-FSH (follicle stimulating hormone) by sequentially conjugating with FI and FSH 33 (peptide domain of FSH). This was achieved using a linker, i.e., sulfosuccinimidyl 4-(N-maleimidomethyl) cyclohexane-1-carboxylate after which acetylation of one amino group was achieved (Figure 12.4). FSH 33 has the strongest binding affinity for the FSHR. The prepared conjugate system specifically targets the over-expressed FSHR cells, sparing the immature follicles. Cellular uptake studies were performed using OVCAR-3 and SKOV-3 cell lines in 2D cell culture, which proved that the dendrimers were highly specific in targeting FSHR cells. In addition, 3D ex-vivo organ culture was performed, which proves that the dendrimer conjugate system acts as the FSHR in ovarian cancer (Modi et al. 2014).

12.2.3 Conjugation with Antibodies

Jain et al. synthesised monoclonal antibody conjugated modified half-generation poly(propylene imine) (PPI) (G 4.5) dendrimers encapsulating paclitaxel (PTX) for targeting ovarian cancer. The G 4.5 PPI dendrimers possessing a cyanide terminal were first modified into carboxylic acid terminated dendrimers persuaded by monoclonal antibody conjugation and encapsulation of PTX. These immunodendrimers significantly reduced the tumour volume and considerably reduced hepato-, haemolytic and nephrotoxicity (Jain et al. 2015).

Otis et al. constructed an imaging agent using gold nanoparticles (AuNPs) and gadolinium. The G 5.0 PAMAM dendrimer structure was modified using PEG and also with chelating agent, 1,4,7,10-tetraazacyclododecane-1,4,7,10-tetraacetic acid (DOTA). This modified dendrimer encapsulated the AuNPs and Gd inside the cavity. Furthermore, humanised anti-HER-2 (human oestrogen receptor) antibody, *i.e.,* herceptin conjugate was prepared using the aforesaid dendrimer. The click reaction was used which uses an azide-alkyne group using copper as the catalyst (Figure 12.5). This herceptin dendrimer conjugate served as a targeted nanotherapeutic agent for treatment and premature detection of cancer (Otis et al. 2016).

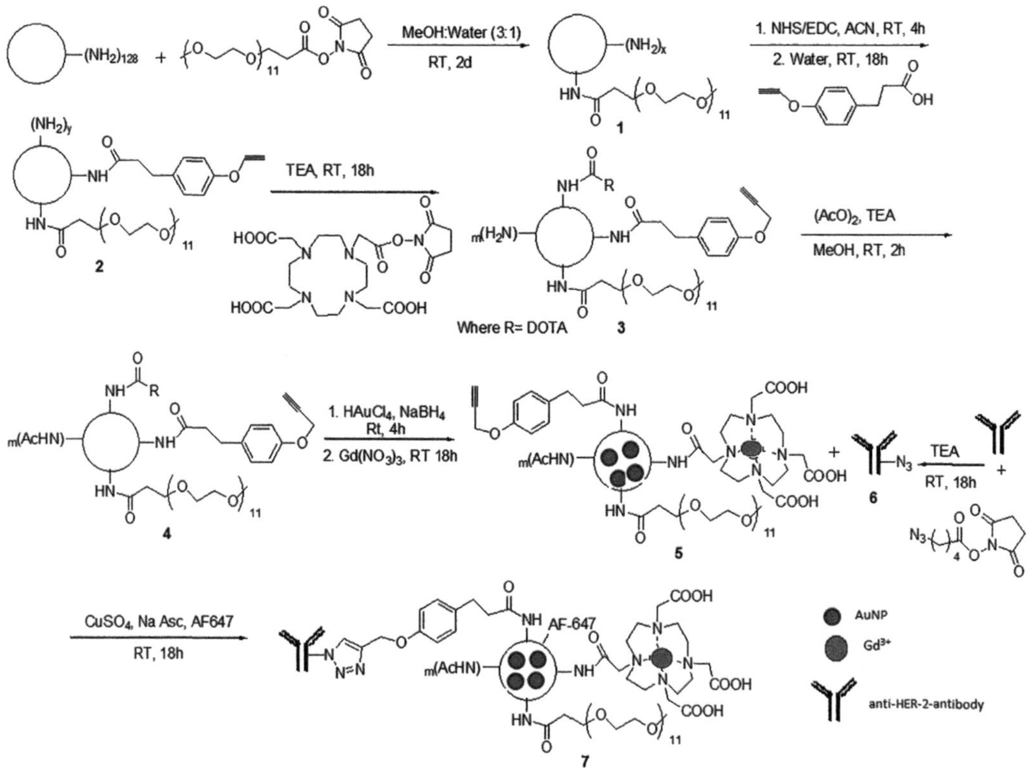

FIGURE 12.5 Schematic representation of preparation of the Au-G5-Gd-herceptin conjugate (Adapted with permission from Otis et al., 2016).

12.2.4 Conjugation with Nucleic acids

Yuan et al. reported the synthesis of epidermal growth factor (EGF) conjugated PAMAM G 4.0 dendrimer nanoparticles in two steps a) introduction of triglycine spacer and then b) coupling of EGF to the PAMAM dendrimer with the help of spacer. The prepared dendrimer was then labelled with quantum dots, which can be used for imaging and effectively delivering shRNA plasmids and siRNA at the targeted site (Figure 12.6) (Yuan et al. 2010).

Choi et al. conjugated G 5.0 PAMAM dendrimers to bi-functional moieties, i.e., folic acid (FA) and fluorescein and then linked them together utilising complementary DNA oligonucleotides to synthesised target molecules. First, an amine-terminated, G 5.0 PAMAM dendrimer was acetylated. After the covalent bonding of complementary, 5′-phosphate-modified oligonucleotides, the dendrimer was conjugated with FI or FA, to produce clustered molecules (Figure 12.7) for targeting cancer cells, which over express the immense-affinity folate receptor (Choi et al. 2005).

12.2.5 Conjugation with Ligand Molecules

Dichwalkar et al. synthesised G 4.0 PAMAM-NH$_2$-docosahexaenoic acid (DHA) – PTX by conjugating DHA and PAMAM-PTX with an objective of enhancing anticancer activity of PTX. The prime motive behind the study was to use PTX, a drug with low solubility and low bioavailability and reconstruct it in a manner so as to effectively use it in upper gastrointestinal cancer. The DHA–PTX conjugate was synthesised by reacting PAMAM-DHA and paclitaxel-NHS-ester under dry conditions. So in the first step the terminal amine group of G 4.0 PAMAM-NH$_2$ was conjugated with carboxyl terminals of DHA to form a strong amide bond in the presence of hydroxybenztriazole and

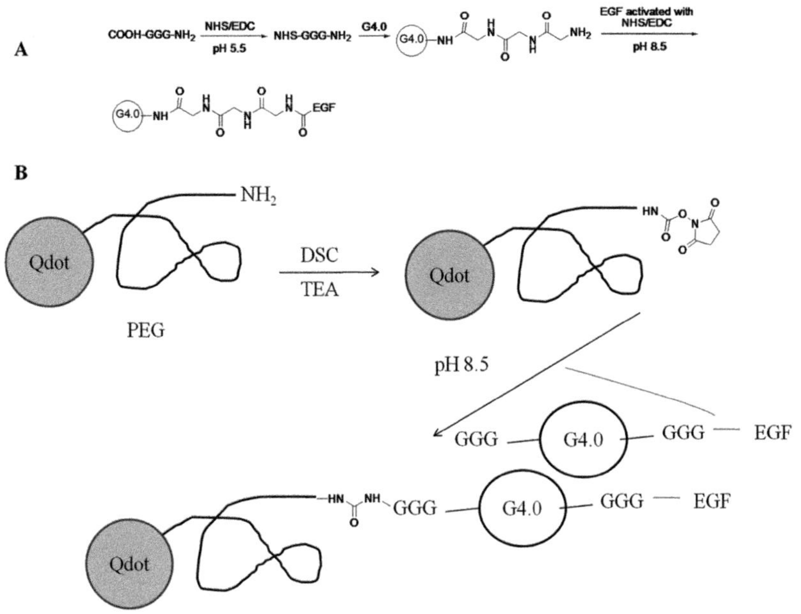

FIGURE 12.6 Synthetic scheme for synthesis of EGF-triglycine-dendrimer conjugates and labeling with quantum dots coated with amine-derivatised PEG of EGF-triglycine-dendrimer conjugates (Adapted with permission from Yuan et al., 2010).

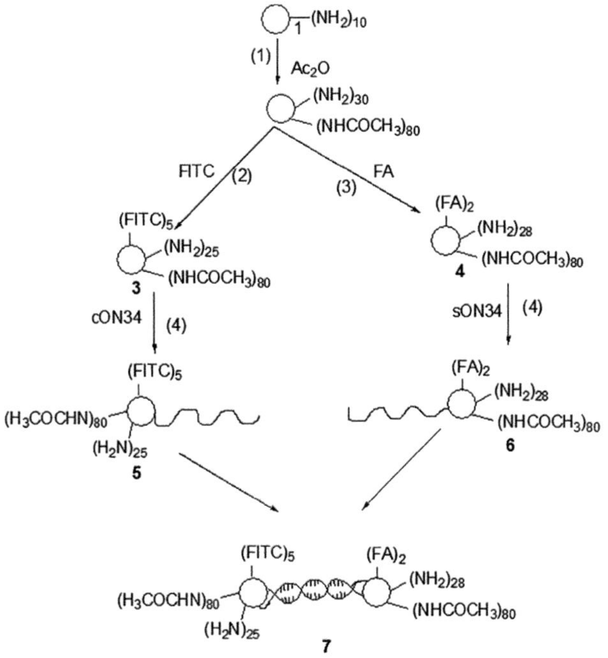

Reagents and conditions: (1) Triethylamine, MeOH, 16h; (2) DMSO; (3) EDC in DMF:DMSO(3:1, v/v); (4) 0.1 M EDC/0.1 M imidazole (pH 6.0) in 0.5 M LiCl; (5) 10mM phosphate buffer (pH 7.4), 150 mM NaCl, annealed at 90°C 10 min the cooled to room temprature over 3 h

FIGURE 12.7 The schematic representation of synthesis of DNA-linked cluster of G5-FITC and G5-FA dendrimers (Adapted with permission from Choi et al., 2005).

O-(benztriazol-1-yl)-N,N,N'-N'-tetramethyluronium hexafluorophosphate as coupling agents. In the second step, PTX-2'-hemisuccinate NHS ester was coupled with amine terminal of PAMAM-DHA in the presence of dry DMF (dimethyl formamide) to synthesise the desired complex (Figure 12.8) and further characterised by ^{1}H NMR spectral means. The study clearly showed that the synthesised conjugate was highly effective against tumour protein P53. This is a critical pathway for cell death especially in case of upper gastrointestinal cancer, in both wild and mutant forms, where mutations are common and there are frequent chances of reduced response to chemotherapy (Kihara et al. 2000, Shi and Gao 2016, Dichwalkar et al. 2017).

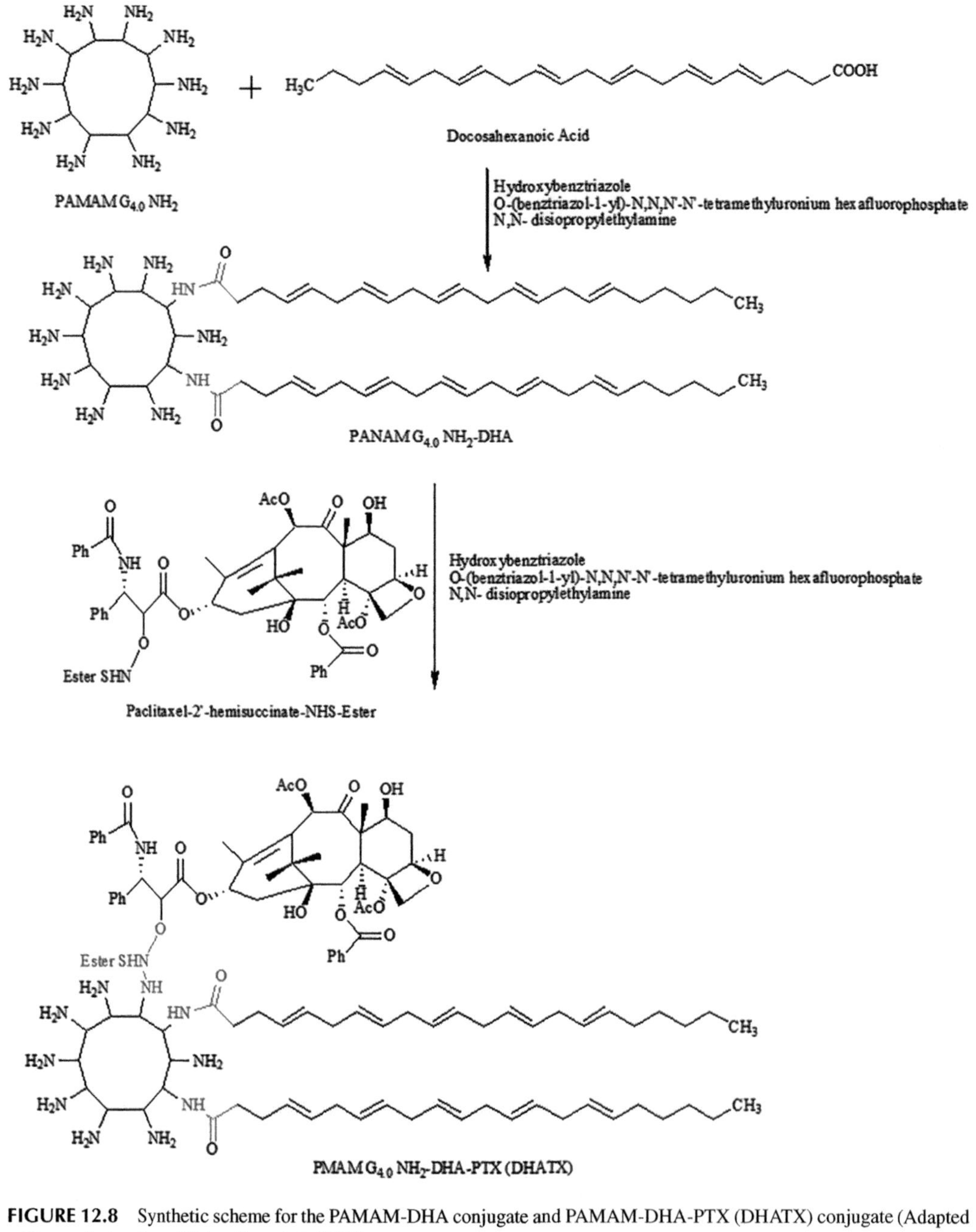

FIGURE 12.8 Synthetic scheme for the PAMAM-DHA conjugate and PAMAM-DHA-PTX (DHATX) conjugate (Adapted with permission from Dichwalkar et al., 2017).

Choi et al. conjugated the G 5.0 PAMAM dendrimer as a targeted carrier with FA ligands for the photo-caged DOX (Figure 12.9). An amide coupling of FA with ethylenediamine resulted in a ligand that targets FA receptor. Then a caged DOX (**5**) was produced after conservation with an *ortho* nitrobenzyl located photocleavable group (**3**) at its primary amine. G 5.0 PAMAM–(glutaric acid) (carboxylic acid-termi-nated dendrimer, (**6**) through amide formation covalently coupled to both (FAR-targeting ligand) and (**5**) (photocaged DOX) for obtaining nanoconjugates. Two fluorescein isothiocyanate (FI)[23]-labelled

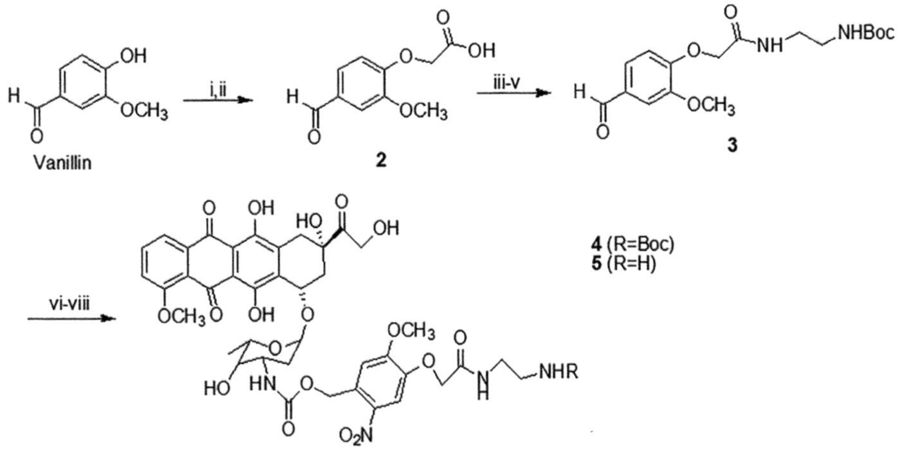

Reagents and conditions- (**i**) ethyl bromoacetate, K_2CO_3, DMF, rt, 17h, 75%; (**ii**) NaOH, THF, MeOH, H_2O,, rt, 33h, 82%; (**iii**) conc. HNO_3, AcOH, 0°C to rt, 26h; (**iv**) N-Boc-1,2-diaminomethane, DCC, DMAP, DMF, 0°C to rt, 36h; (**v**) $NaBH_4$,THF, MeOH, rt, 5h; (**vi**) 4-nitrophenyl chloroformate, DIPEA, THF, $CHCl_3$, rt, 16h; (**vii**) doxorubicin.HCl, Et_3N, DMF, rt, dark, 36h; (**viii**) TFA, $CHCl_3$, rt, 15 min

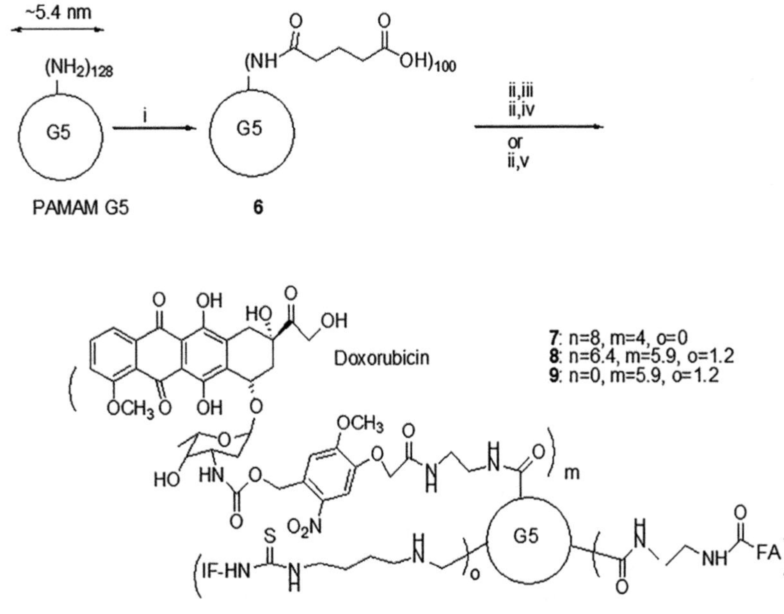

Reagents and conditions: (**i**) glutaric anhydride, Et_3N, MeOH, rt, 48h; (**ii**) NHS, EDC, DMAP, DMF, rt, 36h; (**iii**) FA-$CONHCH_2CH_2NH_2$ (**1**), doxorubicin-photocleavable linker (**5**), Et_3N, DMF, rt, 36h, then H_2O, rt, 4h; (**iv**) 1,5, FITC-$NH(CH_2)_4NH_2$, Et_3N, DMF; then H_2O; (**v**) FITC-$NH(CH_2)_4NH_2$, Et_3N, DMF; then H_2O

FIGURE 12.9 The synthetic scheme for the PAMAM dendrimer conjugate {**7-9**} including G5-FA-doxorubicin **7**, a FA-attached and doxorubicin loaded G5 PAMAM dendrimer in which doxorubicin is caged with a photocleavable ortho-nitro benzyl group and tethered to the dendrimer surface (Adapted with permission from Choi et al., 2010).

PAMAM DOX conjugates were also synthesised either with or without FA ligand attachment for evaluation of cellular binding and uptake studies (Choi et al. 2010).

Fox et al. synthesised two kinds of dendrimers with two different amino acids, i.e., using glycine and β-alanine. PEGylated PLL-CPT (PLL: poly(L-lysine); CPT: camptothecin) dendrimers were formed linked separately with two amino acids for the treatment of cancer. Around a diaminopropane core, two generation two lysine dendrons were constructed and to incorporate two functional handles, i.e., an amine and a carboxylic acid at the periphery of dendrimer was amended with aspartic acid. The carboxylic groups were esterified with glycine-20(O)-CPT conjugate, and the amino group, aspartic acid was PEGylated to yield the PEGylated PLL-CPT conjugate (Fox et al. 2009).

12.2.6 Conjugation with Imaging Agents

Mendoza-Nava et al. synthesised a radio labelled Lutetium (^{177}Lu) G 4.0 PAMAM dendrimer conjugated with folate and bombesin. Inside the cavity of the prepared dendrimer, AuNPs were caged. In aqueous basic medium, first S-2-(4-isothiocyanatobenzyl)-1,4,7,10-tetraazacyclododecane tetra acetic acid (p-SCN-benzyl-DOTA) was conjugated to the dendrimer and then conjugation of Lys^1 Lys^3 dodecane tetra acetic acid - bombesin and FA with dendrimer was accomplished by activation of carboxylate group with O-(7-azabenzotriazol-1-yl)-N,N,N'-N'-tetramethyluronium-hexafluoro phosphate (HATU). The prepared conjugate was then combined with 1% AuCl$_4$ followed by addition of NaBH$_4$. Then this purified conjugate was radio labelled with Lu to form AuNP-folate-bombesin to be effectively used as an imaging agent for the detection of breast tumours. Studies showed that the radio labelled dendrimer had promising detection potential of gastrin-releasing peptide receptors as well as over-expressed breast cancer folate receptors (Mendoza-Nava et al. 2016).

12.3 Diversity in Applications

The diversity of applications of dendrimers is primarily due to the following reasons:

a) Different methods of synthesis of dendrimers, i.e., conventional which include convergent and divergent synthesis; accelerated synthesis as well as orthogonal and chemoselective growth strategies of dendrimers (Syren and Malkoch 2012).

b) Large number of groups provided for the attachment– polyvalency, which ensures better molecular interactions at the biological site (Silva, Menacho and Chorilli 2012).

c) Variety of modifications possible on the surface using different modifiers based on the type of application, e.g., using tumour targeting module to target the tumour site. More specifically, the linkers are being used, and this approach of synthesis can be applied to all generations of dendrimers (Wang et al. 2018).

d) Enhancing the solubility with conjugation and increased circulating time with decreased half-life of the conjugated drug also gives an edge to the dendrimers over other polymers (Vega-Vásquez, Mosier and Irudayaraj 2020).

12.3.1 Dendrimer Conjugation with Quantum Dots

Quantum dots are classified as nanocrystals, which are semiconductors and possess optical characteristics. The optical property of these nanocrystals is widely now explored using advanced imaging techniques particularly by health professionals. Magnetic resonance imaging (MRI) and positron emission tomography (PET) are well-known imaging techniques that provide detailed three-dimensional anatomical information with high resolution. Despite this, there are certain constraints that need further improvisation. For instance, MRI is associated with low sensitivity for identification of cells that are few in number, need individual cell tracking and are sometimes unable to differentiate between benign or malignant types of cancer (Nohroudi et al. 2010). Moreover, PET requires exposure to a high level of radiation (Aicher et al. 2003). Fluorescence, another sensitive method for optical imaging of biological cells, is a promising approach.

Recent advances in single cell imaging have led quantum dots to be utilised with the fluorescence technique. Quantum dots provide certain advantages such as broader absorption and narrow emission spectra, high sensitivity, better optical stability and brightness. Quantum dots allow differentiating certain cancer cells with precision and individual cell tracking in case of stem cell therapy. Along with these advantages, quantum dots are associated mainly with limitations such as low cellular uptake and reduced spreading within the cells due to being trapped in endosomes or lysosomes. Therefore, to establish and make them utilised as a promising alternative approach to other aforesaid imaging techniques, these obstacles need to be resolved. Dendrimers are studied experimentally to entrap semiconductor nanomaterials and targeted to biological tissues. For instance, studies have shown that cationic dendrimers conjugated with quantum dots are the key to overcome the limitations of using quantum dots in cell imaging. Cationic dendrimers such as PAMAM conjugated with quantum dots were studied, and it was observed that the cationic charge on dendrimers facilitates cellular uptake. The possible explanation given for this observation was the electrostatic interaction of positive charge of dendrimers and negative charge of the biological cell membrane. Moreover, the PAMAM dendrimers after cellular uptake facilitate cytoplasmic migration of quantum dots by developing osmotic pressure gradient within endosomes/lysosomes and cytoplasm. The study concluded that the PAMAM dendrimer conjugated with quantum dots improved its internalisation within mesenchymal stem cells and increased the fluorescence intensity (Higuchi et al. 2011).

12.3.2 Dendrimer Conjugation with Carbon Nanotubes (CNTs)

CNTs are atomically sp^2 hybridised fullerene compounds with a layered structure in nanoscale. CNTs possess properties such as a nanosize cylindrical shape that facilitates transmembrane penetration from biological membranes. Due to their hollow structure, CNTs show high drug loading, which demonstrates their utility in drug delivery systems. Moreover, CNTs allow surface functionalisation with different ligands, which further facilitate targeting (Guo et al. 2017). However, CNTs have poor aqueous solubility that limits their applicability in the development of advanced drug delivery systems. Therefore, dendrimers are conjugated with CNTs to provide solubility to CNTs. The properties of dendrimers such as high solubility, dense structure in a nanosize and highly functionalised surface make them an ideal candidate to maximise the pharmaceutical applications of CNTs. Being highly branched and ease of surface modifications with a wide variety of ligands, dendrimers allow CNTs to target at certain biological tissues with high drug loading capacity.

The conjugation of PAMAM dendrimers with CNTs has also been realised in the field of biosensing. A field effect based on the semiconductor of electrolyte-insulator is commonly exploited in various studies for biosensing reactions such as enzymatic, antigen-antibody and detection of certain biological macromolecules particularly in cancer (Poghossian and Schöning 2014). For instance, in a study the PAMAM dendrimer was complexed with single walled CNTs to fabricate a semiconductor chip wherein the PAMAM dendrimer provides a highly porous structure, and single walled CNTs impart a highly conducting system. The advantage of this conjugation lies in the morphology of complex, *i.e.,* highly porous structure increases the surface area which is necessary for enzyme immobilisation thereby enhancing the sensitivity for biosensing the activity. A layer-by-layer technique was adopted wherein the PAMAM dendrimer and CNT complex was fabricated in layer form and optimised for modulating ionic transport. The study explained the potential of this particular complex in a three bilayered form of film in diagnostic applications (Sousa et al. 2017).

12.4 Challenges

Although there are numerous uses and a variety of applications of dendrimers in the present scenario, there are certain shortcomings which are being faced while using dendrimers. These are mainly due to the size of the dendrimers which usually higher generations of dendrimers have. A low clearance rate and intracellular trafficking restrict the use of dendrimers in certain conditions. Moreover, cytotoxicity is another related problem being faced, but has been overcome to some extent by the use of biodegradable dendrimer conjugates.

Another issue related to the synthesis of dendrimers is its heterogeneity. Due to the large number of groups available for the synthesis, sometimes specificity is lost resulting in the production of multiple polymer mixtures with difference in drug loading (Ekladious, Colson and Grinstaff 2019). One of the recently faced problems with dendrimer conjugated nanomedicines is its scale-up production in the pharmaceutical industries. It is a major challenge being faced nowadays including the characterisation of dendrimers as these require highly sophisticated analytical techniques for the complex and large structures of dendrimers (Kim, Park and Na 2018).

12.5 Concluding Remarks and Future Perspective

The present chapter entails in detail the synthesis of the dendrimer and the bioactives with the help of various studies carried out in the decade. The synthesis details have been meticulously been explained diagrammatically giving a clear view, which can be reproducible efficiently. Although many challenges are being faced by the dendrimers these days, still these systems hold a large amount of patents and clinical applications to its name. Apart from being matured as the delivery systems, dendrimer-conjugates have entered into clinical trials in a decade proving their substantial growth in the forthcoming years.

Abbreviations

AuNPs	Gold nanoparticles
CNTs	Carbon nanotubes
CPT	Camptothecin
DHA	Docosahexaenoic acid
DMF	Dimethylformamide
DOTA	1,4,7,10-tetraazacyclododecane-1,4,7,10-tetraacetic acid
DOX	Doxorubicin
EGF	Epidermal growth factor
FA	Folic acid
FI	Fluorescein isothiocyanate
FSH	Follicle stimulating hormone
FSHR	Follicle stimulating hormone receptor
GFLG-DOX	Gly-Phe-Leu-Gly and doxorubicin
HATU	O-{7-azabenzotriazol-1-yl}-N,N,N'-N'-tetramethyluronium-hexafluoro phosphate
HBTU	O-{benztriazol-1-yl}-N,N,N'-N'-tetramethyluronium hexafluorophosphate
HER	Human estrogen receptor
HOBt	Hydroxybenztriazole
mPEG	Methoxy poly(ethylene glycol)
MRI	Magnetic resonance imaging
PAMAM	Polyamidoamine
PEG	Polyethylene glycol
PET	Positron emission tomography
PLL	Poly (L-lysine)
PPI	Poly (propylene imine)
RGD	Arginine-glycine-aspartic acid
TEM	Transmission electron microscope

REFERENCES

Abbasi, Elham, Sedigheh Fekri Aval, Abolfazl Akbarzadeh, Morteza Milani, Hamid Tayefi Nasrabadi, Sang Woo Joo, Younes Hanifehpour, Kazem Nejati-Koshki, and Roghiyeh Pashaei-Asl. 2014. "Dendrimers: synthesis, applications, and properties." *Nanoscale Research Letters* no. 9 (1):247–247. doi:10.1186/1556-276X-9-247.

Aicher, A., W. Brenner, M. Zuhayra, C. Badorff, S. Massoudi, B. Assmus, T. Eckey, E. Henze, A. M. Zeiher, and S. Dimmeler. 2003. "Assessment of the tissue distribution of transplanted human endothelial progenitor cells by radioactive labeling." *Circulation* no. *107* (16):2134–2139. doi:10.1161/01. CIR.0000062649.63838.C9.

Astruc, Didier, Elodie Boisselier, and Cátia Ornelas. 2010. "Dendrimers designed for functions: from physical, photophysical, and supramolecular properties to applications in sensing, catalysis, molecular electronics, photonics, and nanomedicine." *Chemical Reviews* no. *110* (4):1857–1959. doi:10.1021/cr900327d.

Bhyrappa, P., James K. Young, Jeffrey S. Moore, and Kenneth S. Suslick. 1996. "Dendrimer-metalloporphyrins: synthesis and catalysis." *Journal of the American Chemical Society* no. *118* (24):5708–5711. doi:10.1021/ja953474k.

Bonam, Srinivasa Reddy, Aparna Areti, Prashanth Komirishetty, and Sylviane Muller. 2020. "10 - Dendrimers in immunotherapy and hormone therapy." In *Pharmaceutical Applications of Dendrimers*, edited by Abhay Chauhan and Hitesh Kulhari, 233–249. Elsevier.

Bose, C. K. 2008. "Follicle stimulating hormone receptor in ovarian surface epithelium and epithelial ovarian cancer." *Oncology Research* no. *17* (5):231–238. doi:10.3727/096504008786111383.

Bosman, A. W., H. M. Janssen, and E. W. Meijer. 1999. "About Dendrimers: structure, physical properties, and applications." *Chemical Reviews* no. *99* (7):1665–1688. doi:10.1021/cr970069y.

Buhleier, Egon, Winfried Wehner, and Fritz VÖGtle. 1978. "'Cascade'- and 'Nonskid-Chain-like' syntheses of molecular cavity topologies." *Synthesis* no. *1978* (2):155–158. doi:10.1055/s-1978-24702.

Caminade, Anne-Marie, Deyue Yan, and David K. Smith. 2015. "Dendrimers and hyperbranched polymers." *Chemical Society Reviews* no. *44* (12):3870–3873. doi:10.1039/C5CS90049B.

Canas-Arranz, R., M. Forner, and S. Defaus. 2020. "A bivalent B-cell epitope dendrimer peptide can confer long-lasting immunity in swine against foot-and-mouth disease." doi:10.1111/tbed.13497.

Choi, S. K., T. Thomas, M. H. Li, A. Kotlyar, A. Desai, and J. R. Baker, Jr. 2010. "Light-controlled release of caged doxorubicin from folate receptor-targeting PAMAM dendrimer nanoconjugate." *Chemical Communications (Cambridge)* no. *46* (15):2632–2634. doi:10.1039/b927215c.

Choi, S. K., T. P. Thomas, M. H. Li, A. Desai, A. Kotlyar, and J. R. Baker, Jr. 2012. "Photochemical release of methotrexate from folate receptor-targeting PAMAM dendrimer nanoconjugate." *Photochemical and Photobiological Sciences* no. *11* (4):653–660. doi:10.1039/c2pp05355a.

Choi, Y., T. Thomas, A. Kotlyar, M. T. Islam, and J. R. Baker, Jr. 2005. "Synthesis and functional evaluation of DNA-assembled polyamidoamine dendrimer clusters for cancer cell-specific targeting." *Chemistry & Biology* no. *12* (1):35–43. doi:10.1016/j.chembiol.2004.10.016.

Chonco, L., M. Pion, E. Vacas, B. Rasines, M. Maly, M. J. Serramia, L. Lopez-Fernandez, J. De la Mata, S. Alvarez, R. Gomez, and M. A. Munoz-Fernandez. 2012. "Carbosilane dendrimer nanotechnology outlines of the broad HIV blocker profile." *Journal of Controlled Release* no. *161* (3):949–958. doi:10.1016/j. jconrel.2012.04.050.

Chou, Chih-Ming, and Hsing-Lung Lien. 2011. "Dendrimer-conjugated magnetic nanoparticles for removal of zinc (II) from aqueous solutions." *Journal of Nanoparticle Research* no. *13* (5):2099–2107. doi:10.1007/s11051-010-9967-5.

Denkewalter, Robert G., Jaroslav Kolc, and William J. Lukasavage. 1979. Macromolecular highly branched homogeneous compound based on lysine units. United States.

Dichwalkar, T., S. Patel, S. Bapat, P. Pancholi, N. Jasani, B. Desai, V. K. Yellepeddi, and V. Sehdev. 2017. "Omega-3 fatty acid grafted PAMAM-paclitaxel conjugate exhibits enhanced anticancer activity in upper gastrointestinal cancer cells." *Macromolecular Bioscience* no. *17* (8). doi:10.1002/mabi.201600457.

van Dongen, Mallory A., Rahul Rattan, Justin Silpe, Casey Dougherty, Nicole L. Michmerhuizen, Margaret Van Winkle, Baohua Huang, Seok Ki Choi, Kumar Sinniah, Bradford G. Orr, and Mark M. Banaszak Holl. 2014. "Poly(amidoamine) Dendrimer–methotrexate conjugates: the mechanism of interaction with folate binding protein." *Molecular Pharmaceutics* no. *11* (11):4049–4058. doi:10.1021/mp500608s.

Ekladious, I., Y. L. Colson, and M. W. Grinstaff. 2019. "Polymer-drug conjugate therapeutics: advances, insights and prospects." *Nature Reviews Drug Discovery* no. *18* (4):273–294. doi:10.1038/s41573-018-0005-0.

Fox, Megan E., Steve Guillaudeu, Jean M. J. Fréchet, Katherine Jerger, Nichole Macaraeg, and Francis C. Szoka. 2009. "Synthesis and in vivo antitumor efficacy of PEGylated poly(l-lysine) Dendrimer–Camptothecin conjugates." *Molecular Pharmaceutics* no. *6* (5):1562–1572. doi:10.1021/mp9001206.

Freestone, Ian, Nigel Meeks, Margaret Sax, and Catherine Higgitt. 2007. "The Lycurgus Cup—A Roman nanotechnology." *Gold Bulletin* no. *40* (4):270–277. doi:10.1007/BF03215599.

Gilks, C. B. 2010. "Molecular abnormalities in ovarian cancer subtypes other than high-grade serous carcinoma." *Journal of Oncology* no. *2010*:740968. doi:10.1155/2010/740968.

Guo, Q., X. T. Shen, Y. Y. Li, and S. Q. Xu. 2017. "Carbon nanotubes-based drug delivery to cancer and brain." *Journal of Huazhong University of Science and Technology* no. *37* (5):635–641. doi:10.1007/s11596-017-1783-z.

Hayati, B., A. Maleki, F. Najafi, H. Daraei, F. Gharibi, and G. McKay. 2017. "Super high removal capacities of heavy metals (Pb(2+) and Cu(2+)) using CNT dendrimer." *Journal of Hazardous Materials* no. *336*:146–157. doi:10.1016/j.jhazmat.2017.02.059.

He, X., C. S. Alves, N. Oliveira, J. Rodrigues, J. Zhu, I. Banyai, H. Tomas, and X. Shi. 2015. "RGD peptide-modified multifunctional dendrimer platform for drug encapsulation and targeted inhibition of cancer cells." *Colloids and Surfaces B: Biointerfaces* no. *125*:82–89. doi:10.1016/j.colsurfb.2014.11.004.

Higuchi, Y., C. Wu, K. L. Chang, K. Irie, S. Kawakami, F. Yamashita, and M. Hashida. 2011. "Polyamidoamine dendrimer-conjugated quantum dots for efficient labeling of primary cultured mesenchymal stem cells." *Biomaterials* no. *32* (28):6676–6682. doi:10.1016/j.biomaterials.2011.05.076.

Jain, N. K., M. S. Tare, V. Mishra, and P. K. Tripathi. 2015. "The development, characterization and in vivo anti-ovarian cancer activity of poly(propylene imine) (PPI)-antibody conjugates containing encapsulated paclitaxel." *Nanomedicine* no. *11* (1):207–218. doi:10.1016/j.nano.2014.09.006.

Kihara, C., T. Seki, Y. Furukawa, H. Yamana, Y. Kimura, P. van Schaardenburgh, K. Hirata, and Y. Nakamura. 2000. "Mutations in zinc-binding domains of p53 as a prognostic marker of esophageal-cancer patients." *Japanese Journal of Cancer Research* no. *91* (2):190–198. doi:10.1111/j.1349-7006.2000.tb00931.x.

Kim, Yejin, Eun Park, and Dong Hee Na. 2018. "Recent progress in dendrimer-based nanomedicine development." *Archives of Pharmacal Research* no. *41*. doi:10.1007/s12272-018-1008-4.

Klajnert, B., and M. Bryszewska. 2001. "Dendrimers: properties and applications." *Acta Biochimica Polonica* no. *48* (1):199–208.

Li, Ning, Na Li, Qiangying Yi, Kui Luo, Chunhua Guo, Dayi Pan, and Zhongwei Gu. 2014. "Amphiphilic peptide dendritic copolymer-doxorubicin nanoscale conjugate self-assembled to enzyme-responsive anti-cancer agent." *Biomaterials* no. *35* (35):9529–9545. doi:10.1016/j.biomaterials.2014.07.059.

Liu, M., and J. M. Fréchet. 1999. "Designing dendrimers for drug delivery." *Pharmaceutical Science & Technology Today* no. *2* (10):393–401. doi:10.1016/s1461-5347(99)00203-5.

Longmire, Michelle R., Mikako Ogawa, Peter L. Choyke, and Hisataka Kobayashi. 2011. "Biologically optimized nanosized molecules and particles: more than just size." *Bioconjugate Chemistry* no. *22* (6):993–1000. doi:10.1021/bc200111p.

McCarthy, T. D., P. Karellas, S. A. Henderson, M. Giannis, D. F. O'Keefe, G. Heery, J. R. Paull, B. R. Matthews, and G. Holan. 2005. "Dendrimers as drugs: discovery and preclinical and clinical development of dendrimer-based microbicides for HIV and STI prevention." *Molecular Pharmaceutics* no. *2* (4):312–318. doi:10.1021/mp050023q.

Mendoza-Nava, Héctor, Guillermina Ferro-Flores, Flor de María Ramírez, Blanca Ocampo-García, Clara Santos-Cuevas, Liliana Aranda-Lara, Erika Azorín-Vega, Enrique Morales-Avila, and Keila Isaac-Olivé. 2016. "[177]Lu-Dendrimer conjugated to Folate and Bombesin with gold nanoparticles in the dendritic cavity: a potential theranostic radiopharmaceutical." *Journal of Nanomaterials* no. *2016*:1039258. doi:10.1155/2016/1039258.

Modi, Dimple A., Suhair Sunoqrot, Jason Bugno, Daniel D. Lantvit, Seungpyo Hong, and Joanna E. Burdette. 2014. "Targeting of follicle stimulating hormone peptide-conjugated dendrimers to ovarian cancer cells." *Nanoscale* no. *6* (5):2812–2820. doi:10.1039/C3NR05042D.

Newkome, G. R., E. He, and C. N. Moorefield. 1999. "Suprasupermolecules with novel properties: metallodendrimers." *Chemical Reviews* no. *99* (7):1689–1746. doi:10.1021/cr9800659.

Nohroudi, K., S. Arnhold, T. Berhorn, K. Addicks, M. Hoehn, and U. Himmelreich. 2010. "In vivo MRI stem cell tracking requires balancing of detection limit and cell viability." *Cell Transplant* no. *19* (4):431–441. doi:10.3727/096368909X484699.

Otis, James B., Hong Zong, Alina Kotylar, Anna Yin, Somnath Bhattacharjee, Han Wang, James R. Baker, Jr., and Su He Wang. 2016. "Dendrimer antibody conjugate to target and image HER-2 overexpressing cancer cells." *Oncotarget* no. *7* (24):36002–36013. doi:10.18632/oncotarget.9081.

Poghossian, Arshak, and Michael Schöning. 2014. "Label-free sensing of biomolecules with field-effect devices for clinical applications." *Electroanalysis* no. *26*:1197–1213. doi:10.1002/elan.201400073.

Shi, Wen-Jia, and Jin-Bo Gao. 2016. "Molecular mechanisms of chemoresistance in gastric cancer." *World Journal of Gastrointestinal Oncology* no. *8* (9):673–681. doi:10.4251/wjgo.v8.i9.673.

Silva, J. R., F. P. Menacho, and Marlus Chorilli. 2012. "Dendrimers as potential platform in nanotechnology-based drug delivery systems." *IOSR Journal of Pharmacy* no. *2*:23–30.

Sousa, M. A. M., J. R. Siqueira, A. Vercik, M. J. Schöning, and O. N. Oliveira. 2017. "Determining the optimized layer-by-layer film architecture with Dendrimer/carbon nanotubes for field-effect sensors." *IEEE Sensors Journal* no. *17* (6):1735–1740. doi:10.1109/JSEN.2017.2653238.

Swager, Timothy, and Alberto Concellón. 2019. "Sweet inhibition of Zika and Dengue via [60]Fullero-Dendrimers." *Synfacts* no. *15*. doi:10.1055/s-0039-1691267.

Syren, Marie, and Michael Malkoch. 2012. "ChemInform abstract: simplifying the synthesis of Dendrimers: accelerated approaches." *Chemical Society Reviews* no. *41*:4593–4609. doi:10.1039/c2cs35062a.

Tang, Mary X., Carl T. Redemann, and Francis C. Szoka. 1996. "In vitro gene delivery by degraded polyamidoamine Dendrimers." *Bioconjugate Chemistry* no. *7* (6):703–714. doi:10.1021/bc9600630.

Thompson, David. 2007. "Michael Faraday's recognition of ruby gold: the birth of modern nanotechnology." *Gold Bulletin* no. *40* (4):267–269. doi:10.1007/BF03215598.

Tomalia, D. A., H. Baker, J. Dewald, M. Hall, G. Kallos, S. Martin, J. Roeck, J. Ryder, and P. Smith. 1985. "A new class of polymers: starburst-Dendritic macromolecules." *Polymer Journal* no. *17* (1):117–132. doi:10.1295/polymj.17.117.

Trollsås, Mikael, and James L. Hedrick. 1998. "Dendrimer-like star polymers." *Journal of the American Chemical Society* no. *120* (19):4644–4651. doi:10.1021/ja973678w.

Twyman, Lance J., Anthony E. Beezer, R. Esfand, Martin J. Hardy, and John C. Mitchell. 1999. "The synthesis of water soluble dendrimers, and their application as possible drug delivery systems." *Tetrahedron Letters* no. *40* (9):1743–1746. doi:10.1016/S0040-4039(98)02680-X.

Tyssen, David, Scott A. Henderson, Adam Johnson, Jasminka Sterjovski, Katie Moore, Jennifer La, Mark Zanin, Secondo Sonza, Peter Karellas, Michael P. Giannis, Guy Krippner, Steve Wesselingh, Tom McCarthy, Paul R. Gorry, Paul A. Ramsland, Richard Cone, Jeremy R. A. Paull, Gareth R. Lewis, and Gilda Tachedjian. 2010. "Structure activity relationship of Dendrimer microbicides with dual action antiviral activity." *PLOS ONE* no. *5* (8):e12309. doi:10.1371/journal.pone.0012309.

Vega-Vásquez, Pablo, Nathan S. Mosier, and Joseph Irudayaraj. 2020. "Nanoscale drug delivery systems: from medicine to agriculture." *Frontiers in Bioengineering and Biotechnology* no. *8*:79–79. doi:10.3389/fbioe.2020.00079.

Vögtle, Fritz, Gabriele Richardt, and Nicole Werner. 2009. *"Dendrimer chemistry: concepts, syntheses, properties, applications."* *Dendrimer Chemistry: Concepts, Syntheses, Properties, Applications* 1–342. doi:10.1002/9783527626953.

Vunain, E., A. K. Mishra, and B. B. Mamba. 2016. "Dendrimers, mesoporous silicas and chitosan-based nanosorbents for the removal of heavy-metal ions: a review." *International Journal of Biological Macromolecules* no. *86*:570–586. doi:10.1016/j.ijbiomac.2016.02.005.

Wang, Tao, Yaozhong Zhang, Longfei Wei, Yuhan G. Teng, Tadashi Honda, and Iwao Ojima. 2018. "Design, synthesis, and biological evaluations of asymmetric bow-tie PAMAM Dendrimer-based conjugates for tumor-targeted drug delivery." *ACS Omega* no. *3* (4):3717–3736. doi:10.1021/acsomega.8b00409.

Xu, J., H. Wang, L. Xu, Y. Chao, C. Wang, X. Han, Z. Dong, H. Chang, R. Peng, Y. Cheng, and Z. Liu. 2019. "Nanovaccine based on a protein-delivering dendrimer for effective antigen cross-presentation and cancer immunotherapy." *Biomaterials* no. *207*:1–9. doi:10.1016/j.biomaterials.2019.03.037.

Yoo, Hoon, Peter Sazani, and R. L. Juliano. 1999. "PAMAM Dendrimers as delivery agents for antisense oligonucleotides." *Pharmaceutical Research* no. *16* (12):1799–1804. doi:10.1023/A:1018926605871.

Yuan, Q., E. Lee, W. A. Yeudall, and H. Yang. 2010. "Dendrimer-triglycine-EGF nanoparticles for tumor imaging and targeted nucleic acid and drug delivery." *Oral Oncology* no. *46* (9):698–704. doi:10.1016/j.oraloncology.2010.07.001.

Zhang, C., D. Pan, K. Luo, W. She, C. Guo, Y. Yang, and Z. Gu. 2014a. "Peptide dendrimer-Doxorubicin conjugate-based nanoparticles as an enzyme-responsive drug delivery system for cancer therapy." *Advanced Healthcare Materials* no. *3* (8):1299–1308. doi:10.1002/adhm.201300601.

Zhang, Chengyuan, Dayi Pan, Kui Luo, Ning Li, Chunhua Guo, Xiuli Zheng, and Zhongwei Gu. 2014b. "Dendrimer–doxorubicin conjugate as enzyme-sensitive and polymeric nanoscale drug delivery vehicle for ovarian cancer therapy." *Polymer Chemistry* no. *5* (18):5227–5235. doi:10.1039/C4PY00601A.

Zhu, Wen-Ping, Jie Gao, Shi-Peng Sun, Sui Zhang, and Tai-Shung Chung. 2015. "Poly(amidoamine) dendrimer (PAMAM) grafted on thin film composite (TFC) nanofiltration (NF) hollow fiber membranes for heavy metal removal." *Journal of Membrane Science* no. *487*:117–126. doi:https://doi.org/10.1016/j.memsci.2015.03.033.

Zong, Hong, Thommey P. Thomas, Kyung-Hoon Lee, Ankur M. Desai, Ming-Hsin Li, Alina Kotlyar, Yuehua Zhang, Pascale R. Leroueil, Jeremy J. Gam, Mark M. Banaszak Holl, and James R. Baker. 2012. "Bifunctional PAMAM Dendrimer conjugates of folic acid and methotrexate with defined ratio." *Biomacromolecules* no. *13* (4):982–991. doi:10.1021/bm201639c.

13

Smart Functionalised-Dendrimeric Medicine in Cancer Therapy

Vijay Mishra, Manvendra Singh and Pallavi Nayak

CONTENTS

13.1 Introduction

Nowadays, cancer is the most common disease spreading all over the world, causing a high mortality and morbidity rate. The World Cancer Survey performed in 2014 has reported that 70% of cancer-related deaths occur in low and middle-income countries, while in 2010, the total annual economic cost of cancer was about US$ 1.16 trillion (Ambekar et al. 2020).

Cancer is responsible for one in six deaths worldwide. The rate of reported incidents and deaths tends to rise due to increased life expectancy as well as demographic and epidemiological changes. An additional 18.1 million new diagnoses and 9.6 million cancer deaths were recorded in 2018, with the most commonly detected cancer being lung (11.6% of all patients), led by female breast (11.6%) and colorectal cancer (10.2%), respectively. Lung cancer is the leading cause of cancer mortality (18.4% of all deaths), accompanied by colorectal (9.2%) and stomach cancer (8.2%) (Figure 13.1) (WHO 2020).

Radiation, chemotherapy and surgery are conventional treatment methods, which cannot specifically target the cancer cell and show toxicity to healthy cell/tissues. Therefore, nanotechnology has been developed to solve these hurdles. This novel technology guides the medication to show its action on a specific site and reduces the toxicity problem (Chaurasiya and Mishra 2018). Food and Drug Administration (FDA)-approved latest drugs or a combination of drugs for the treatment of cancer are tabulated in Table 13.1 (FDA 2020).

Cancer nanotechnology is a major research field, in which nanotools are used to diagnose and treat cancer with unparalleled precision. A targeted drug delivery system (TDDS) is an ideal key for more successful cancer treatment. In the past few years, nanocarriers have been widely used to enhance the precision, selectivity

TABLE 13.1

FDA-approved latest drugs or combination of drug for the treatment of various cancer

Drug	Cancer Type and Patient Condition	Approval Date
Nivolumab + Ipilimumab and two cycles of platinum-doublet chemotherapy	• Recurrent non-small cell lung cancer (NSCLC) • No epidermal growth factor receptor (EGFR) • No anaplastic lymphoma kinase (ALK) genomic tumour aberrations	May 26, 2020
Brigatinib	• Adult patients with ALK positive metastatic NSCLC	May 22, 2020
Olaparib	• Adult patients with somatic homologous recombination repair (HRR) gene-mutated metastatic castration-resistant prostate cancer (mCRPC)	May 19, 2020
Atezolizumab	• Adult patients with NSCLC • Tumours with high PD-L1 expression • No EGFR or ALK genomic tumour aberrations	May 18, 2020
Ripretinib	• Adult patients with advanced gastrointestinal stromal tumour • Patients must receive prior treatment with three or more kinase inhibitors, including imatinib.	May 15, 2020
Rucaparib	• mCRPC • Patients have been treated with androgen receptor-directed therapy and a taxane-based chemotherapy	May 15, 2020
Capmatinib	• Adult patients with NSCLC	May 6, 2020
Daratumumab and hyaluronidase-fihj	• Adult patients with newly diagnosed or relapsed/ refractory multiple myeloma	May 1, 2020
Niraparib	• Adult patients with advanced epithelial ovarian, fallopian tube or primary peritoneal cancer	April 29, 2020
Sacituzumab govitecan	• Adult patients with metastatic triple-negative breast cancer • Patients must receive at least two prior therapies for metastatic disease	April 22, 2020

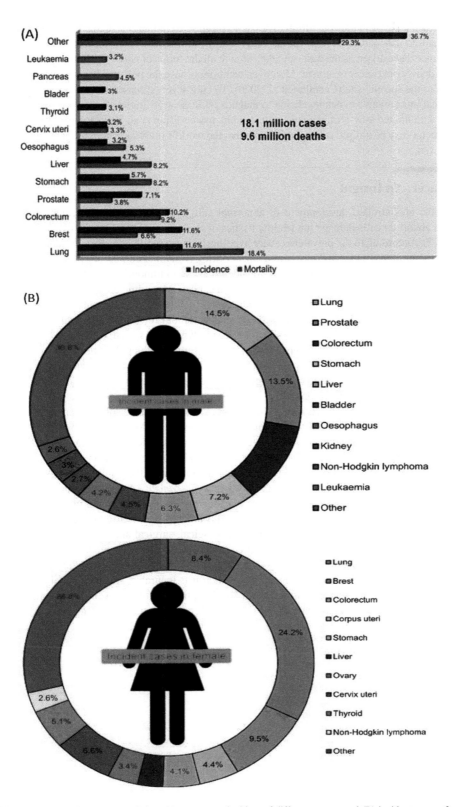

FIGURE 13.1 Worldwide cancer statistics (A) percentage incident of different cancer and (B) incident cases of different cancer in male and female.

and regulated release of therapeutic drugs. They hold promise to mitigate systemic toxicity by production of functionalised nanoparticles (NPs) for targeted therapy (Patil et al. 2019). Recent studies demonstrated that dendrimers have arisen as the best candidate among all the forms of nanocarriers for optimised and targeted drug delivery in cancer treatment. Moreover, dendrimers have the potential to carry multiple and large drug loads to the specific site (Carvalho et al. 2020). To cure severe diseases, the medicinal field involves technological innovations i.e. nanomedicine to diagnose disease at its initial stage and rectify quickly and accurately with limited long-term damage to the healthy tissues (Tran et al. 2017; Juliano 2016). The nanomedicine sector may be utilised strategically to combat the world's most dangerous ailment 'cancer'.

13.2 Cancer: An Insight

Cancer is the uncontrolled development of abnormal cells with a faster rate as compared to normal cells. Such abnormal cells damage the healthy tissues of the body and promote the expansion of cancer (Figure 13.2). According to the prevalent cancer hypothesis, normal cells transform to cancerous cells via irregular structural, molecular and biochemical networking (Rajani et al. 2020). The group of cancerous cells forming an abnormal mass of tissues is considered as 'solid tumour', while the abnormal cells that do not form any mass of tissue are 'leukaemia' or blood cancer. Depending on their cell type, tumours are classified as malignant and benign tumours. A malignant tumour is a cancerous tumour, which can invade its surrounding tissues and spread all over the body. A new tumour forms, when a malignant tumour separated out from the parent tumour and passes through the lymphatic system (or blood) from their main site to another site. The benign tumour is a non-cancerous tumour, which does not grow and spread like malignant tumours. If the infected cells exert pressure on the vital organs, the benign tumour can become more harmful. Gene mutation caused by heritage or non-heritage factors is responsible for the growth and proliferation of anomalous cells (Rajani et al. 2020).

Hippocrates, the Greek physician (460–370 BC) known as the 'Father of Medicine', has examined many diseases which involved the formation of lumps or masses (onkos). He was the first scientist who had used two major terms to define atumour i.e. the non-healing and ulcerative lumps as '*karkinos*' while the non-ulcerative lumps as '*karkinomas*' (Faguet 2015). He denoted atumour as Greek word '*crab*' because it can travel and its fingers tend to propagate projection. Later, the Roman doctor Aulus Cornelius Celsus (28–50 BC) converted '*crab*' term into the Latin term '*cancer*' (American Cancer Society 2018).

Earliest Egyptian papyri showed the footprints of cancerous cells discovered in the fossils of dinosaur and human skull as reported in 1500–1600 BC. In the nineteenth century, George Ebers and Edwin Smith published a papyrus that describes the pharmacological, surgical and mystical treatments of cancer. Imhotep, an architectural-physician, has written Smith papyrus. He was the first researcher who has described chest tumours with reference of breast cancer. Moreover, he cautioned that if this tumour is swelling, feeling cold and spread all over the breast then it is incurable. In 1000 BC, the development of cancerous cells was also observed in Peruvian and Egyptian mummies. Almost 2700 years ago, the Scythian monarch (around 40–50 years in age) who lived in Southern Siberian steppes was the earliest and scientifically recorded case of disseminated tumour/cancer (Faguet 2015).

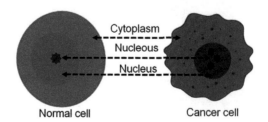

FIGURE 13.2   Pictorial representation of normal and cancer cells.

13.3 Dendrimers: A Versatile Nanocarrier

Dendrimers are nanosized artificial macromolecules with definite, symmetrical and mono-dispersed architecture. Due to their highly branched three-dimensional (3D) cascade, dendrimers are denoted as trees with three major architectural elements such as a core, branching layer (called as generation) consisting of recurring units around the inner core and an outer layer. The generation number is denoted as 'G', which states the number of key elements from the inner shell to the outer shell. Furthermore, it is used in the determination of molecular weight and total number of functional end groups (Lee et al. 2005). The G0 to G3.0 of dendrimers are identical to a natural organic molecule. These are tiny and flexible without great homogeneity or a certain 3D structure and have asymmetrical shapes with an open structure. After G4.0 was developed, dendrimers with 3D structures began to become spherical. By G5.0, they form specific and consistent 3D structures. After G5.0, they are very structured spheres (Munir et al. 2016). As they grow to the peripheries, dendrimers form a spherical membrane-like structure, tightly packed (Figure 13.3). Dendrimers cannot extend out, when the critical branched state is obtained owing to the lack of space. This state is called the 'starburst effect' (Shinde et al. 2014). Depending on the generation, dendrimers have a diameter of 2–10 nm (Garg et al. 2011). The core molecule includes several reaction sites, like ammonia (the smallest one), which comprises three reaction sites. The core molecule is connected to the monomer molecules via the reaction sites to build the first generation. Each unreacted end group of bonded monomers provides a reaction site that combines it to build higher generations with several molecules (Yousefi et al. 2020).

Drugs can either be encapsulated within the cavity of dendrimers or attached to their outer surfaces via hydrophobic, electrostatic or covalent bonding. Their vast properties contribute to the fabrication of different dendrimers, such as micellar, chiral, poly(propylene imine) (PPI), amphiphilic and

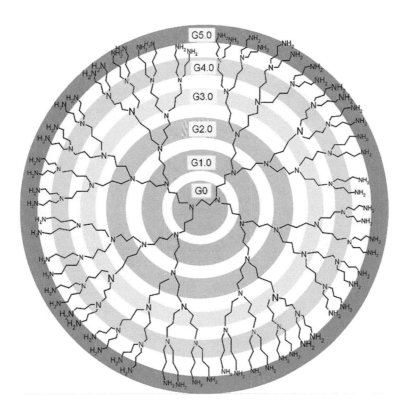

FIGURE 13.3 Pictorial presentation of different generations of dendrimers.

poly(amido amine) (PAMAM). Their unique characteristics allow dendrimers to be employed in various biomedical applications.

The modern imaging technologies help in early detection of the molecular progressions of cancer. Various types of NPs are employed and demonstrated in cancer theranostic in which multipurpose dendrimer-based NPs may be the best choice. This cancer theranostic is one of the dynamic techniques that diagnose and treat patients at the initial stage. Nanocarriers have the ability to passively target the tumours *via* the enhanced permeation and retention (EPR) effect (Fang et al. 2011). However, delivery of an anti-cancer drug is accomplished through active molecular targeting in which different targeting molecules are engineered on the surface of dendrimers (Saluja et al. 2019).

Dendrimers are classified on the basis of their terminal functional groups, structure and inner cavities (Figure 13.4) (Shinde et al. 2010). Simple dendrimers consist of 4, 10, 22 and 46 benzene rings associated correspondingly with 45 Å of molecular diameters. The chirality of chiral dendrimers is based upon the formation of constitutional branches but chemically alike branches to an achiral core e.g. pentaerythritol. Liquid crystalline dendrimers are composed of mesogenic monomer (e.g. mesogenic carbosilane dendrimers). Basically, micellar dendrimers are unimolecular arrangement of micelles in the dendrimers (Shinde et al. 2010). Hybrid dendrimers are composed of linear and dendritic polymers, arranged in a copolymer or hybrid manner. Metallodendrimers are connected through metal ions and lead to the formation of a complex either in a peripheral or interior core. Amphiphilic dendrimers are the form of globular dendrimers, having irregular and extremely restricted division of the end-terminal structure (Shinde et al. 2010; Baig et al. 2015).

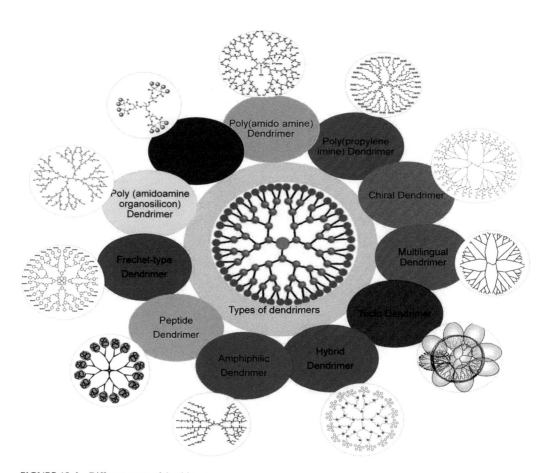

FIGURE 13.4   Different types of dendrimers.

13.4 Characterisation Techniques of Dendrimers

Different characterisation techniques are used for the identification and determination of dendrimers. Some of the techniques are discussed as follows:

13.4.1 Electrical Techniques

Electron paramagnetic resonance is employed for quantitative identification of substitution efficiency on the outer surface of PAMAM dendrimers. Electrochemistry offers data based upon the chances of interaction between electronically active terminal groups. Electrophoresis helps in the evaluation of homogeneity and purity of different water-soluble dendrimers (Francese et al. 2003; Kukowska-Latallo et al. 1996; Tabakovic et al. 1997).

13.4.2 Microscopy Methods

Transmission electron microscopy (TEM) is mainly used to determine the size and shape as well as size distribution of specific molecules of synthesised PAMAM dendrimers. The scanning electron microscopy (SEM) technique observes and determines the surface morphology of dendrimers (Gensch et al. 1999; Hofkens et al. 1998). The principle of size exclusion chromatography (SEC) or molecular sieve chromatography (MSC) is based on the high-performance liquid chromatography (HPLC) separation method. This is employed to estimate the molecular weight of dendrimers (Zeng et al. 2002).

13.4.3 Rheology and Physical Properties

Intrinsic viscosity is an analytical tool used to determine the morphological architecture of dendrimers. Dielectric spectroscopy helped in the determination of the molecule based dynamic methods (α, β). Depending on the chain arrangement, molecular weight and entanglement of polymers, differential scanning calorimetry (DSC) detects the temperature of glass transition (Trahasch et al. 1999; Mourey et al. 1992; Matos et al. 2000; Dantras et al. 2002).

13.4.4 Scattering Techniques

The small angle X-ray scattering (SAXS) technique is employed for solutions to measure the radius of gyration (Rg). Furthermore, their scattering intensity determines the order of polymeric segments. In contrast to SAXS, the small angle neutron scattering (SANS) technique provides a more precise Rg value and identifies the site of the end-terminal groups *e.g.* PAMAM and PPI dendrimers. The laser light scattering (LLS) method measures the hydrodynamic radius and also detects the aggregates of dendrimers (Achar and Puddephatt 1994; Chu and Hsiao 2001; Prosa et al. 1997; Rietveld and Smit 1999; Topp et al. 1999).

13.4.5 Spectroscopy and Spectrometry Methods

The fluorescence method quantifies the defects that came during the fabrication of dendrimers. Ultraviolet-visible spectroscopy helps to identify the dendrimer synthesis method. Generally, the absorption band intensity is directly proportional to the quantity of chromophoric units. Infra-red spectroscopy is used to detect different functional groups engineered during surface modification of dendrimers. Nuclear magnetic resonance spectroscopy (NMR) analyses various functional groups present on plain as well as surface-modified dendrimers. Near-infrared spectroscopy characterises the delocalised interaction of π-π stacking between the end-terminal groups in PAMAM dendrimers. In mass spectroscopy, chemical ionisation is utilised for evaluation of cargo dendrimers having molecular weight <3000 Da. In addition, the electrospray ionisation technique is used for dendrimers that can form a stable compound (multi-charged). X-ray diffraction spectroscopy determines the chemical configuration, size, shape and architecture of dendrimers (Larre et al. 1998; Miller et al. 1997; Wilken and Adams 1997; Hummelen et al. 1997; Sakthivel et al. 1998).

13.4.6 Miscellaneous

The titrimetric technique analyses the number of NH_2 groups present on the PAMAM dendrimer. The sedimentation technique is utilised for lactosylated PAMAM dendrimers and measures the dipole moments. X-ray photoelectron spectroscopy (XPS) characterises different layers of dendrimers and chemicals, for example, poly(aryl ether) dendrons or PAMAM dendrimers.

13.5 Applications of Dendrimers

Dendrimers have various applications ranging from drug solubilisation to drug delivery as well as treatment of different diseases such as diabetes (Zhuo et al. 1999), glaucoma (Wooley et al. 1993), brain tumour and ovarian cancer (Pavlov et al. 2001; Mishra et al. 2019; Mishra and Jain 2014;Saluja et al. 2019; Mishra and Kesharwani 2016; Dwivedi et al. 2016; Jain et al. 2015; Kesharwani et al. 2015; Tunki et al. 2020).Figure 13.5 presents various applications of dendrimers.

13.6 Surface-Functionalised Dendrimers

Therapeutic drugs can either be engineered on the surface or encapsulated inside the core of the dendrimers chemically or mechanically, respectively. In chemical bonding, a strong bond can be obtained if functional groups are enabled before coupling. Primary amine (NH_2), hydroxyl (OH), guanidino, carboxyl (COOH) and thiol (SH) are some common functional groups found in polymers and drug molecules (Mishra et al. 2009).

13.6.1 Peptide-Functionalised Dendrimers

Peptide dendrimers are wedge or radial-like branched dendrimers with peptidyl branching in the core and covalently modified functional groups on the surface. The dendritic core is composed of irregular

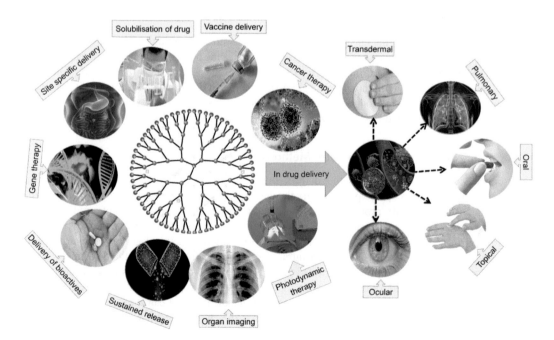

FIGURE 13.5 Overview of different applications of dendrimers.

branching of lysine attached with 2–8 units of a tetra-peptide or an octa-peptide. Therefore, peptide dendrimers are used on the microbial surface for selective action. In contrast to linear polymeric analogues, peptide dendrimers are less toxic, highly water soluble, stable and proteolytic in nature. The lysine core-based dendrimers were designed with arginine residues on their surface to develop an anti-angiogenetic agent, while azobenzene residues engineered on the surface of lysine dendrimers were developed for a photo-responsive drug delivery system (DDS) (Mishra et al. 2009).

13.6.2 PEGylated Dendrimers

In the DDS, poly(ethylene glycol) (PEG) is a commonly employed polymer. It is highly water soluble, non-toxic and non-immunogenic in nature. Thus, PEG is engineered on the surface of dendrimers with an ultimate goal to change their properties, which results in reduced immunogenicity and toxicity. These PEGylated dendrimers can also act as dendritic unimolecular micelles. The PAMAM dendrimers have cationic charge (due to peripherally attached NH_2 group) which resulted in the leakage of drug, haemolysis, cytotoxicity and uptake of reticuloendothelial system (RES). The PEG-functionalised DDS overcomes all these shortcomings and results in improved drug loading, macrophage absorption, half-life and reduced haemolytic toxicity, RES uptake and drug leakage (Mishra et al. 2009).

13.6.3 Glycodendrimers

Glycodendrimer is a versatile term used for varied structures of dendrimers, which integrate carbohydrates in their structure. These are categorised as carbohydrate-coated, carbohydrate-centred and fully carbohydrate-based dendrimers. The glycol dendrimers are synthesised by modification of commercially available dendrimers and synthesised de novo dendrimers. Many research studies have been performed in which largely investigations are focussed on PAMAM dendrimers (Starburst™ dendrimers) and PPI dendrimers (Astramol™ dendrimers). Both the classes of dendrimers consist of tertiary amine-dependent architectures with primary amine groups present peripherally. In addition, the formation of thiourea and amide, photo-addition to allyl ethers, glycosylation and reductive amination are the main coupling reactions involved in their fabrication methods (Mishra et al. 2009).

13.6.4 Galabiose-Functionalised Dendrimers

Carbosilane galabiose (Galα1-4Gal) dendrimers were designed for artificial inhibitor purpose against the Shiga toxins (Stxs), produced from *Escherichia coli*. The galabiose-functionalised dendrimers were composed of 3,4,6 galabiosis units at the peripheral axis of dendrimers. The galabiose was formed when the penta-O-acetyl-β-d-galactopyranose was paired with carbosilane dendrimers, having three-shapes that resulted in acetyl-protected glycol dendrimers in virtuous yields. The cluster of de-O-acetylation was formed in the presence of sodium methoxide, and subsequently aqueous sodium hydroxide was added, which resulted in three desired shapes of galabiose-coated carbosilane dendrimers (Mishra et al. 2009).

13.6.5 Miscellaneous Surface-Engineered Dendrimers

Dendrimers such as tecto dendrimers, amphiphilic, hybrid, chiral dendrimers, liquid crystalline and poly(amidoamine-organosilicon) (PAMAMOS) dendrimers are versatile in nature and smartly tailored for various DDS-based applications. It was observed that using the convergent approach, the polyaryl ether dendrimer was synthesised with two functional groups present on their surface. One functional group was utilised to bind PEG on the surface of dendrimers to increase the water solubility of assembly, while other functional groups were employed to bind the hydrophobic drugs (Mishra et al. 2009). In a current report, Xiong et al fabricated zwitterion-carboxy betaine acrylamide (CBAA) dendrimer (PAMAM G5.0) entrapped gold (Au) NPs for serum-improved gene delivery to obstruct the metastasis of cancer cells. The gene delivery capacities of morpholine-engineered AuNPs and Morpholine-free AuNPs in the serum medium were found to be 1.4 and 1.7 times greater than the equivalent vector present in serum-free medium, respectively (Xiong et al. 2019).Therefore, dendrimers have been explored as excellent drug delivery vehicles for different routes of administration.

13.7 Supramolecular Architectures of Dendrimers in Cancer Therapy

The structure of dendrimers provides unique prospects through chemical conjugation. Thus, their supramolecular structure is designed, when various interactions occurred as hydrogen/hydrophobic and electrostatic-bond or encapsulation, inside the core and/or in numerous channels amid the dendrons. These molecules have wide applications, ranging from the DDS, in which dendrimers NPs are associated with drugs and directed to target the gene or tissues delivery that performs correspondingly to the DDS. Nonetheless, binding the dendrimers NPs using nucleic acids provides alternative and advanced results, like a drug nanocarrier because of specific physicochemical properties, for example, net charge, polarity, solubility in water, etc. (Castro et al. 2018).

Recent developments show that PAMAM dendrimer is an effective tool in gene [pDNA, small interfering ribonucleic acid (siRNA)] and delivery of anticancer drugs. The PAMAM dendrimers are employed to overcome the multidrug resistance of tumour by hybrid NPs and loaded or linked in other NPs (Li et al. 2018). Zhu and co-authors synthesised AuNP-entrapped dendrimers engineered with arginyl-glycyl-aspartic acid (RGD) peptide and α-tocopheryl succinate (α-TOS) for targeted chemotherapy. The authors observed that Au-TOS-RGD DENPs have not shown any significant toxicity to healthy cells at 5.12 μM (highest concentration selected) with 75% cell viability. Their results concluded that the developed multifunctional dendrimers could hold great potential as a theranostic tool in the treatment of cancer (Zhu et al. 2015).

Some investigators evaluated bio-reducible polymer-based dendrimers for delivery of anti-VEGF siRNA to the cancer cell lines like human hepatocarcinoma (Huh-7), human fibrosarcoma (HT1080 cells) and human lung adenocarcinoma (A549). The bio-reducible polymer-based dendrimers and siRNA resulted in polyplex formation having 116 nm of average diameter and +24.6 mV of charge (Nam et al. 2015).

Gardikis et al introduced a novel DDS for cancer treatment, in which the combined dendrimeric and liposomal technology (liposomal locked-in dendrimers) was utilised. In this study, G1.0 and G2.0 dendrimers with OH-end groups were fabricated and 'sealed' inside the liposomes. The observations showed high drug release and encapsulation rate (Gardikis et al. 2010)·

13.8 Dendrimer-Based Nanomedicines in Cancer Therapy

Dendrimers are one of the potential nanocarriers, which have attracted the researchers worldwide for the treatment of different types of cancers (Figure 13.6). Among the members of the dendrimer family, widely explored dendrimers are PAMAM, PPI, poly(L-lysine) (PLL) and carbosilane dendrimers. The role of dendrimers in delivery of various anticancer bioactives is well documented. The targeted delivery of various anticancer bioactive/drugs by ligand anchored PAMAM dendrimers is listed in Table 13.2 (Tunki et al. 2020).

13.8.1 Prostate Cancer

Epithelium of PC showed over-expression of prostate-specific membrane antigen (PSMA) in comparison to benign hyperplasia and normal prostate tissue. Lesniak et al. have constructed NH_2-terminated PAMAM dendrimer (G4.0) NPs conjugated with two dodecane tetra-acetic acid (DOTA) molecules (for radiolabelling), three rhodamines (for optical evaluation) and ten Lys-Glu-urea (low molecular weight) PSMA-targeted ligands. The remaining end-amine groups were interacted with glycidol to eliminate the positive charge of the surface. Thus, the non-targeted action (cellular uptake) and toxicity of formulated G4.0-PSMA NPs were reduced (Lesniak et al. 2019).

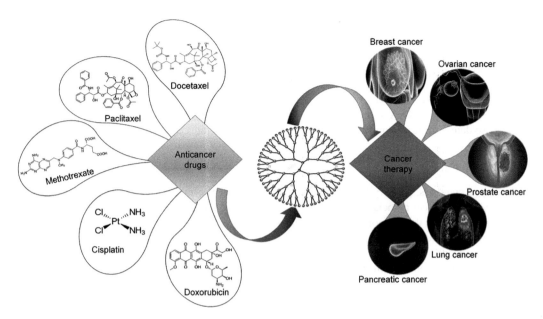

FIGURE 13.6 Dendrimers in the treatment of different cancer.

Kavosi et al. developed an ultrasensitive electrochemical immunosensor for detection of the PSMA biomarker in PC cells employing AuNPs/PAMAM dendrimers loaded with enzyme associated aptamers as the combined triple signal amplification approach (Kavosi et al. 2015).

Self-assembled amphiphilic peptide dendrimer-dependent nanovectors were designed by Dong et al. for successful delivery of siRNA therapy in PC treatment. The cytotoxicity assays of

TABLE 13.2

Targeted delivery of anticancer drugs by ligand anchored PAMAM dendrimers

Bioactive/Drug	Targeting Ligand	Type of Cancer
Curcumin	Hyaluronic acid	Pancreatic cancer
Methotrexate	Folic acid	Ovarian cancer
Docetaxel (DTX)	pHBA	Brain tumour
DTX	Antibody	Breast cancer cells
Paclitaxel (PTX)	Antibody	Colon cancer
PTX	Antibody	Lung cancer
PTX	Antibody	Ovarian cancer
10-hydroxycamptothecin	c(RGDfK)	Prostate and breast cancer cells
cis-diaminodichloro platinum	Folic acid	Lung cancer
Cisplatin	Biotin	Ovarian cancer cells
Doxorubicin (DOX)	Glycodendrimers	Hepatic cancer
DOX	Transferrin	Brain tumour
DOX	Transferrin, WGA	Brain tumour
DOX	CREKA, LyP-1	Prostate cancer (PC), MDA-MB-435 cancer cells
DOX	Folic acid	Brain tumour
DOX	LFC131	Brain tumour
DOX	RGD	Brain tumour

Source: Tunki et al. (2020).

1,2-distearoyl-sn-glycer-3-phosphoethanolamine (DSPEKK$_2$) (siRNA) were used in vitro with the help of MTT and lactate dehydrogenase (LDH) assays. Results of the MTT assay demonstrated that the cells normally grew under the siRNA delivery conditions after treatment with either DSPEKK$_2$ or scramble siRNA/DSPE-KK$_2$, signalling that no metabolic toxicity was detected, while neither DSPE-KK$_2$ nor scramble/DSPE-KK$_2$ complexes showed significant LDH release with regard to the positive control of the lysis buffer, and has no release of LDH, implying no damage to the membrane. Such findings suggest that the availability of siRNA dependent on DSPE-KK$_2$ in PC-3 cells is not cytotoxic. Further haemolysis studies also demonstrate its non-toxic properties even at elevated rates (Figure 13.7) (Dong et al. 2020).

13.8.2 Lung Cancer

In developed countries, primary lung cancer mainly of epithelial origin is one of the main reasons for about 23% of all cancer associated deaths. Kaminskas et al. synthesised a DOX-conjugated dendrimer and administered it in the pulmonary route for enhanced drug delivery to lung metastases and improved cancer therapy. Twice a week intratracheal administration of dendrimers showed >95% decrease in lung tumour subsequently two weeks as compared with DOX solution administered intravenously reduced 30–50% of lung tumour (Kaminskas et al. 2012). Ayatollahi et al. designed a polyplex (functionalised PAMAM dendrimers/aptamer/plasmid) for potent delivery of RNAi-related genes for lung cancer treatment. The PAMAM dendrimers were functionalised using PEG and 10-bromodecanoic acid. Their outcome indicated 25% of gene silencing effect and induced 14% of late apoptosis effect with greater selectivity in targeted cancer cells (Ayatollahi et al. 2017).

Almuqbil et al developed a dendrimer-based nanoDDS that enhances and correlates the penetration of DOX in the 3D in vitro model of the lung tumour with its effectiveness. The basic components of the extracellular matrix (ECM), recognised as a physical barrier to the transport of DOX, are shown to generate spheroids produced with the human adenocarcinoma cells. DOX was conjugated to G4.0 succinamic acid-terminated-PAMAM dendrimers (G4SA) through an enzyme-liable tetrapeptide (G4SA-GFLG-DOX). MTT assay used to evaluate 2D cell viability by using free DOX, G4SA or G4SA-GFLG-DOX conjugates after 72 h of incubation. The G4SA-GFLG-DOX cell viability curve follows a very comparable DOX free profile, but it is upward with the half maximal inhibitory concentration (IC$_{50}$) of 1.11 µM. It indicates that DOX was produced from the conjugate after it was internalised in the tumour cells and stored in the nucleus. The apoptosis test was conducted after 24 h incubation in both A549 cultural spheroids and co-culture A549/3T3 spheroids, and the finding suggested that the DOX and G4SAGFFLG-DOX tests for single culture spheroids have the same effect on apoptosis in all taken concentrations. The G4SA-GFLG-DOX nanocarriers, however, displayed enhanced behaviour in the 3D coculture system compared with free DOX at a concentration of 10 µM DOXeq, indicating an improved conjugate performance with more complicated and abounding ECM in the system. This research has shown that DOX-G4SA combination facilitates the penetration of DOX in an ECM-producing model of lung cancer co-culture and that this enhanced penetration is associated with an improved cytotoxic activity compared with free DOX (Almuqbil et al. 2020).

The mitochondria known as the powerhouse of cell regulate many biological mechanisms including cell metabolism and cell death. Recently, let-7b has been observed as tumour suppressor microRNA in human cell mitochondria targeting many genes of the respiratory chain associated with mitochondria. Maghsoudnia et al. fabricated nanocarriers of let-7b-PAMAM (G5)-TPP (triphenylphosphonium cation) and let-7b-PAMAM (G5)-TPP-HA (hyaluronic acid) to supply let-7b miRNA mimics in NSCLC mitochondria cells as a fascinating way of inhibiting cancer cells. The study found that no cell cytotoxicity in HA treated cells was observed compared to nontreated cells. Furthermore, the viability of the cells treated with various HA concentrations was not substantially different. The reduced cell toxicity of let-7b-PAMAM-TPP-HA NPs was lower in comparison with Let-7b-PAMAM-TPP NPs to inhibit cancer cells. The introduction of TPP molecules to PAMAM dendrimers will increase the cytotoxicity of the cell compared with unmodified dendrimers, since TPP itself may be destructive of PAMAM (G5)-TPP biological membranes as an effective method of generation of the genes to supply microRNAs with mitochondria to offer a completely new way in NSC (Maghsoudnia et al. 2020).

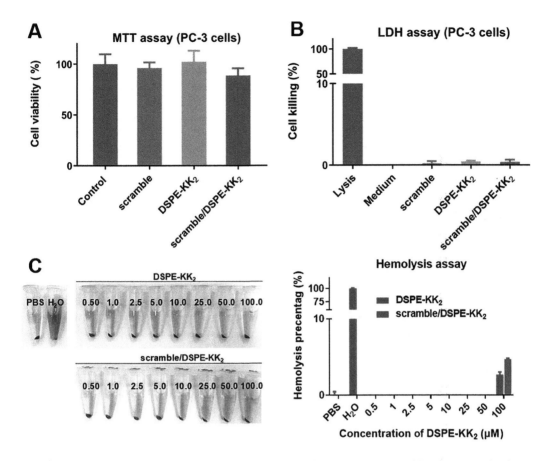

FIGURE 13.7 Safety assessment of siRNA/AmPPD complexes both in vitro and in vivo. Toxicity assessment using the (A) MTT assay, (B) LDH assay on PC-3 prostate cancer cells (50 nM scramble siRNA, N/P ratio of 10) (mean ± SD, n = 3) and (C) haemolysis assay of the DSPE-KK2 (0.5, 1, 2.5, 5, 10, 25, 50 and 100 μM) and siRNA/DSPE-KK2complexes (N/P ratio of 10) (mean ± SD, n = 3) (Reproduced with permission from Dong et al. 2020).

13.8.3 Ovarian Cancer

Ovarian cancer is an extremely lethal gynaecologic disease, with the high-grade serous subtype predominantly associated with poor survival rates. Lack of early diagnostic biomarkers and prevalence of post-treatment recurrence create substantial challenges in the treatment of ovarian cancer. Liu et al revealed lipid-based dendrimer (PAMAM G4.0) hybrid nanocarriers as an innovative DDS for PTX in therapy of the ovarian cancer. The resulting efficiency of drug encapsulation in the lipid-based dendrimer system was about 78.0±2.1% with 37-fold increased drug potency (Liu et al. 2015). Luong and others fabricated a ligand-decorated nanoarchitecture dendrimer for enhanced solubility and specified delivery of anticancer flavonoid analogues to the cancer cells/tissues. The researchers used folate as a ligand, which was decorated on the surface of PAMAM dendrimers for ultimate goal of enhanced water solubility of 3,4-difluorobenzylidene diferuloylmethane (CDF), highly hydrophobic but effective anticancer flavonoid (Luong et al. 2016).

Cruz et al. developed a new formulation using polyurea (PURE) dendrimers with the use of L-buthionine sulphoximine (L-BSO) (L-BSO@PURE$_{G4}$-FA$_2$) to test in vitro (GSH) synthesis on the restoration of ovarian cell cancer susceptibility to carboplatin, in the form of a folate-functional NP. Free L-BSO effectively decreases GSH bioavailability, impairing carboplatin tolerance. This effect of L-BSO has also been observed in the in vivo model of ovarian cancer, significantly reducing the size of subcutaneous tumours and GSH levels and peritoneal dissemination. L-BSO@PURE$_{G4}$-FA$_2$ nanoformulation is more successful in causing death of ovarian cancer cells than free L-BSO; and ovarian cancer cells are more vulnerable to L-BSO@PURE$_{G4}$-FA$_2$ than non-cancer squamous cells (HaCaT), confirming abdominal-mediated putative treatment (Cruz et al. 2020).

The results of selenium containing chryside (SeChry) in three separate ovarian cell lines (ES2, OVCAR3 and OVCAR8) and in two non-malignant cell lines (HaCaT and HK2) were examined by Santos et al. Findings demonstrate that SeChry does not affect cysteine uptake as well as being highly cytotoxic, but it increases GSH depletion and SeChry can induce oxidative stress.SeChry@PURE$_{G4}$-FA NPs have increased the specificity of SeChry delivery to ovarian cancer cells, significantly reducing the toxicity to non-malignant cells. The study concluded that the SeChry@PURE$_{G4}$-FA NPs as a potential method for enhancing ovarian cancer care underlie SeChry cytotoxicity in the case of GSH depletion and cystathionine β-synthase inhibition (Santos et al. 2019).

13.8.4 Oral Cancer

Oral and oropharyngeal cancer is the sixth most common cancer worldwide. The treatment outcome for oral cancer remains poor. Liu et al. investigated the anticancer activity of PAMAM dendrimer-linked short hairpin RNA (shRNA) in contrast to hTERT in oral cancer, which resulted in induced cell growth and apoptosis of cancer cells (Liu et al. 2011). Other investigators employed G3.0 PAMAM dendrimers and dimethylaminododecyl methacrylate (DMADDM) to design a biofilm adhesive for anti-caries action with enhanced re-mineralisation capabilities and biofilm regulation. This study showed no adverse effects on the dentin bond strength (Ge et al. 2017).

13.8.5 Breast Cancer

Breast cancer is the second important reason of women death worldwide. The triple negative breast cancer (TNBC) accounts for about 10–15% in women. Pourianazar and Gunduz fabricated CpG oligodeoxynucleotide (ODN)-loaded on PAMAM dendrimers coated with magnetic NPs to deliver the oligonucleotides, genes and drugs and promoted cell death in breast cancer cells. These NPs with a magnetic core have 40±10 nm average size and efficiently bind to the CpG-ODN molecules. Thus, it resulted in induced apoptosis of tumour cells (SKBR3 and MDA-MB231) and was considered as an effective targeted DDS for CpG-ODN in the biomedical applications (Pourianazar and Gunduz, 2016). In another study, Ghosh et al. developed and analysed a targeting non-viral vector that efficiently delivered the specific gene and diagnosed TNBC. The carbon quantum dots were prepared using peels of sweet lemon and then conjugated with various generations of PAMAM dendrimers, which resulted in a promising tool for the treatment of TNBC (Ghosh et al. 2019).

Torres-Pérez et al. fabricated new one-step PAMAM dendrimers loaded with methotrexate (MTX) and D-glucose (OS-PAMAM-MTX-GLU) and evaluated in TNBC cell line, (MDA-MB-231). The findings reveal that OS-PAMAM-MTX-GLU and controls have the primary and secondary amides characteristic of PAMAM dendrimers. The OS-PAMAM-MTX-GLU decreases the cell viability of MDA-MB-231 cells up to 20% and is considerably higher than free MTX, without significantly affecting HaCaT cells. Cell uptake analysis found that glycosylation enhanced the internalisation of OS-PAMAM conjugates in cancer cells relative to non-cancer cells. The uptake of OS-PAMAM-MTX-GLU inhibits MDA-MB-231 selectively, which is a desirable strategy for targeted breast cancer cell therapy (Torres-Pérez et al. 2020).

13.9 Pancreatic Cancer

Pancreatic cancer is one of the most lethal of the solid malignancies. However, surgery is safer and less invasive. Advances in radiation therapy have resulted in less toxicity. Optiz et al. designed a unique strategy to measure the mass transfer performance of fluorescent PAMAM dendrimers into the mammalian cells. The resultant effective mass transfer coefficient was obtained as 0.054±0.043 µM/min that resembled the rate constant of 0.035±0.023 min^{-1} for uptake of the fluorescent PAMAM dendrimers into the pancreatic cells (Opitz et al. 2013).

Lin and co-workers designed and reported advanced ultrasound-targeting microbubble devastation (UTMD)-related DDS based on dendrimers-encapsulated AuNPs for co-delivery of miR-21 inhibitor and drug, gemcitabine. The outcomes represented that the cellular uptake of engineered DDS was facilitated by UTMD with 0.4 W/cm^2 ultrasound power to improve the permeability of cells. Additionally, the co-delivery of the inhibitor and drug with or without UTMD therapy exhibited 13-fold and 82-fold lesser IC$_{50}$ values than free drug, respectively (Lin et al. 2018).

13.9.1 Liver Cancer

Liver cancer is listed as the second most common cause of tumour associated deaths in East Asian countries. The most common subtype comprising over 90% of human liver cancers is hepatocellular carcinoma (HCC). In a recent study, Jedrzak et al. synthesised multifunctional nanocarriers using PAMAM dendrimers G4.0, G5.0 and G6.0 attached with polydopamine-coated magnetite NPs (Fe$_3$O$_4$). The fabricated nanoplatform was used as smart DDS for cancer treatment (Jędrzak et al. 2019). Medina et al. reported N-acetylgalactosamine (NAc Gal) coupling of PAMAM-NH$_2$ dendrimers (G5.0) (thiourea and peptide bond) to formulate NAc Gal-targeted nanocarriers. It was examined that a sialoglycoprotein receptor-mediated endocytosis has contributed in cellular uptake of G5.0 NAcGal conjugate, which resulted in targeted delivery of anticancer drugs to the hepatic cancerous cells (Medina et al. 2011).

13.10 Toxicological Profiling of Dendrimers in Cancer Treatment

Dendrimers can be used as a drug delivery platform through in vitro and in vivo assessments needed to predict biocompatibility, cytotoxicity and biodistribution of dendrimers. When exposed to dendrimers, cell viability, haemolysis, protein content and other cellular parameters promote further dendrimer studies as carriers for drug delivery.

Sharma et al. synthesised a surface engineered dendrimer for the cancer targeted DDS. PAMAM dendrimers (G4.0) were employed and conjugated with gallic acid. The research revealed that the formulation targeted MCF-7 cells and performed synergistic action with anti-cancer drugs. Hence, conjugated gallic acid-dendrimers may be an effective nanocarrier in diagnosing and targeting the cancer (Sharma et al. 2011).

Lai et al. developed a promising approach to reduce the cytotoxicity and increase the efficacy of drug. The anticancer drug, DOX, was conjugated with PAMAM dendrimers using sensitive and insensitive-pH linkers and then combined with various photochemical internalisation strategies to estimate their cytotoxicity. The outcome showed that photochemical internalisation-based strategies could significantly enhance the drug cytotoxicity on Ca9-22 cells at larger concentrations (Lai et al. 2007)

Chen et al have employed PAMAM-OH derivative (PAMSPF) for co-delivery of MDM2 inhibitor (RG7388) and p53 plasmid to the cancer/tumour-specific site as well as assessed the synergistic anti-tumour activity of RG7388 and p53 plasmid. The particle size of PAMSPF was found to be 200 nm. Furthermore, the observations revealed 92.5% of encapsulation capacity of RG7388 in PAMSPF/p53/RG (Chen et al. 2019).

Dib and other collaborators sculptured an approach based on the complexes of dendrimers-guest for the enhancement of the solubility of phenazine N, N′-dioxide derivative with anti-tumour properties. It is found that phenazine anionic drug association with PPI dendrimers diminished the cytotoxicity of polycationic carriers by masking the charge of the surface (Dib et al. 2019).

13.11 Conclusion

This chapter has addressed the current advancement in various dendrimer-based platforms to design different strategies for cancer therapy. The peripheral axis of dendrimers can be functionalised with zwitterions, PEG, glycosyl and targeting agents to enhance their therapeutic efficacy. The inherent cationic amines present on the dendrimer surface enable various modifications and make them probable to condense the nucleic acid for targeted gene delivery. Mainly, dendrimers have been employed in conjugated cancer therapy to encourage the synergistic therapeutic effects. Consequently, dendrimers have been proved as a novel platform for fabrication of various multifunctional nanocarriers in cancer therapy-based applications. Many dendrimer-based DDSs have been explored with the aim of optimised and enhanced cancer treatment. But more research needs to be performed for greater biocompatibility, treatment efficacy and metabolism rate. Although the clinical translation of these exciting dendrimers has not been progressed yet, still, there is extended way to drive for the effective advancement of dendrimer-based nanomedicines. The enhancements in drug uptake can be obtained through bottom less knowledge of the microenvironment, metabolism and metastasis.

Abbreviations

3D	Three dimensional
A549	Human lung adenocarcinoma
ALK	Anaplastic lymphoma kinase
CBAA	Zwitterion-carboxy betaine acrylamide
CBS	Cystathionine β-synthase
CDF	3,4-difluorobenzylidene diferuloylmethane
DDS	Drug delivery system
DMADDM	Dimethylaminododecyl methacrylate
DOTA	Dodecane tetra-acetic acid
DOX	Doxorubicin
DSC	Differential scanning calorimetry
DSPEKK$_2$	1,2-distearoyl-sn-glycer-3-phosphoethanolamine
ECM	Extracellular matrix
EGFR	Epidermal growth factor receptor
EPR	Electron paramagnetic resonance
EPR	Enhanced permeation and retention
FDA	Food and drug administration
Galα1-4Gal	Carbosilane galabiose
GIST	Gastrointestinal stromal tumor
GSH	Glutathione
HA	Hyaluronic acid
HCC	Hepatocellular carcinoma
HPLC	High performance liquid chromatography
HRR	Homologous recombination repair
HT1080 cells	Human fibrosarcoma
Huh-7	Human hepatocarcinoma
IC$_{50}$	The half maximal inhibitory concentration
L-BSO	L-buthionin sulfoximine
LDH	Lactate dehydrogenase
LLS	Laser light scattering
mCRPC	Metastatic castration-resistant prostate cancer
MSC	Molecular sieve chromatography
NAc Gal	N-acetylgalactosamine
NMR	Nuclear magnetic resonance spectroscopy
NPs	Nanoparticles
NSCLC	Non-small cell lung cancer
ODN	Oligodeoxynucleotide
PAMAM	Poly(amido amine)
PAMAMOS	Poly(amidoamine-organosilicon)
PC	Prostate cancer

PDA	Polydopamine
PEG	Poly(ethylene glycol)
PLL	Poly(L-lysine)
PPI	Poly(propylene imine)
PSMA	Prostate-specific membrane antigen
PTX	Paclitaxel
PURE	Polyurea
RES	Reticulo endothelial system
RGD	Arginylglycylaspartic acid
SANS	Small angle neutron scattering
SAXS	Small angle X-ray scattering
SEC	Size exclusion chromatography
SeChry	Selenium containing chryside
SEM	Scanning electron microscopy
shRNA	Short hairpin RNA
siRNA	Small interfering ribonucleic acid
Stxs	Shiga toxins
TDDS	Targeted drug delivery system
TEM	Transmission electron microscopy
TNBC	Triple negative breast cancer
TPP	Triphenylphosphonium
UTMD	Ultrasound-targeting microbubble devastation
XPS	X-ray photoelectron spectroscopy
α-TOS	α-tocopheryl succinate

REFERENCES

Achar, S. and Puddephatt, R.J., 1994. Organoplatinum dendrimers formed by oxidative addition. *AngewandteChemie International Edition in English*, *33*(8): 847–849.

Almuqbil, R.M., Heyder, R.S., Bielski, E.R., Durymanov, M., Reineke, J.J. and da Rocha, S.R., 2020. Dendrimer conjugation enhances tumour penetration and efficacy of doxorubicin in extracellular matrix-expressing 3d lung cancer models. *Molecular Pharmaceutics*, *17*(5): 1648–1662.

Ambekar, R.S., Choudhary, M. and Kandasubramanian, B., 2020. Recent advances in dendrimer-based nanoplatform for cancer treatment: A review. *European Polymer Journal*, *126*: 109546.

American Cancer Society,2018. *Early history of cancer*. Available:www.cancer.org/cancer/cancer-basics/history-of-cancer/what-is-cancer.html; (Accessed on June 18, 2020).

Ayatollahi, S., Salmasi, Z., Hashemi, M., Askarian, S., Oskuee, R.K., Abnous, K. and Ramezani, M., 2017. Aptamer-targeted delivery of Bcl-xLshRNA using alkyl modified PAMAM dendrimers into lung cancer cells. *The International Journal of Biochemistry & Cell Biology*, *92*: 210–217.

Baig, T., Nayak, J., Dwivedi, V., Singh, A., Srivastava, A. and Tripathi, P.K., 2015. A review about dendrimers: synthesis, types, characterization and applications. *International Journal of Advances in Pharmacy, Biology and Chemistry*, *4*: 44–59.

Carvalho, M.R., Reis, R.L. and Oliveira, J.M., 2020. Dendrimer nanoparticles for colorectal cancer applications. *Journal of Materials Chemistry B*, *8*(6): 1128–1138.

Castro, R.I., Forero-Doria, O. and Guzman, L., 2018. Perspectives of dendrimer-based nanoparticles in cancer therapy. *Anais da Academia Brasileira de Ciências*, *90*(2): 2331–2346.

Chaurasiya, S. and Mishra, V., 2018. Biodegradable nanoparticles as theranostics of ovarian cancer: an overview. *Journal of Pharmacy and Pharmacology*, *70*(4): 435–449.

Chen, K., Xin, X., Qiu, L., Li, W., Guan, G., Li, G., Qiao, M., Zhao, X., Hu, H. and Chen, D., 2019. Co-delivery of p53 and MDM2 inhibitor RG7388 using a hydroxyl terminal PAMAM dendrimer derivative for synergistic cancer therapy. *ActaBiomaterialia*, *100*: 118–131.

Chu, B. and Hsiao, B.S., 2001. Small-angle X-ray scattering of polymers. *Chemical Reviews*, *101*(6): 1727–1762.

Cruz, A., Mota, P., Ramos, C., Pires, R.F., Mendes, C., Silva, J.P., Nunes, S.C., Bonifácio, V.D. and Serpa, J., 2020. Polyurea dendrimer folate-targeted nanodelivery of l-buthionine sulfoximine as a tool to tackle ovarian cancer chemoresistance. *Antioxidants, 9*(2): 133.

Dantras, E., Dandurand, J., Lacabanne, C., Caminade, A.M. and Majoral, J.P., 2002. Enthalpy relaxation in phosphorus-containing dendrimers. *Macromolecules, 35*(6): 2090–2094.

Dib, N., Fernández, L., Santo, M., Otero, L., Alustiza, F., Liaudat, A.C., Bosch, P., Lavaggi, M.L., Cerecetto, H. and González, M., 2019. Formation of dendrimer-guest complexes as a strategy to increase the solubility of a phenazine N, N′-dioxide derivative with antitumour activity. *Heliyon, 5*(4): e01528.

Dong, Y., Chen, Y., Zhu, D., Shi, K., Ma, C., Zhang, W., Rocchi, P., Jiang, L. and Liu, X., 2020. Self-assembly of amphiphilic phospholipid peptide dendrimer-based nanovectors for effective delivery of siRNA therapeutics in prostate cancer therapy. *Journal of Controlled Release, 322*: 416–425.

Dwivedi, N., Shah, J., Mishra, V., Mohd Amin, M.C.I., Iyer, A.K., Tekade, R.K. and Kesharwani, P., 2016. Dendrimer-mediated approaches for the treatment of brain tumour. *Journal of Biomaterials Science, Polymer Edition, 27*(7): 557–580.

Faguet, G.B., 2015. A brief history of cancer: age-old milestones underlying our current knowledge database. *International Journal of Cancer, 136*(9): 2022–2036.

Fang, J., Nakamura, H. and Maeda, H., 2011. The EPR effect: unique features of tumour blood vessels for drug delivery, factors involved, and limitations and augmentation of the effect. *Advanced Drug Delivery Reviews, 63*(3): 136–151.

Food and Drug Administration (FDA). 2020. *Hematology/Oncology (Cancer) Approvals & Safety Notifications,* https://www.fda.gov/drugs/resources-information-approved-drugs/hematologyoncology-cancer-approvals-safety-notifications, (Accessed on June 7, 2020).

Francese, G., Dunand, F.A., Loosli, C., Merbach, A.E. and Decurtins, S., 2003. Functionalization of PAMAM dendrimers with nitronylnitroxide radicals as models for the outer-sphere relaxation in dentritic potential MRI contrast agents. *Magnetic Resonance in Chemistry, 41*(2): 81–83.

Gardikis, K., Hatziantoniou, S., Bucos, M., Fessas, D., Signorelli, M., Felekis, T., Zervou, M., Screttas, C.G., Steele, B.R., Ionov, M. and Micha-Screttas, M., 2010. New drug delivery nanosystem combining liposomal and dendrimeric technology (liposomal locked-in dendrimers) for cancer therapy. *Journal of Pharmaceutical Sciences, 99*(8): 3561–3571.

Garg, T., Singh, O., Arora, S. and Murthy, R., 2011. Dendrimer—A novel scaffold for drug delivery. *International Journal of Pharmaceutical Sciences Review and Research, 7*(2): 211–220.

Ge, Y., Ren, B., Zhou, X., Xu, H.H., Wang, S., Li, M., Weir, M.D., Feng, M. and Cheng, L., 2017. Novel dental adhesive with biofilm-regulating and remineralization capabilities. *Materials, 10*(1): 26.

Gensch, T., Hofkens, J., Heirmann, A., Tsuda, K., Verheijen, W., Vosch, T., Christ, T., Basché, T., Müllen, K. and De Schryver, F.C., 1999. Fluorescence detection from single dendrimers with multiple chromophores. *Angewandte Chemie International Edition, 38*(24): 3752–3756.

Ghosh, S., Ghosal, K., Mohammad, S.A. and Sarkar,K., 2019. Dendrimer functionalized carbon quantum dot for selective detection of breast cancer and gene therapy. *Chemical Engineering Journal, 373*: 468–484.

Hofkens, J., Verheijen, W., Shukla, R., Dehaen, W. and De Schryver, F.C., 1998. Detection of a single dendrimer macromolecule with a fluorescent dihydropyrrolopyrroledione (DPP) core embedded in a thin polystyrene polymer film. *Macromolecules, 31*(14): 4493–4497.

Hummelen, J.C., Van Dongen, J.L. and Meijer, E.W., 1997. Electrospray mass spectrometry of poly (propylene imine) dendrimers—the issue of dendritic purity or polydispersity. *Chemistry–A European Journal, 3*(9): 1489–1493.

Jain, N.K., Tare, M.S., Mishra, V. and Tripathi, P.K., 2015. The development, characterization and in vivo anti-ovarian cancer activity of poly (propylene imine) (PPI)-antibody conjugates containing encapsulated paclitaxel. *Nanomedicine: Nanotechnology, Biology and Medicine, 11*(1): 207–218.

Jędrzak, A., Grześkowiak, B.F., Coy, E., Wojnarowicz, J., Szutkowski, K., Jurga, S., Jesionowski, T. and Mrówczyński, R., 2019. Dendrimer based theranostic nanostructures for combined chemo-and photo-thermal therapy of liver cancer cells in vitro. *Colloids and Surfaces B: Biointerfaces, 173*: 698–708.

Juliano, R.L., 2016. Nanomedicine: Promises and challenges. *In Nanomedicines*, 281–289.

Kaminskas, L.M., McLeod, V.M., Kelly, B.D., Sberna, G., Boyd, B.J., Williamson, M., Owen, D.J. and Porter, C.J., 2012. A comparison of changes to doxorubicin pharmacokinetics, antitumour activity, and toxicity mediated by PEGylated dendrimer and PEGylated liposome drug delivery systems. *Nanomedicine: Nanotechnology, Biology and Medicine, 8*(1): 103–111.

Kavosi, B., Salimi, A., Hallaj, R. and Moradi, F., 2015. Ultrasensitive electrochemical immunosensor for PSA biomarker detection in prostate cancer cells using gold nanoparticles/PAMAM dendrimer loaded with enzyme linked aptamer as integrated triple signal amplification strategy. *Biosensors and Bioelectronics*, *74*: 915–923.

Kesharwani, P., Mishra, V. and Jain, N.K., 2015. Generation dependent hemolytic profile of folate engineered poly(propylene imine) dendrimer. *Journal of Drug Delivery Science and Technology*, *28*: 1–6.

Kukowska-Latallo, J.F., Bielinska, A.U., Johnson, J., Spindler, R., Tomalia, D.A. and Baker, J.R., 1996. Efficient transfer of genetic material into mammalian cells using Starburst polyamidoamine dendrimers. *Proceedings of the National Academy of Sciences*, *93*(10): 4897–4902.

Lai, P.S., Lou, P.J., Peng, C.L., Pai, C.L., Yen, W.N., Huang, M.Y., Young, T.H. and Shieh, M.J., 2007. Doxorubicin delivery by polyamidoamine dendrimer conjugation and photochemical internalization for cancer therapy. *Journal of Controlled Release*, *122*(1): 39–46.

Larre, C., Bressolles, D., Turrin, C., Donnadieu, B., Caminade, A.M. and Majoral, J.P., 1998. Chemistry within megamolecules: Regio specific functionalization after construction of phosphorus dendrimers. *Journal of the American Chemical Society*, *120*(50): 13070–13082.

Lee, C.C., MacKay, J.A., Fréchet, J.M. and Szoka, F.C., 2005. Designing dendrimers for biological applications. *Nature Biotechnology*, *23*(12): 1517–1526.

Lesniak, W., Ray, S., Boinapally, S., Azad, B.B. and Pomper, M., 2019. Evaluation of a PSMA-targeted PAMAM dendrimer in an experimental model of prostate cancer. *Journal of Nuclear Medicine*, *60*: 548–548.

Li, J., Liang, H., Liu, J. and Wang, Z., 2018. Poly (amidoamine)(PAMAM) dendrimer mediated delivery of drug and pDNA/siRNA for cancer therapy. *International Journal of Pharmaceutics*,*546*(1–2): 215–225.

Lin, L., Fan, Y., Gao, F., Jin, L., Li, D., Sun, W., Li, F., Qin, P., Shi, Q., Shi, X. and Du, L., 2018. UTMD-promoted co-delivery of gemcitabine and miR-21 inhibitor by dendrimer-entrapped gold nanoparticles for pancreatic cancer therapy. *Theranostics*, *8*(7): 1923.

Liu, X., Huang, H., Wang, J., Wang, C., Wang, M., Zhang, B. and Pan, C., 2011. Dendrimers-delivered short hairpin RNA targeting hTERT inhibits oral cancer cell growth in vitro and in vivo. *Biochemical Pharmacology*, *82*(1): 17–23.

Liu, Y., Ng, Y., Toh, M.R. and Chiu, G.N., 2015. Lipid-dendrimer hybrid nanosystem as a novel delivery system for paclitaxel to treat ovarian cancer. *Journal of Controlled Release*,*220*: 438–446.

Luong, D., Kesharwani, P., Killinger, B.A., Moszczynska, A., Sarkar, F.H., Padhye, S., Rishi, A.K. and Iyer, A.K., 2016. Solubility enhancement and targeted delivery of a potent anticancer flavonoid analogue to cancer cells using ligand decorated dendrimer nano-architectures. *Journal of Colloid and Interface Science*, *484*: 33–43.

Maghsoudnia, N., BaradaranEftekhari, R., NaderiSohi, A., Norouzi, P., Akbari, H., Ghahremani, M.H., Soleimani, M., Amini, M., Samadi, H. and Dorkoosh, F.A., 2020. Mitochondrial delivery of microRNA Mimic let-7b to NSCLC cells by PAMAM-based nanoparticles. *Journal of Drug Targeting*, *28*(7–8): 810–830.

Matos, M.S., Hofkens, J., Verheijen, W., De Schryver, F.C., Hecht, S., Pollak, K.W., Fréchet, J.M.J., Forier, B. and Dehaen, W., 2000. Effect of core structure on photophysical and hydrodynamic properties of porphyrin dendrimers. *Macromolecules*, *33*(8): 2967–2973.

Medina, S.H., Tekumalla, V., Chevliakov, M.V., Shewach, D.S., Ensminger, W.D. and El-Sayed, M.E., 2011. N-acetylgalactosamine-functionalized dendrimers as hepatic cancer cell-targeted carriers. *Biomaterials*, *32*(17): 4118–4129.

Miller, L.L., Duan, R.G., Tully, D.C. and Tomalia, D.A., 1997. Electrically conducting dendrimers. *Journal of the American Chemical Society*, *119*(5): 1005–1010.

Mishra, V., Gupta, U. and Jain, N.K., 2009. Surface-engineered dendrimers: a solution for toxicity issues. *Journal of Biomaterials Science, Polymer Edition*, *20*(2): 141–166.

Mishra, V. and Jain, N.K., 2014. Acetazolamide encapsulated dendritic nano-architectures for effective glaucoma management in rabbits. *International Journal of Pharmaceutics*, *461*(1–2): 380–390.

Mishra, V. and Kesharwani, P., 2016. Dendrimer technologies for brain tumour. *Drug Discovery Today*, *21*(5): 766–778.

Mishra, V., Yadav, N., Saraogi, G.K., Tambuwala, M.M. and Giri, N., 2019. Dendrimer based nanoarchitectures in diabetes management: An overview. *Current Pharmaceutical Design*, *25*(23): 2569–2583.

Mourey, T.H., Turner, S.R., Rubinstein, M., Fréchet, J.M.J., Hawker, C.J. and Wooley, K.L., 1992. Unique behavior of dendritic macromolecules: intrinsic viscosity of polyether dendrimers. *Macromolecules*, *25*(9): 2401–2406.

Munir, M., Hanif, M. and Ranjha, N.M., 2016. Dendrimers and their applications: a review article. *Pakistan Journal of Pharmaceutical Research*, 2(1): 55–66.

Nam, J.P., Nam, K., Jung, S., Nah, J.W. and Kim, S.W., 2015. Evaluation of dendrimer type bio-reducible polymer as a siRNA delivery carrier for cancer therapy. *Journal of Controlled Release*, 209: 179–185.

Opitz, A.W., Czymmek, K.J., Wickstrom, E. and Wagner, N.J., 2013. Uptake, efflux, and mass transfer coefficient of fluorescent PAMAM dendrimers into pancreatic cancer cells. *Biochimica et Biophysica Acta-Biomembranes*, 1828(2): 294–301.

Patil, A., Mishra, V., Thakur, S., Riyaz, B., Kaur, A., Khursheed, R., Patil, K. and Sathe, B., 2019. Nanotechnology derived nanotools in biomedical perspectives: An update. *Current Nanoscience*, 15(2): 137–146.

Pavlov, G.M., Errington, N., Harding, S.E., Korneeva, E.V. and Roy, R., 2001. Dilute solution properties of lactosylated polyamidoamine dendrimers and their structural characteristics. *Polymer*, 42(8): 3671–3678.

Pourianazar, N.T. and Gunduz, U., 2016. CpG oligodeoxynucleotide-loaded PAMAM dendrimer-coated magnetic nanoparticles promote apoptosis in breast cancer cells. *Biomedicine & Pharmacotherapy*, 78: 81–91.

Prosa, T.J., Bauer, B.J., Amis, E.J., Tomalia, D.A. and Scherrenberg, R., 1997. A SAXS study of the internal structure of dendritic polymer systems. *Journal of Polymer Science Part B: Polymer Physics*, 35(17): 2913–2924.

Rajani, C., Borisa, P., Karanwad, T., Borade, Y., Patel, V., Rajpoot, K. and Tekade, R.K., 2020. Cancer-targeted chemotherapy: Emerging role of the folate anchored dendrimer as drug delivery nanocarrier. In *Pharmaceutical Applications of Dendrimers: Micro and Nano Technologies*, ed Chauhan, A. and Kulhari, H. Elsevier, Netherlands, 151–198.

Rietveld, I.B. and Smit, J.A., 1999. Colligative and viscosity properties of poly (propylene imine) dendrimers in methanol. *Macromolecules*, 32(14): 4608–4614.

Sakthivel, T., Toth, I. and Florence, A.T., 1998. Synthesis and physicochemical properties of lipophilic polyamide dendrimers. *Pharmaceutical Research*, 15(5): 776–782.

Saluja, V., Mankoo, A., Saraogi, G.K., Tambuwala, M.M. and Mishra, V., 2019. Smart dendrimers: Synergizing the targeting of anticancer bioactives. *Journal of Drug Delivery Science and Technology*, 52: 15–26.

Santos, I., Ramos, C., Mendes, C., Sequeira, C.O., Tomé, C.S., Fernandes, D.G., Mota, P., Pires, R.F., Urso, D., Hipólito, A. and Antunes, A.M., 2019. Targeting glutathione and cystathionine β-synthase in ovarian cancer treatment by selenium–chrysinpolyurea dendrimer nanoformulation. *Nutrients*, 11(10): 2523.

Sharma, A., Gautam, S.P. and Gupta, A.K., 2011. Surface modified dendrimers: Synthesis and characterization for cancer targeted drug delivery. *Bioorganic & Medicinal Chemistry*, 19(11): 3341–3346.

Shinde, G.V., Bangale, G.S., Umalkar, D.G., Rathinaraj, B.S., Yadav, C.S. and Yadav, P., 2010. Dendrimers. *J Pharm Biomed Sci*, 3(03): 1–8.

Shinde, M.S., Bhalerao, M.B., Thakre, S., Franklin, J. and Jain, A., 2014. Dendrimers-An excellent polymer for drug delivery system. *Asian Journal of Pharmaceutical Research and Development*, 2(2): 24–34.

Tabakovic, I., Miller, L.L., Duan, R.G., Tully, D.C. and Tomalia, D.A., 1997. Dendrimers peripherally modified with anion radicals that form π-dimers and π-stacks. *Chemistry of Materials*, 9(3): 736–745.

Topp, A., Bauer, B.J., Klimash, J.W., Spindler, R., Tomalia, D.A. and Amis, E.J., 1999. Probing the location of the terminal groups of dendrimers in dilute solution. *Macromolecules*, 32(21): 7226–7231.

Torres-Pérez, S.A., del Pilar Ramos-Godínez, M. and Ramón-Gallegos, E., 2020. Glycosylated one-step PAMAM dendrimers loaded with methotrexate for target therapy in breast cancer cells MDA-MB-231. *Journal of Drug Delivery Science and Technology*, 58: 101769.

Trahasch, B., Stühn, B., Frey, H. and Lorenz, K., 1999. Dielectric relaxation in carbosilane dendrimers with perfluorinated end groups. *Macromolecules*, 32(6): 1962–1966.

Tran, S., DeGiovanni, P.J., Piel, B. and Rai, P., 2017. Cancer nanomedicine: a review of recent success in drug delivery. *Clinical and Translational Medicine*, 6(1): 44.

Tunki, L., Kulhari, H., Sistla, R. and Pooja, D., 2020. Dendrimer-based targeted drug delivery. *Pharmaceutical Applications of Dendrimers*, 107–129.

Wilken, R. and Adams, J., 1997. End-group dynamics of fluorescently labeled dendrimers. *Macromolecular Rapid Communications*, 18(8): 659–665.

Wooley, K.L., Hawker, C.J. and Frechet, J.M., 1993. Unsymmetrical three-dimensional macromolecules: preparation and characterization of strongly dipolar dendritic macromolecules. *Journal of the American Chemical Society*, 115(24): 11496–11505.

World Health Organization, 2020. WHO report on cancer: setting priorities, investing wisely and providing care for all. https://www.who.int/publications/i/item/who-report-on-cancer-setting-priorities-investing-wisely-and-providing-care-for-all (Accessed on June 18, 2020).

Xiong, Z., Alves, C.S., Wang, J., Li, A., Liu, J., Shen, M., Rodrigues, J., Tomás, H. and Shi, X., 2019. Zwitterion-functionalized dendrimer-entrapped gold nanoparticles for serum-enhanced gene delivery to inhibit cancer cell metastasis. *Acta Biomaterialia, 99*: 320–329.

Yousefi, M., Narmani, A. and Jafari, S.M., 2020. Dendrimers as efficient nanocarriers for the protection and delivery of bioactive phytochemicals. *Advances in Colloid and Interface Science, 278*: 102125.

Zeng, F., Zimmerman, S.C., Kolotuchin, S.V., Reichert, D.E. and Ma, Y., 2002. Supramolecular polymer chemistry: design, synthesis, characterization, and kinetics, thermodynamics, and fidelity of formation of self-assembled dendrimers. *Tetrahedron, 58*(4): 825–843.

Zhu, J., Fu, F., Xiong, Z., Shen, M. and Shi, X., 2015. Dendrimer-entrapped gold nanoparticles modified with RGD peptide and alpha-tocopheryl succinate enable targeted theranostics of cancer cells. *Colloids and Surfaces B: Biointerfaces, 133*: 36–42.

Zhuo, R.X., Du, B. and Lu, Z.R., 1999. In vitro release of 5-fluorouracil with cyclic core dendritic polymer. *Journal of Controlled Release, 57*(3): 249–257.

14

Drug Delivery Potential of Dendrimeric Formulation

Musarrat Husain Warsi, Mohammad Akhlaquer Rahman, Mohammad Yusuf, Abuzer Ali, Abdul Muheem, Saima Amin and Javed Ahmad

CONTENTS

14.1 Introduction

Toxicity, poor solubility and stability are the main issues associated with maximum therapeutic molecules that restrict them for limited clinical applications despite having outstanding potency (Madaan et al. 2014). To overcome these problems, a suitable and effective carrier system skilled with efficient drug delivery is desired. Numerous polymers have been explored as a vehicle for drug delivery (Brannon-Peppas and Blanchette 2004), but inappropriate chemical structure (associated to molecular weight and polydispersity of the polymers) is a key challenge with them. To solve these problems, several investigations have been carried out.

Nanotechnology has emerged as a better approach to overcome these challenges by improving physicochemical and biological characteristics of such therapeutically active molecules (Gradishar and Tjulandin 2005; Ko et al. 2013; Awada et al. 2014). Different nanotechnology centred medicinal products are commercially available, and many are in clinical trials (Northfelt et al. 1998; Gradishar et al. 2009). Nanocarrier-based drug delivery and targeting have been established for promising prospects in order to improve safety and reduce drug-related toxicity. Additionally, in this direction recent advancement was an approach of a multifunctional platform for a single system that should be capable of targeting and delivering of therapeutic as well as diagnostic agents (Madaan et al. 2014). Amongst numerous nanocarriers, dendrimers have attracted great attention of investigators owing to their idiosyncratic physical, chemical and mechanical characteristics (Lee et al. 2005; Svenson and Tomalia 2012; Hsu et al. 2017).

Dendrimers are a well-defined hyper-branched, nanosize range 3D structure (Sampathkumar and Yarema 2007; Abbasi et al. 2014). Dendrimers attracted terrific interest in the field of nanomedicine because of having several functional groups on their exterior which provides versatility of dendrimers. They also provide a vital platform for personalised medicine systems (Jain 2017). First effort was made by Buhleier et al. (1978) to design and synthesise dendrimers (Buhleier et al. 1978). Originally, these were called 'cascade molecules'. Later, a new class of cascade molecules allowing a similar structure but comprising amides was reported by Tomalia et al. (Tomalia 2005). Tomalia et al. named such developed highly branched macro-molecules as 'dendrimers'. The dendrimer term originated from Greek words 'dendros', it means 'tree or branch' and 'meros' meaning 'part' (Madaan et al. 2014; de Brabander-van den Berg and Meijer 1993). Fabrication of dendrimers deals with gradual and careful controlled synthesis with the help of monomers, so they show dual characteristic features of molecular chemistry as well as polymer chemistry (Gottis et al. 2013; Kaur et al. 2016). As compared to conventional polymers, dendrimers show characteristic properties. Furthermore, due to the nanoarchitectures of dendrimers, they have promising applications during nanomedicine research. They have emerged as an effective delivery system for therapeutically active molecules (Madaan et al. 2014). The first dendritic marketed medicinal product was VivaGel® (Buhleier et al. 1978). The typical architecture of dendrimers provides distinct scenarios for multivalent host–guest interfaces (Boris and Rubinstein 1996; Spataro et al. 2010).

14.2 Dendrimer-Based Drug Delivery System

The controlled release of therapeutics with specific distribution within the desired site is highly valuable to eliminate the archetypal weaknesses of traditional delivery of drugs. Therefore, a drug delivery approach must be designed to achieve the site-specific drug delivery. In the last decade, site-specific delivery of therapeutic molecules with enhanced pharmacokinetics and proper biodistribution of the therapeutics at the particular site has been attained with polymeric drug delivery (Allen and Cullis 2004). Dendrimers have been established as an alternative of the polymeric drug delivery system and attained extensive consideration in biological applications because of their certain properties such as high water solubility (Soto-Castro et al. 2012), polyvalency (Patton et al. 2006), biocompatibility (Duncan and Izzo 2005) and accurate molecular weight (Tomalia 2005).

In this sequence, Kitchens et al. developed the cationic polyamidoamine (PAMAM)-NH (G0–G4) dendrimers. They assessed their penetrability as a function of dendrimer generation. They use Caco-2 cell monolayers for the permeability study. They examined different parameters such as transepithelial resistance, C-mannitol transportability and seepage of lactate dehydrogenase. It was found that developed amine terminated PAMAM dendrimers were able to traverse the biological membranes probably by endocytosis and paracellular routes (Kitchens et al. 2005).

Side effects of anticancer drugs are very critical in terms of conventional therapy. These drugs also accumulate in different vital organs such as kidneys, liver and spleen and cause toxicity. Therefore, targeted delivery of such agents is essential to diminish their side effects. The site specific targeted drug

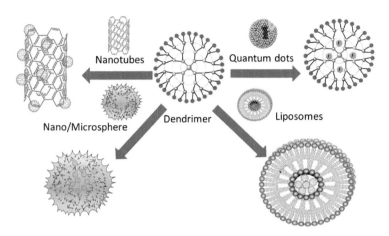

FIGURE 14.1 Schematic illustration of the dendrimer-based nanohybrid system

delivery might be attained by dendrimeric surface modification. In this consequence, different targeting moieties such as folic acid (FA), sugar group peptides and monoclonal antibodies have been employed by modification of the branching units and surface groups of dendrimers (Kesharwani et al. 2014). Targeted delivery of methotrexate was successfully achieved by Patri et al. In this study, FA was conjugated with G5 PAMAM dendrimers in order to achieve site specific delivery of methotrexate. Patri et al., found that the developed conjugated dendrimer showed greater specificity for KB cells overexpressing folate receptors by receptor mediated delivery of drug with slower drug release (Patri et al. 2005). In another attempt, a conjugated system of colchicine was developed with glycodendrimers, which exhibited selective inhibition (20–100 times) of HeLa cells as compared to normal cells (Lagnoux et al. 2005). Impact of surface modification of dendrimers in terms of protein interaction was evaluated by Ciolkowski et al. Poly(propylene imine) (PPI) G4 and G3.5 PAMAM dendrimers were used as a drug vehicle and white lysozyme as a model protein obtained from hen egg. It was observed that by escalating maltose concentration on the PPI dendrimer surface, surface charge of the dendrimers diminished and hence lesser interaction of protein (Ciolkowski et al. 2012).

Dendrimers can be multifunctional oscillating from solubility enhancement to drug targeting. In recent progress towards dendrimers technology, dendrimers have been explored for multifunctional capability by enhancing solubility, dissolution, stability, high drug payload, enhanced GIT permeability, bioavailability and controlled release of drugs in a single formulation. Numerous interrelated approaches are explored concurrently to launch dendrimers as a multifunctional delivery system.

Recently, many researchers have reported duo delivery systems employing dendrimers along with different carriers; viz. microspheres (Fu et al. 2010; Biradar et al. 2011), liposomes (Khopade et al. 2002), NPs (Li et al. 2010a) and CNTs (Qin et al. 2011). As compared to the typical nanocarriers (microsphere, liposomes and NPs), dendrimers are smaller in size. Hence, they can be effectively encapsulated for nanohybrid formation as well as be able to engross on the surface of CNTs. Remarkably, quantum dots (QDs) could be simply lodged within the dendrimer. Figure 14.1 shows the schematic illustration of the dendrimer-based nanohybrid system.

14.3 Dendrimer-Based Nanohybrid System

A nanohybrid system could be an effective approach to pact with the various obstacles related to drug delivery and other biomedical matters. A lot of studies have been reported based on dendrimer-nanohybrid systems in conjunction with other nanocarriers with multipurpose functions in medicine and non-medical fields. In this section, a brief discussion was made based on different developed dendrimer-based nanohybrids.

14.3.1 Hybrid System of Dendrimer-Liposome

Liposomes are one of the popular nanocarrier systems, which are composed of a phospholipid bilayer. Liposomes are nonimmunogenic, biocompatible and biodegradable nanocarriers, and the size of liposomes differs on the basis of selection of lipids during formulation. It has been reported that liposomes are a good choice for targeted delivery (Chandrasekaran and King 2014). Several literature studies have been reported on the development of dendrimers but only limited products based on the dendrimer are available commercially. This might be due to the cytotoxicity issues triggered by the presence of cations on the surface (Parajapati et al. 2016), which confines their clinical applications (Roberts et al. 1996). Different approaches have been employed to diminish the toxicity allied with the dendrimeric cationic charge. The dendrimers based on novel approaches are biodegradable dendrimers (Agrawal et al. 2007), melamine dendrimers (Neerman et al. 2004), poly-L-lysine dendrimers (Jain et al. 2014), polyester dendrimers (Gillies et al. 2005) and triazine dendrimers (Wang et al. 2015). Surface modification of dendrimers is another approach to reduce toxicity such as PEGylation (Bhadra et al. 2003), acetylation (Hu et al. 2011), carbohydrate coating (Cousin and Cloninger 2015) and peptide conjugation (Agashe et al. 2006).

14.3.2 Hybrid System of Dendrimer-Nanoparticles (NPs)

NPs are a very common nanocarrier having a size range of 10–200 nm. Quantum size effects and a higher surface area are the main characteristics of NPs which make them the foremost nanocarriers for effective drug delivery. Initially, gold (Au) and cadmium (Cd) sulphide NPs were developed by using polyethylene glycol (PEG)-grafted PAMAM-NH_2 dendrimers. It was further proposed that star polymers can stabilise polydispersed NPs and could be used as the template. They can also enhance miscibility in diverse organic solvents. Correspondingly, to immobilise albumin protein more competently, dendrimer-grafted NPs were employed. For the same, PAMAM dendrimers were synthesised on the outer surface of customised aminosilane NPs. By increasing dendrimer generation, BSA immobilisation was increased. However, this increase was not exponential due to the large size of BSA and inadequate surface area of NPs. Furthermore, 3.9–7.7 fold higher immobilisation was observed as compared to amino silane modified NPs (Pan et al. 2005).

14.3.3 Hybrid System of Dendrimer-QDs

QDs are a spherical shape (radius in the range of 10–100 A°) semi-conducting material, and their physical size is responsible for different properties of QDs. QDs have been reported for bioimaging (Chan and Nie 1998), fluorescence, readily tunable spectral properties, higher chemical stability, high photostability and narrow as well as broad emission and excitation (Bruchez et al. 1998), as well as in various vivo applications (Larson et al. 2003). It has been reported that the hybrid of dendrimer–QD was capable of reducing the toxicity related to QDs as well as to improve the aqueous solubility profile of QDs and their quantum yield. Dendrimers can be used as a tool to maximise quantum yield by acting as a stabiliser in QD synthesis with excellent photostability. G4-OH, G6-OH and G8-OH PAMAM dendrimers have been employed for encapsulating the Cadmium sulphide (CdS) QDs. It was found that there was reduction in encapsulating size of the dendrimer. The developed nanocomposite was equipped with all the above-stated characteristics (Lemon and Crooks 2000). Moreover, a PAMAM G4 dendrimer hybrid was developed having epidermal growth factor (EGF) labelled with 525 ITKTM(PEG) QDs. The developed system offered imaging modality yellow fluorescent protein (YFP) siRNA as well as site-specific delivery of the vimentin shRNA plasmid. Aptitude of targeting efficiency of the EGF-conjugated dendrimer was arbitrated by epithelial growth factor receptor (EGFR) against NIH3T3 and HN12 cells. An overexpression of this was found in the cell lines. The conjugate system was internalised into endosomes along with the cytoplasm and nuclear membrane in the existence of EGFR. Furthermore, due to the negatively charged cell and positively charged EGF, NPs were energetic in receptor-binding action. The above conjugate system does not trigger activation of proliferation in the EGFR-expressing cells but substantial reduction of vimentin and YFP was observed in HN12 after treating with dendrimer-conjugated shRNA and siRNA, respectively. The fabricated system was valuable for targeted delivery of growth factor (Yuan et al. 2010). Correspondingly, the QD–PAMAM G4-aptamer GBI-10 complex also has been constructed with higher water solubility along with targeted delivery. An improved binding efficiency of Apt-QD nanoprobes was found for U251 glioblastoma cells. GBI-10 was able to identify tenacin-C present on the human U251 glioblastoma cell surface, and the nanocomplex had extreme affinity with the cells (Li et al. 2010b). It also has been investigated that arginine–glycine–aspartic- acid (RGD) peptide can enhance the imaging as well as targeting properties of dendrimer–QD hybrids (Li et al. 2010c).

14.3.4 Hybrid System of Dendrimer-Carbon Nanotubes (CNTs)

CNTs are made up of mainly carbon nanomaterials (sp2 hybridised carbons) and architecturally demarcated as three-dimensional hexagonal sheets. CNTs have been reported as auspicious nanocarriers for numerous biomedical applications. Lipophilic nature, non-dispersibility as well as non-biodegradability and toxicity are the key drawbacks of CNTs, which restricts their clinical values. To overcome the limitation of CNTs, different strategies have been implemented it includes functionalisation and fabrication of conjugated systems (Kesharwani et al. 2012). CNTs have upper hand by having some extraordinary features like biosensing (Joshi et al. 2005) and bio-catalysis. Poor dispersibility of CNTs in solvents, mainly in water, restricts their application. Several literature studies have been published in terms of enhancing the aqueous dispersibility of CNTs. Various approaches have been employed to improve dispersibility like surface modifications (partial oxidation) (Kesharwani et al. 2015) and incorporation of hydrophilic groups or modification with hydrophilic polymers (Bottini et al. 2005). However,, these alterations were not able to make enough hydrophilic CNTs. Conjugation of CNTs with dendrimers was also reported in order to enhance aqueous dispersibility.

14.3.5 Hybrid System of Dendrimer–Hydrogel

Hydrogels have important application in the biomedical field. Gelation characteristics of amphiphilic dendrimers were investigated in order to form hydrogels for enhancing drug encapsulation and to obtain controlled release of therapeutics (Nummelin et al. 2015). Afterwards, the impact of PAMAM dendrimers on the hydrogel properties of erythromycin was explored by Wro' blewska and Winnicka. They observed that the release of erythromycin from dendrimer formulations was rapid and improved and depends on concentration and generation. A typical shear-thinning performance was reported. Additionally, dendrimers demonstrated an increment in the antimicrobial action (Wróblewska and Winnicka 2015). Biomedical utilisations of hydrogels rely upon functional groups, which are available at the surface because dendrimers may manipulate it. Dendrimer-based hydrogels were constructed by structuring dendrimers, designed with propargylamine-determined hyaluronic acid having an azido ring. The investigation on conduct and construction of hydrogels indicated that dendrimers did not impact the rheological habit of hyaluronic acid-based hydrogel and demonstrated as promising thermo-responsive vehicles for different pharmaceutical applications (Seelbach et al. 2014).

14.3.6 Hybrid System of Dendrimer-Clays

Clay minerals, composed of particles with a diameter below 2 μm, are defined as hydrous aluminosilicates that are dominant in the clay-sized fraction of soil. There are two structural features in their crystal structures–tetrahedron and octahedron, differing in number and arrangement of oxygen ions. Organo-clays with incorporated dendrimers find many applications. However, the possibility of their biomedical use seems particularly interesting. Both dendrimers and clay minerals are tested as drug carriers (Massaro et al. 2017). PAMAM dendrimer-functionalised halloysite is a suitable material for effective drug delivery. Such a system is capable of increased adsorption and/or reduced release rate of a given drug compared to pristine clay (Kurczewska et al. 2018). This area of scientific research is not as widely developed as the previously described applications, but it has great potential for development in the nearest future.

14.4 Drug Delivery Application of Dendrimer-Based Formulation

Dendrimers are characterised by the distinctive properties such as monodispersity, internal cavities and flexible surface modification. Besides these properties, other important features which make them a promising drug delivery system include their exceptional cellular uptake, several functionalities and ability to bind with high molecular weight compounds and increased their retention time. Furthermore, nanosize of dendrimer promotes the passive targeting of therapeutics to cancerous tissues through the improved permeation and retention effect (Kumar et al. 2015; Najlah et al. 2007; Saovapakhiran et al. 2009). The increased retention and permeability effect facilitates them for targeted drug delivery of macromolecules. Kitchens et al, in one of his permeability studies on Caco-2 cell using cationic dendrimers like PAMAM-NH_2 (G_0–G_4), recommended that the PAMAM cationic dendrimers with –NH_2 terminal groups may effectively passage via biological membranes almost certainly through both endocytosis and paracellular channels (Kitchens et al. 2005). Patri et al established the targeting attributes of FA enabled cationic G_5-PAMAM dendrimer using anticancer drug 'methotrexate' as therapeutic molecules. A receptor-mediated drug carrier system exhibiting elevated selectivity for KB cells with sustain release of drug was investigated (Patri et al. 2005). Due to specific nonstructural characteristics and monitored particle size, dendrimers have turned into fascinating bits and pieces for biochemical applications. In recent times, consolidating the distinguishing properties of dendrimer with various NPs like magnetic nanoparticles (MNPs) has been worked out to accomplish enhanced therapeutics and biomedical applications (Taheri-Kafrani et al. 2017).

Dendrimers have shown promise in various medical applications including diagnosis, drug delivery, transfection and therapy. A dendrimer-based cancer treatment, DTXSPL8783, is being investigated in clinical trials; a phase 1 study of this agent is under way in patients with advanced cancer (Caster et al. 2017). A dendrimeric antiviral/antibiotic compound, Vivagel (Starpharma), is in phase 3 clinical trials for bacterial vaginosis (BV). This unique nanodrug incorporates naphthalene disulphate groups on the surface of dendrimers. Phase 2 data have indicated high rates of clinical and pathologic cure of BV, as evidenced by symptomatic improvement and clear laboratory results, respectively. However, phase 3 data have been equivocal, with high rates of symptomatic improvement but lower rates of clinical laboratory cure being observed. Vivagel has also exhibited potent *in vitro* activity against HIV and herpes simplex virus. Phase 1 studies have indicated that vaginal use of this nanoformulation is well tolerated and that antiviral activity is retained by cervicovaginal fluids in most patients up to 24 hours after administration (Price et al. 2011). Vivagel is available in Australia as a condom lubricant. A list of dendrimer-based products that were approved and currently under clinical trial is summarised in Table 14.1.

TABLE 14.1

List of dendrimer-based products currently approved or under clinical trial

Category	Brand Name	Application	Status	Proprietor
Diagnostic agent	Stratus® CS acute care diagnostic system	Measurement of cardiac biomarkers	Marketed	Siemens Healthcare Diagnostics
Transfection agent	Superfect®	Cell transfection	Marketed	Qiagen
Transfection agent	Priostar®	Cell transfection	Marketed	EMD
Contraceptive	VivaGel® condoms	Prevention of sexually transmitted infection	Marketed in Australia and Canada	Starpharma
Therapeutic agent	VivaGel®	Antiviral agent	In Phase I clinical study	Starpharma
Therapeutic agent	VivaGel® BV	Treatment and symptomatic relief of bacterial vaginosis	Approved for marketing in Europe	Starpharma
Therapeutic agent	VivaGel® BV	Prevention of recurrent bacterial Vaginosis	Phase III clinical study completed	Starpharma
Drug-delivery system	DEP™ docetaxel	Anticancer agent	In Phase I clinical study	—

14.4.1 Transdermal Drug Delivery System

Transdermal delivery systems of drugs have been helpful to surmount the gastrointestinal and renal adverse effect of various non-steroidal anti-inflammatory drugs. Besides that, they show an extended drug release effect by controlling blood levels over a longer period. Application of dendrimers in the transdermal drug delivery system has been studied by various researchers by incorporating the drug into suitable vehicles. Chauhan and co-authors in one of the studies established the transdermal transport by employing indomethacin-loaded cationic PAMAM dendrimers (Kawaguchi et al. 1995). Manikkath et al. demonstrated the collective impacts of dendrimers associated with arginine-based peptide and moderate frequency ultrasound on the transdermal penetration of ketoprofen. The amalgamation of ultrasound and peptide dendrimers showed a collective effect and showed a high drug plasma concentration compared to passive delivery in an animal model. Administration of ketoprofen through the transdermal route using dendrimers reported the same assimilation and plasma drug concentration as was with oral delivery (Manikkath et al. 2017).

14.4.2 Oral Drug Delivery System

Oral delivery is most applicable among all others because of ease of administration and the patient compliance. Due to lipophilicity and low permeability, anticancer drugs restrict their ingestion via the oral route, still they are commonly used as they are economic and also promote the use of more chronic treatment regimens (Csaba et al. 2006; Malingré et al. 2001). To overcome the drawbacks associated with these drugs, promising results were found in various studies when dendrimers were used as the drug delivery vehicles. Jevprasesphant et al studied the permeation of PAMAM dendrimers and surface modified PAMAM dendrimers across the Caco-2 cell monolayers and accomplished that PAMAM dendrimers and lauroyl conjugated PAMAM dendrimers could proficiently pass through epithelial pathways via transcellular and paracellular pathways (Gothwal et al. 2015). In another study, modified PAMAM dendrimers with propranolol were studied for transport through Caco-2 cell line. It was showed that propranolol-PAMAM dendrimers might diminish the influence of P-glycoprotein (P-gp) transport on the absorption of propranolol in intestine. Conclusion referred that P-gp efflux transport can be avoided by dendrimers and hence improved the oral drug administration (D'emanuele et al. 2004). Najlah et al. formulated naproxen dendrimer conjugates using two different linkers (diethylene glycol and lactate ester) and examined the transepithelial permeability of the conjugates through oral delivery. The authors also investigated the stability of these conjugates in 80% human plasma and 50% liver homogenate. The results demonstrated that both these linker conjugated dendrimers showed considerably significant impact on the stability of the developed system. Compared to diethylene glycol like conjugates, lactate ester linked conjugates showed more stability in blood plasma and displayed reduced hydrolysis in liver homogenate, whereas the diethylene glycol linker-based conjugate showed high chemical stability with rapid release of therapeutic molecules in both plasma and liver homogenate. The conclusion established that the dendritic conjugate of naproxen can augment the oral bioavailability and lactate ester linker conjugate may act as a potential carrier for monitored release of drug (Najlah et al. 2007).

14.4.3 Ocular Drug Delivery System

The topical usage of active therapeutics for the therapy of ocular diseases is the main route. In general, topical ocular delivery of drugs has very low bioavailability due to the loss of drugs by tear turnover, blinking of eye and nasolacrimal drainage of fluid. Prolonged retention and enhanced corneal permeation are the major factors to improve ocular availability of drugs (Nanjwade et al. 2009; D'emanuele et al. 2004). Dendrimers have been effectively employed for the ocular administration of drugs. PAMAM dendrimers have been successfully applied with hydroxyl or carboxyl end functionalities for ophthalmic administration of drugs. Such dendrimers enhance the corneal contact time of pilocarpine (Vandamme and Brobeck 2005). Ocular absorption of dexamethasone (DEX) was found to be improved by conjugating it with PAMAM dendrimers. MTT assay demonstrated that all the groups showed in cell viability as compared to DEX solution. An increase transcorneal permeation was also reported after applying PAMAM-DEX (Yavuz et al. 2015). Different studies have been performed in order to assess the proficiency of dendrimers as a carrier of therapeutics (Tolia and Choi 2008; Oliveira et al. 2010).

14.4.4 Drug Delivery and Targeting to Bone

Dendrimers in bone targeting have been also used as potential delivery systems. Yamashita and coworkers developed PEG-conjugated cationic G3-PAMAM dendrimers for bone targeting in order to treat bone diseases. PAMAM backbones were conjugated with different carboxylic groups (succinic acid, aspartic acid, aconitic acid and glutamic acid). Four different types of PAMAM dendrimers were developed. PEG was used for PEGylation, and surface modified PEGylated carboxylic acid-PAMAM dendrimers were synthesised. An intra-bone delivery study revealed that fluorescein isothiocyanate-labelled PEGylated (5)-Aspartic acid (Asp)-PAMAM amassed in large quantity on the bone surfaces (quiescent and eroded). These surfaces are accountable for bone disorders such as osteoporosis and rheumatoid arthritis (Yamashita et al. 2017).

14.5 Therapeutic and Biomedical Applications of Dendrimer-Based Formulation

Dendrimers have various points of interest as they may be utilised as vehicles or platforms for diagnosis and treatment. Moreover, dendrimers preclude the drug instability as being monomolecular polymer micelles. Then again, because of flexible size (varied according to generation) of dendrimers, they have vast application in biomedical engineering.

14.5.1 Cancer Therapy

Due to the well-defined architecture (multivalences) of dendrimers, drugs can be attached covalently to their periphery. This property of dendrimers was very fruitful to develop it as a potential carrier for anticancer agents.

Zhang et al. have reported the development of pH-sensitive multifunctional doxorubicin (DOX) conjugated PAMAM dendrimers as a unique platform for targeted cancer chemotherapy (Zhang et al. 2018). One of the potent anticancer drugs, paclitaxel, has poor aqueous solubility. After conjugating anticancer drug (paclitaxel) with PAMAM G4 dendrimer having hydroxyl end groups, an enhanced solubility and cytotoxicity were observed. Tenfold enhanced cytotoxicity was reported in terms of conjugated paclitaxel than the non-conjugated anti-cancer drug (paclitaxel) (Jones et al. 2012; Khandare et al. 2006). In another study, cancerous tissue targeting efficiency of anti-Human growth factor receptor-2 (HER2) monoclonal antibodies was reported. HER2 specific (monoclonal antibody) was conjugated with PAMAM G5 dendrimer. The developed dendrimer–antibody conjugate demonstrated fast and effective cellular internalisation devoid of discrepancies in selectivity of targeting during internalisation process with free antibody. Animal studies data also revealed the targeting of HER2-expressing tumours by the conjugated monoclonal antibody (Shukla et al. 2006).

Chittasupho and co-authors developed a doxycycline conjugated PAMAM dendrimers. The effects of doxycycline conjugated PAMAM were investigated on cytotoxicity, cellular binding and migration of BT-549-Luc and T47D breast cancer cells. The drug conjugated dendrimers were modified with LFC131 peptide. It has capability to distinguish CXCR4, which is expressed on the surface of breast tumour cells. The developed formulation (LFC131-DOX-D4) displayed considerably better cellular toxicity in vitro, as related to non-targeting dendrimers (Chittasupho et al. 2017). In another study four different poloxamer (F108, F68, P123 and F127) were conjugated with PAMAM G4, and 5-fluorouracil was taken as a model drug. Higher drug payload was observed with poloxamer P123 conjugated PAMAM G4 compared with other poloxamer grades. The optimised drug-loaded dendrimer exhibited higher antiproliferative action against MCF-7 breast cancer cell line (Nguyen et al. 2017). Dendrimers as a carrier for herbal drugs have been established by numerous investigators. A natural alkaloid, Berberine, showing promising anticancer action is less studied because of its low pharmacokinetic performance. Gupta et al. developed berberine conjugated G4 PAMAM dendrimer for effective delivery of berberine. The outcomes of MTT assay showed substantially better antiproliferative activity against MDA-MB-468 and MCF-7 breast cancer cells with PAMAM berberine conjugate. Additionally, the developed conjugate formulation was reported biocompatible and safe. *In vivo* studies exhibited remarkable improvement in pharmacokinetic parameters (area under the curve and half-life) (Gupta et al. 2017). A nucleoside analogue, cytosine arabinoside (Ara-C), is one of the naturally active agents having promising anti-tumour activity, but its use is limited due to insufficient uptake and accretion of therapeutic molecules within the tumours. Szulc et al. developed maltose-modified glycodendrimers (PPI-m OS) containing Ara-C triphosphate (Ara-CTP) for the treatment of leukaemia. The developed Ara-CTP dendrimers conjugate showed enhanced cytotoxic activity against 1301 leukemic cells equated to pure Ara-C and Ara-CTP. Authors further reported that such improved uptake as well as cytotoxicity of Ara-CTP-dendrimers with blocked human equilibrative nucleoside transporter (hENT1) might be applied as potential therapy for acute lymphoblastic leukemia cells (Szulc et al. 2016).

A dendritic polyester system based on monomers 2,2-bis(hydroxymethyl)propionic acid attached to DOX or hydroxyl-terminated generation 4 PAMAM in conjugation with PTX through a union with succinic acid has shown great anticancer activity against ovarian cancer cells (Castro et al. 2018). However, at present the dendrimers used as drug-carriers do not satisfactorily meet the necessary characteristic of an ideal dendrimer for targeted drug delivery. However, the development and study of new dendrimers drug-carriers continues to be an important tool in the cancer therapy (Prasad and Srivastava 2020).

14.5.2 Gene, Enzyme and Protein Carrier System

Vectors are responsible for the transfer of genes into the nucleus. Various studies are being carried out by utilising dendrimers as a carrier or vector without harming or disabling the nucleic acid such as DNA. A well-known dendrimer, PAMAM dendrimer has been explored as vectors for gene therapy. The amino groups present on the surface of PAMAM react and form complexes with the phosphate groups of nucleic acids. Furthermore, the complex dendrimers are able to transport a higher quantity of biological material such as DNA, RNA, protein and enzymes, contrasted to that of viruses (Klajnert and Bryszewska 2001).

Nitrogen core dendrimer containing polypropyl ether imine (PETIM) has been reported by Lakshminarayanan et al., which have complexation with DNA and delivery of gene abilities. Quantitative luciferase assay showed hundred fold gene transfection related to poly(ethylene imine) branched polymer comprising equal figure of cationic sites as the dendrimer (Lakshminarayanan et al. 2013). Lungs delivery of siRNA is very challenging resulting very lower bioavailability. Additionally, the formulation development, stability and maintaining the activity of free siRNA after pulmonary administration by inhalation is a difficult task. In vitro transfection efficacy of a triphenylphosphonium altered PAMAM G4 dendrimer (G4NH2-TPP) was studied by Bielski E et al. They developed dendriplex conjugates comprising 12 TPP molecules complexed with siRNA on the surface. The highest in vitro gene knockdown proficiency was observed. Furthermore, 12 TPP-dendriplex-loaded mannitol microparticles were formulated to establish the efficacy of TPP-dendriplexes for pulmonary applications. A deep lung deposition of dendriplexes was found deprived of any impact on the *in-vitro* gene giveaway effectiveness of the siRNA (Bielski et al. 2017). Wang X et al. studied the performance of large therapeutic molecules like proteins by incorporating into dendrimers. They developed a series of conjugates (in various molar ratios) by the active ester method containing therapeutically active protein like streptokinase (SK) and PAMAM G3.5 dendrimers. It was observed that the conjugate having equimolar (1:1) ratio exhibited significant enzymatic activity. However, SK-PAMAM conjugates with high molar ratios (1:20 and 1:10 of PAMAM: SK) displayed lesser enzymatic activities but prolonged thrombolytic action in plasma. Therefore, it was concluded that based on the molar ratio of protein and PAMAM, enzymatic activity of therapeutic molecules could be altered as per need and thus employs the possible application of dendrimers (Wang et al. 2007).

14.5.3 Efficient Delivery of Drug to Brain

An increasing number of hybrid nanocarriers based on dendrimers have popped up for brain drug delivery to fathom and resolve obstacles in drug delivery and other biomedical issues such as side effects, non-negligible biotoxicity, constrained penetration and inadequate drug loading capacity. Recently, dendrimer-dationised-albumin (D-Alb) has been synthesised following the carboxyl activation technique (Muniswamy et al. 2019). The synthesised D-Alb was then coated on DOX-loaded poly(lactic-co-glycolic acid) (PLGA) NPs (D-Alb@NP-DOX) to produce a neoteric hybrid nanoformulation for the treatment of brain tumours. It is significant to note that D-Alb@NP-DOX provided exceptional antitumour activity of DOX in glioblastoma cells while extensively improved its blood–brain barrier permeability. The dendrimer was also hybridised with QDs to improve its water solubility and quantum yield and to turn downsize its cytotoxicity effect. Aiming at efficiently stipulating the accurate dopamine concentration in a customizable manner for assessing Parkinson's disease, a glass surface was modified on a QD-encapsulated dendrimer, forming a hybrid biosensor to evaluate the dopamine concentration. Such chemically modified dendrimer-QDs are also good at identifying and tracking neural stem cells as they move around (Zhu et al. 2018).

NPs of size (>100 nm) are characterised with weak tumour penetration but encouraging pharmacokinetics, while NPs of small size (<20 nm) result in poor tumour retention but strong tumour penetration. Tumour microenvironment-responsive dendrimer-gelatin hybrid NPs or multistage-responsive hybrid NPs based on dendrimers may favour the NP accumulation in brain tumours, may release small dendrimer NPs to enhance drug penetration in tissue and may improve brain tumour treatment (Hu et al. 2018a, 2018b). Inflammation involves immune cells, blood vessels and molecular mediators directed against detrimental stimuli, so biomimetic NPs that mimic immune cells could help deliver drugs to these inflammatory sites specifically to the brain (Jin et al. 2018). Covering dendrimer-based NPs with immune cell membranes may constrain upcoming expansion for improving brain drug delivery (Li et al. 2018). Hence, designing a single hybrid nanosystem using different components may result in the generation of multifunctional NPs with perfect structural and biological features. Such NPs not only acquire advantageous properties of all the bulk materials but can also be enhanced in conditions of structural and functional moieties to put forward carriers with enhanced brain-targeting efficacy.

14.5.4 Photodynamic Therapy (PDT)

Dendrimers have potential applications in PDT. Dendrimers have been explored as a photosensitiser. A study reported development of polymeric micelles in complexation with dendrimer phthalocyanine. They reported enhanced photodynamic effects of the developed system (Herlambang et al. 2011). Dendrimers also have been applied for effective delivery of 5-aminolevulinic acid (a natural photosensitiser protoporphyrin 1X). An increased deposition of porphyrin was found in cells which consequently ensued in toxicity (Battah et al. 2006; Di Venosa et al. 2006). Furthermore, a photosensitive drug delivery system was synthesised for PDT consisting of G3 PAMAM grafted porous silica NPs. A conjugation of gluconic acid was made to this developed system. The PAMAM-functionalised outer surface consists of numerous amino groups enabled high loading of aluminium phthalocyanine tetrasulphonate. The entire system is irradiated in combination of light established active generation of singlet oxygen, which triggered noteworthy mutilation to the cancer cells (Tao et al. 2013).

14.5.5 Boron Neutron Capture Remedy

Treatment of cancers by boron therapy is a promising therapy based on the boron capture reaction (Barth et al. 1992). The PAMAM dendrimers have been employed in neutron capture therapy for the effective delivery of anticancer agents. The efficiency of a boronated monoclonal antibody was studied for an intracerebral glioma, after applying the boron neutron capture therapy (Barth et al. 2004; Yang et al. 2009). In this study, functionalised G5 PAMAM dendrimers were synthesised and conjugated with an epidermal growth factor receptor (EGFR) inhibitor, Cetuximab. These starburst dendrimers carried approximately 1100 boron atoms. In vivo study data revealed 10 fold greater deposition of conjugates in brain tumour cells compared to healthy brain cells.

14.5.5.1 Tissue Regeneration

Tissue engineering has emerged as a new strategy mainly in terms of healing process and organ transplantations over the past few periods (Langer 1993; Langer and Tirrell 2004). This deals with restoration of indigenous tissue by augmenting the usual restorative mechanism of the body or the establishment of whole organs for replacement. The primary phase in creating tissue structure is the proper choice of a suitable scaffold system. Scaffolds might be simple (a 2D surface), on which cells can grow, or urbane scaffolds (capture more cell categories in 3D) (Lutolf and Hubbell 2005). Generally, scaffolds are porous in structures which helps them in diffusion of nutrients. The main objective of tissue engineering is to restore innate extracellular matrix (ECM) and ultimately swap the scaffold. Hence, the scaffold should be degraded at a level that harmonises the biosynthesis of a new-fangled ECM. Polymer-based scaffolds can be categorised into natural and synthetic types based on the compositions. Natural scaffolds are mainly synthesised from carbohydrates, proteins or glycoproteins. Collagen is most favoured protein for the creation of scaffolds (Glowacki and Mizuno 2008). Fibrin, due to its capability to gather into mesh-like networks, is also employed for scaffold construction (Ahmed et al. 2008). Hyaluronic acid (Nettles et al. 2004) and chondroitin sulphate (Li et al. 2004) are crucial fundamental materials and extensively engaged in tissue engineering scaffolds. Due to their aptitude to form hydrated networks, different carbohydrates such as chitosan (Laurencin et al. 2008), alginate (Augst et al. 2006) and dextran (Bajgai et al. 2008) are also applied as scaffold stuffs. There are various synthetic polymers, such as PEG, poly(glycolic acid) (PGA), poly(lactic acid) and poly(caprolactone) which have been utilised broadly as scaffold materials (Ifkovits and Burdick 2007; Barnes et al. 2007).

14.5.6 Diagnostic Application

Dendrimers have been utilised as a carrier to carry different varieties of small bioactive molecules for diagnosis and therapy. The interior cavities and amphiphilic characteristic of dendrimers provide them a useful tool to encapsulate hydrophobic as well as hydrophilic drugs based on components of dendrimers. Furthermore, dendrimers as being monomolecular polymer micelles provide the stability of drug formulations. Another important advantage of dendrimers is having the linear correlation with generation size. Hence, they can be designated for particular biomedical purposes with appropriate size. For instance, dendrimers about 5 nm size are employed as magnetic resonance imaging (MRI) contrast agents for the diagnosis of lymphatic systems (Kobayashi and Choyke 2006). The multivalent, highly branched characteristics of dendrimers give itself ideal candidateship for employing in tissue engineering as well as in surface chemistry and act as key elements in scaffolds that simulate the extracellular matrices.

14.5.6.1 MRI

MRI is now widely used for diagnosis of different disorders because of its property to provide high-quality 3D images without utilising the harmful ionising radiation. In clinical applications, several contrast agents are known that are superparamagnetic iron oxide particles, gadolinium chelates and hepatobiliary contrast agents. The prime set of MRI contrast agents are gadolinium chelates and are believed to be reliable and safe. Low toxicity, biocompatibility and high relaxivities are the key features of MRI contrast agents. It was observed that MRI contrast agents with low molecular weight diffuse swiftly from blood vessels and reach into the interstitial domain followed by their fast excretion from the body (Jang et al. 2009). Paramagnetic metal chelates [Gd(III)-N,N′,N″,N‴-tetracarboxymethyl-1,4,7,10-tetraazacyclododecane (Gd(III)-DOTA), Gd(III)-diethylenetriamine pentaacetic acid (Gd(III)-DTPA)] and their derivatives are employed as contrast agents for MRI. These metallic chelates have the ability to increase the relaxation kinetics of adjacent water protons (Hay et al. 2004). Nevertheless, the disadvantages of such low molecular weight contrast agents are little circulation times inside the body and ineffective prejudice between diseased and regular tissues. Later, to enhance image contrast enrichment macromolecular Gd(III) complexes was developed. It was achieved by conjugation of Gd(III) chelates to polymers, containing poly(amino acids), polysaccharides, and proteins. These macromolecular mediators have been established excellent advancement in contrast for the imaging of blood pool and tumours in animal models. Inappropriately, having an improved imaging, macromolecular agents have limited clinical application because of their sluggish excretion which consequences their accumulation inside the body. It was also reported that prolong residence of MRI agents augments the hazard of toxicity associated with release of Gd(III) ions due to the metabolism of such agents (Caravan et al. 1999; Franano et al. 1995). Wiener et al. established a newer class of MRI contrast agents (Gd(III)-DTPA-based

PAMAM dendrimers). It has characteristics of bulky proton relaxation enrichments with high molecular relaxivities. These were categorised as sixth generation PAMAM dendrimers, which own 192 sensitive terminal amines and can be conjugated via a thiourea linkage with the chelating ligand 2-(4-isothiocy-anato-benzyl)-6-methyl-(DTPA). They found six fold higher relaxivity in comparison to the free Gd(III)-DTPA complex. *In vivo* studies performed using rabbits model revealed brilliant MRI images of blood vessels with prolong circulation times (>100 min) after i.v. injection of Gd(III)-DTPA-based PAMAM dendrimers (Weiner et al. 1994). Furthermore, Kobayashi et al. developed small dendrimer-based MRI contrast agents. They examined the connection between relaxivity and generation of dendrimer by applying Gd(III)-DTPA based PAMAM dendrimers. They reported that there was increase in relaxivities along with the generation of dendrimer, but beyond seventh generation no further significant improvement in relaxivities (Kobayashi et al. 2003).

14.5.7 Theranostic Application

The theranostic application has attracted great attention in recent years due to the advantage of offering the concurrent diagnosis and treatment for certain diseases. This is a proposal which assimilates therapeutic molecules with an imaging agent and collectively called theranostic strategies (Figure 14.2). Generally, it becomes difficult to observe the distribution and function of the administered drug molecule as well as the exposure of targeted tissue. The theranostic devices can concurrently follow and treat the diseases. It have been critically studied in the earlier years (McCarthy and Weissleder 2008). The core of the dendrimer is unoccupied for the drug encapsulation and the functional groups present peripherally are accessible for the alteration of therapeutic molecules either via covalent bonds or by the formation of complex.

Recently, remarkable interest has been shown in the fabrication of dendrimers ranging from imaging to drug delivery platforms (Chen et al. 2013; Zhu and Shi 2013; Sk et al. 2013) (Soundararajan et al. 2009). Some other investigators have presented the construction of pH-responsive multifunctional dendrimer-based theranostic nanosystem (in which Gold NPs conjugated with DOX were entrapped in dendrimer) to be utilised for simultaneous chemotherapy and computed tomography (CT) imaging of various types of cancer cells (Zhu et al. 2018). The polyhedral oligomeric silsesquioxane (POSS) core of the dendrimers have been developed for biomedical applications. The POSS core of the dendrimer has ensured structural advantages over traditional dendrimers. The developed POSS core dendrimers have a large content of functional groups on their surface and definite sizes in nanometre range with compact globular constructions. Due to presence of reactive termini (eight in number) of the POSS core, surface functional groups are available in abundance which helps in binding of hydrophobic antineoplastic agents and imaging agents together with the targeting agents (Mintzer and Grinstaff 2011). Fluorescence sensing technology has a wide range of applications in the field of diagnostics and the biological sciences (Giepmans et al. 2006; Lakowicz 2013). Majoros and co-workers developed a multifunctional PAMAM dendrimer as a theranostic device. It contains an anticancer agent (Methotrexate), fluorescent agent (fluorescence) and

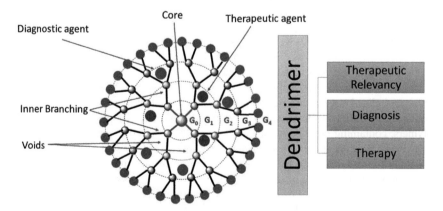

FIGURE 14.2 Schematic illustration of dendrimeric formulation as theranostics.

site specific targeting agent (folic acid) (Majoros et al. 2005). In this consequence, He et al. developed a G5 PAMAM dendrimer conjugated with polyethylene glycol, arginyl-glycyl-aspartic acid peptide and FI. The developed multifunctional G5 PAMAM dendrimer was employed for the effective delivery of DOX to integrin αvβ3-overexpressing glioblastoma U87MG cancer cells. It was found that a stable encapsulation takes place between the inner lipophilic spaces of the dendrimers. The formed system showed prolonged release with particular internalisation inside the cells through receptor-mediated endocytosis (He et al. 2015).

It has been reported that for T2-weighted MRI super paramagnetic iron oxide NPs (SPIONs) are an appropriate contrast agent (Thorek et al. 2006; Lee et al. 2009). The potential application of SPIONs in the delivery of drug and gene has been investigated (Morishita et al. 2005). Chang et al developed a hybrid nanomaterial composite of a pH-responsive DOX conjugated PEGylated dendrimer (G2.5)/DOX complexed with SPIONs. The release profile for the developed hybrid system was established at different pH values. The developed system demonstrated enhanced release of DOX at the pH 5.2 compared to pH 7.4. It was employed as an MRI contrast agent for the diagnosis of cancer (Chang et al. 2011).

In another study, a synchronised system was developed consisting of siRNA and MRI contrast agents (SPIO) employing dendrimers for specific delivery to cancer cells. SPIONs were combined with siRNA and a G5 PPI dendrimer. Furthermore, the developed complexes were altered with targeting agent by PEG carrying LHRH peptide to offer a tumour-specific targeting moiety and hinder non-specific interactions. This hybrid dendrimeric system containing PEG-LHRH has improved internalisation in the cancer cells and augmented the suppression of the targeted gene *in vitro*. Furthermore, the hybrid NPs could be exploited for the delivery of the anticancer agent 'cisplatin', which displayed improved *in vivo* anticancer activity (Taratula et al. 2011). The researchers also demonstrated the synthesis of lactobionic acid (LA)-modified Au DENPs. It was found that the developed system was stable at different pH and temperatures. It was employed for CT imaging of human hepatocellular carcinoma and has been established that the multifunctional Au DENPs were safe to normal cells, but cytotoxicity was observed to the targeted hepatocellular carcinoma cells (Liu et al. 2014).

In a study, conjugation was made by FI, PEGylated α-TOS and PEGylated FA with G5 amine-terminated PAMAM dendrimer. It was utilised as a prototype for the Au DENPs development. Further, adjustment of the Au DENPs with FA empowered it for proficient targeting of tumours by the developed nanodevice. It successfully targeted the CT imaging of tumours *in vitro* and the induced *in vivo* tumour model. The developed hybrid dendrimeric system demonstrated inhibition of targeted tumours in the mouse model (Zhu et al. 2014).

14.6 Future Prospects

Hybrid nanomaterials are always a key approach in the arena of nanobased medicines, which are being studied for the effective transport of different bioactive. A hybrid nanosystem constructed by encompassing meso-porous silica NPs with grafting of carbosilane dendron exhibited outstanding DNA-binding characteristics and transfection proficiency. The amplified transfection efficiency of the developed hybrid system of NPs and grafted dendrons ensued in ascribed to a porous network of silica NPs and dendrimers having cationic charge, which confer high binding and transfection efficiency as well as biocompatibility (Martínez et al. 2015). From several studies, it was concluded that nanohybrids may have some vital benefits in biomedical applications with respect to the amalgamation designed. Such developed systems are available in the nanometric range and therefore able to interact with intracellular modules and could considerably alter the functions of essential organs (Cancino et al. 2013). The developed dendrimer-based nanohybrid system should be assessed crucially before any clinical applications. Recently, various researchers have proposed a guideline to scientists working in the area of dendrimers which was based on several translational prerequisite to move for an Investigational New Drug application, which is an assessment of the safety profile before starting clinical trials. Furthermore, Mignani and Majoral have demarcated the dendrimer space idea as an strategy that offers a new paradigm for medicinal scientists and unlocks innovative and auspicious paths for the identification of unique and biocompatible dendrimers (Mignani et al. 2013; Mintzer and Grinstaff 2011; Leiro et al. 2015).

14.7 Concluding Remarks

The multiple features like high branching, multivalency, flexible chemical composition, well-defined structures and high biological compatibility provide dendrimers potent and diverse candidates for drug delivery, imaging and tissue engineering. Simplicity in modifying surface of dendrimers and their capability to interact with functional groups established them as an attractive tool in drug discovery process. Furthermore, dendrimers are also defined as a different category of MRI contrast agents. Dendrimers also useful in the advancement of new devices for assisting in the diagnosis of ailments in a rapid time period deprived of the pre-treatment of sample. Even though dendrimers offer various pluses such as effective carries for therapeutic molecules and complexing with contrast agents due to multifunctional surface, the clinical aspects are constrained due to the toxicity issues and excessive cost. Thus, to overcome such issues and to make sure clinical suitability, there is still need of different approaches.

Abbreviations

AUC	Area under the curve
BSA	Bovine serum albumin
CNTs	Carbon nanotubes
DOX	Doxorubicin
DTPA	Diethylenetriamine pentaacetic acid
EGFR	Epithelial growth factor receptor
FA	folic acid
GIT	Gastrointestinal tract
IND	Investigational new drug application
MNPs	Magnetic nanoparticles
MRI	Magnetic resonance imaging
MTT	3-(4,5-dimethylthiazol-2-yl)-2,5-diphenyl-2H-tetrazolium bromide
NPs	Nanoparticles
PAMAM	Polyamidoamine
PDT	Photodynamic therapy
PEG	Polyethylene glycol
P-gp	P-glycoprotein
POSS	Polyhedral oligomeric silsesquioxane
PPI	Poly(propylene imine)
QDs	Quantum dots
SPION	Supramagnetic iron oxide NPs

REFERENCES

Abbasi, E., Aval, S. F., Akbarzadeh, A., et al. 2014. Dendrimers: synthesis, applications, and properties. *Nanoscale Research Letters 9* (1): 247.

Agashe, H. B., Dutta, T., Garg, M., et al. 2006. Investigations on the toxicological profile of functionalized fifth-generation poly (propylene imine) dendrimer. *Journal of Pharmacy and Pharmacology 58* (11): 1491–1498.

Agrawal, P., Gupta, U., and Jain, N. 2007. Glycoconjugated peptide dendrimers-based nanoparticulate system for the delivery of chloroquine phosphate. *Biomaterials 28* (22): 3349–3359.

Ahmed, T. A., Dare, E. V., and Hincke, M. 2008. Fibrin: a versatile scaffold for tissue engineering applications. *Tissue Engineering Part B: Reviews 14* (2): 199–215.

Allen, T. M., and Cullis, P. R. 2004. Drug delivery systems: entering the mainstream. *Science 303* (5665): 1818–1822.

Augst, A. D., Kong, H. J., and Mooney, D. J. 2006. Alginate hydrogels as biomaterials. *Macromolecular Bioscience 6* (8): 623–633.

Awada, A., Bondarenko, I., Bonneterre, J., et al. 2014. A randomized controlled phase II trial of a novel composition of paclitaxel embedded into neutral and cationic lipids targeting tumour endothelial cells in advanced triple-negative breast cancer (TNBC). *Annals of Oncology 25* (4): 824–831.

Bajgai, M. P., Aryal, S., Bhattarai, S. R., et al. 2008. Poly (ε-caprolactone) grafted dextran biodegradable electrospun matrix: A novel scaffold for tissue engineering. *Journal of Applied Polymer Science 108* (3): 1447–1454.

Barnes, C. P., Sell, S. A., Boland, E. D., et al. 2007. Nanofiber technology: designing the next generation of tissue engineering scaffolds. *Advanced Drug Delivery Reviews 59* (14): 1413–1433.

Barth, R. F., Soloway, A. H., Fairchild, R. G., et al. 1992. Boron neutron capture therapy for cancer. *Realities and prospects. Cancer 70* (12): 2995–3007.

Barth, R. F., Wu, G., Yang, W., et al. 2004. Neutron capture therapy of epidermal growth factor (+) gliomas using boronated cetuximab (IMC-C225) as a delivery agent. *Applied Radiation and Isotopes 61* (5): 899–903.

Battah, S., O'Neill, S., Edwards, C., et al. 2006. Enhanced porphyrin accumulation using dendritic derivatives of 5-aminolaevulinic acid for photodynamic therapy: an in vitro study. *The International Journal of Biochemistry & Cell Biology 38* (8): 1382–1392.

Bhadra, D., Bhadra, S., Jain, S., et al. 2003. A PEGylated dendritic nanoparticulate carrier of fluorouracil. *International Journal of Pharmaceutics 257* (1–2): 111–124.

Bielski, E., Zhong, Q., Mirza, H., et al. 2017. TPP-dendrimer nanocarriers for siRNA delivery to the pulmonary epithelium and their dry powder and metered-dose inhaler formulations. *International Journal of Pharmaceutics 527* (1–2): 171–183.

Biradar, A. V., Biradar, A. A., and Asefa, T. 2011. Silica–dendrimer core–shell microspheres with encapsulated ultrasmall palladium nanoparticles: efficient and easily recyclable heterogeneous nanocatalysts. *Langmuir 27* (23): 14408–14418.

Boris, D., and Rubinstein, M. 1996. A self-consistent mean field model of a starburst dendrimer: dense core vs dense shell. *Macromolecules 29* (22): 7251–7260.

Bottini, M., Tautz, L., Huynh, H., et al. 2005. Covalent decoration of multi-walled carbon nanotubes with silica nanoparticles. *Chemical Communications 14* (6): 758–760.

Brannon-Peppas, L., and Blanchette, J. O. 2004. Nanoparticle and targeted systems for cancer therapy. *Advanced Drug Delivery Reviews 56* (11): 1649–1659.

Bruchez, M., Moronne, M., Gin, P., et al. 1998. Semiconductor nanocrystals as fluorescent biological labels. *Science 281* (5385): 2013–2016.

Buhleier, E., Wehner, W., Voegtle, F. 1978. "Cascade" and "Nonskid-Chain-Like" syntheses of molecular cavity topologies. *Heterocyclic Compounds 9* (25): 7–11.

Cancino, J., Paino, I., Micocci, K., et al. 2013. In vitro nanotoxicity of single-walled carbon nanotube–dendrimer nanocomplexes against murine myoblast cells. *Toxicology Letters 219* (1): 18–25.

Caravan, P., Ellison, J. J., McMurry, T. J., et al. 1999. Gadolinium (III) chelates as MRI contrast agents: structure, dynamics, and applications. *Chemical Reviews 99* (9): 2293–2352.

Caster, J. M., Patel, A. N., Zhang, T., et al. 2017. Investigational nanomedicines in 2016: a review of nanotherapeutics currently undergoing clinical trials. *Wiley Interdisciplinary Reviews: Nanomedicine and Nanobiotechnology 9* (1): e1416.

Castro, R. I., Forero-Doria, O., and Guzman, L. 2018. Perspectives of dendrimer-based nanoparticles in cancer therapy. *Anais da Academia Brasileira de Ciências 90* (2): 2331–2346.

Chan, W. C., and Nie, S. 1998. Quantum dot bioconjugates for ultrasensitive nonisotopic detection. *Science 281* (5385): 2016–2018.

Chandrasekaran, S., and King, M. R. 2014. Microenvironment of tumour-draining lymph nodes: opportunities for liposome-based targeted therapy. *International Journal of Molecular Sciences 15* (11): 20209–20239.

Chang, Y., Meng, X., Zhao, Y., et al. 2011. Novel water-soluble and pH-responsive anticancer drug nanocarriers: Doxorubicin–PAMAM dendrimer conjugates attached to superparamagnetic iron oxide nanoparticles (IONPs). *Journal of Colloid and Interface Science 363* (1): 403–409.

Chen, Q., Li, K., Wen, S., et al. 2013. Targeted CT/MR dual mode imaging of tumours using multifunctional dendrimer-entrapped gold nanoparticles. *Biomaterials 34* (21): 5200–5209.

Chittasupho, C., Anuchapreeda, S., and Sarisuta, N. 2017. CXCR4 targeted dendrimer for anti-cancer drug delivery and breast cancer cell migration inhibition. *European Journal of Pharmaceutics and Biopharmaceutics 119*: 310–321.

Ciolkowski, M., Pałecz, B., Appelhans, D., et al. 2012. The influence of maltose modified poly (propylene imine) dendrimers on hen egg white lysozyme structure and thermal stability. *Colloids and Surfaces B: Biointerfaces 95*: 103–108.

Cousin, J. M., and Cloninger, M. J. 2015. Glycodendrimers: tools to explore multivalent galectin-1 interactions. *Beilstein Journal of Organic Chemistry 11* (1): 739–747.

Csaba, N., Garcia-Fuentes, M., and Alonso, M. J. 2006. The performance of nanocarriers for transmucosal drug delivery. *Expert Opinion on Drug Delivery 3* (4): 463–478.

D'emanuele, A., Jevprasesphant, R., Penny, J., et al. 2004. The use of a dendrimer-propranolol prodrug to bypass efflux transporters and enhance oral bioavailability. *Journal of Controlled Release 95* (3): 447–453.

de Brabander-van den Berg, E. M. M., and Meijer, E. 1993. Poly (propylenimin)-Dendrimere: Synthese in größerem Maßstab durch heterogen katalysierte Hydrierungen. *Angewandte Chemie 105* (9): 1370–1372.

Di Venosa, G. M., Casas, A. G., Battah, S., et al. 2006. Investigation of a novel dendritic derivative of 5-aminolaevulinic acid for photodynamic therapy. *The International Journal of Biochemistry & Cell Biology 38* (1): 82–91.

Duncan, R., and Izzo, L. 2005. Dendrimer biocompatibility and toxicity. *Advanced Drug Delivery Reviews 57* (15): 2215–2237.

Franano, F. N., Edwards, W. B., Welch, M. J., et al. 1995. Biodistribution and metabolism of targeted and nontargeted protein-chelate-gadolinium complexes: evidence for gadolinium dissociation in vitro and in vivo. *Magnetic Resonance Imaging 13* (2): 201–214.

Fu, H.-L., Li, Y.-Q., Shao, L., et al. 2010. Gene expression mediated by dendrimer/DNA complexes encapsulated in biodegradable polymer microspheres. *Journal of Microencapsulation 27* (4): 345–354.

Giepmans, B. N., Adams, S. R., Ellisman, M. H., et al. 2006. The fluorescent toolbox for assessing protein location and function. *Science 312* (5771): 217–224.

Gillies, E. R., Dy, E., Fréchet, J. M., et al. 2005. Biological evaluation of polyester dendrimer: poly (ethylene oxide)"bow-tie" hybrids with tunable molecular weight and architecture. *Molecular Pharmaceutics 2* (2): 129–138.

Glowacki, J., and Mizuno, S. 2008. Collagen scaffolds for tissue engineering. *Biopolymers: Original Research on Biomolecules 89* (5): 338–344.

Gothwal, A., Kesharwani, P., Gupta, U., et al. 2015. Dendrimers as an effective nanocarrier in cardiovascular disease. *Current Pharmaceutical Design 21* (30): 4519–4526.

Gottis, S., Rodriguez, L.-I., Laurent, R., et al. 2013. Janus carbosilane/phosphorhydrazone dendrimers synthesized by the 'click' Staudinger reaction. *Tetrahedron Letters 54* (50): 6864–6867.

Gradishar, W., and Tjulandin, S. 2005. Davidson n, Shaw H, Desai n, Bhar P, Hawkins M, o'Shaughnessy J. Phase III trial of nanoparticle albumin-bound paclitaxel compared with polyethylated castor oilbased paclitaxel in women with breast cancer. *Journal of Clinical Oncology 23*: 7794–7803.

Gradishar, W. J., Krasnojon, D., Cheporov, S., et al. 2009. Significantly longer progression-free survival with nab-paclitaxel compared with docetaxel as first-line therapy for metastatic breast cancer. *Journal of Clinical Oncology 27* (22): 3611–3619.

Gupta, L., Sharma, A. K., Gothwal, A., et al. 2017. Dendrimer encapsulated and conjugated delivery of berberine: A novel approach mitigating toxicity and improving in vivo pharmacokinetics. *International Journal of Pharmaceutics 528* (1–2): 88–99.

Hay, B. P., Werner, E. J., and Raymond, K. N. 2004. Estimating the number of bound waters in Gd (III) complexes revisited. Improved methods for the prediction of q-values. *Bioconjugate Chemistry 15* (6): 1496–1502.

He, X., Alves, C. S., Oliveira, N., et al. 2015. RGD peptide-modified multifunctional dendrimer platform for drug encapsulation and targeted inhibition of cancer cells. *Colloids and Surfaces B: Biointerfaces 125*: 82–89.

Herlambang, S., Kumagai, M., Nomoto, T., et al. 2011. Disulfide crosslinked polyion complex micelles encapsulating dendrimer phthalocyanine directed to improved efficiency of photodynamic therapy. *Journal of Controlled Release 155* (3): 449–457.

Hsu, H. J., Bugno, J., Lee, S. r., et al. 2017. Dendrimer-based nanocarriers: a versatile platform for drug delivery. *Wiley Interdisciplinary Reviews: Nanomedicine and Nanobiotechnology 9* (1): e1409.

Hu, C., Yang, X., Liu, R., et al. 2018a. Coadministration of iRGD with multistage responsive nanoparticles enhanced tumour targeting and penetration abilities for breast cancer therapy. *ACS Applied Materials & Interfaces 10* (26): 22571–22579.

Hu, C., Cun, X., Ruan, S., et al. 2018b. Enzyme-triggered size shrink and laser-enhanced NO release nanoparticles for deep tumour penetration and combination therapy. *Biomaterials 168*: 64–75.

Hu, J., Su, Y., Zhang, H., et al. 2011. Design of interior-functionalized fully acetylated dendrimers for anticancer drug delivery. *Biomaterials 32* (36): 9950–9959.

Ifkovits, J. L., and Burdick, J. A. 2007. Photopolymerizable and degradable biomaterials for tissue engineering applications. *Tissue Engineering 13* (10): 2369–2385.

Jain, K. 2017. Dendrimers: smart nanoengineered polymers for bioinspired applications in drug delivery. In *Biopolymer-Based Composites: Drug Delivery and Biomedical Applications* 169–220.

Jain, K., Gupta, U., and Jain, N. K. 2014. Dendronized nanoconjugates of lysine and folate for treatment of cancer. *European Journal of Pharmaceutics and Biopharmaceutics 87* (3): 500–509.

Jang, W.-D., Selim, K. K., Lee, C.-H., et al. 2009. Bioinspired application of dendrimers: from bio-mimicry to biomedical applications. *Progress in Polymer Science 34* (1): 1–23.

Jin, K., Luo, Z., Zhang, B., et al. 2018. Biomimetic nanoparticles for inflammation targeting. *Acta Pharmaceutica Sinica B 8* (1): 23–33.

Jones, C. F., Campbell, R. A., Franks, Z., et al. 2012. Cationic PAMAM dendrimers disrupt key platelet functions. *Molecular Pharmaceutics 9* (6): 1599–1611.

Joshi, P. P., Merchant, S. A., Wang, Y., et al. 2005. Amperometric biosensors based on redox polymer– carbon nanotube– enzyme composites. *Analytical Chemistry 77* (10): 3183–3188.

Kaur, D., Jain, K., Mehra, N. K., et al. 2016. A review on comparative study of PPI and PAMAM dendrimers. *Journal of Nanoparticle Research 18* (6): 146.

Kawaguchi, T., Walker, K. L., Wilkins, C. L., et al. 1995. Double exponential dendrimer growth. *Journal of the American Chemical Society 117* (8): 2159–2165.

Kesharwani, P., Ghanghoria, R., and Jain, N. K. 2012. Carbon nanotube exploration in cancer cell lines. *Drug Discovery Today 17* (17–18): 1023–1030.

Kesharwani, P., Jain, K., and Jain, N. K. 2014. Dendrimer as nanocarrier for drug delivery. *Progress in Polymer Science 39* (2): 268–307.

Kesharwani, P., Mishra, V., and Jain, N. K. 2015. Validating the anticancer potential of carbon nanotube-based therapeutics through cell line testing. *Drug Discovery Today 20* (9): 1049–1060.

Khandare, J. J., Jayant, S., Singh, A., et al. 2006. Dendrimer versus linear conjugate: influence of polymeric architecture on the delivery and anticancer effect of paclitaxel. *Bioconjugate Chemistry 17* (6): 1464–1472.

Khopade, A. J., Caruso, F., Tripathi, P., et al. 2002. Effect of dendrimer on entrapment and release of bioactive from liposomes. *International Journal of Pharmaceutics 232* (1–2): 157–162.

Kitchens, K. M., El-Sayed, M. E., and Ghandehari, H. 2005. Transepithelial and endothelial transport of poly (amidoamine) dendrimers. *Advanced Drug Delivery Reviews 57* (15): 2163–2176.

Klajnert, B., and Bryszewska, M. 2001. Dendrimers: properties and applications. *Acta Biochimica Polonica 48* (1): 199–208.

Ko, A., Tempero, M., Shan, Y., et al. 2013. A multinational phase 2 study of nanoliposomal irinotecan sucrosofate (PEP02, MM-398) for patients with gemcitabine-refractory metastatic pancreatic cancer. *British Journal of Cancer 109* (4): 920–925.

Kobayashi, H., and Choyke, P. 2006. Methods for tumour treatment using dendrimer conjugates: U.S. Patent Application No. 11/371780.

Kobayashi, H., Kawamoto, S., Jo, S.-K., et al. 2003. Macromolecular MRI contrast agents with small dendrimers: pharmacokinetic differences between sizes and cores. *Bioconjugate Chemistry 14* (2): 388–394.

Kumar, A. S. P., Latha, S., and Selvamani, P. 2015. Dendrimers: multifunctional drug delivery carriers. *International Journal of Technology Research Engine 2*: 1569–1575.

Kurczewska, J., Cegłowski, M., Messyasz, B., et al. 2018. Dendrimer-functionalized halloysite nanotubes for effective drug delivery. *Applied Clay Science 153*: 134–143.

Lagnoux, D., Darbre, T., Schmitz, M. L., et al. 2005. Inhibition of mitosis by glycopeptide dendrimer conjugates of colchicine. *Chemistry–A European Journal 11* (13): 3941–3950.

Lakowicz, J. R. 2013. *Principles of Fluorescence Spectroscopy*: Springer Science & Business Media.

Lakshminarayanan, A., Ravi, V. K., Tatineni, R., et al. 2013. Efficient dendrimer–DNA complexation and gene delivery vector properties of nitrogen-core poly (propyl ether imine) dendrimer in mammalian cells. *Bioconjugate Chemistry 24* (9): 1612–1623.

Langer, R. 1993. JP Vacanti. Tissue engineering. *Science 260* (5110): 920–926.

Langer, R., and Tirrell, D. A. 2004. Designing materials for biology and medicine. *Nature 428* (6982): 487–492.

Larson, D. R., Zipfel, W. R., Williams, R. M., et al. 2003. Water-soluble quantum dots for multiphoton fluorescence imaging in vivo. *Science 300* (5624): 1434–1436.

Laurencin, C. T., Jiang, T., Kumbar, S. G., et al. 2008. Biologically active chitosan systems for tissue engineering and regenerative medicine. *Current Topics in Medicinal Chemistry 8* (4): 354–364.

Lee, C. C., MacKay, J. A., Fréchet, J. M., et al. 2005. Designing dendrimers for biological applications. *Nature Biotechnology 23* (12): 1517–1526.

Lee, J. H., Lee, K., Moon, S. H., et al. 2009. All-in-one target-cell-specific magnetic nanoparticles for simultaneous molecular imaging and siRNA delivery. *Angewandte Chemie International Edition 48* (23): 4174–4179.

Leiro, V., Garcia, J. P., Tomás, H., et al. 2015. The present and the future of degradable dendrimers and derivatives in theranostics. *Bioconjugate Chemistry 26* (7): 1182–1197.

Lemon, B. I., and Crooks, R. M. 2000. Preparation and characterization of dendrimer-encapsulated CdS semiconductor quantum dots. *Journal of the American Chemical Society 122* (51): 12886–12887.

Li, C., Li, D., Zhao, Z.-S., et al. 2010a. Platinum nanoparticles from hydrosilylation reaction: Carbosilane dendrimer as capping agent. *Colloids and Surfaces A: Physicochemical and Engineering Aspects 366* (1–3): 45–49.

Li, Z., Huang, P., He, R., et al. 2010b. Aptamer-conjugated dendrimer-modified quantum dots for cancer cell targeting and imaging. *Materials Letters 64* (3): 375–378.

Li, Z., Huang, P., Lin, J., et al. 2010c. Arginine-glycine-aspartic acid-conjugated dendrimer-modified quantum dots for targeting and imaging melanoma. *Journal of Nanoscience and Nanotechnology 10* (8): 4859–4867.

Li, Q., Williams, C. G., Sun, D. D., et al. 2004. Photocrosslinkable polysaccharides based on chondroitin sulfate. *Journal of Biomedical Materials Research Part A 68* (1): 28–33.

Li, R., He, Y., Zhu, Y., et al. 2018. Route to rheumatoid arthritis by macrophage-derived microvesicle-coated nanoparticles. *Nano Letters 19* (1): 124–134.

Liu, H., Wang, H., Xu, Y., et al. 2014. Lactobionic acid-modified dendrimer-entrapped gold nanoparticles for targeted computed tomography imaging of human hepatocellular carcinoma. *ACS Applied Materials & Interfaces 6* (9): 6944–6953.

Lutolf, M., and Hubbell, J. 2005. Synthetic biomaterials as instructive extracellular microenvironments for morphogenesis in tissue engineering. *Nature Biotechnology 23* (1): 47–55.

Madaan, K., Kumar, S., Poonia, N., et al. 2014. Dendrimers in drug delivery and targeting: Drug-dendrimer interactions and toxicity issues. *Journal of Pharmacy & Bioallied Sciences 6* (3): 139.

Majoros, I. J., Thomas, T. P., Mehta, C. B., et al. 2005. Poly (amidoamine) dendrimer-based multifunctional engineered nanodevice for cancer therapy. *Journal of Medicinal Chemistry 48* (19): 5892–5899.

Malingré, M. M., Beijnen, J. H., and Schellens, J. H. 2001. Oral delivery of taxanes. *Investigational New Drugs 19* (2): 155–162.

Manikkath, J., Hegde, A. R., Kalthur, G., et al. 2017. Influence of peptide dendrimers and sonophoresis on the transdermal delivery of ketoprofen. *International Journal of Pharmaceutics 521* (1–2): 110–119.

Martínez, Á., Fuentes-Paniagua, E., Baeza, A., et al. 2015. Mesoporous Silica Nanoparticles Decorated with Carbosilane Dendrons as New Non-viral Oligonucleotide Delivery Carriers. *Chemistry–A European Journal 21* (44): 15651–15666.

Massaro, M., Lazzara, G., Milioto, S., et al. 2017. Covalently modified halloysite clay nanotubes: synthesis, properties, biological and medical applications. *Journal of Materials Chemistry B 5* (16): 2867–2882.

McCarthy, J. R., and Weissleder, R. 2008. Multifunctional magnetic nanoparticles for targeted imaging and therapy. *Advanced Drug Delivery Reviews 60* (11): 1241–1251.

Mignani, S., El Kazzouli, S., Bousmina, M., et al. 2013. Expand classical drug administration ways by emerging routes using dendrimer drug delivery systems: a concise overview. *Advanced Drug Delivery Reviews 65* (10): 1316–1330.

Mintzer, M. A., and Grinstaff, M. W. 2011. Biomedical applications of dendrimers: a tutorial. *Chemical Society Reviews 40* (1): 173–190.

Morishita, N., Nakagami, H., Morishita, R., et al. 2005. Magnetic nanoparticles with surface modification enhanced gene delivery of HVJ-E vector. *Biochemical and Biophysical Research Communications 334* (4): 1121–1126.

Muniswamy, V. J., Raval, N., Gondaliya, P., et al. 2019. 'Dendrimer-Cationized-Albumin' encrusted polymeric nanoparticle improves BBB penetration and anticancer activity of doxorubicin. *International Journal of Pharmaceutics 555*: 77–99.

Najlah, M., Freeman, S., Attwood, D., et al. 2007. Synthesis and assessment of first-generation polyamidoamine dendrimer prodrugs to enhance the cellular permeability of P-gp substrates. *Bioconjugate Chemistry 18* (3): 937–946.

Nanjwade, B. K., Bechra, H. M., Derkar, G. K., et al. 2009. Dendrimers: emerging polymers for drug-delivery systems. *European Journal of Pharmaceutical Sciences 38* (3): 185–196.

Neerman, M. F., Zhang, W., Parrish, A. R., et al. 2004. In vitro and in vivo evaluation of a melamine dendrimer as a vehicle for drug delivery. *International Journal of Pharmaceutics 281* (1–2): 129–132.

Nettles, D. L., Vail, T. P., Morgan, M. T., et al. 2004. Photocrosslinkable hyaluronan as a scaffold for articular cartilage repair. *Annals of Biomedical Engineering 32* (3): 391–397.

Nguyen, T. T. C., Nguyen, C. K., Nguyen, T. H., et al. 2017. Highly lipophilic pluronics-conjugated polyamidoamine dendrimer nanocarriers as potential delivery system for hydrophobic drugs. *Materials Science and Engineering: C 70*: 992–999.

Northfelt, D. W., Dezube, B. J., Thommes, J. A., et al. 1998. Pegylated-liposomal doxorubicin versus doxorubicin, bleomycin, and vincristine in the treatment of AIDS-related Kaposi's sarcoma: results of a randomized phase III clinical trial. *Journal of Clinical Oncology 16* (7): 2445–2451.

Nummelin, S., Liljeström, V., Saarikoski, E., et al. 2015. Self-Assembly of Amphiphilic Janus Dendrimers into Mechanically Robust Supramolecular Hydrogels for Sustained Drug Release. *Chemistry–A European Journal 21* (41): 14433–14439.

Oliveira, J. M., Salgado, A. J., Sousa, N., et al. 2010. Dendrimers and derivatives as a potential therapeutic tool in regenerative medicine strategies—a review. *Progress in Polymer Science 35* (9): 1163–1194.

Pan, B., Gao, F., Ao, L., et al. 2005. Controlled self-assembly of thiol-terminated poly (amidoamine) dendrimer and gold nanoparticles. *Colloids and Surfaces A: Physicochemical and Engineering Aspects 259* (1–3): 89–94.

Parajapati, S. K., Maurya, S. D., Das, M. K., et al. 2016. Potential application of dendrimers in drug delivery: A concise review and update. *Journal of Drug Delivery and Therapeutics 6* (2): 71–88.

Patri, A. K., Kukowska-Latallo, J. F., and Baker Jr, J. R. 2005. Targeted drug delivery with dendrimers: comparison of the release kinetics of covalently conjugated drug and non-covalent drug inclusion complex. *Advanced Drug Delivery Reviews 57* (15): 2203–2214.

Patton, D., Sweeney, Y. C., McCarthy, T., et al. 2006. Preclinical safety and efficacy assessments of dendrimer-based (SPL7013) microbicide gel formulations in a nonhuman primate model. *Antimicrobial Agents and Chemotherapy 50* (5): 1696–1700.

Prasad, P., and Srivastava, A. 2020. A beacon for gynecological cancers patients: pH-sensitive. *Nanomedicine. Clinics in Oncology 5*: 1681–1689

Price, C. F., Tyssen, D., Sonza, S., et al. 2011. SPL7013 Gel (VivaGel®) retains potent HIV-1 and HSV-2 inhibitory activity following vaginal administration in humans. *PLoS One 6* (9): e24095.

Qin, W., Yang, K., Tang, H., et al. 2011. Improved GFP gene transfection mediated by polyamidoamine dendrimer-functionalized multi-walled carbon nanotubes with high biocompatibility. *Colloids and Surfaces B: Biointerfaces 84* (1): 206–213.

Roberts, J. C., Bhalgat, M. K., and Zera, R. T. 1996. Preliminary biological evaluation of polyamidoamine (PAMAM) StarburstTM dendrimers. *Journal of Biomedical Materials Research: An Official Journal of The Society for Biomaterials and The Japanese Society for Biomaterials 30* (1): 53–65.

Sampathkumar, S. G., and Yarema, K. J. 2007. Dendrimers in cancer treatment and diagnosis. *Nanotechnologies for the Life Sciences* [Online]. doi:10.1002/9783527610419.ntls0071

Saovapakhiran, A., D'Emanuele, A., Attwood, D., et al. 2009. Surface modification of PAMAM dendrimers modulates the mechanism of cellular internalization. *Bioconjugate Chemistry 20* (4): 693–701.

Seelbach, R. J., Fransen, P., Peroglio, M., et al. 2014. Multivalent dendrimers presenting spatially controlled clusters of binding epitopes in thermoresponsive hyaluronan hydrogels. *Acta Biomaterialia 10* (10): 4340–4350.

Shukla, R., Thomas, T. P., Peters, J. L., et al. 2006. HER2 specific tumour targeting with dendrimer conjugated anti-HER2 mAb. *Bioconjugate Chemistry 17* (5): 1109–1115.

Sk, U. H., Dixit, D., and Sen, E. 2013. Comparative study of microtubule inhibitors–estramustine and natural podophyllotoxin conjugated PAMAM dendrimer on glioma cell proliferation. *European Journal of Medicinal Chemistry 68*: 47–57.

Soto-Castro, D., Cruz-Morales, J. A., Apan, M. T. R., et al. 2012. Solubilization and anticancer-activity enhancement of Methotrexate by novel dendrimeric nanodevices synthesized in one-step reaction. *Bioorganic Chemistry 41*: 13–21.

Soundararajan, A., Bao, A., Phillips, W. T., et al. 2009. [186Re] Liposomal doxorubicin (Doxil): in vitro stability, pharmacokinetics, imaging and biodistribution in a head and neck squamous cell carcinoma xenograft model. *Nuclear Medicine and Biology 36* (5): 515–524.

Spataro, G., Malecaze, F., Turrin, C.-O., et al. 2010. Designing dendrimers for ocular drug delivery. *European Journal of Medicinal Chemistry 45* (1): 326–334.

Svenson, S., and Tomalia, D. A. 2012. Dendrimers in biomedical applications—reflections on the field. *Advanced Drug Delivery Reviews 64*: 102–115.

Szulc, A., Pulaski, L., Appelhans, D., et al. 2016. Sugar-modified poly (propylene imine) dendrimers as drug delivery agents for cytarabine to overcome drug resistance. *International Journal of Pharmaceutics 513* (1–2): 572–583.

Taheri-Kafrani, A., Shirzadfar, H., and Tavassoli-Kafrani, E. 2017. Dendrimers and dendrimers-grafted superparamagnetic iron oxide nanoparticles: synthesis, characterization, functionalization, and biological applications in drug delivery systems. In *Nano-and Microscale Drug Delivery Systems*: Elsevier, 75–94.

Tao, X., Yang, Y.-J., Liu, S., et al. 2013. Poly (amidoamine) dendrimer-grafted porous hollow silica nanoparticles for enhanced intracellular photodynamic therapy. *Acta Biomaterialia 9* (5): 6431–6438.

Taratula, O., Garbuzenko, O., Savla, R., et al. 2011. Multifunctional nanomedicine platform for cancer specific delivery of siRNA by superparamagnetic iron oxide nanoparticles-dendrimer complexes. *Current Drug Delivery 8* (1): 59–69.

Thorek, D. L., Chen, A. K., Czupryna, J., et al. 2006. Superparamagnetic iron oxide nanoparticle probes for molecular imaging. *Annals of Biomedical Engineering 34* (1): 23–38.

Tolia, G. T., and Choi, H. H. 2008. The role of dendrimers in topical drug delivery. *Pharmaceutical Technology 32* (11): 88–98.

Tomalia, D. A. 2005. Birth of a new macromolecular architecture: dendrimers as quantized building blocks for nanoscale synthetic polymer chemistry. *Progress in Polymer Science 30* (3–4): 294–324.

Vandamme, T. F., and Brobeck, L. 2005. Poly (amidoamine) dendrimers as ophthalmic vehicles for ocular delivery of pilocarpine nitrate and tropicamide. *Journal of Controlled Release 102* (1): 23–38.

Wang, X., Inapagolla, R., Kannan, S., et al. 2007. Synthesis, characterization, and in vitro activity of dendrimer– Streptokinase conjugates. *Bioconjugate Chemistry 18* (3): 791–799.

Wang, Y., Li, L., Shao, N., et al. 2015. Triazine-modified dendrimer for efficient TRAIL gene therapy in osteosarcoma. *Acta Biomaterialia 17*: 115–124.

Weiner, E., Brechbiel, M., Brothers, H., et al. 1994. Dendrimer-based metal chelates: a new class of magnetic resonance imaging contrast agents. *Magnetic Resonance in Medicine 31*: 1–8.

Wróblewska, M., and Winnicka, K. 2015. The effect of cationic polyamidoamine dendrimers on physicochemical characteristics of hydrogels with erythromycin. *International Journal of Molecular Sciences 16* (9): 20277–20289.

Yamashita, S., Katsumi, H., Hibino, N., et al. 2017. Development of PEGylated carboxylic acid-modified polyamidoamine dendrimers as bone-targeting carriers for the treatment of bone diseases. *Journal of Controlled Release 262*: 10–17.

Yang, W., Barth, R. F., Wu, G., et al. 2009. Boron neutron capture therapy of EGFR or EGFRvIII positive gliomas using either boronated monoclonal antibodies or epidermal growth factor as molecular targeting agents. *Applied Radiation and Isotopes 67* (7–8): S328–S331.

Yavuz, B., Pehlivan, S. B., Vural, İ., et al. 2015. In vitro/in vivo evaluation of dexamethasone—PAMAM dendrimer complexes for retinal drug delivery. *Journal of Pharmaceutical Sciences 104* (11): 3814–3823.

Yuan, Q., Lee, E., Yeudall, W. A., et al. 2010. Dendrimer-triglycine-EGF nanoparticles for tumour imaging and targeted nucleic acid and drug delivery. *Oral Oncology 46* (9): 698–704.

Zhang, M., Zhu, J., Zheng, Y., et al. 2018. Doxorubicin-conjugated PAMAM dendrimers for pH-responsive drug release and folic acid-targeted cancer therapy. *Pharmaceutics 10* (3): 162.

Zhu, J., and Shi, X. 2013. Dendrimer-based nanodevices for targeted drug delivery applications. *Journal of Materials Chemistry B 1* (34): 4199–4211.

Zhu, J., Wang, G., Alves, C. S., et al. 2018. Multifunctional dendrimer-entrapped gold nanoparticles conjugated with doxorubicin for pH-responsive drug delivery and targeted computed tomography imaging. *Langmuir 34* (41): 12428–12435.

Zhu, J., Zheng, L., Wen, S., et al. 2014. Targeted cancer theranostics using alpha-tocopheryl succinate-conjugated multifunctional dendrimer-entrapped gold nanoparticles. *Biomaterials 35* (26): 7635–7646.

15

Biomedical Applications of Dendrimers

Sonali Batra and Sumit Sharma

CONTENTS

15.1 Introduction

Structurally dendrimers are symmetrical and highly branched polymeric nanomaterials with three distinct features, *i.e.,* inner core, repeated units, which signify branching and multiple functionalities at the surface. The interior of dendrimers comprises a high density of positive charge due to amine groups. This density provides extra buffering capacity, which allows endosomal escape of dendrimers and stability to loaded sensitive drugs and DNA/RNA. Moreover, the multi-valency at the surface gives multifunctional attributes to the dendrimers such as pH sensitivity, charge neutralisation, specific receptor-mediated targeting, controlled release and biocompatibility, which further widen the range of biomedical applications of dendrimers. Dendrimers have been extensively utilised in drug delivery, as a non-vector for gene delivery, diagnostic applications, tissue restoration, etc. In this chapter, we have discussed the diverse modifications of dendrimers that have been explored in wide era of biomedical applications mechanistically.

15.2 Dendrimers as Drug Carriers

Amalgamation of nanotechnology and unique properties of dendrimers make dendrimers a potential choice of nanodrug delivery of sensitive molecules such as proteins and peptides, controlled drug delivery, targeting hydrophobic molecules and enhancing permeation. Dendrimers have 3-D architectural design with an internal cavity and external variable functional dendritic surface, which allows heavy payload as compared to other nanodrug delivery systems. In regard to toxicity, cationic dendrimers are observed to be highly cytotoxic and haemolytic due to terminally attached positively charged functional groups (Ziemba et al. 2012). This issue can be resolved by cross linking with certain neutral molecules such as polyethylene glycol (Astruc, Boisselier and Ornelas 2010) and oligosaccharides (Appelhans et al. 2009) thereby reducing the cytotoxicity and increasing the biocompatibility. Moreover, an ideal drug delivery system should be free from toxicity and non-immunogenicity.

All the above-mentioned attributes of dendrimers were explored in various studies of novel drug delivery systems. For instance, fourth generation (G4.0) dendrimers was fabricated using poly(propylene imine) to facilitate the permeation of activated hydrophilic triphosphate form of cytarabine. Cytarabine is a nucleoside analogue (inactive form) which is administered as prodrug because the active form is hydrophilic in nature and hence cannot permeate the cell membrane. The inactive form can pass through cell membrane and is supposed to convert into active form (cytarabine triphosphate) by kinases inside the cancer cells. The drawback of this approach is that due to the limited amount of kinases in cancer cells, inefficient cell uptake through nucleoside specific transporter and poor biodistribution renders the cytarabine treatment inefficient and develops drug resistance. Therefore, the active form of cytarabine was loaded into the poly(propylene imine) based dendrimers. The surface of dendrimers was neutralised by maltose which renders the dendrimers themselves non-toxic to normal cells as well as cancer cells, thereby reducing non-specific cell toxicity. On the other hand, cytarabine triphosphate complexed with dendrimers was observed to be more toxic against 1301 and HL-60 cell lines than cytarabine and cytarabine triphosphate. Moreover, human equilibrative nucleoside transporter 1 (HENT-1) is supposed to mediate the cytarabine cell uptake. The decreased expression of this transporter leads to the limitation of drug resistance against cancer. The dendrimers as a nanocarrier system dodge this mechanism of uptake and were found to be independent of HENT-1 transporter. All these findings from cytarabine triphosphate-loaded dendrimers have reduced the probability of drug resistance in chemotherapy, enhanced the cellular uptake and circumvent the prodrug conversion by kinases (Szulc et al. 2016).

Dendrimers based on polyamidoamine (PAMAM), poly(propylene imine), poly (glycerol-co-succinic acid), poly-l-lysine, melamine, triazine, poly(glycerol), poly[2,2-bis (hydroxymethyl) propionic acid] and poly(ethylene glycol) are widely studied as drug delivery dendrimers for various drugs. Among them, PAMAM and poly(propylene imine) are highly appreciated because of their pH sensitivity which facilitates drug targeting, controlled release and improving aqueous solubility of hydrophobic drugs (Zhou, D'Emanuele and Attwood 2013). Dendrimers are conjugated with folic acid in various studies in order to target the cancerous cells specifically. The theory behind this is that in cancer cells the folate receptors are overexpressed and dendrimers have multiple sites at the surface for the conjugation of folic acid. As a consequence, the dendrimers conjugated with folic acid target the cancer cells specifically and lead to enhanced internalisation of these nanocarriers loaded with anti-cancer drugs.

For instance, Zhang and co-workers have fabricated pH sensitive and folic acid conjugated PAMAM fifth generation dendrimers to target doxorubicin to the cancer cells. pH sensitivity was achieved by conjugating cis-aconityl (as a linker between dendrimers and doxorubicin) at the periphery. This allows maximum drug release at pH around 5–6, which is close to the microenvironment of cancer cells whereas restricts the drug release at pH around 7.4 (normal blood pH). Moreover, folic acid linkage at an average number of 5 was observed to be sufficient to increase the internalisation of dendrimers into the cancer cells through folate receptor mediated endocytosis. In conclusion, it was postulated that dendrimers as a nanodrug carrier provide prospects of multi-functional as targeting and controlled release of drugs (Zhang et al. 2018). Table 15.1 represents various drug molecules that were delivered smartly either by targeting through conjugation and/or complexation with ligands or increasing their aqueous solubility for the treatment of diseases.

15.3 Dendrimers in Gene Delivery

A therapeutic gene delivery is important as it is used to treat certain inherited disorders, viral infections and metabolic disorders. Therapeutic delivery of genetic material (nucleic acids, plasmids, etc.) always remains a challenge. Gene delivery can be made either by viral vectors or non-viral vectors. The problems associated with non-viral vectors are immunogenicity, safety and highly sophisticated methods for commercial production which ultimately add to the cost of the treatment. Dendrimers as a non-viral vector are

TABLE 15.1

Drug molecules delivered through dendrimers as a drug carrier

Drug molecules	Type of dendrimers	Description	Application	Reference
Methotrexate	PAMAM	Alendronate was covalently bonded to PEGylated PAMAM dendrimers conjugate and loaded with methotrexate	• Alendronate, a biphosphonate has high affinity for hydroxyapatite (component of bones) and facilitates targeting in the bones. • Targeting results in the enhanced tissue distribution of methotrexate for the treatment of bone metastasis.	Yamashita et al. (2018)
Doxorubicin	PAMAM	5.0G PAMAM dendrimers conjugated with lactobionic acid using polyethylene glycol as a linker and neutralisation of amine sites by acetylation.	• Enhanced specific targeting to overexpressed ASGPR liver cancer cells than non-cancer cells. • Polyethylene glycol linkage potentiated the receptor mediated cellular uptake by lactobionic acid conjugated dendrimers.	Fu et al. (2014)
Paclitaxel	PAMAM	Diblock copolymer of poly(butylene oxide) and poly(ethylene oxide) conjugated with 2.0G PAMAM dendrimers	• Micellar encapsulation in 2% w/v diblock copolymer conjugated dendrimers shown 3700 fold increase in solubility of paclitaxel in phosphate buffer (pH 7.4). • Micellar structure of dendrimers provides multiple encapsulation sites for loading hydrophobic drugs.	Zhou, D'Emanuele, and Attwood (2013)
Amphotericin B	Poly (propyleneimine)	5.0G Poly (propyleneimine) dendrimer conjugated with muramyl dipeptide by divergent technique.	• Macrophages were targeted by ampho-tericin B loaded dendrimers and released maximum payload at endosomal pH i.e. around 5.5. • Low haemolytic and cytotoxcity due to neutralisation of cationic charge of dendrimers by muramyl dipeptide.	Jain et al. (2015)
Tamoxifen	PAMAM	4.0G PAMAM dendrimers conjugated with transferrin at the periphery and tamoxifen loaded in the inner core. Doxorubicin linked through hydrazone linkage which imparts pH sensitivity to the dendrimers.	• Transferrin facilitates uptake of drug loaded dendrimers through overexpressed transferring receptors on cancer cells and brain capillary endothelium. • PEGylation was done to reduce cationic cytotoxicity. • pH sensitivity allows maximum drug release at pH 4.5 mimicking the microenvironment of brain glioma cells.	Li et al. (2012)

extensively explored for transfection of the genetic material and provide the advantage of biocompatibility and flexibility in modification with different functional ligands. Apart from all the above-mentioned advantages, dendrimers have to face certain drawbacks in efficient gene transfection for instance, degradation of DNA by cytosolic nuclease, polyplex instability and inefficient cellular uptake. Moreover, earlier anionic and neutral polyplexes were preferred for fabrication of dendrimers and gene transfection. However due to complexity in production and more time consumption, their use is limited, and the attention of researchers then shifted to cationic dendrimers. It is reported that cationic polymers have the potential not only to deliver DNA plasmids but can also deliver antisense oligonucleotides (Bielinska et al. 1996). The cationic polymers are functionally modified with ligands such as lipids, nucleic acids, neutral moieties such as polyethylene glycol and sugar moieties in order to reduce cytotoxicity, increase polyplex stability and increase specific cellular accumulation (Liu et al. 2016). PAMAM and poly (propyleneimine) are highly reported in the literature for targeting genes to the several organs and cancer cells (Dufes et al. 2005, Wada et al. 2005).

Amino acids such as arginine and lysine have positive charge and on conjugation with cationic dendrimers increase the charge density at the surface, which ultimately provides higher DNA condensation and polyplex stability to the genetic material. The performance of ligands is also correlated with the type of dendrimer. For example, the transfection efficiency of DNA and siRNA using arginine was observed to be more efficient when conjugated with PAMAM than with poly (propyleneimine). Similarly, lysine modified poly(propyleneimine) showed efficient transfection than arginine modified PAMAM dendrimers.

Hydrophobicity also plays an important role in increasing cellular uptake of dendrimers loaded with genetic material. Cholesterol and fatty acids like palmitic acid and myristic acid are observed to have excellent fusogenic activity into cell membrane *via* caveolted mediated endocytosis and circumvent endosomal degradation as a result provide enhanced gene transfection. However, it is important to balance the level of hydrophobicity as it reduces aqueous solubility and ultimately obstructs gene transfection (Yang et al. 2015). For instance, alkyls and fluoroalkyls show higher transfection through increased cellular uptake and endosomal escape but considering the effect of hydrophobicity it was observed that short chain fluoroalkyl groups are preferred for efficient gene delivery (Liu et al. 2016, Wang et al. 2014).

15.4 Dendrimers in Tissue Engineering

The surface of dendrimers has wide acceptability for body defence cells; their excellent cross linking properties due to the presence of multiple end groups and flexible mechanical properties make them potential candidates in various tissue engineering applications. Highly modified surface chemistry is the key component, which provides an additional advantage to dendrimers over other nanocompounds such as liposomes, nanoemulsion and niosomes for tissue engineering applications. Moreover, the primary goals of dendrimers in tissue regeneration are to support and mimic the extracellular matrix and hasten the natural wound healing process. In tissue reconstruction, the porous and nanoscaffolds are engineered using biocompatible natural or synthetic polymers and commonly explored (Batra and Sharma 2019, Sharma and Batra 2019, Kesharwani et al. 2017). The cascade of tissue reconstruction requires adequate exposure to the external environment for the diffusion of growth factors, chemical mediators and other nutrients along with sufficient mechanical support to allow adherence and limited movement of encapsulated cells involve in tissue repair. Dendrimers are the pharmaceutical delivery system which comprise all these properties and therefore holds a good candidacy as an ideal pharmaceutical aid for tissue repair. Biomaterials are designed to manufacture scaffolds that will mimic the extra cellular matrix such as the three-dimensional network and optimum mechanical strength to support cell adherence, accumulation of growth factors and little movement of cells for remodelling in excised or injured tissue. Chan and co-workers designed a scaffold from extracellular matrix derived with cholecyst and functionalise with amine groups using PAMAM dendrimers (Generation 1; G1.0). The elastic modulus of cross linked scaffolds with PAMAM was compared with that of the non-cross linked scaffolds. It was observed that cross linked scaffolds with PAMAM showed reduced value of elastic modulus with no significant change in tensile strength as compared to non-cross linked scaffolds. This signifies that cross linking with PAMAM dendrimers has increased elasticity without changing the tensile strength much and maintaining the desired mechanical strength of the scaffold for soft tissues. This slight increase in extensibility of scaffold allows increased cellular seeding, reducing undesired stiffness of the scaffold and little movement of fibroblasts within the scaffold. These properties are said to be very advantageous for scaffolds employed for tissue remodelling (Chan et al. 2008).

Another exclusive property of dendrimers was exploited by Feng and co-workers. PAMAM based dendrimers were grafted with diazirine and developed a UV activated in situ bioadhesive hydrogel formulation. The properties of developed hydrogel bioadhesive are (1) variable and tunable mechanical and elastic properties and (2) formation of a covalent bond with tissue water content in the extracellular matrix and enabling a rapid adhesion without interfering with structural conformation of proteins. Diazirine provides carbene groups on slight activation by ultraviolet which forms strong bonds at the hydrated tissue interface and act as a biocompatible sealant. The major advantage of using PAMAM is that it provides comparatively more number of reactive sites for diazirine. By virtue of which the elastic modulus and mechanical properties of hydrogel can be modified by tailoring the diazirine concentration. It was observed that the rheological properties and elastic modulus depend on the number of carbene formed C-H bonds with water. The biomedical utility of developing such bioadhesive dendrimers based hydrogel was realised in major surgeries (like highly vascular organ incisions and vascular anastomosis) where incised tissue needs to be stapled or sealed. The developed formulation act as bioadhesive sealant and fasten the process of tissue healing and avoid exsanguinations (Feng et al. 2016). Certain reports in the literature have been reported for tissue regeneration claiming the nanodendrimers formulations comprising growth factors, proteins and chemically modified genetic material complexed with biocompatible scaffolds using carbohydrates, collagen, chitosan, etc. The natural polymers are observed to be naturally active and promote cell attachment (Gorain et al. 2017, Joshi and Grinstaff 2008, Elangovan et al. 2014).

The advantage of loading protein compounds in the dendrimers is that it does not change the protein structure also provide comparatively extra drug loading. A three dimensional architectural design of dendrimeric scaffold mimic the physical structure of extracellular matrix and allows the seeded cells, stem cells and growth factors constantly available for the tissue regeneration. Dendrimeric scaffolds provide a natural requisite microenvironment and facilitate the cascade of tissue healing. Given the advantages of dendrimers in tissue regeneration such as biocompatibility, biodegradability, tunable mechanical integrity, elasticity, supporting the progenitor cells to grow and preserving the structural integrity of loaded proteins and peptide research has focused to develop functional constructs like dendrimers for tissue regeneration. Research approaches have been performed for tissue engineering and found utility particularly in osteoarthritis, bone surgeries, vascular tissue regeneration, antimicrobial bandages, burns and wounds.

15.5 Dendrimers in Diagnostics

Medical diagnosis is a process for determination of the diseased condition. Early disease detection is important and has gained a boom with the help of radio imaging in the field of nuclear medicine. Dendrimers have been recognised as good candidates for diagnosis due to several advantages these offer. First, their low poly-dispersed nature helps to synthesis precise product with consistent results. Second, possibilities of variety of modifications that can be done on dendrimers provide the required characteristics and pharmacokinetics. Third, different generations of dendrimers exhibit unique morphology and thus, this can be effectively used as contrast agents in different organs depending upon the size of veins of organs and its permeability (Barrett et al. 2009, Zhao et al. 2017a).

Since dendrimers themselves are not fluorescent, these are therefore conjugated with fluorophores to the externally protruding amine groups, which helps detection by imaging. Use of dendrimers as contrast agents have been largely exploited in various diagnostics like X-ray-computed tomography (CT), magnetic resonance imaging (MRI), positron emission tomography (PET) (Barrett et al. 2009), single photon emission computed tomography and as biosensors (Rai et al. 2020).

15.5.1 Dendrimers as CT Contrast Agents

CT is one of the crucial medical imaging tools used for the diagnosis of the disease. This is based on the difference in the X-ray absorption of various organs. For higher contrast imaging in CT, agents that provide higher X-ray attenuation coefficients are preferred. This is one of the prime requirements so as to differentiate between the normal and the diseased tissues (Barrett et al. 2009). Other major requirements include the specific targeting, biocompatibility and extreme sensitivity. Conventionally adopted, iodine based CT agents use a high dose of iodine which can lead to acute kidney damage. Hence to overcome such problems, development of nanoparticle based agents using metals like bismuth and gold were investigated. These provided low-dose usage of the metals but stability was still matter of concern (Sk Ugir and Kojima 2015). Here thus, the macromolecule-based technique involving dendrimers played an excellent role as CT contrast agents. Dendrimers are either conjugated with metal nanoparticles or the metal ions are fabricated into the core surrounded by modified dendrimers (Rai et al. 2020). Generation 2 PAMAM dendrimers modified using folic acid were used to prepare gold nanoparticles. These gold-containing dendrimers stabilised nanoparticles showed much higher attenuation coefficient than conventionally used contrast agent, i.e., omniplaque when given intravenously. These low generation dendrimers were able to successfully target the human carcinoma cell lines and hence can be potentially used as CT contrast agent (Liu et al. 2013). In another study, G4 dendrimers were stabilised with bismuth ions to form Bi_2S_3 and formulated into nanoparticles. These fabricated dendrimers stabilised nanoparticles were non-toxic to cells as proved by 3-(4,5-Dimethylthiazol-2-yl)-2,5-diphenyltetrazolium bromide assay. Hence, these can be efficiently used as contrast agents in CT imaging (Fang et al. 2013).

15.5.2 Dendrimers as MRI Contrast Agents

MRI is a non-invasive technique which is used for the diagnosis as well as treatment of medical condition. It has been a more reliable technique as it provides 3-D images and uses magnetic moment of the protons. Mainly the protons of the water molecules are involved in these are used as contrast agents when magnetic field is applied to the tissue of interest. Paramagnetic substances are thus used in this technique which aligns itself into certain direction on exposure to the magnetic radiations (Krause et al. 2000). In MRI, high molecular contrast agents are preferred as they effectively differentiate between the normal and the diseased tissue. Dendrimers are thus the ideal high molecular weight carriers to be used as contrast agents in MRI, but as they are themselves not paramagnetic; hence these are combined with other metal ions. Gadolinium is one of the most widely used metal ions, the complexation of which with the dendrimers is now being used for over a decade (Sk Ugir and Kojima 2015, Nagpal et al. 2018). Fourth generation PAMAM dendrimers complexed with iminodiacetic acid derivatives of gadolinium has proven affinity towards liver cells. Due to its good imaging and minimised side effects, the complex is used as a hepatoproic MRI contrast agent (Markowicz-Piasecka et al. 2015). In addition, gadolinium-conjugated nanoclusters have been synthesised for early detection of cancerous cells when used as a MRI contrast agent. However, gadolinium has longer circulation and residence time and is thus associated with nephrogenic systemic fibrosis (Cheng, Thorek and Tsourkas 2010). To overcome such a problem, biodegradable polysulphide linkages were made in the dendrimers to modify gadolinium conjugated nanoclusters. These exhibited smaller degradation products of gadolinium complex, thereby reducing its circulation time and thus its long-term toxicity (Huang et al. 2012).

15.5.3 Dendrimers as PET Contrast Agents

Among various used imaging techniques, the PET technique is one of the widely accepted imaging tools of nuclear medicine. Due to the use of radio nuclides, this technique has the advantage of monitoring the pharmacokinetics and of course has better sensitivity as compared to other diagnostic techniques. Generally, radio nuclides having short half-life ($t_{1/2}$) are used as positron emitters in this technique. [11]C, ^{13}N, ^{15}O, ^{18}F and ^{68}Ga are few among clinically used among them (Zhao et al. 2017b, 2017a).

15.6 Other Applications

Dendrimers have been used in diverse areas enormously, especially in recent times. The never ending uses and applications of the dendrimers have made them blockbuster compounds being used nowadays. Dendrimers have been used as mimicking agents to mimic the hardest known material of the human body, i.e., dental enamel. Also studies have been performed which show that collagen mimicking has also been done in which trimesic acid or G1 PAMAM core has been used in different studies along with trimer of amino acids, which has been proved to be highly useful and substitute to the biodegraded collagen with better thermo-responsive properties (Ray et al. 2018).

Applications of dendrimers have not only been restricted towards the gene diagnostics but also as therapeutics, as a result of which these are better known as theranostics. These play a vital role in cancer especially in case of pancreas, where it is difficult to deliver the drug using conventional chemotherapy. Yalcin and co-workers formulated PAMAM coated magnetic nanoparticles loaded with gemcitabine and retinoic acid which delivered the drug at the targeted site of pancreatic tumours (Yalcin et al. 2014).

Apart from these, dendrimers have been formulated into stable sensor systems, which detect the presence of metal ions in water. Florescence-emitting dendrimers were used by Grabchev and co-workers which electively detected metal ions like Cu^{2+}, Li^+, Fe^{3+} and Zn^{2+} in water at very low concentrations as other sensor devices (Grabchev, Staneva and Betcheva 2012).

Moreover, an elaborated use of dendrimers as biosensors has been compiled in detail by Bahadir and Sezginturk. The review states how dendrimers can be widely applied in biosensor technology in the pharmaceutical, medical and food industries. These sensors are used to determine various marker molecules like in cardiac disease and cancer. Also, these can easily sense the target DNA molecule, proteins and its metabolites and even the pesticides and heavy metals through electrochemical sensing (Bahadir and Sezginturk 2016).

15.7 Conclusion and Future Perspectives

The applications and future perspectives of dendrimers and their modifications are very diverse and in abundance. The variety of amendments that can be played with dendrimers depends entirely on the researcher's framework of mind and the objective behind it. Looking at the present scenario, it seems that dendrimers still have long journey and an era to be captured owing to their applications and none the less the large number of patents to their name.

REFERENCES

Appelhans, D., H. Komber, M. A. Quadir, S. Richter, S. Schwarz, J. van der Vlist, A. Aigner, M. Muller, K. Loos, J. Seidel, K. F. Arndt, R. Haag, and B. Voit. 2009. "Hyperbranched PEI with various oligosaccharide architectures: synthesis, characterization, ATP complexation, and cellular uptake properties." *Biomacromolecules* no. *10* (5):1114–1124. doi:10.1021/bm801310d.

Astruc, D., E. Boisselier, and C. Ornelas. 2010. "Dendrimers designed for functions: from physical, photophysical, and supramolecular properties to applications in sensing, catalysis, molecular electronics, photonics, and nanomedicine." *Chem Rev* no. *110* (4):1857–1959. doi:10.1021/cr900327d.

Bahadir, E. B., and M. K. Sezginturk. 2016. "Poly(amidoamine) (PAMAM): an emerging material for electrochemical bio(sensing) applications." *Talanta* no. *148*:427–438. doi:10.1016/j.talanta.2015.11.022.

Barrett, Tristan, Gregory Ravizzini, Peter L. Choyke, and Hisataka Kobayashi. 2009. "Dendrimers in medical nanotechnology." *IEEE Engineering in Medicine and Biology Magazine: The Quarterly Magazine of the Engineering in Medicine & Biology Society* no. *28* (1):12–22. doi:10.1109/MEMB.2008.931012.

Batra, Sonali, and Sumit Sharma. 2019. "Chapter 14 - Trend of nanofibers in dental regeneration: perspectives and challenges." In *Materials for Biomedical Engineering*, edited by Valentina Grumezescu and Alexandru Mihai Grumezescu, 463–484. Elsevier.

Bielinska, A., J. F. Kukowska-Latallo, J. Johnson, D. A. Tomalia, and J. R. Baker, Jr. 1996. "Regulation of in vitro gene expression using antisense oligonucleotides or antisense expression plasmids transfected using starburst PAMAM dendrimers." *Nucleic Acids Res* no. *24* (11):2176–2182. doi:10.1093/nar/24.11.2176.

Chan, J. C., K. Burugapalli, H. Naik, J. L. Kelly, and A. Pandit. 2008. "Amine functionalization of cholecyst-derived extracellular matrix with generation 1 PAMAM dendrimer." *Biomacromolecules* no. *9* (2):528–536. doi:10.1021/bm701055k.

Cheng, Zhiliang, Daniel L. J. Thorek, and Andrew Tsourkas. 2010. "Gadolinium-conjugated dendrimer nano-clusters as a tumor-targeted T1 magnetic resonance imaging contrast agent." *Angewandte Chemie (International ed. in English)* no. *49* (2):346–350. doi:10.1002/anie.200905133.

Dufes, C., W. N. Keith, A. Bilsland, I. Proutski, I. F. Uchegbu, and A. G. Schatzlein. 2005. "Synthetic anticancer gene medicine exploits intrinsic antitumor activity of cationic vector to cure established tumors." *Cancer Res* no. *65* (18):8079–8084. doi:10.1158/0008-5472.CAN-04-4402.

Elangovan, Satheesh, Sheetal Reginald R. D'mello, Anh-Vu T. Do, Liu Hong, Behnoush Khorsand-Sourkohi, Aliasger K. Salem, and Michael Kormann. 2014. *RNA based Biomaterial for Tissue Engineering Applications.* United States.

Fang, Yi, Chen Peng, Rui Guo, Linfeng Zheng, Jinbao Qin, Benqing Zhou, Mingwu Shen, Xinwu Lu, Guixiang Zhang, and Xiangyang Shi. 2013. "Dendrimer-stabilized bismuth sulfide nanoparticles: synthesis, characterization, and potential computed tomography imaging applications." *Analyst* no. *138* (11):3172–3180. doi:10.1039/C3AN00237C.

Feng, G., I. Djordjevic, V. Mogal, R. O'Rorke, O. Pokholenko, and T. W. Steele. 2016. "Elastic light tunable tissue adhesive Dendrimers." *Macromol Biosci* no. *16* (7):1072–1082. doi:10.1002/mabi.201600033.

Fu, F., Y. Wu, J. Zhu, S. Wen, M. Shen, and X. Shi. 2014. "Multifunctional lactobionic acid-modified dendrimers for targeted drug delivery to liver cancer cells: investigating the role played by PEG spacer." *ACS Appl Mater Interfaces* no. *6* (18):16416–16425. doi:10.1021/am504849x.

Gorain, Bapi, Muktika Tekade, Prashant Kesharwani, Arun K. Iyer, Kiran Kalia, and Rakesh Kumar Tekade. 2017. "The use of nanoscaffolds and dendrimers in tissue engineering." *Drug Discov Today* no. *22* (4):652–664. doi:10.1016/j.drudis.2016.12.007.

Grabchev, I., D. Staneva, and R. Betcheva. 2012. "Fluorescent Dendrimers as sensors for biologically important metal cations." *Curr Med Chem* no. *19* (29):4976–4983. doi:10.2174/0929867311209024976.

Huang, Ching-Hui, Kido Nwe, Ajlan Al Zaki, Martin W. Brechbiel, and Andrew Tsourkas. 2012. "Biodegradable polydisulfide Dendrimer nanoclusters as MRI contrast agents." *ACS Nano* no. *6* (11):9416–9424. doi:10.1021/nn304160p.

Jain, K., A. K. Verma, P. R. Mishra, and N. K. Jain. 2015. "Characterization and evaluation of amphotericin B loaded MDP conjugated poly(propylene imine) dendrimers." *Nanomedicine* no. *11* (3):705–713. doi:10.1016/j.nano.2014.11.008.

Joshi, Neel, and Mark Grinstaff. 2008. "Applications of Dendrimers in tissue engineering." *Curr Top Med Chem* no. 8:1225–1236. doi:10.2174/156802608785849067.

Kesharwani, Prashant, Mohd Cairul Iqbal Mohd Amin, Namita Giri, Ashay Jain, and Virendra Gajbhiye. 2017. "Chapter 11 - Dendrimers in targeting and delivery of drugs." In *Nanotechnology-Based Approaches for Targeting and Delivery of Drugs and Genes*, edited by Vijay Mishra, Prashant Kesharwani, Mohd Cairul Iqbal Mohd Amin and Arun Iyer, 363–388. Academic Press.

Krause, W., N. Hackmann-Schlichter, F.K. Maier, and R. Müller. 2000. "Dendrimers in diagnostics." In *Dendrimers II. Topics in Current Chemistry*, edited by Vögtle F. Berlin: Springer.

Li, Y., H. He, X. Jia, W. L. Lu, J. Lou, and Y. Wei. 2012. "A dual-targeting nanocarrier based on poly(amidoamine) dendrimers conjugated with transferrin and tamoxifen for treating brain gliomas." *Biomaterials* no. *33* (15):3899–3908. doi:10.1016/j.biomaterials.2012.02.004.

Liu, H., H. Chang, J. Lv, C. Jiang, Z. Li, F. Wang, H. Wang, M. Wang, C. Liu, X. Wang, N. Shao, B. He, W. Shen, Q. Zhang, and Y. Cheng. 2016. "Screening of efficient siRNA carriers in a library of surface-engineered dendrimers." *Sci Rep* no. 6:25069. doi:10.1038/srep25069.

Liu, H., Y. Xu, S. Wen, Q. Chen, L. Zheng, M. Shen, J. Zhao, G. Zhang, and X. Shi. 2013. "Targeted tumor computed tomography imaging using low-generation dendrimer-stabilized gold nanoparticles." *Chemistry* no. *19* (20):6409–6416. doi:10.1002/chem.201204612.

Markowicz-Piasecka, Magdalena, Joanna Sikora, Szyma, Pawel Szymabski, Oliwia Kozak, Micha Studniarek, and Elhbieta Mikiciuk-Olasik. 2015. "PAMAM Dendrimers as potential carriers of gadolinium complexes of iminodiacetic acid derivatives for magnetic resonance imaging." *J Nanomater* no. *2015*:11. doi:10.1155/2015/394827.

Nagpal, Kalpana, Anand Mohan, Sourav Thakur, and Pradeep Kumar. 2018. "Dendritic platforms for biomimicry and biotechnological applications." *Artif. Cell Nanomed Biotechnol* no. *46* (sup1):861–875. doi:10.1080/21691401.2018.1438451.

Rai, Divya Bharti, Nitin Gupta, Deep Pooja, and Hitesh Kulhari. 2020. "13 - Dendrimers for diagnostic applications." In *Pharmaceutical Applications of Dendrimers*, edited by Abhay Chauhan and Hitesh Kulhari, 291–324. Elsevier.

Ray, Sayoni, Zhao Li, Chao-Hsiung Hsu, Lian-Pin Hwang, Ying-Chih Lin, Pi-Tai Chou, and Yung-Ya Lin. 2018. "Dendrimer- and copolymer-based nanoparticles for magnetic resonance cancer theranostics." *Theranostics* no. *8* (22):6322–6349. doi:10.7150/thno.27828.

Sharma, Sumit, and Sonali Batra. 2019. "Chapter 5 - Recent advances of chitosan composites in artificial skin: the next era for potential biomedical application." In *Materials for Biomedical Engineering*, edited by Alina-Maria Holban and Alexandru Mihai Grumezescu, 97–119. Elsevier.

Sk Ugir, Hossain, and Chie Kojima. 2015. "Dendrimers for theranostic applications." *Biomol Concepts* no. *6* (3):205–217. doi: 10.1515/bmc-2015-0012.

Szulc, Aleksandra, Lukasz Pulaski, Dietmar Appelhans, Brigitte Voit, and Barbara Klajnert-Maculewicz. 2016. "Sugar-modified poly(propylene imine) dendrimers as drug delivery agents for cytarabine to overcome drug resistance." *Int J Pharm* no. *513* (1):572–583. doi:10.1016/j.ijpharm.2016.09.063.

Wada, K., H. Arima, T. Tsutsumi, F. Hirayama, and K. Uekama. 2005. "Enhancing effects of galactosylated dendrimer/alpha-cyclodextrin conjugates on gene transfer efficiency." *Biol Pharm Bull* no. *28* (3):500–505. doi:10.1248/bpb.28.500.

Wang, Mingming, Hongmei Liu, Lei Li, and Yiyun Cheng. 2014. "A fluorinated dendrimer achieves excellent gene transfection efficacy at extremely low nitrogen to phosphorus ratios." *Nat Commun* no. *5* (1):3053. doi:10.1038/ncomms4053.

Yalcin, S., M. Erkan, G. Unsoy, M. Parsian, J. Kleeff, and U. Gunduz. 2014. "Effect of gemcitabine and retinoic acid loaded PAMAM dendrimer-coated magnetic nanoparticles on pancreatic cancer and stellate cell lines." *Biomed Pharmacother* no. *68* (6):737–743. doi:10.1016/j.biopha.2014.07.003.

Yamashita, S., H. Katsumi, T. Sakane, and A. Yamamoto. 2018. "Bone-targeting dendrimer for the delivery of methotrexate and treatment of bone metastasis." *J Drug Target* no. *26* (9):818–828. doi:10.1080/1061186X.2018.1434659.

Yang, J., Q. Zhang, H. Chang, and Y. Cheng. 2015. "Surface-engineered dendrimers in gene delivery." *Chem Rev* no. *115* (11):5274–5300. doi:10.1021/cr500542t.

Zhang, Mengen, Jingyi Zhu, Yun Zheng, Rui Guo, Shige Wang, Serge Mignani, Anne-Marie Caminade, Jean-Pierre Majoral, and Xiangyang Shi. 2018. "Doxorubicin-conjugated PAMAM Dendrimers for pH-responsive drug release and folic acid-targeted cancer therapy." *Pharmaceutics* no. *10* (3):162. doi:10.3390/pharmaceutics10030162.

Zhao, Lingzhou, Xiangyang Shi, and Jinhua Zhao. 2017b. "Dendrimer-based contrast agents for PET imaging." *Drug Deliv* no. *24* (2):81–93. doi:10.1080/10717544.2017.1399299.

Zhao, Lingzhou, Meilin Zhu, Yujie Li, Yan Xing, and Jinhua Zhao. 2017a. "Radiolabeled Dendrimers for Nuclear Medicine Applications." *Molecules (Basel, Switzerland)* no. *22* (9):1350. doi:10.3390/molecules22091350.

Zhou, Z., A. D'Emanuele, and D. Attwood. 2013. "Solubility enhancement of paclitaxel using a linear-dendritic block copolymer." *Int J Pharm* no. *452* (1–2):173–179. doi:10.1016/j.ijpharm.2013.04.075.

Ziemba, B., I. Halets, D. Shcharbin, D. Appelhans, B. Voit, I. Pieszynski, M. Bryszewska, and B. Klajnert. 2012. "Influence of fourth generation poly(propyleneimine) dendrimers on blood cells." *J Biomed Mater Res A* no. *100* (11):2870–2880. doi:10.1002/jbm.a.34222.

16

Biodistribution, Toxicity and Regulatory Considerations of Dendrimers

Anisha D'Souza and Pratikkumar Patel

CONTENTS

16.1 Introduction

Dendrimers are nanosized three-dimensional, repeated scaffolds of polymeric macromolecules (Grayson and Frechet 2001). Dendrimers radiate from a central core (focal point) with repeated unit layers or are also called generations, ranging from 1 to 12. Thus overall, architecturally, they have a distinguishable core, generations and terminal functional units present at the exterior (Jain et al. 2010). They are used in the delivery of drugs or genetic materials (Zhang and Smith 2000), as discussed in the earlier chapters of this book. Figure 16.1a shows a schematic representation of a dendrimer structure.

Due to their unique chemical composition, dendrimers have been fabricated into multipurpose carriers for delivering DNA, RNA, nucleic acids and small molecules and in diagnostics. Though Flory was the first to publish a paper on branched polymers in 1948 (Moura et al. 2019), the first successful dendritic synthesis was carried out in 1978 by Vogtle, followed by Tomalia and a group from Dow Chemicals.

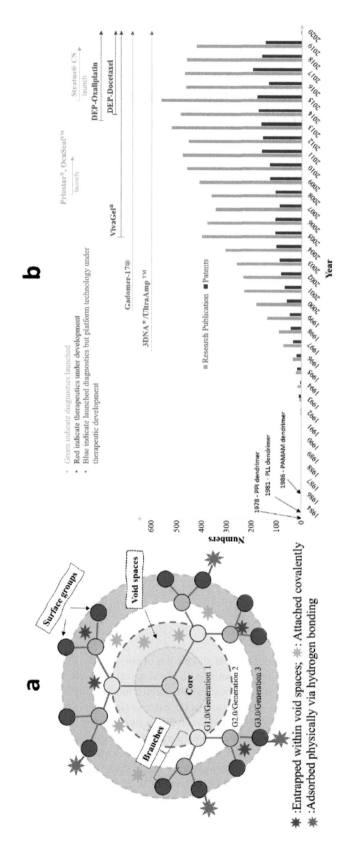

FIGURE 16.1 Representation of dendrimers having different generations and year wise publications. (a) A typical schematic representation of dendrimer with different generations. The drug is entrapped either within void spaces or attached covalently to dendrimers or physically adsorbed via hydrogen bonding. (b) Year-wise publications, patents and clinical success of dendrimers-based products for therapeutics and diagnostics. Over the years, there has been a gradual increase in the research publications and filing of patents (Source: Scifinder accessed on January 2, 2020).

Polyamidoamine (PAMAM) was the first starburst dendrimer with ammonia as the core. Subsequently, poly(amidoamine-organosilicon) (the first commercial dendrimer), diaminobutane (DAB) dendrimers comprising poly(propyleneimine) (PPI) and poly(propylene amine) tecto-dendrimers, chiral dendrimers, amphiphilic dendrimers, micellar dendrimers, multiple antigen peptide dendrimers, Frechet type dendrimers and so on were developed (Surendra and Das 2013).

The first commercially available therapeutic dendrimer was VivaGel®, a topical gel containing astodrimer sodium, SPL7013 and polyanionic poly(L-lysine) (PLL) for the treatment of bacterial vaginosis developed by StarPharma Holding Limited (Melbourne, Australia) (Kannan et al. 2014; Moura et al. 2019). The gel has received phase III approval for bacterial vaginosis therapy. In a similar line, another dendrimer based on poly(lysine) containing docetaxel (DEP® docetaxel) has advanced to Phase II trials for solid types of cancer.

Many dendrimer-related research publications and their clinical applications have been growing exponentially since the last couple of decades (Figure 16.1b). Despite a plethora of publications and patents, there has not been substantial clinical, regulatory or commercial approvals of dendrimers. Between 1974 and 2015, about 359 investigational new drug (IND) applications were submitted related to nanomaterials. Amongst them, approximately 65% of applications were for cancer treatment using liposomal formulations followed by nanocrystals, emulsions, iron–polymer complexes and micelles (Mignani et al. 2018). The progress of dendrimers has been hampered to their fullest applicability due to the inherent cytotoxicity and haematological toxicity of cationic dendrimers. Nevertheless, the potential of their therapeutic applications cannot be overlooked and remains a considerable interest to support the development (Nijhara and Balakrishnan 2006). If the herculean barrier of dendrimer toxicity needs to be minimised, the impact of various types of dendrimers on human biochemical pathways needs to be clearly understood. Based on these interactions, new strategies for the synthesis and functionalisation of dendrimers are necessary to make them a safer delivery carrier. The current chapter would discuss the biodistribution, toxicology and the translational prerequisites of dendrimers to move towards the IND application to reach clinical and diagnostic applications.

16.2 Biodistribution of Dendrimers

Besides the anatomy and physiology characteristics of the biological system, the architecture and physicochemical properties of dendrimers influence its pharmacokinetics and biodistribution. Upon intravenous administration, dendrimers are rapidly bioavailable into the systemic circulation. The dendrimers are then cleared from the systemic circulation (plasma clearance) by diffusion to extravascular space or specific organs for elimination. Dendrimers with molecular weight 30–50 kDa and 5–6 nm in size are readily excreted via glomerular filtration, while higher molecular weight undergoes the first-pass effect (Wijagkanalan, Kawakami and Hashida 2011). Moreover, positively charged dendrimers can be adsorbed electrostatically to any negatively charged cells especially of liver and kidneys (Brenner, Hostetter and Humes 1978; Gerlowski and Jain 1983). Polyanionic molecules undergo non-parenchymal uptake in the liver. In the active- or ligand-targeted delivery, the carriers undergo specific receptor-mediated uptake. It is not only dependent on ligand affinity but also on the physicochemical properties of molecules such as hydrophilicity, steric hindrance and saturation dependency. In addition to this, plasma clearance also plays a significant role in tissue uptake. Hence to achieve organ-specificity and maximum therapeutic efficiency, the biodistribution of dendrimers should be optimised (Wijagkanalan, Kawakami and Hashida 2011).

The critical nanoscale parameters such as size, surface chemistry, shape, flexibility/rigidity, elemental composition and architecture have been used to predict the dendrimers' properties, especially their biodistribution, pharmacokinetics, endocytosis and toxicology. These patterns are known to be incremental to changes in these nanoscale parameters. The individual parameters influencing the biodistribution, toxicology and pharmacokinetics are often difficult to identify distinctly. Biodistribution and toxicology have been impacted majorly by surface chemistry and size rather than the interior compositions. Exceptions are the lower generation dendrimers whose biodistribution and toxicology are dependent on their interior compositions due to their architectural flexibility. A detailed review of the same is given in Tomalia,

Christensen and Boas (2012) and Shcharbin et al. (2014). The influence of nanoscale parameters on dendrimers' *in vivo* behaviour is discussed in detail below.

16.2.1 Particle Size

The particle size of dendrimers is directly proportional to the number of generations, i.e., higher the generation, larger is the size of the dendrimers. Though the molecular weight could double, the particle size would rarely change more than 20–25% of the earlier generation. Biodistribution of dendrimers is very sensitive to the particle size. Changes as small as 1–3 nm alter the distribution.

Cationic PAMAM dendrimers were synthesised (G1.0–G10.0, 1–935 kDa, 1–15 nm) using ethylenediamine and an ammonia core conjugated to diethylenetriamine penta-acetic acid chelator (used for chelating gadolinium (Gd)) and were studied for biodistribution. An increase in the particle size (those with G5.0–G8.0 and molecular weight 88–954 kDa) resulted in prolonged systemic circulation and reduced glomerular elimination (Tomalia, Reyna and Svenson 2007; Longmire, Choyke and Kobayashi 2008). Further increase in a generation (G9.0) resulted in the first-pass effect by the reticuloendothelial system (RES) and failed to be retained in the systemic circulation. An increase of more than 10 nm in particle size causes protein opsonisation, followed by RES uptake (Kobayashi et al. 2006).

Anionic charged [153]Gd-radiolabelled PAMAM dendrimers (PAMAM-Gd) increased the molecular weight of G1.0–G10.0 PAMAM dendrimers up to 7–3820 kDa. Intravenous injection of these dendrimers exhibited rapid systemic clearance and accumulation in the liver, spleen and kidneys according to their respective size and molecular weight. G5.0 PAMAM-Gd exhibited partial glomerular filtration despite higher molecular weight due to the compact structure of dendrimers. G6.0 PAMAM-Gd showed the lowest renal excretion of approximately 15% of the dose (Kobayashi, Sato et al. 2001; Kobayashi, Kawamoto, Saga, Sato, Hiraga and Konishi et al. 2001b; Bryant et al. 2002), while G6.0–G8.0 showed retention in the systemic circulation. Thus, lower generation or smaller particle size dendrimers, preferably below G6.0, showed rapid systemic clearance and nonspecific accumulation (Kannan et al. 2014). In leaky tumour vasculature, G7.0 PAMAM-Gd dendrimers are retained more than G8.0 PAMAM-Gd dendrimers (Kobayashi, Kawamoto, Saga, Sato, Hiraga and Konishi et al. 2001b; Sarin et al. 2009).

G8.0 PAMAM-Gd exhibited lower permeability and had restricted distribution to the lymphatic vessels after intracutaneous injection (Kobayashi, Kawamoto and Choyke et al. 2003; Kobayashi et al. 2006). The biodistribution of dendrimers with various particle sizes and generations is depicted in Figure 16.2. Nevertheless, permeability studies of G4.0 PAMAM dendrimers were higher than G1.0 or G0.0 > G3.0 > G2.0 PAMAM dendrimers in Madin–Darby Canine Kidney cell lines (Tajarobi et al. 2001). Preliminary studies of G4.0 PAMAM dendrimers were unable to cross the human placenta *ex vivo* (Menjoge et al. 2011).

Smaller sized particles are thus often cleared by the biliary or glomerular filtration (Kobayashi, Watanabe and Choyke 2013), while larger sized dendrimers show lower systemic and renal clearance but a higher accumulation in lymph nodes (Kobayashi, Kawamoto and Star et al. 2003). Thus, to summarise, G2.0–G4.0 would be ideal for kidney/bladder targeting while G4.0–G7.0 for targeting the normal vasculature and G8.0 onwards for targeting the lymph circulation or tumour vasculature. Under tumour or cancer conditions, smaller sized dendrimers could also extravasate from the defective blood vessels and show non-specific distribution and renal excretion except for PEGylated molecules (Moura et al. 2019). For enhanced permeability and retention (EPR) effect, particles with neutral charge and a range of particle sizes are known to be profoundly responsible for better pharmacokinetics/pharmacodynamics of the nanoparticles. (Mignani et al. 2018). The designing of the delivery vehicle can accordingly be determinant of particle size. Moreover, the side effects of these dendrimers can thus be considerably reduced with lower generation dendrimers.

Dendrimers have also been explored via the skin for certain cosmetic and topical uses. Higher generation dendrimers > G6.0 being a poor skin permeator can be used for cosmetic purposes and other topical applications. These G2.0–G5.0 dendrimers can lead to better penetration through skin appendages and stratum corneum and can be localised up to the viable epidermis layer. Neutral dendrimers can be explored for release within epidermal interstitium, cationic dendrimers for intracellular release, while G1.0–G4.0 dendrimers can be used for transdermal administration for a systemic effect when combined with permeation strategies especially iontophoresis and ultrasound (Dave and Krishna Venuganti 2017).

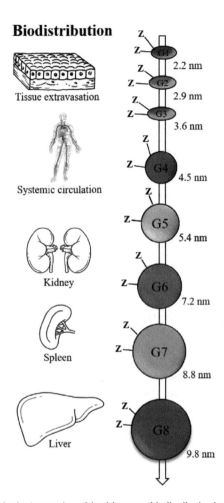

FIGURE 16.2   Influence of particle size/generation of dendrimers on biodistribution in the systemic circulation. Depending on the size, dendrimers can be used to target various organs. The lower generation dendrimers get extravasated into tissues while higher generations end up in the liver. Z indicates a functional group. As the generation increases in number, the surface density increases.

16.2.2 Surface Charge and Surface Chemistry

Cationic dendrimers are an ideal choice for gene transfection such as DNA or siRNA. However, these dendrimers exhibit rapid systemic clearance and non-specific distribution in the liver, spleen, lungs, kidneys and pancreas (Malik et al. 2000; Tomalia, Reyna and Svenson 2007).

Neutral and anionic dendrimers show higher systemic levels compared to cationic dendrimers.

Small and uncharged dendrimers with size less than 25 kDa are rapidly excreted through urinary excretion. Although dendrimers greater than G6.0 accumulate in the liver, G5.0–G7.0 radiolabelled poly(ester) dendrimers with hydroxyl groups are excreted via kidneys (Parrott et al. 2009).

Amine-terminated G5.0 PAMAM dendrimers fabricated with gold nanocomposites of size 5 nm were studied for the impact of surface charge on biodistribution of dendrimers in tumours and normal vasculature. Different charges were induced on the surface of the polyionic particles in dilute aqueous solutions at different biological pH values. In aqueous solutions, the dendrimers were positively charged due to the protonation of terminal amines. For instance, dendrimers conjugated to gold with positive charge $(Au^0)_9$-PAMAM-$E_5(NH_2)_{110}$ having pH 7.0, partially positive charge (PAMAM-$E_5(NH_2)_{44}$ (NHCOCH$_3$)$_{66}$ having pH 7.0, and neutrally charged $(Au^0)_9$-PAMAM-E_5.(NHCOCH$_3$)$_{110}$ and (PAMAM-E_5(NHCOCH$_3$)$_{110}$. It was found that positively charged dendrimers with and without an Au nanocomposite showed toxicity upon intravenous administration, while dendrimers without an Au nanocomposite caused death in B16 melanoma mice. The neutrally charged and partially positively charged Au nanocomposite showed lower accumulation in brain tissues and heart. The neutrally charged Au nanocomposite dendrimers were accumulated in the liver and spleen of the mice (Khan et al. 2005). Technetium ($^{99m}TcO_4^-$) radiolabelled lactosylated/ mannosylated and unmodified G5.0 PPI dendrimers were studied for biodistribution in mice. The unmodified G5.0 PPI dendrimers were excreted through kidneys. However, the glycosylated dendrimers showed a higher preference for liver uptake with a faster clearance rate. The clearance pattern was found to be dependent upon its concentration and was biphasic in nature (Agashe et al. 2007).

Injection of G4.0 PAMAM dendrimers into rabbits with cerebral palsy showed the accumulation of dendrimers in astrocytes and microglia of the injured brain parts in the rabbit (Greish et al. 2012). A similar observation was made in the canine model as well (Mishra et al. 2014). However, no accumulation of dendrimers was detected after 24 h of dendrimer injection into the cerebrospinal fluid of a healthy rabbit (Kannan et al. 2012).

Intraventricular injection of polar dendrimer-amino-terminated G4.0 PAMAM dendrimers and apolar lipophilic G4.0 PAMAM dendrimers (G4-C_{12} PAMAM) was studied for brain and central nervous system (CNS) delivery. Amino-terminated G4.0 PAMAM dendrimers freely diffused into the cerebrospinal fluid and permeated the brain parenchyma through the ependymal cell borer. Localisation could be seen up to the outer cortical layers. The lipophilic G4-C_{12} PAMAM dendrimers were retained in the proximity with no accumulation in brain parenchyma. The lipidic C_{12} moieties accumulated within the ependymal cell layer lining the ventricles. Subarachnoid injection of G4.0 PAMAM dendrimers too diffused up to the CNS dural surface and escaped the ependyma. G4.0 thus can be used for targeting deep brain parenchymal cells. Intraparenchymal injection of G4.0 completely diffused from the site of injection due to the strong interaction of G4.0 with surrounding membranes. The diffusion of G4-C_{12} was completely blocked within place probably due to the hydrophobicity of C_{-12} lipid (Albertazzi et al. 2013).

Oral delivery of G6.5 PAMAM dendrimers could permeate the dendrimers through the gut epithelial barrier of CD-1 mice during the evaluation of acute oral toxicity (Thiagarajan et al. 2013). The maximum tolerated dose was 10-fold higher for anionic dendrimers (G7.0 hydroxylated PAMAM dendrimers) compared to cationic dendrimers (10–200 mg/kg) (G7.0 amino-terminated PAMAM dendrimers) (Thiagarajan, Greish and Ghandehari 2013). Permeability of G2.0 PAMAM dendrimers across Caco-2 cells was in the following sequence: cationic > anionic > uncharged or PEGylated (Figure 16.3a). Moreover, microvilli of Caco-2 were disrupted in the presence of amino-terminated G4.0 PAMAM dendrimers compared to no cellular change observed with carboxylate terminated G3.5 PAMAM dendrimers at the same concentration (Kitchens et al. 2007). Similar results were reported for cationic charged dendrimers in another study (Jevprasesphant et al. 2003).

Similarly, the influence of change in the particle size on the penetration of dendrimers during transdermal delivery has been observed and is depicted in Figure 16.3b. For percutaneous absorption, the order was found to be cationic > uncharged > anionic (Kaminskas, Boyd and Porter 2011). The large molecular weight of dendrimers does not ideally fit for transdermal and skin penetration; however, their physicochemical properties could make it possible. Cationic dendrimers can penetrate the deeper layers of skin by virtue of positive charge interaction with the negative skin surface and intercellular skin lipids. The appendages having a wider opening up to 70 microns occupy a lesser percentage for the skin surface. G4.0 PAMAM dendrimers can penetrate the intercellular matrix with limited penetration of anionic dendrimers due to charge. Lower generation dendrimers could penetrate better than higher generation PAMAM dendrimers (Cevc and Vierl 2010).

16.2.3 Hydrophobicity

The nature of the central core also plays an important role in the biodistribution. During the clearance study of equivalent generation sized PPI dendrimers and PAMAM dendrimers, it was found that PPI dendrimers were systemically cleared at a higher rate than PAMAM dendrimers due to the hydrophobic interiors. The elimination, however, was found to be dependent on hepatic clearance. Likewise, G5.0 PPI dendrimers derivatised from DAB and G4.0 PAMAM dendrimers of a similar molecular weight led to more accumulation of G5.0 PPI into the lymphatic system due to hydrophobicity conferred by the DAB (Kobayashi et al. 2006). In another study, DAB used in PPI dendrimers exhibited more rapid systemic clearance with a two-fold increase in liver accumulation than PAMAM dendrimers of similar surface functionality and generation (Kobayashi, Kawamoto and Jo et al. 2003).

16.2.4 Particle Shape

The lower generation dendrimers from G1.0 to G4.0 are planar and elliptical, and G5.0–G10.0 are dense spherical shaped dendrimers (Moura et al. 2019) (Figure 16.2). Radially symmetric dendrimers showed higher entrapment efficiency; however, optimisation of the dendrimers' symmetry and density needs to be balanced out for better activity. Elongated and flexible nanocarriers are reported to have better drug delivery potential than the spherically dense ones (Svenson 2015).

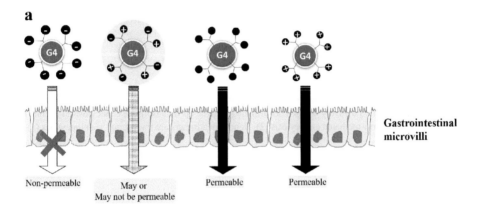

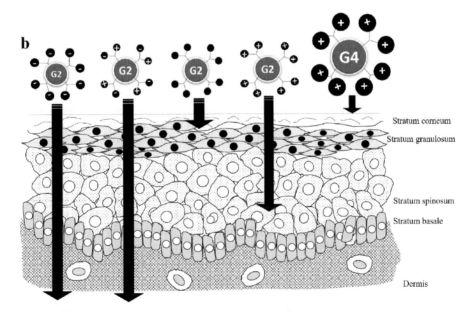

FIGURE 16.3 Pathways of dendrimers across different barriers. (a) Permeation of different charge dendrimers across gastroendothelial microvilli. Anionically charged dendrimers do not permeate across the microvilli, while neutrally charged and positively charged dendrimers permeate through gastroendothelial microvilli. (b) Permeation of dendrimers having different particle sizes and charges across the skin layers. Positively charged G4.0 does not permeate across stratum corneum while positively charged lower generation G2.0 permeates up to stratum spinosum. The negatively and neutrally charged polar dendrimers G2.0 permeate up to deeper layers of the skin, i.e., dermis, while non-polar dendrimers G2.0 reach up to stratum corneum. *(Continued)*

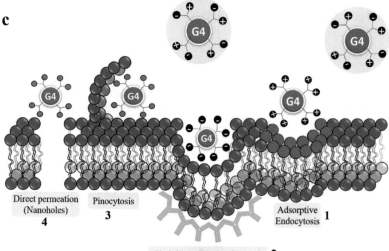

FIGURE 16.3 (Continued)   Pathways of dendrimers across different barriers. (c) Cellular uptake mechanism and influence of surface charge of dendrimers. Positively charged dendrimers are taken up via adsorptive endocytosis (1), while negatively charged are taken up via clathrin-mediated endocytosis (2). Neutral dendrimers are transported via pinocytosis (3) or permeated directly (4).

16.3 Toxicity of Dendrimers

Toxicity is a complex sequence of undesirable events, and the severity depends upon nanoscale design parameters of dendrimers, dose, route of administration and physiological interactions (Vega-Villa et al. 2008). Clinical development of dendrimers is possible only if the dendrimers are non-toxic and safe to the patients. Nanosized formulations by virtue of their size are associated with a risk of interacting at the cellular and sub-cellular levels (Jain et al. 2010). Hence, toxicity needs to be evaluated in *in vitro* and *in vivo* models for cell viability, immunogenicity, biocompatibility and haemocompatibility (Jain et al. 2010). The toxicokinetic study should be designed carefully based on earlier clinical experience for similar carriers/vectors for proposed clinical use. The non-selective uptake of nanosized particles is a concern for cytotoxicity. Different cellular uptake mechanisms for any nanosized carriers are depicted in Figure 16.3c. *In vitro,* cytotoxicity has been studied on different cell lines at different concentrations and different times of incubation using different methods of the assay. Cytotoxicity is dependent on the generation and functional group chemistry on the surface of the dendrimers (Roberts, Bhalgat, Zera 1996).

Dendrimers are suggested to undergo paracellular transportation across epithelial membranes as studied with a polyamine, ornithine and arginine conjugated to amine terminals of G4.0 PAMAM dendrimers (Pisal et al. 2008). Kitchens and co-authors have investigated that G0–G4.0 amino-terminated PAMAM dendrimers crossed the biological membranes. The potential pathways could be either paracellular or endocytic, with increased permeability of ^{14}C-mannitol along with the release of lactate dehydrogenase (Roberts, Bhalgat, Zera 1996). Nevertheless, molecular dynamics of PAMAM dendrimers along with 1,2-dimyristoyl-sn-glycero3-phosphocholine bilayers suggested that the spheroidal dendrimers enhanced pore formation and membrane disruption. The release of lactate dehydrogenase was enhanced with the concentration of dendrimers and with the increase in its charge densities (Lee and Larson 2008). A similar concept of nanohole formation and lysis of cells has been reported in biological and living cell membranes upon interaction with positively charged dendrimers (Jain et al. 2010). Some reports indicate that the internalisation pathway and transfection of genetic materials by dendrimers vary in different cells (Manunta et al. 2004, 2006). Manunta et al. (2006) have reported cholesterol-dependent pathways via membrane rafts for uptake of dendrimer-mediated gene delivery in endothelial cells compared to those observed in HepG2 and HeLa cell lines. Disruption of biological cellular integrity raises the concern of dendrimer toxicity. In general, the toxicity of dendrimers can be reduced by complexing with guest molecules via electrostatic interaction (Chauhan, Jain and Diwan 2010). The toxicity of dendrimers is influenced by surface charge and surface chemistry and discussed in the following section.

16.3.1 Surface Charge and Surface Chemistry

Terminal functional groups of dendrimers are crucial to determine their toxicity. Positively charged dendrimers such as PPI, PLL and PAMAM are popular as non-viral carriers in transfection but exhibit significant cytotoxicity *in vitro*. The positive charge is essential for the formation of the optimum dendrimer–gene complex and interacts with the biological membranes electrostatically and promotes intracellular delivery (Moura et al. 2019). However, in the process, they disrupt membrane integrity inducing toxicity and cell death (Jain et al. 2010). This poor biocompatibility is shown despite their smaller size and hydrophilicity, limiting their application *in vivo* (Mignani et al. 2018). The density/concentration of free primary amine terminal groups of G5.0 PPI on cytotoxicity in COS-7 and HepG2 cells has been demonstrated (Agashe et al. 2006). Cytotoxicity of amine-terminated PAMAM dendrimers also increases with an increase in the generation level (greater than G5.0). Surface-modified amino dendrimers are more cytotoxic than negatively charged dendrimers modified like hydroxyl, carboxylate, acetamido, phosphonate or sulphonate functionalised or neutral surfaces of PEGylated dendrimers. G4.0 PAMAM dendrimers with hydroxyl groups depicted no hepatocellular injury or change in renal function or any neurobehavioural change (Kannan et al. 2012).

However, some positively charged PAMAM and PPI-DAB dendrimers failed to induce apoptosis in NIH/3T3 mouse fibroblast cells and BNL CL.2 mouse liver cells compared to macrophage RAW 264.7 cells. Apoptosis in macrophage cells involves a caspase-dependent pathway (Kuo, Jan and Chiu 2005). Apoptosis of apolar lipophilic G4-C_{12} PAMAM dendrimers was higher with decreased viability of neural cultures compared to unmodified amino-terminated G4.0 PAMAM dendrimers. Though G4.0 PAMAM exhibited more diffusion, it showed lower apoptotic cell death and failed to activate the microglia or the blood–brain barrier integrity. The authors have emphasised that the toxicity is lower when the concentrations of dendrimers are at nanoconcentrations, albeit other research papers studying at micromolar concentrations (Albertazzi et al. 2013). G2.0 amino-terminated PAMAM dendrimers are effective as an antibacterial agent, but they exhibited slight toxicity in human gastric epithelial cells. However, they did not induce any antibiotic resistance in bacteria (Xue et al. 2013). Substituted amino-terminated G3.0 and G5.0 PAMAM dendrimers with polyethylene glycol (PEG) chains retained the antibacterial effect against *S. aureus* and *P. aeruginosa* with low toxicity. Similarly, amino-substituted G4.0 PAMAM with nitric oxide-releasing moieties revealed higher antibacterial activity with low mammalian cell toxicity compared to unmodified dendrimers (Lu et al. 2013; Worley, Schilly and Schoenfisch 2015).

Positively charged dendrimers are also known to cause haemotoxicity resulting in haemolysis and subsequently altered haematological parameters. As seen in cytotoxicity for positively charged amino-terminated dendrimers, haemotoxicity is also generation- and concentration-dependent. Positively charged G7.0 PAMAM dendrimers caused *in vitro* platelet disruption and blood coagulation due to fibrinogen and albumin aggregation at 100 µg/mL (Jones, Campbell and Franks et al. 2012).

Hydroxyl- and carboxylic-terminated dendrimers were nontoxic in mice up to an oral dose of 500 mg/kg (Jones, Campbell and Brooks et al. 2012; Jones, Campbell and Franks et al. 2012; Greish et al. 2012). Amino-terminated PAMAM dendrimers were lethal to mice due to platelet disruption compared to hydroxyl-terminated neutral dendrimers and carboxyl terminated PAMAM dendrimers, which did not impair them (Jones, Campbell and Franks et al. 2012). Poly(ethylene oxide) (PEO)-grafted carbosilane with diaminoethane (DAE) as the core and polyether dendrimers neither exhibited cytotoxicity nor haemolysis in CCRF and HepG2 cells. However, cytotoxicity with the lower generation of PEO-grafted carbosilane dendrimers was observed in B16F10 cells at higher concentrations. Toxicity of PEO-grafted carbosilane reduced with an increase in branching. Increased branching rendered the dendrimer core inaccessible to the cells. Hence, the branching of PEO resulted in increased dendrimer biocompatibility. Polyether dendrimers functionalised with malonate and carboxylate groups were non-haemolytic at 1 h but were haemolytic after 24 h, which was not observed in anionic PAMAM dendrimers. Additionally, positively charged PAMAM dendrimers were lesser toxic compared to the cationic DAB and DAE-PPI dendrimers with an equivalent number of amino groups (Malik et al. 2000). Functionalisation of G5.0 PPI dendrimers with mannose, tert-butyloxycarbonyl and tuftsin masked the primary amino acids and decreased the G5.0 PPI toxicity (Dutta et al. 2008). Likewise, cytotoxicity of PPI dendrimers in human umbilical vein endothelial cells could be decreased by neutrally charged acetamide derivatised PPI dendrimers and PEGylated dendrimers (Stasko et al. 2007).

Despite the cytotoxicity and haematotoxicity of dendrimers, the immunogenic potential of either humoral or antibody titre has not been reported. PAMAM dendrimers by themselves are non-immunogenic (Araujo et al. 2018). Unmodified, as well as modified PPI, also showed no immunogenicity (Dutta et al. 2008). However, magnetic resonance imaging (MRI) contrast conjugated dendrimers G4D-(1B4M-Gd)$_{62}$ could become immunogenic, which can be overcome by PEG (Kobayashi, Kawamoto, Saga, Sato, Hiraga and Ishimori et al. 2001a). This could be considered a positive aspect of the development of dendrimers. However, the molecular level of interaction between dendrimers, especially cationic surfaces with the cellular components, cannot be overlooked. They have binding affinity to biological systems, lipids, nucleic acids, vitamins, proteins and bile acids, which may stimulate the loss of their activities.

16.3.2 Particle Size

Injection of G3.0, G5.0 and G7.0 PAMAM dendrimers into mice at a dose of 10 mg/kg was found to be non-toxic up to the G5.0 PAMAM generations. G7.0 cationic PAMAM dendrimers also exhibited lower viability (~10%) in Chinese hamster lung fibroblast cells at 100 nM as against 10 μM of G5.0 dendrimers and 1 mM of G3.0 dendrimers after exposing to 4 h and 24 h. However, all the generations were non-immunogenic (Roberts, Bhalgat, Zera 1996; Bourne et al. 2000).

G7.0 PAMAM dendrimers demonstrated potential biological complications in Swiss-Webster mice amongst G3.0, G5.0 and G7.0 PAMAM dendrimers. However, the authors concluded that these dendrimers could be used in biological applications as they failed to exhibit any properties which could preclude their use. Moreover, the biodistribution needs to be evaluated (Roberts, Bhalgat, Zera 1996).

Intraperitoneal administration of melamine dendrimers at doses of 2.5, 10, 40 and 160 mg/kg was studied during the acute toxicity in mice. Complete mortality was observed within 6–12 h post-dosing with 160 mg/kg. Liver enzyme activity indicated hepatotoxicity in the group dosed at 40 mg/kg. The other doses exhibited no major mortality or any damage (Neerman et al. 2004). Amongst G4.0–G7.0 PAMAM dendrimers, G6.0 and G7.0 dendrimers induced leaky fusion and had a higher membrane disruptive potential (Zhang and Smith 2000). The nanoholes created by G7.0 PAMAM dendrimers were between 15 and 40 nm in diameter. G5.0 PAMAM dendrimers by themselves failed to induce any nanoholes; however, they expanded the size of any pre-existing defects of the bilayer (Mecke et al. 2004). Influence of G4.0, G5.0 and G6.0 PAMAM dendrimers was studied on SW480, primary colon adenocarcinoma cells and HaCaT, human keratinocytes. The increase in generation correlated with the increase in the reactive oxygen species (ROS) formed, increase in lysosomal activity, apoptotic induction and increase in DNA damage. After cellular uptake, PAMAM dendrimers were found to be in mitochondria. The timescales and the oxygen species were generation-dependent (Mukherjee, Davoren and Byrne 2010; Mukherjee, Lyng et al. 2010). The effect of G4.0 dendrimers on the photosynthetic microorganisms was also generation-dependent. The ROS was formed only with hydroxyl-terminated PAMAM dendrimers and strictly restricted to only mitochondria (Gonzalo et al. 2015).

G2.0 and G3.0 amino-terminated PAMAM dendrimers were studied for topical toxicity upon applications of 0.3, 3, 6, 30 and 300 mg/mL on rat skin. Epithelial cells showed morphological changes of cytoplasmic vacuolisation at concentrations ≥6 mg/mL. Granulocyte infiltration was observed in the upper dermis. The increase in nuclear immunoreactivity was correlated with the amino-terminated PAMAM dendrimers with no significant difference seen in G2.0 or G3.0. Higher expression of proliferating cell nuclear antigen suggested abnormal cell proliferation, which might lead to neoplastic changes (Winnicka et al. 2015).

The biocompatibility of dendrimers needs judicious modifications. Various strategies have been adopted and studied to decrease toxicity. Conjugation to biorthogonal groups, stimuli-responsive linkers such as disulphide or hydrazine bonds, PEG, folic acid, carbohydrates, pyrrolidone, any other anionic or neutral biocompatible moieties, etc. masks the cationic group (Mehta et al. 2019). Coating the dendrimers with PEG increases systemic circulation time depending upon the molecular weight of PEG and flexibility of dendrimers (Kojima et al. 2010, 2011). PEG having a molecular weight ranging from 2 to 5 kDa is preferred and is known to reduce toxicity in Caco-2 cells, endothelial cells and haemolysis (Wijagkanalan, Kawakami and Hashida 2011).

G4.0 PAMAM dendrimers functionalised with 4-carbomethoxypyrrolidone were non-apoptotic, at 200 μM with no change in ROS levels and mitochondrial membrane potential (Janaszewska et al. 2013, 2015). Carboxymethyl chitosan/ PAMAM dendrimers were internalised in the glioblastoma U87MG GBM cell line and hTERT/E6/E7 human immortalised astrocytes with no change in cell viability up to 48 h upon exposure at 200 and 400 μg/mL. However, a 20% decrease in cellular metabolism was observed up to seven days (Pojo et al. 2013).

Conjugation of maltose to amino-terminated G4.0 PPI dendrimers was effective against bacteria as well as fungi with lower toxicity in the hepatocellular carcinoma cell line (HepG2), Chinese hamster fibroblasts (B14), rat liver cell line (BRL-3A) and mouse neuroblastoma cell line (N2a) (Felczak et al. 2012). Those substituted with 25% of maltose were more effective than those with 100% substituted maltose (Felczak, Wrońska et al. 2012). Amino acid-based dendrimers exhibited antibacterial activity against resistant strains of *P. aeruginosa* and *A. baumanii* with low toxicity (Pires et al. 2015). A polyanionic ether-based lipophilic core and G1.0 dendrimers were effective against *B. subtilis* with low toxicity to human cells (Meyers et al. 2008). Folate-acetylated or folate-conjugated PAMAM dendrimers were less haemolytic (Asthana et al. 2005) and less toxic *in vivo* (Kukowska-Latallo et al. 2005).

Similar other derivatives include PEG-polyesters (Gillies et al. 2005; Lee et al. 2005) and glutamic acid (Miyano et al. 2010). A review paper on different amino acid-based, glycodendrimers, carbosilane and PAMAM dendrimers on their antiviral activity discusses their role in preventing viral entry to cells (Ceña-Díez et al. 2016). Certain polyanions such as sulphonate and carboxylate also prove beneficial by preventing the interaction of human immunodeficiency virus (HIV) envelope proteins with the host cell receptor disrupting the communication (Moura et al. 2019). G5.0 PAMAM dendrimers attached to G2.5 carboxyl-terminated PAMAM tecto-dendrimers (central dendrimers peripherally attached to many dendrimers) showed an anti-melanoma property in the SK-Mel-28 cell line with no toxicity on healthy keratinocytes (HaCaT cell lines) (Schilrreff et al. 2012).

Functionalisation of hydroxyl-terminated G5.0 PAMAM dendrimers with cholesteryl or dodecyl moieties increased the toxicity in pig lens epithelial cells and Chinese Hamster ovarian cells. The unmodified dendrimers were non-toxic, followed by cholesteryl-modified dendrimers, which were less toxic in the epithelial cells (Setaro et al. 2014).

Cationic polymelamine dendrimers upon injecting intraperitoneally (10 mg/kg) induced liver toxicity with increased levels of serum alanine transaminase (Neerman et al. 2004). PEGylated cationic polymelamine dendrimers and PEG–polyester dendrimers were well tolerated up to a dose of 1 g/kg following intraperitoneal or intravenous injection (Padilla De Jesús et al. 2002; Chen et al. 2004). Melamine-functionalised cationic dendrimers with amino, guanidino, carboxylate, phosphonate or sulphonate groups also concluded the cytotoxicity of cationic dendrimers than the PEG or anionic charged dendrimers (Chen et al. 2004). Acetylation or amidation too decreased cytotoxicity ten-fold lower than unmodified dendrimers in Caco-2 cells with no impact on permeability (Kolhatkar et al. 2007). Dose-dependent release of serum alanine transaminase was also reported along with the intravenous injection of G6.0 PAMAM dendrimers (Okuda et al. 2006).

Another technique to have biocompatible polymeric dendrimers is to have the repeated subunits from the units, which are produced during different metabolic pathways. Details of the same can be found in the literature (Jain et al. 2010). A brief of the cytotoxicity and haematological studies performed with the surface engineered dendrimers is given in Table 16.1.

TABLE 16.1

Effect of various dendrimer generations on cytotoxicity and haematological parameters

Dendrimers	Cytotoxicity	Haemolysis (%)	Haematological Parameters					References
			RBC	WBC	Hb	HCT	MCH	
G2.0 PAMAM	>700 µM (Caco-2 cell line)	Not Reported			Not Reported			(Jevprasesphant et al. 2003)
Anionic or G2.5 PAMAM	Non-cytotoxic up to 1mM (Caco-2)	Not Reported						
G3.0 PAMAM	Cytotoxic at all concentration (Caco-2 cell line)	Haemolytic above 1 mg/mL			Not Reported			(Malik et al. 2000, Jevprasesphant et al. 2003)
Anionic or G3.5 PAMAM	Non-cytotoxic up to 1 mM (Caco-2)	Not Reported			Not Reported			(Jevprasesphant et al. 2003)
G3.0 PPI	Not Reported	Haemolytic above 1 mg/mL			Not Reported			(Malik et al. 2000)
G4.0 PAMAM	Cytotoxic at all concentration (Caco-2 cell line)	18%			Not Reported			(Malik et al. 2000, Jevprasesphant et al. 2003)
	Not Reported	~15.3 – 17.3%			Not Reported			(Bhadra et al. 2003)
G4.0 Poly-L-lysine	Not Reported	14.1 ± 1.02%	7.5 ± 0.3	15.2 ± 0.4	Not Reported			(Agrawal, Gupta and Jain 2007)
G4.0 Poly-L-Lysine (Galactose coated)	7.3 ± 2.8%	Not Reported	8.8 ± 0.2	11.3 ± 0.4	Not Reported			(Agrawal, Gupta and Jain 2007)
G4.0 PPI	Not Reported	35.7%			Not Reported			(Bhadra et al. 2003)
G5.0 PPI	66.8% (0.001 mg/mL) in HepG2 71.5% (0.001 mg/mL) in COS-7	86.2 ± 0.6% (1 mg/mL)	4.43 ± 1.15 (1 mg/mL)	12.01 ± 1.23	8.26 ± 1.39	27.06 ± 3.16	18.82 ± 1.06	(Agashe et al. 2006)
Glycine-coated G5.0 PPI	Not Reported	49.2%			Not Reported			(Bhadra et al. 2003)
	4.9 ± 0.2% (1 mg/mL)	95.37% (1 mg/mL) (HepG2 cells) 96.1% (1 mg/mL) (COS-7 cells)	7.01 ± 1.41	9.01 ± 1.02	12.76 ± 1.31	38.71 ± 3.18	18.15 ± 0.78	(Agashe et al. 2006)

(Continued)

TABLE 16.1 (*Continued*)

Effect of various dendrimer generations on cytotoxicity and haematological parameters

Dendrimers	Cytotoxicity	Haemolysis (%)	Haematological Parameters					References
			RBC	WBC	Hb	HCT	MCH	
Phenylalanine-coated G5.0 PPI	3.3 ± 0.4% (1 mg/mL)	96.2% (1 mg/mL) (HepG2 cells) 93.0% (1 mg/mL) (COS-7 cells)	6.99 ± 1.22	8.99 ± 1.36	12.51 ± 1.35	38.38 ± 3.14	18.01 ± 1.08	(Agashe et al. 2006)
Mannose-coated G5.0 PPI	2.9 ± 0.6% (1mg/mL)	95.2% (1mg/mL) (HepG2 cells) 98.1% (1mg/mL) (COS-7 cells)	7.36 ± 1.01	8.24 ± 1.89	13.46 ± 1.78	41.82 ± 4.32	18.41 ± 0.53	(Agashe et al. 2006)
Lactose-coated G5.0 PPI	2.2 ± 0.3% (1 mg/mL)	94.7% (1 mg/mL) (HepG2 cells) 97.71% (1 mg/mL) (COS-7 cells)	7.24 ± 1.24	8.37 ± 1.46	13.01 ± 1.59	41.02 ± 4.62	18.15 ± 1.21	(Agashe et al. 2006)
G5.0 PPI with neutral acetamide groups	Not Reported	Decrease in cytotoxicity for HUVEC cells			Not Reported			(Stasko et al. 2007)
G5.0 PPI with carboxylic acid groups	Non-hemolytic up to 2 mg/mL	Non-toxic up to 1 mg/mL (B16F10 cells)			Not Reported			(Malik et al. 2000)

RBC: red blood cells; WBC: White blood cells; Hb: haemoglobin; HCT: hematocrit; MCH: mean corpuscular haemoglobin.

16.4 Regulatory

Nanomedicine exhibit a molecular level interaction and thus encounters unique challenges in addition to the traditional hurdle faced during any IND submission (Nijhara and Balakrishnan 2006). All nanomedicines must be evaluated and approved by the US Food and Drug Administration (FDA) as IND submission followed by new drug application (NDA). The major concerns of the FDA are the characterisation of the entity, controls taken to maintain and preserve the quality of the entity, safety and efficacy for the proposed use, risks and benefits of the entity. Approval of any nanomedicine is usually considered as a new molecular entity due to the enhanced qualities such as better absorption, enhanced bioavailability, probable dose reduction, altered toxicity, etc., which do not qualify them as bioequivalent to the original drug (Moradi 2004). They are therefore considered as new, and the safety and efficacy need to be evaluated, which is a time consuming and cost-driven activity. With the continuously emerging nanomedicine concepts, the demand for regulations to ensure the safety and efficacy of the product has increased (Prashant et al. 2014).

Teva Pharmaceuticals and the New York Academy of Sciences together had held a conference in 2013 to emphasise on the regulatory and scientific challenges during the development of nanomedicine based on the learnings of Doxil (Tinkle et al. 2014). The unanimous outcome was the lack of knowledge of nanomedicines attributed to the non-homogeneous and complex nature of the nanomaterial. Besides, a thorough understanding of full characterisation for nanomedicines is lacking. The commercial success is also limited, to a certain extent, due to difficulties in controlling the manufacturing processes. The cost involved in manufacturing is related to increasing complexity. Harmonisation of regulatory guidelines and adapting the same are needed urgently. FDA and European Medicine Agency have published a draft guidelines which discussed the potential challenges and evaluation techniques for few nanotechnologies based products e.g. liposomes (EMA/CHMP/806058/2009/Rev. 02), iron-based colloids (EMA/CHMP/SWP/620008/2012), block copolymer micelles (EMA/CHMP/13099/2013), surface coatings used in nanomedicines and novel nanodevices (Hafner et al. 2014).

A major hurdle during the development of innovative nanomedicines is the limited thought or knowledge to develop the nanocarrier quality as a profile, meeting the regulatory and industrial requirements. Probably before entering Phase I of a clinical trial, the following non-clinical reports need to be kept ready beforehand (Patri 2012, 2014).

The requirements could be broadly classified under various headings as follows:

16.4.1 CMC of Dendrimers – Chemistry, Manufacturing and Controls

To ensure the robustness of synthesis and methodology for scale-up, all manufacturing practices must comply with the current USFDA's regulations. A novel technology or innovative manufacturing process has its unique challenges and difficulty in complying with regulations. For instance, as per 21CFR211.72 2019, every person involved in the manufacturing of drug products must be trained, educated and experienced to perform the particular activity besides current good manufacturing practices (cGMP) continuingly. The requirement of well-trained and well-educated staff to understand the process, understand the issues related to the manufacturing of nanomedicines and operate quality control, in addition, to be well versed with FDA requirements is a common issue among industries.

The equipment used in the manufacturing of nanomaterials or nanodevices should be inert to the materials so that the critical quality attributes of the drug product are not altered (21CFR211.72 2019). To prevent cross-contamination of products, proper sanitisation and cleaning validation should be in place to prevent any adverse effects. For nano-related work, microscopic examination becomes a challenging aspect as chances of unexpected reaction with other molecules are high (Mignani et al. 2018).

Manufacturing standards for nanomedicine with precision to produce the exact particle size maintaining its unique characteristics cost-effectively during scalability is another question. Dendrimers are a well-defined framework with surface functionalities that can be mathematically defined with synthetic chemistry; they should therefore not pose an issue. Full hazard assessment of toxicity such as the disposal method, count of manufacturing and distribution, appropriate measures to maintain the environment and human exposures well below the mandated thresholds should be established (Mignani et al. 2018).

16.4.2 *In vitro* Characterisation of Physical, Chemical, Biological and Physiological Properties

The characterisation methods should be adequate to ensure the quality and safety of the products. These *in vitro* studies include size, surface functionality, purity, encapsulation efficiency, release profiles, degradation, stability, cell viability, polymer excretion mode and mechanism of cellular uptake.

16.4.3 Release and Therapeutic Profiles

The characterisation protocols for biodistribution, pharmacokinetics, efficacy and safety for both placebo and drug-loaded dendrimers should be ensured to be adequate. The simple protocols of traditional small molecules must be expanded to evaluate the risk/benefit ratio of the nanoplatform. These may include evaluation of redox activity, complement activation, membrane disruption, photoactivation, generation of ROS and toxicity measurement. Any off-target effects or vehicle-related responses such as inflammation should be assessed (Kannan et al. 2014). Because the nanoparticle properties are dependent on their critical nanoscale design parameters such as surface chemistry, shape, flexibility/rigidity, elemental composition and size, their influence during clinical translation such as bioaccumulation, pharmacokinetic profile and toxicology is of importance (Kannan et al. 2014). As discussed in the earlier section, a paradigm built on critical nanoscale design parameters should provide a lead to further development of dendrimers predicting the excretion pattern, biodistribution and EPR effect as well as transdermal delivery (Kannan et al. 2014).

16.4.4 Safety Studies

Any associated environmental issues should also be covered. A review of the safety of dendrimers on the non-target organisms and ecotoxicity has been reported (E. García-Calvo 2011). The review discusses the effect on PAMAM dendrimers on aquatic organisms (*Daphnia magna, Vibrio fischeri, Thamnocephalus platyurus* and fish cell lines) on exposure to different generations of dendrimers. The embryonic zebrafish was found to be an indicator of dendrimers' cytotoxicity (Pryor, Harper and Harper 2014). If the dendrimers are intended for topical or dermal application, dermal toxicity and skin irritation studies for prolonged application needs to be performed with toxicity studies even at the cellular level (Winnicka et al. 2015).

16.4.5 Toxicology Study/Toxicokinetics

Carcinogenicity, reproductive toxicology and chronic toxicology or any other toxicity issues or special safety due to the nanosize of dendrimers should be studied (Singh 2018).

16.4.6 Good Intellectual Property Position for Patents

Though basic research does not mandate good laboratory practice, cGMP, quality system regulations and International Council for Harmonisation of Technical Requirements for Pharmaceuticals for Humans (ICH) Q8, Q9 and Q10 (Quality Guidelines), but eventually compliance to these regulations is required for any IND approval (Mignani et al. 2018), for instance, elemental impurities, residual solvents, etc., complying with ICH guidelines. These regulations are law binding, and violations include issuing a warning letter, seizure of a product or court intervention, prolonged IND approval, etc. Once this list of the basic requirements for dendrimers is prepared, it enables the industrial scientists to decide to 'go' or 'no-go' for the IND submission of the dendrimers.

16.5 Clinical Trials and Commercialisation

StarPharma is proactively involved in the dendrimeric application. StarPharma's VivaGel ® is currently in Phase III (NCT01577537) as a topical vaginal virocide for the treatment of bacterial vaginosis in women (https://starpharma.com/drug_delivery 2019). It is also effective against HIV-1, herpes simplex viruses type-1, HIV-2 and human papillomavirus. The lead nanostructure has 32 amino groups and an anionic, naphthalene disulphonate conjugated to the amine of G4.0 PLL dendrimers. The gel is a water-based carbopol gel containing SPL7013 (3% wt/wt) (Moura et al. 2019). This SPL7013 dendrimer is approved in Japan as a protective coating in condoms and is effective against genital herpes and HIV (Kannan et al. 2014). It seems that the virocide activity was statistically significantly; however, the clinical cure was not met at 2–3 weeks after treatment (https://starpharma.com/news/139 2012). VivaGel was well tolerated in sexually abstinent and healthy women without any systemic toxicity (O'Loughlin et al. 2010). However, sexually active women showed low-grade genital adverse events in comparison to placebo gel (McGowan et al. 2011). Poor patient compliance for vaginal related formulations comprises the validity of such outcomes.

Dendrimer enhanced product (DEP®) DEP has also been under evaluation of clinical trials (https://starpharma.com/drug_delivery 2019). Phase II trial of docetaxel–dendrimer conjugate (DEP- Docetaxel, DTX-SPL8783) with a dose of 200 mg administered intravenously establishing the pharmacokinetics and efficacy in candidates with non-small cell lung cancer is almost at completion (https://www.clinicaltrialsregister.eu/ctr-search/search?query=docetaxel 2019). DEP-docetaxel exhibited a 40-fold higher tumour-targeting than free docetaxel (Taxotere). DEP-cabazitaxel has received the ethics and regulatory approval to begin with the clinical trials for tumour targeting (https://starpharma.com/news/356 2018). Dendrimers as a vaccine against tumour-associated carbohydrate antigens are a prospective candidate in Phase I clinical studies developed by Institute Pasteur (Ganneau et al. 2016).

16.6 Diagnostics and Theranostics

The advantages of dendrimers to permit attachment at the multiple sites have extended the utility of dendrimers in diagnostics as superior in terms of sensitivity and highly stable upon attachment. Many of them have been successfully commercialised for the detection of pathogens (Le Berre et al. 2003; Trévisiol et al. 2003; Singh 2007; Senescau et al. 2018) They have also been used in hospitals for immunoassay screening, protein detection (3DNA and UltraAmp developed by Genisphere, Inc.), gene transfection vectors for siRNA delivery such as Superfect from Qiagen and Priofect from StarPharma and diagnostics such as the StratusCS Acute Care diagnostic system (Siemens HealthCare) have been used as cardio diagnostics (https://www.siemens-healthineers.com/cardiac/cardiac-systems/stratus-cs-acute-care 2019). They have been commercially available since the late 1990s. The main challenge in the success of nanodevices has been in the identification of affected and crucial biochemical parameters, proper identification of biomarkers, identification and validation of target related to human disease (Chen and Du 2007). Ocular or surgical adhesives based on dendrimer technology (OcuSeal/Adherus) developed by HyperBranch Medical Technology, Inc, USA, have also been commercialised (Kannan et al. 2014). As per the US FDA, Section 201(h), a medical device includes any machine, instrument or *in vitro* reagent either used as a part of a component or as an accessory intended for diagnosis of disease (https://www.fda.gov/medical-devices/classify-your-medical-device/how-determine-if-your-product-medical-device 2019).

As per the Food, Drug and Cosmetic act of the FDA, diagnostic devices can be broadly divided as follows:

Class I: These are the lowest risk devices and have general controls.

Class II: A bit risky and requires special controls.

Class III: Greatest risk and needs effectiveness and safety reviews.

For new medical devices, clinical trials can be initiated after submitting an 'investigational device exemption.' Class II and Class III devices mandate the manufacturer to file an FDA 510k application or Premarket Notification with the FDA. A 'substantially equivalent device to the already marketed one' does not require a review as a Class III device and can obtain 510(k) approval. Gadomer-17, a MRI contrast linked dendrimer, is under clinical trials for blood pool imaging agents by Bayer Schering Pharma AG, Germany. The imaging agent is similar to the diagnostic agent Gd-DTPA-PLL prototype

(DTPA-diethylene triamine penta-acetic acid; PLL-polylysine) but with a superior rate of elimination (Kannan et al. 2014).

Salimi et al. studied the biodistribution, pharmacokinetics and toxicity of G4.0 PAMAM dendrimers coated with iron oxide nanoparticles after intraperitoneal administration in BALB/c mice bearing tumours. After 24 h, higher iron content was detected in kidneys, liver and lungs, while administering at double dose concentrations (10 mg/mL) caused elevated bilirubin and blood urea nitrogen levels. Liver tissues also showed histopathological abnormalities (Salimi et al. 2018).

16.7 Conclusion

Despite 40 years of research in dendrimers, there are not many clinically approved products. In contrast, 20 years of liposome research ended up Doxil and many more in the market. Another reason for the slow translation of dendrimers is the perception that the synthesis of dendrimers is slow, complex and expensive and difficult to reproduce as per regulatory agencies. Research on dendrimers should focus more on the number of functional groups needed for required pharmacological efficacy, activity and safety for a delivery. Demonstrating minimised toxicity and side effects would not be sufficient to convince the adoption of new technology (Svenson 2015).

Unlike the other nanosystems, the major advantage of dendrimers is the reproducibility in manufacturing them with diverse multifunctionalities and structural precision at the required sites with control on the molecules (Duncan 2014). This has made dendrimers an ideal carrier moiety. Besides, it is easier to correlate the *in vivo* biological behaviour of dendrimers with the specific and critical nanoscale design parameters, ensuring efficient and safe delivery, which could prevent failure during clinical trials. The latter needs to be improved and requires more understanding with an exhaustive physical, chemical and biological characterisation. Though certain limitations of toxicity have been reported, with dendrimers, nevertheless, tuning this toxicity with derivatised dendrimers provides a more safety/benefit ratio creating a success towards commercialisation. These anionic or neutral dendrimers could prove to be of clinical significance. However, *in vitro* studies cannot always be related to the *in vivo* effect. Duncan et. al developed a battery of tests both *in vitro* and *in vivo,* which could be used as screening parameters before initiating any preclinical studies (Duncan and Izzo 2005). 3-(4,5-Dimethylthiazol-2-yl)-2,5-diphenyltetrazolium bromide assay in different cell lines is commonly used for *in vitro* cytotoxicity followed by permeability and *in vivo* related assays such as haematocompatibility assay involving red blood cell (RBC) lysis and complement activation. Short-term and long-term biodistribution, immunogenic responses and study of dendrimer metabolism are crucial for *in vivo* toxicity studies.

Toxicokinetic aspects in representative and reliable models should also be evaluated. Most of the biodistribution has been studied using plain dendrimers without any active load, which may change in the presence of any active load due to the change in the overall properties. The metabolic by-product of the dendrimers may also be involved in redistribution, which should also be considered (Wijagkanalan, Kawakami and Hashida 2011). Empirical toxicity testing of every dendrimer is necessary because a simple guess on the toxicity of molecules on the basis of scientific principles may not always work well as reflected by the studies carried out by some of the authors (Khan et al. 2005).

Focussed and persistent efforts from cross-functional disciplines and acquired expertise are needed to provide a building block or lego approach to offer the best platforms for dendrimers as a carrier. Besides, the functioning of these disciplines should be nurtured too. Not only the manufacturers but also the federal government has been continuously dealing with the new regulatory challenges posed as the number of NDAs related to nanomedicine is being added upon. Close working with the regulatory divisions of the FDA could help in advancement in these particular areas.

Cost-effectiveness of dendrimers has been debated over time. A rough estimate says Vivagel for bacterial vaginosis can be manufactured at the cost of $1500 per kg, which makes them 15 cents per dose (Halford 2005). Moreover, the bench side to clinical phase transition also requires enormous funding, technical and scientific expertise at basics and for moving towards clinical phases along with incentives. An interesting review of the strategic and financial decisions to build and maintain competitiveness and satisfactory share in the market has been published (Morigi et al. 2012). Interest from major companies and start-ups- provides an additional impetus or bonanza. Mutual cooperation between pharma companies and researchers could boost up the translation. The concept of de-risking should be removed. Lots of vistas need to be connected. Dendrimers have the full potential to be developed as a medicine. It is high time to have dendrimers translated to the clinic.

Abbreviations

CFR	Code of federal regulation
cGMP	Current good manufacturing practices
CMC	Chemistry, manufacturing and controls
CNS	Central nervous system
Da	Dalton
DAB	Diaminobutane dendrimers
DAE	Diaminoethane
DMPC	1,2-dimyristoyl-Sn-glycero3-phosphocholine
DNA	Deoxyribose nucleic acid
DTPA	Diethylene triamine pentacetic acid
EPR	Enhanced permeability and retention
FDA	Food and Drug Administration
FDC	Food, drug and cosmetic
G	Generation
GLP	Good laboratory practice
Hb	Hemoglobin
HCT	Hematocrit
ICH	International Council for Harmonisation
IND	Investigational new drug
MCH	Mean corpuscular hemoglobin
MRI	Magnetic resonance imaging
MTT	3-(4,5-Dimethylthiazol-2-yl)-2,5-diphenyltetrazolium
NDA	New drug application
PAMAM	Polyamidoamine
PAMAM-Gd	Polyamidoamine ^{153}Gd-radiobelled
PAMAMOS	Poly(amidoamine-organosilicon)
PEG	Polyethylene glycol
PEO	Polyethylene oxide
PKPD	Pharmacokinetic/pharmacodynamic
PLL	Poly-l-lysine
POPAM	Poly(propylene amine)
PPI	Poly(propyleneimine)
QSR	Quality system regulations
RBC	Red blood cells
RES	Reticuloendothelial system
ROS	Reactive oxygen species
siRNA	Small interfering ribose nucleic acid
WBC	White blood cells

REFERENCES

Agashe, H. B., A. K. Babbar, S. Jain, R. K. Sharma, A. K. Mishra, A. Asthana, M. Garg, T. Dutta, and N. K. Jain. 2007. "Investigations on biodistribution of technetium-99m-labeled carbohydrate-coated poly(propylene imine) dendrimers." *Nanomedicine 3* (2):120–127. doi:10.1016/j.nano.2007.02.002.

Agashe, H. B., T. Dutta, M. Garg, and N. K. Jain. 2006. "Investigations on the toxicological profile of functionalized fifth-generation poly (propylene imine) dendrimer." *J Pharm Pharmacol 58* (11):1491–1498. doi:10.1211/jpp.58.11.0010.

Agrawal, P., U. Gupta, and N. K. Jain. 2007. "Glycoconjugated peptide dendrimers-based nanoparticulate system for the delivery of chloroquine phosphate." *Biomaterials 28* (22):3349–3359. doi:10.1016/j.biomaterials.2007.04.004.

Albertazzi, L., L. Gherardini, M. Brondi, S. Sulis Sato, A. Bifone, T. Pizzorusso, G. M. Ratto, and G. Bardi. 2013. "In vivo distribution and toxicity of PAMAM dendrimers in the central nervous system depend on their surface chemistry." *Mol Pharm 10* (1):249–260. doi:10.1021/mp300391v.

Araujo, R. V., S. D. S. Santos, E. Igne Ferreira, and J. Giarolla. 2018. "New advances in general biomedical applications of PAMAM Dendrimers." *Molecules 23* (11). doi:10.3390/molecules23112849.

Asthana, A., A. S. Chauhan, P. V. Diwan, and N. K. Jain. 2005. "Poly(amidoamine) (PAMAM) dendritic nanostructures for controlled site-specific delivery of acidic anti-inflammatory active ingredient." *AAPS PharmSciTech 6* (3):E536–E542. doi:10.1208/pt060367.

Le Berre, Véronique, Emmanuelle Trévisiol, Adilia Dagkessamanskaia, Serguei Sokol, Anne-Marie Caminade, Jean Pierre Majoral, Bernard Meunier, and Jean François. 2003. "Dendrimeric coating of glass slides for sensitive DNA microarrays analysis." *Nucleic Acids Res 31* (16):e88–e88. doi:10.1093/nar/gng088.

Bhadra, D., S. Bhadra, S. Jain, and N. K. Jain. 2003. "A PEGylated dendritic nanoparticulate carrier of fluorouracil." *Int J Pharm 257* (1–2):111–124. doi:10.1016/s0378-5173(03)00132-7.

Bourne, N., L. R. Stanberry, E. R. Kern, G. Holan, B. Matthews, and D. I. Bernstein. 2000. "Dendrimers, a new class of candidate topical microbicides with activity against herpes simplex virus infection." *Antimicrob Agents Chemother 44* (9):2471–2474. doi:10.1128/aac.44.9.2471-2474.2000.

Brenner, B. M., T. H. Hostetter, and H. D. Humes. 1978. "Glomerular permselectivity: barrier function based on discrimination of molecular size and charge." *Am J Physiol 234* (6):F455–F460. doi:10.1152/ajprenal.1978.234.6.F455.

Bryant, L. H., Jr., E. K. Jordan, J. W. Bulte, V. Herynek, and J. A. Frank. 2002. "Pharmacokinetics of a high-generation dendrimer-Gd-DOTA." *Acad Radiol 9* (Suppl 1):S29–S33. doi:10.1016/s1076-6332(03)80390-2.

Ceña-Díez, Rafael, Daniel Sepúlveda-Crespo, Marek Maly, and Mª Angeles Muñoz-Fernández. 2016. "Dendrimeric based microbicides against sexual transmitted infections associated to heparan sulfate." *RSC Adv 6* (52):46755–46764. doi:10.1039/C6RA06969J.

Cevc, G., and U. Vierl. 2010. "Nanotechnology and the transdermal route: A state of the art review and critical appraisal." *J Control Release 141* (3):277–299. doi:10.1016/j.jconrel.2009.10.016.

Chauhan, Abhay Singh, Narender Kumar Jain, and Prakash Vamanrao Diwan. 2010. "Pre-clinical and behavioural toxicity profile of PAMAM dendrimers in mice." *Proc Math Phys Eng Sci 466* (2117):1535–1550.

Chen, H. T., M. F. Neerman, A. R. Parrish, and E. E. Simanek. 2004. "Cytotoxicity, hemolysis, and acute in vivo toxicity of dendrimers based on melamine, candidate vehicles for drug delivery." *J Am Chem Soc 126* (32):10044–10048. doi:10.1021/ja048548j.

Chen, X. P., and G. H. Du. 2007. "Target validation: A door to drug discovery." *Drug Discov Ther 1* (1):23–29.

21CFR211.72. 2019. Current Good Manufacturing Practice (CGMP) Regulations (Accessed on 18 Apr 2020).

Dave, K., and V. V. Krishna Venuganti. 2017. "Dendritic polymers for dermal drug delivery." *Ther Deliv 8* (12):1077–1096. doi:10.4155/tde-2017-0091.

Duncan, R. 2014. "Polymer therapeutics: Top 10 selling pharmaceuticals - what next?" *J Control Release 190*:371–380. doi:10.1016/j.jconrel.2014.05.001.

Duncan, R., and L. Izzo. 2005. "Dendrimer biocompatibility and toxicity." *Adv Drug Deliv Rev 57* (15):2215–2237. doi:10.1016/j.addr.2005.09.019.

Dutta, Tathagata, Minakshi Garg, Vaibhav Dubey, Dinesh Mishra, Kanhaiya Singh, Deepti Pandita, Ajeet K. Singh, Alok K. Ravi, Thirumurthy Velpandian, and Narendra K. Jain. 2008. "Toxicological investigation of surface engineered fifth generation poly (propyleneimine) dendrimers in vivo." *Nanotoxicology 2* (2):62–70. doi:10.1080/17435390802105167.

EMA/CHMP/13099/2013. European Medicines Agency (Accessed on 18 Apr 2020).

EMA/CHMP/806058/2009/Rev. 02 European Medicines Agency (Accessed on 18 Apr 2020).

EMA/CHMP/SWP/620008/2012. European Medicines Agency (Accessed on 18 Apr 2020).

Felczak, Aleksandra, Natalia Wrońska, Anna Janaszewska, Barbara Klajnert, Maria Bryszewska, Dietmar Appelhans, Brigitte Voit, Sylwia Różalska, and Katarzyna Lisowska. 2012. "Antimicrobial activity of poly(propylene imine) dendrimers." *New J Chem 36* (11):2215–2222. doi:10.1039/C2NJ40421D.

Ganneau, C., C. Simenel, E. Emptas, T. Courtiol, Y. M. Coic, C. Artaud, E. Deriaud, F. Bonhomme, M. Delepierre, C. Leclerc, R. Lo-Man, and S. Bay. 2016. "Large-scale synthesis and structural analysis of a synthetic glycopeptide dendrimer as an anti-cancer vaccine candidate." *Org Biomol Chem 15* (1):114–123. doi:10.1039/c6ob01931e.

García-Calvo, E., I.J. Suarez, R. Rosal, A. Rodriguez, A. Ucles, A.R. Fernandez-Albab, M.D. Hernando. 2011. "Chemical and ecotoxicological assessment of poly(amidoamine) dendrimers in the aquatic environment." *TrAC Trends Anal Chem 30* (3).

Gerlowski, L. E., and R. K. Jain. 1983. "Physiologically based pharmacokinetic modeling: principles and applications." *J Pharm Sci 72* (10):1103–1127. doi:10.1002/jps.2600721003.

Gillies, Elizabeth R., Edward Dy, Jean M. J. Fréchet, and Francis C. Szoka. 2005. "Biological evaluation of polyester Dendrimer: poly(ethylene oxide) "Bow-Tie" hybrids with tunable molecular weight and architecture." *Mol Pharm 2* (2):129–138. doi:10.1021/mp049886u.

Gonzalo, S., I. Rodea-Palomares, F. Leganes, E. Garcia-Calvo, R. Rosal, and F. Fernandez-Pinas. 2015. "First evidences of PAMAM dendrimer internalization in microorganisms of environmental relevance: A linkage with toxicity and oxidative stress." *Nanotoxicology 9* (6):706–718. doi:10.3109/17435390.2014.969345.

Grayson, S. M., and J. M. Frechet. 2001. "Convergent dendrons and dendrimers: from synthesis to applications." *Chem Rev 101* (12):3819–3868. doi:10.1021/cr990116h.

Greish, K., G. Thiagarajan, H. Herd, R. Price, H. Bauer, D. Hubbard, A. Burckle, S. Sadekar, T. Yu, A. Anwar, A. Ray, and H. Ghandehari. 2012. "Size and surface charge significantly influence the toxicity of silica and dendritic nanoparticles." *Nanotoxicology 6* (7):713–723. doi:10.3109/17435390.2011.604442.

Hafner, A., J. Lovric, G. P. Lakos, and I. Pepic. 2014. "Nanotherapeutics in the EU: an overview on current state and future directions." *Int J Nanomedicine 9*:1005–1023. doi:10.2147/ijn.S55359.

Halford, Bethany. 2005. "Dendrimers Branch Out." *Chemical & Engineering News Archive 83* (24):30–36. doi:10.1021/cen-v083n024.p030.

https://starpharma.com/drug_delivery. 2019. *Dendrimer Drug Delivery (DEP®)* (Accessed on 18 April 2020).

https://starpharma.com/news/139. 2012. *VivaGel phase 3 study results* (Accessed on 18 April 2020).

https://starpharma.com/news/356. 2018. (Accessed on 18 April 2020).

https://www.clinicaltrialsregister.eu/ctr-search/search?query=docetaxel. 2019. *Clinical trials for docetaxel* (Accessed on 18 April 2020).

https://www.fda.gov/medical-devices/classify-your-medical-device/how-determine-if-your-product-medical-device. 2019. (Accessed on 18 April 2020).

https://www.siemens-healthineers.com/cardiac/cardiac-systems/stratus-cs-acute-care. 2019. (Accessed on 18 April 2020).

Jain, K., P. Kesharwani, U. Gupta, and N. K. Jain. 2010. "Dendrimer toxicity: Let's meet the challenge." *Int J Pharm 394* (1–2):122–142. doi:10.1016/j.ijpharm.2010.04.027.

Janaszewska, A., M. Ciolkowski, D. Wrobel, J. F. Petersen, M. Ficker, J. B. Christensen, M. Bryszewska, and B. Klajnert. 2013. "Modified PAMAM dendrimer with 4-carbomethoxypyrrolidone surface groups reveals negligible toxicity against three rodent cell-lines." *Nanomedicine 9* (4):461–464. doi:10.1016/j.nano.2013.01.010.

Janaszewska, A., M. Studzian, J. F. Petersen, M. Ficker, J. B. Christensen, and B. Klajnert-Maculewicz. 2015. "PAMAM dendrimer with 4-carbomethoxypyrrolidone--in vitro assessment of neurotoxicity." *Nanomedicine 11* (2):409–411. doi:10.1016/j.nano.2014.09.011.

Jevprasesphant, R., J. Penny, R. Jalal, D. Attwood, N. B. McKeown, and A. D'Emanuele. 2003. "The influence of surface modification on the cytotoxicity of PAMAM dendrimers." *Int J Pharm 252* (1–2):263–266. doi:10.1016/s0378-5173(02)00623-3.

Jones, C. F., R. A. Campbell, A. E. Brooks, S. Assemi, S. Tadjiki, G. Thiagarajan, C. Mulcock, A. S. Weyrich, B. D. Brooks, H. Ghandehari, and D. W. Grainger. 2012. "Cationic PAMAM dendrimers aggressively initiate blood clot formation." *ACS Nano 6* (11):9900–9910. doi:10.1021/nn303472r.

Jones, C. F., R. A. Campbell, Z. Franks, C. C. Gibson, G. Thiagarajan, A. Vieira-de-Abreu, S. Sukavaneshvar, S. F. Mohammad, D. Y. Li, H. Ghandehari, A. S. Weyrich, B. D. Brooks, and D. W. Grainger. 2012. "Cationic PAMAM dendrimers disrupt key platelet functions." *Mol Pharm 9* (6):1599–1611. doi:10.1021/mp2006054.

Kaminskas, L. M., B. J. Boyd, and C. J. Porter. 2011. "Dendrimer pharmacokinetics: the effect of size, structure and surface characteristics on ADME properties." *Nanomedicine (Lond) 6* (6):1063–1084. doi:10.2217/nnm.11.67.

Kannan, R. M., E. Nance, S. Kannan, and D. A. Tomalia. 2014. "Emerging concepts in dendrimer-based nano-medicine: from design principles to clinical applications." *J Intern Med 276* (6):579–617. doi:10.1111/joim.12280.

Kannan, Sujatha, Hui Dai, Raghavendra S. Navath, Bindu Balakrishnan, Amar Jyoti, James Janisse, Roberto Romero, and Rangaramanujam M. Kannan. 2012. "Dendrimer-based postnatal therapy for neuroinflam-mation and cerebral palsy in a rabbit model." *Sci Transl Med 4* (130):130ra46–130ra46. doi:10.1126/scitranslmed.3003162.

Khan, M. K., S. S. Nigavekar, L. D. Minc, M. S. Kariapper, B. M. Nair, W. G. Lesniak, and L. P. Balogh. 2005. "In vivo biodistribution of dendrimers and dendrimer nanocomposites -- implications for cancer imaging and therapy." *Technol Cancer Res Treat 4* (6):603–613. doi:10.1177/153303460500400604.

Kitchens, K. M., A. B. Foraker, R. B. Kolhatkar, P. W. Swaan, and H. Ghandehari. 2007. "Endocytosis and interaction of poly (amidoamine) dendrimers with Caco-2 cells." *Pharm Res 24* (11):2138–2145. doi:10.1007/s11095-007-9415-0.

Kobayashi, H., S. Kawamoto, M. Bernardo, M. W. Brechbiel, M. V. Knopp, and P. L. Choyke. 2006. "Delivery of gadolinium-labeled nanoparticles to the sentinel lymph node: comparison of the sentinel node visual-ization and estimations of intra-nodal gadolinium concentration by the magnetic resonance imaging." *J Control Release 111* (3):343–351. doi:10.1016/j.jconrel.2005.12.019.

Kobayashi, H., S. Kawamoto, P. L. Choyke, N. Sato, M. V. Knopp, R. A. Star, T. A. Waldmann, Y. Tagaya, and M. W. Brechbiel. 2003. "Comparison of dendrimer-based macromolecular contrast agents for dynamic micro-magnetic resonance lymphangiography." *Magn Reson Med 50* (4):758–766. doi:10.1002/mrm.10583.

Kobayashi, H., S. Kawamoto, S. K. Jo, H. L. Bryant, Jr., M. W. Brechbiel, and R. A. Star. 2003. "Macromo-lecular MRI contrast agents with small dendrimers: pharmacokinetic differences between sizes and cores." *Bioconjug Chem 14* (2):388–394. doi:10.1021/bc025633c.

Kobayashi, H., S. Kawamoto, T. Saga, N. Sato, A. Hiraga, T. Ishimori, J. Konishi, K. Togashi, and M. W. Brech-biel. 2001a. "Positive effects of polyethylene glycol conjugation to generation-4 polyamidoamine den-drimers as macromolecular MR contrast agents." *Magn Reson Med 46* (4):781–788. doi:10.1002/mrm.1257.

Kobayashi, H., S. Kawamoto, T. Saga, N. Sato, A. Hiraga, J. Konishi, K. Togashi, and M. W. Brechbiel. 2001b. "Micro-MR angiography of normal and intratumoural vessels in mice using dedicated intravascular MR contrast agents with high generation of polyamidoamine dendrimer core: reference to pharmacokinetic properties of dendrimer-based MR contrast agents." *J Magn Reson Imaging 14* (6):705–713. doi:10.1002/jmri.10025.

Kobayashi, H., S. Kawamoto, R. A. Star, T. A. Waldmann, Y. Tagaya, and M. W. Brechbiel. 2003. "Micro-magnetic resonance lymphangiography in mice using a novel dendrimer-based magnetic resonance imaging contrast agent." *Cancer Res 63* (2):271–276.

Kobayashi, H., N. Sato, A. Hiraga, T. Saga, Y. Nakamoto, H. Ueda, J. Konishi, K. Togashi, and M. W. Brech-biel. 2001. "3D-micro-MR angiography of mice using macromolecular MR contrast agents with poly-amidoamine dendrimer core with reference to their pharmacokinetic properties." *Magn Reson Med 45* (3):454–460. doi:10.1002/1522-2594(200103)45:3<454::aid-mrm1060>3.0.co;2-m.

Kobayashi, H., R. Watanabe, and P. L. Choyke. 2013. "Improving conventional enhanced permeability and retention (EPR) effects; what is the appropriate target?" *Theranostics 4* (1):81–89. doi:10.7150/thno.7193.

Kojima, C., C. Regino, Y. Umeda, H. Kobayashi, and K. Kono. 2010. "Influence of dendrimer generation and polyethylene glycol length on the biodistribution of PEGylated dendrimers." *Int J Pharm 383* (1–2):293–296. doi:10.1016/j.ijpharm.2009.09.015.

Kojima, C., B. Turkbey, M. Ogawa, M. Bernardo, C. A. Regino, L. H. Bryant, Jr., P. L. Choyke, K. Kono, and H. Kobayashi. 2011. "Dendrimer-based MRI contrast agents: the effects of PEGylation on relaxivity and pharmacokinetics." *Nanomedicine 7* (6):1001–1008. doi:10.1016/j.nano.2011.03.007.

Kolhatkar, R. B., K. M. Kitchens, P. W. Swaan, and H. Ghandehari. 2007. "Surface acetylation of polyamido-amine (PAMAM) dendrimers decreases cytotoxicity while maintaining membrane permeability." *Bio-conjug Chem 18* (6):2054–2060. doi:10.1021/bc0603889.

Kukowska-Latallo, J. F., K. A. Candido, Z. Cao, S. S. Nigavekar, I. J. Majoros, T. P. Thomas, L. P. Balogh, M. K. Khan, and J. R. Baker, Jr. 2005. "Nanoparticle targeting of anticancer drug improves therapeutic response in animal model of human epithelial cancer." *Cancer Res 65* (12):5317–5324. doi:10.1158/0008-5472.Can-04-3921.

Kuo, J. H., M. S. Jan, and H. W. Chiu. 2005. "Mechanism of cell death induced by cationic dendrimers in RAW 264.7 murine macrophage-like cells." *J Pharm Pharmacol 57* (4):489–495. doi:10.1211/0022357055803.

Lee, C. C., M. Yoshida, J. M. Frechet, E. E. Dy, and F. C. Szoka. 2005. "In vitro and in vivo evaluation of hydrophilic dendronized linear polymers." *Bioconjug Chem 16* (3):535–541. doi:10.1021/bc0497665.

Lee, Hwankyu, and Ronald G. Larson. 2008. "Lipid bilayer curvature and pore formation induced by charged linear polymers and dendrimers: the effect of molecular shape." *J Phys Chem B 112* (39):12279–12285. doi:10.1021/jp805026m.

Longmire, M., P. L. Choyke, and H. Kobayashi. 2008. "Clearance properties of nano-sized particles and molecules as imaging agents: considerations and caveats." *Nanomedicine (Lond) 3* (5):703–717. doi:10.2217/17435889.3.5.703.

Lu, Y., D. L. Slomberg, A. Shah, and M. H. Schoenfisch. 2013. "Nitric oxide-releasing amphiphilic poly(amidoamine) (PAMAM) dendrimers as antibacterial agents." *Biomacromolecules 14* (10):3589–3598. doi:10.1021/bm400961r.

Malik, N., R. Wiwattanapatapee, R. Klopsch, K. Lorenz, H. Frey, J. W. Weener, E. W. Meijer, W. Paulus, and R. Duncan. 2000. "Dendrimers: relationship between structure and biocompatibility in vitro, and preliminary studies on the biodistribution of 125I-labelled polyamidoamine dendrimers in vivo." *J Control Release 65* (1–2):133–148. doi:10.1016/s0168-3659(99)00246-1.

Manunta, M., B. J. Nichols, P. H. Tan, P. Sagoo, J. Harper, and A. J. George. 2006. "Gene delivery by dendrimers operates via different pathways in different cells, but is enhanced by the presence of caveolin." *J Immunol Methods 314* (1–2):134–146. doi:10.1016/j.jim.2006.06.007.

Manunta, Maria, Peng Hong Tan, Pervinder Sagoo, Kirk Kashefi, and Andrew J. T. George. 2004. "Gene delivery by dendrimers operates via a cholesterol dependent pathway." *Nucleic Acids Res 32* (9):2730–2739. doi:10.1093/nar/gkh595.

McGowan, Ian, Kailazarid Gomez, Karen Bruder, Irma Febo, Beatrice A. Chen, Barbra A. Richardson, Marla Husnik, Edward Livant, Clare Price, Cindy Jacobson, and M. T. N. Protocol Team. 2011. "Phase 1 randomized trial of the vaginal safety and acceptability of SPL7013 gel (VivaGel) in sexually active young women (MTN-004)." *AIDS (London, England) 25* (8):1057–1064. doi:10.1097/QAD.0b013e328346bd3e.

Mecke, A., S. Uppuluri, T. M. Sassanella, D. K. Lee, A. Ramamoorthy, J. R. Baker, Jr., B. G. Orr, and M. M. Banaszak Holl. 2004. "Direct observation of lipid bilayer disruption by poly(amidoamine) dendrimers." *Chem Phys Lipids 132* (1):3–14. doi:10.1016/j.chemphyslip.2004.09.001.

Mehta, Piyush, Shivajirao Kadam, Atmaram Pawar, and C. Bothiraja. 2019. "Dendrimers for pulmonary delivery: current perspectives and future challenges." *New J Chem 43* (22):8396–8409. doi:10.1039/C9NJ01591D.

Menjoge, A. R., A. L. Rinderknecht, R. S. Navath, M. Faridnia, C. J. Kim, R. Romero, R. K. Miller, and R. M. Kannan. 2011. "Transfer of PAMAM dendrimers across human placenta: prospects of its use as drug carrier during pregnancy." *J Control Release 150* (3):326–338. doi:10.1016/j.jconrel.2010.11.023.

Meyers, S. R., F. S. Juhn, A. P. Griset, N. R. Luman, and M. W. Grinstaff. 2008. "Anionic amphiphilic dendrimers as antibacterial agents." *J Am Chem Soc 130* (44):14444–14445. doi:10.1021/ja806912a.

Mignani, S., J. Rodrigues, H. Tomas, R. Roy, X. Shi, and J. P. Majoral. 2018. "Bench-to-bedside translation of dendrimers: Reality or utopia? A concise analysis." *Adv Drug Deliv Rev 136–137*:73–81. doi:10.1016/j.addr.2017.11.007.

Mishra, M. K., C. A. Beaty, W. G. Lesniak, S. P. Kambhampati, F. Zhang, M. A. Wilson, M. E. Blue, J. C. Troncoso, S. Kannan, M. V. Johnston, W. A. Baumgartner, and R. M. Kannan. 2014. "Dendrimer brain uptake and targeted therapy for brain injury in a large animal model of hypothermic circulatory arrest." *ACS Nano 8* (3):2134–2147. doi:10.1021/nn404872e.

Miyano, T., W. Wijagkanalan, S. Kawakami, F. Yamashita, and M. Hashida. 2010. "Anionic amino acid dendrimer-trastuzumab conjugates for specific internalization in HER2-positive cancer cells." *Mol Pharm 7* (4):1318–1327. doi:10.1021/mp100105c.

Moradi, Michael. 2004. *Nano-enabled Drug Delivery Systems Market*, Sterling, VA: NanoMarkets, LC.

Morigi, Valentina, Alessandro Tocchio, Carlo Bellavite Pellegrini, Jason H. Sakamoto, Marco Arnone, and Ennio Tasciotti. 2012. "Nanotechnology in medicine: from inception to market domination." *J Drug Deliv 2012*:389485–389485. doi:10.1155/2012/389485.

Moura, Liane I. F., Alessio Malfanti, Carina Peres, Ana I. Matos, Elise Guegain, Vanessa Sainz, Mire Zloh, María J. Vicent, and Helena F. Florindo. 2019. "Functionalized branched polymers: promising immunomodulatory tools for the treatment of cancer and immune disorders." *Mater Horiz 6* (10):1956–1973. doi:10.1039/C9MH00628A.

Mukherjee, S. P., M. Davoren, and H. J. Byrne. 2010. "In vitro mammalian cytotoxicological study of PAMAM dendrimers - towards quantitative structure activity relationships." *Toxicol In Vitro 24* (1):169–177. doi:10.1016/j.tiv.2009.09.014.

Mukherjee, S. P., F. M. Lyng, A. Garcia, M. Davoren, and H. J. Byrne. 2010. "Mechanistic studies of in vitro cytotoxicity of poly(amidoamine) dendrimers in mammalian cells." *Toxicol Appl Pharmacol 248* (3):259–268. doi:10.1016/j.taap.2010.08.016.

Neerman, M. F., W. Zhang, A. R. Parrish, and E. E. Simanek. 2004. "In vitro and in vivo evaluation of a melamine dendrimer as a vehicle for drug delivery." *Int J Pharm 281* (1–2):129–132. doi:10.1016/j.ijpharm.2004.04.023.

Nijhara, R., and K. Balakrishnan. 2006. "Bringing nanomedicines to market: regulatory challenges, opportunities, and uncertainties." *Nanomedicine 2* (2):127–136. doi:10.1016/j.nano.2006.04.005.

O'Loughlin, J., I. Y. Millwood, H. M. McDonald, C. F. Price, J. M. Kaldor, and J. R. Paull. 2010. "Safety, tolerability, and pharmacokinetics of SPL7013 gel (VivaGel): a dose ranging, phase I study." *Sex Transm Dis 37* (2):100–104. doi:10.1097/OLQ.0b013e3181bc0aac.

Okuda, T., S. Kawakami, T. Maeie, T. Niidome, F. Yamashita, and M. Hashida. 2006. "Biodistribution characteristics of amino acid dendrimers and their PEGylated derivatives after intravenous administration." *J Control Release 114* (1):69–77. doi:10.1016/j.jconrel.2006.05.009.

Padilla De Jesús, Omayra L., Henrik R. Ihre, Lucie Gagne, Jean M. J. Fréchet, and Francis C. Szoka. 2002. "Polyester dendritic systems for drug delivery applications: In vitro and in vivo evaluation." *Bioconjug Chem 13* (3):453–461. doi:10.1021/bc010103m.

Parrott, M. C., S. R. Benhabbour, C. Saab, J. A. Lemon, S. Parker, J. F. Valliant, and A. Adronov. 2009. "Synthesis, radiolabeling, and bio-imaging of high-generation polyester dendrimers." *J Am Chem Soc 131* (8):2906–2916. doi:10.1021/ja8078175.

Patri, A. 2012. "Clinical Translation of Dendrimer Based Nanomedicines: Challenges and Solution." Biodendrimer symposium, Toledo (Spain).

Patri, A. 2014. "Dendrimer Design and Preclinical Testing for Successful Clinical Translation." Biodendrimer symposium, Lugano (Switzerland).

Pires, J., T. N. Siriwardena, M. Stach, R. Tinguely, S. Kasraian, F. Luzzaro, S. L. Leib, T. Darbre, J. L. Reymond, and A. Endimiani. 2015. "In Vitro Activity of the Novel Antimicrobial Peptide Dendrimer G3KL against Multidrug-Resistant *Acinetobacter baumannii* and *Pseudomonas aeruginosa*." *Antimicrob Agents Chemother 59* (12):7915–7918. doi:10.1128/aac.01853-15.

Pisal, D. S., V. K. Yellepeddi, A. Kumar, R. S. Kaushik, M. B. Hildreth, X. Guan, and S. Palakurthi. 2008. "Permeability of surface-modified polyamidoamine (PAMAM) dendrimers across Caco-2 cell monolayers." *Int J Pharm 350* (1–2):113–121. doi:10.1016/j.ijpharm.2007.08.033.

Pojo, M., S. R. Cerqueira, T. Mota, A. Xavier-Magalhães, S. Ribeiro-Samy, J. F. Mano, J. M. Oliveira, R. L. Reis, N. Sousa, B. M. Costa, and A. J. Salgado. 2013. "In vitro evaluation of the cytotoxicity and cellular uptake of CMCht/PAMAM dendrimer nanoparticles by glioblastoma cell models." *J Nanopart. Res. 15* (5):1621. doi:10.1007/s11051-013-1621-6.

Prashant, Kesharwani, Keerti Jain, Narendra Kumar Jain. 2014. "Dendrimer as nanocarrier for drug delivery." *Prog Polym Sci 39* (2).

Pryor, J. B., B. J. Harper, and S. L. Harper. 2014. "Comparative toxicological assessment of PAMAM and thiophosphoryl dendrimers using embryonic zebrafish." *Int J Nanomedicine 9*:1947–1956. doi:10.2147/ijn.S60220.

Roberts, Jeanette C., Mahesh K. Bhalgat, Richard T. Zera. 1996. "Preliminary biological evaluation of polyamidoamine (PAMAM) StarburstTM dendrimers." *J Biomed Mater Res 30* (1).

Salimi, M., S. Sarkar, S. Fathi, A. M. Alizadeh, R. Saber, F. Moradi, and H. Delavari. 2018. "Biodistribution, pharmacokinetics, and toxicity of dendrimer-coated iron oxide nanoparticles in BALB/c mice." *Int J Nanomedicine 13*:1483–1493. doi:10.2147/ijn.S157293.

Sarin, H., A. S. Kanevsky, H. Wu, A. A. Sousa, C. M. Wilson, M. A. Aronova, G. L. Griffiths, R. D. Leapman, and H. Q. Vo. 2009. "Physiologic upper limit of pore size in the blood-tumour barrier of malignant solid tumours." *J Transl Med 7*:51. doi:10.1186/1479-5876-7-51.

Schilrreff, Priscila, Cecilia Mundiña-Weilenmann, Eder Lilia Romero, and Maria Jose Morilla. 2012. "Selective cytotoxicity of PAMAM G5 core--PAMAM G2.5 shell tecto-dendrimers on melanoma cells." *Int J Nanomedicine 7*:4121–4133. doi:10.2147/IJN.S32785.

Senescau, A., T. Kempowsky, E. Bernard, S. Messier, P. Besse, R. Fabre, and J. M. Francois. 2018. "Innovative DendrisChips((R)) technology for a syndromic approach of in vitro diagnosis: application to the respiratory infectious diseases." *Diagnostics (Basel) 8* (4). doi:10.3390/diagnostics8040077.

Setaro, F., R. Ruiz-Gonzalez, S. Nonell, U. Hahn, and T. Torres. 2014. "Synthesis, photophysical studies and (1)O(2) generation of carboxylate-terminated zinc phthalocyanine dendrimers." *J Inorg Biochem 136*:170–176. doi:10.1016/j.jinorgbio.2014.02.007.

Shcharbin, D., A. Janaszewska, B. Klajnert-Maculewicz, B. Ziemba, V. Dzmitruk, I. Halets, S. Loznikova, N. Shcharbina, K. Milowska, M. Ionov, A. Shakhbazau, and M. Bryszewska. 2014. "How to study dendrimers and dendriplexes III. Biodistribution, pharmacokinetics and toxicity in vivo." *J Control Release 181*:40–52. doi:10.1016/j.jconrel.2014.02.021.

Singh, Gursharan. 2018. "Chapter 4 - Preclinical Drug Development." In *Pharmaceutical Medicine and Translational Clinical Research*, edited by Divya Vohora and Gursharan Singh, 47–63. Boston: Academic Press.

Singh, P. 2007. "Dendrimers and their applications in immunoassays and clinical diagnostics." *Biotechnol Appl Biochem 48* (Pt 1):1–9. doi:10.1042/ba20070019.

Stasko, N. A., C. B. Johnson, M. H. Schoenfisch, T. A. Johnson, and E. L. Holmuhamedov. 2007. "Cytotoxicity of polypropylenimine dendrimer conjugates on cultured endothelial cells." *Biomacromolecules 8* (12):3853–3859. doi:10.1021/bm7008203.

Surendra, Tripathy, Malay K Das. 2013. "Dendrimers and their applications as novel drug delivery carriers." *J Appl Pharm Sci 9* (3):9.

Svenson, S. 2015. "The dendrimer paradox--high medical expectations but poor clinical translation." *Chem Soc Rev 44* (12):4131–4144. doi:10.1039/c5cs00288e.

Tajarobi, F., M. El-Sayed, B. D. Rege, J. E. Polli, and H. Ghandehari. 2001. "Transport of poly amidoamine dendrimers across Madin-Darby canine kidney cells." *Int J Pharm 215* (1–2):263–267. doi:10.1016/s0378-5173(00)00679-7.

Thiagarajan, G., K. Greish, and H. Ghandehari. 2013. "Charge affects the oral toxicity of poly(amidoamine) dendrimers." *Eur J Pharm Biopharm 84* (2):330–334. doi:10.1016/j.ejpb.2013.01.019.

Thiagarajan, G., S. Sadekar, K. Greish, A. Ray, and H. Ghandehari. 2013. "Evidence of oral translocation of anionic G6.5 dendrimers in mice." *Mol Pharm 10* (3):988–998. doi:10.1021/mp300436c.

Tinkle, S., S. E. McNeil, S. Muhlebach, R. Bawa, G. Borchard, Y. C. Barenholz, L. Tamarkin, and N. Desai. 2014. "Nanomedicines: addressing the scientific and regulatory gap." *Ann N Y Acad Sci 1313*:35–56. doi:10.1111/nyas.12403.

Tomalia, D. A., L. A. Reyna, and S. Svenson. 2007. "Dendrimers as multi-purpose nanodevices for oncology drug delivery and diagnostic imaging." *Biochem Soc Trans 35* (Pt 1):61–67. doi:10.1042/bst0350061.

Tomalia, Donald, Jørn Christensen, and Ulrik Boas. 2012. "Toxicology of dendrimers and dendrons." In *Dendrimers, Dendrons, and Dendritic Polymers Discovery, Applications, and the Future* New York, NY: Cambridge University Press.

Trévisiol, Emmanuelle, Véronique Le Berre-Anton, Julien Leclaire, Geneviève Pratviel, Anne-Marie Caminade, Jean-Pierre Majoral, Jean Marie François, and Bernard Meunier. 2003. "Dendrislides, dendrichips: a simple chemical functionalization of glass slides with phosphorus dendrimers as an effective means for the preparation of biochips." *New J Chem 27* (12):1713–1719. doi:10.1039/B307928G.

Vega-Villa, K. R., J. K. Takemoto, J. A. Yanez, C. M. Remsberg, M. L. Forrest, and N. M. Davies. 2008. "Clinical toxicities of nanocarrier systems." *Adv Drug Deliv Rev 60* (8):929–938. doi:10.1016/j.addr.2007.11.007.

Wijagkanalan, W., S. Kawakami, and M. Hashida. 2011. "Designing dendrimers for drug delivery and imaging: pharmacokinetic considerations." *Pharm Res 28* (7):1500–1519. doi:10.1007/s11095-010-0339-8.

Winnicka, K., M. Wroblewska, K. Sosnowska, H. Car, and I. Kasacka. 2015. "Evaluation of cationic polyamidoamine dendrimers' dermal toxicity in the rat skin model." *Drug Des Devel Ther 9*:1367–1377. doi:10.2147/dddt.S78336.

Worley, B. V., K. M. Schilly, and M. H. Schoenfisch. 2015. "Anti-Biofilm Efficacy of Dual-Action Nitric Oxide-Releasing Alkyl Chain Modified Poly(amidoamine) Dendrimers." *Mol Pharm 12* (5):1573–1583. doi:10.1021/acs.molpharmaceut.5b00006.

Xue, X., X. Chen, X. Mao, Z. Hou, Y. Zhou, H. Bai, J. Meng, F. Da, G. Sang, Y. Wang, and X. Luo. 2013. "Amino-terminated generation 2 poly(amidoamine) dendrimer as a potential broad-spectrum, nonresistance-inducing antibacterial agent." *AAPS J 15* (1):132–142. doi:10.1208/s12248-012-9416-8.

Zhang, Z. Y., and B. D. Smith. 2000. "High-generation polycationic dendrimers are unusually effective at disrupting anionic vesicles: membrane bending model." *Bioconjug Chem 11* (6):805–814. doi:10.1021/bc000018z.

Index

Page numbers in *italic* indicate figures. Page numbers in **bold** indicate tables.

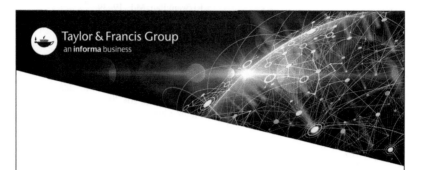